Bird Migration

Thomas Alerstam
Department of Ecology
University of Lund

Translated by David A. Christie
Illustrations by Astrid Ulfstrand

CAMBRIDGE
UNIVERSITY PRESS

Published by the Press Syndicate of the University of Cambridge
The Pitt Building, Trumpington Street, Cambridge CB2 1RP
40 West 20th Street, New York, NY 10011–4211, USA
10 Stamford Road, Oakleigh, Melbourne 3166, Australia

Originally published in Swedish as *Fågelflyttning* by Bokförlaget Signum in 1982 and
© Thomas Alerstam and Bokförlaget Signum 1982

First published in English by Cambridge University Press 1990 as *Bird Migration*
English translation © Cambridge University Press 1990
First paperback edition 1993

Printed in Great Britain by the Bath Press, Avon

British Library cataloguing in publication data

Alerstam, Thomas
Bird migration.
1. Birds. Migration
I. Title   II. Fågelflyttning. *English*
598.2525

Library of Congress cataloguing in publication data

Alerstam, Thomas
[Fågelflyttning. English]
Bird migration / Thomas Alerstam; translated by David A. Christie.
    p.   cm.
Originally published in Swedish as Fågelflyttning by Bokförlaget Signum in 1982–T.p. verso.
Includes bibliographical references.
ISBN 0 521 32865 9
1. Birds–Migration.   I. Title.
QL698.9.A4413   1990
598.2'525–dc20   90-31054   CIP

ISBN 0 521 32865 9 hardback
ISBN 0 521 44822 0 paperback

Contents

Preface to Swedish edition

Preparing this book has taken me three years, 1979–1981. During this time I have also had a number of other tasks to perform, and I have therefore been repeatedly compelled to break off the writing of the book, occasionally for several months. The time restrictions have of course meant that it has not been possible to deal with all different aspects of bird migration. Nor has my aim been to communicate as many cold facts as possible, but instead to provide an insight into the richly varied life of migratory birds and, here and there, to discuss and speculate on the kind of things that exist in the border area between what is known and what are still unsolved mysteries within the branches of bird migration which I personally find to be most exciting.

Following the book's introduction and a background chapter on the earth's climate and vegetation, there are three main parts. The first, 'Summer and winter quarters', deals with the birds' ecology. Different life patterns and migration patterns are presented and discussed in nine separate chapter-sections and the evolution of bird migration is then illustrated in a summary tenth chapter-section. The second main part, 'The migratory journey', deals in a series of chapter-sections with how birds migrate and the fortunes and hazards they encounter at that time. Finally, some light is thrown on the 'Orientation and navigation' of migrants. The various chapters and sections stand by themselves to the extent that the reader can quite easily read only those parts in which he or she is most interested. There are no pictures of different bird species in this book; for those who are unsure of what some species look like and where they live, it may be a good thing to have a general bird guide handy. A globe may also at times be found to be a useful aid.

While working on this book I have received much support and assistance from various quarters; for this I am deeply grateful. The analyses and syntheses which form the basic data for the book I have been able to carry out within the framework of a research grant from the Swedish Natural Science Research Council. The Ecology Department at the Zoological Institute at the University of Lund has been my warmly appreciated place of work. Per Brinck has, as head of department, given me all the support I could have wished for. The Zoological Institute's well-stocked library has been of great use and service with its willingness to help and its generous lending regulations, all largely thanks to Ulla Holmberg and Stig Belfrage. Ornithologist friends and colleagues at the Ecology Building whom I have pursued with questions have offered many valuable points of view, provided tips on literature and lent books and journals. It has been a great source of help every time somebody has stuck his or her head round the door of my study or approached me during a coffee-break: 'I recently read a very interesting article on bird migration which you might perhaps find useful for your book. Have you read it? I have brought along the journal for you so that you can see what you think yourself.' Thank-you Per Andell, Bo Ebenman, Pehr H. Enckell, Paul-Eric Jönsson, Leif Nilsson, Lars-Eric Persson, Gustaf Rudebeck, Bo Svensson, Sören Svensson, Magnus Sylvén and Staffan Ulfstrand (Uppsala). My thanks also to Gunnar Roos for the latest news from Falsterbo.

I wish to thank particularly warmly Göran Högstedt, Johnny Karlsson and Bertil Larsson in Ljungbyhed for their excellent co-operation and great assistance. With Göran Högstedt I have had many and long discussions on birds and ecology during the late hours of the evening, discussions which have provided me with great help and stimulus.

My gratitude is greatest of all for the enormous assistance and encouragement I have received from Astrid Ulfstrand and Inga Rudebeck. Astrid Ulfstrand has drawn all of the book's many illustrations meticulously and with interest. Inga Rudebeck produced the typescript from my handwritten original and furthermore retyped it with amendments and additions up to the final version. Inga Rudebeck in addition read through the whole manuscript and saved me from many factual errors, ambiguities and instances of poor wording.

Thomas Alerstam
Ecology Building, Lund, 29 December 1981

Preface to English edition

This present work on bird migration is an English translation of a book that I wrote in Swedish and which was originally published in 1982. Consequently there are at least two obvious biases. First, the contents are biased towards dealing with the migratory habits of Scandinavian birds, the reason being, of course, that I have felt at ease with discussing and presenting those birds and their migratory habits about which I have some first-hand field experience. Bird migration is a very intrusive and inspiring phenomenon in the natural scenery of northern Europe. I hope that the reader will also be able to feel some of that inspiration when I choose many of my examples from among the north European bird fauna, particularly since the majority of the bird species will be familiar to British ornithologists.

Secondly, the contents are biased in the sense that the very latest findings and studies (those which have been reported during the eight years since the book first appeared) are not included to any important extent. When reading the proofs of the present edition, I have, however, inserted a small number of updating corrections and additions, including recent population estimates and notes on some of the most important findings during the 1980s, and added some twenty new references. I hope that this will be considered sufficient as a minimal level of updating the original Swedish text. Still, this edition is in all important respects basically the same book, with the same text, illustrations and tables, as the Swedish edition.

It has been a great pleasure to have my book translated by David Christie. I am deeply impressed by the way he has completed a very extensive work in the finest possible way. David Christie has the unique combination of having mastered the Swedish and English languages and, in addition, having expert ornithological knowledge and being a keen birdwatcher; hence, he has contributed additions and updatings of population numbers and records from Britain, for which I am extremely grateful.

I also wish to thank the biological editors of Cambridge University Press, Martin Walters and Robin Smith, for their kind and positive determination to realise this English translation, in spite of the fact that it took several years before we were able to find a suitable translator.

Finally, both David Christie and I would like to express our sincere thanks to Karin Fancett for her meticulous copy-editing of the English translation.

Thomas Alerstam
Lund, February 1990

1 *The journeys of birds and the ideas of man*

Strictly speaking, it is cheating a bit to begin with these lines. The fact is that they were not written first at all, but just the reverse, at a very late stage of my work on the book, one day when – as so many times before – I had been at the library to look up information on bird migration. Every time that I visit the library, my gratitude and respect increases for all the striving for knowledge and enlightenment that is documented there in masses of books and scientific journals. Without all these books, papers and reports that ornithologists have published, it would of course have been impossible even to attempt to produce a book on bird migration.

Why are so many people prepared to put in such hard work to give accurate accounts of their observations and analyses and to develop their ideas in scientific literature? We live in an age of knowledge-seeking. Since the Renaissance, Western culture has to a great extent revolved around the ideal of seeking objective knowledge. Knowledge, not only for practical use but also for its own sake, in order to satisfy man's interest and imagination. The sciences have developed in order to cultivate and systematise these efforts. The result has been a tremendous evolving of knowledge. During the last 300 years, ever since Galileo and Newton, our conception of the world has been revolutionised time after time. And the pursuit of understanding of the world around us, as well as of our own situation and history, goes on. Up to now, science has succeeded in defending its stand against its most serious enemy – dogmatism. What we know, or think we know, will not be sanctioned without constant questioning. All knowledge is provisional and incomplete. We can always be wrong. Today, there are more scientists, working at universities and other research institutes, than there have been during the whole development of science put together.

The knowledge of birds – ornithology – is of course only one very small, in many respects minor, part of the natural sciences. Nevertheless, it has its obvious *raison d'être*. I find it hard to understand people who pick and choose among the sciences and regard some as uninteresting, sometimes even lacking any useful purpose, and others as of value. For my own part, I am convinced that I should be able to learn to appreciate the most diverse subject areas if I acquired a deeper insight into them. The whole world is remarkable, and the most remarkable thing of all is that we human beings can get to know it with our thoughts and ideas. Why is our conceptual ability such that we are capable of discussing the organisation and evolution of the universe, the properties of elementary particles, time and space, energy and mass, the symbolic harmonies of mathematics, the molecules that direct and run life and – why not? – the fantastic journeys across the earth of migratory birds?

The tools of research are the same in all the subject areas: our *imagination* and our *ideas*, combined with *criticism* and *controls* using repeated observations and experiments. There

1

really is no more to it. One often hears people who believe that research into animals and nature is more or less incompatible with natural experiences. The scientist is supposed to endeavour to measure and record objectively; to be led astray by subjective admiration of nature should not be permitted. This is a capital misconception. Nobody will convince me that imagination, richness of ideas and logic, the basis of research, create obstacles to one's imbibing to the full and being captivated by the spectacle of nature. Quite the opposite: admiration and respect for nature, indeed for the whole of our surroundings, are the chief driving force behind inquisitiveness and the striving for knowledge.

I do not, however, wish to disguise the fact that one can be thoroughly disheartened when in certain cases research becomes debased. There are a fair number of research reports that lack both new ideas and meaningful observations. Such articles are of course totally worthless from a scientific viewpoint. They are written in order to confer concessionary advantages in the hard competition on the academic career ladder. Shame to say, it has to be admitted that, in the academic context, good research is all too often equated with frequent 'production of papers'. The caricature of the academic has, I suppose, always had, and unfortunately I imagine always will have, its models.

There are also, however, thank heavens, plenty of examples of studies carried forward by intractable curiosity and an honest desire to understand. It brings joy to the heart and respect for nature to read and to learn from these research results.

To our knowledge of birds, indeed to our knowledge in general of the life of animals and plants – ecology – all who are interested can contribute. Important discoveries can still be made in this field without complicated devices or mathematical methods of analysis. Our knowledge of the universal patterns or rules that govern the organisation of life forms on the earth, the number of species, population sizes, ecological niches, migration patterns etc, is very superficial and tentative. The development of knowledge in this respect has been slow and irregular.

People who live in primitive hunting and gathering communities have naturally always been intimate with nature and birds. Their knowledge has rested on personal experience and has been supplemented with tales which have been handed down from generation to generation. Such knowledge – it may in many cases be characterised by great wisdom and clarity of vision – has its obvious limitations. For us to be able to talk of science requires that knowledge be administered so that it can constantly be developed. This in turn demands that it be stored on a larger scale than is possible using simply our memories. When put into print, knowledge can be accumulated and at the same time become generally available for checking and reappraisal. The specialist libraries are storehouses for our knowledge. Here can be found in stupendous variety ideas, speculations, observations and reports on experiments.

When knowledge about birds and bird migration was first put into print a good 2000 years ago, by Aristotle, it was not much to boast about. There are about 8700 species of birds on the earth, but Aristotle knew no more than 140 of them. Even in the 1700s, our knowledge was very fragmentary; Linnaeus could distinguish only about 500 different birds, and in his treatise on the migration of birds, *Migrationes Avium* (1757), he still propagated the delusion that the swallows hibernate at the bottom of lakes. Not until the 1800s do the outlines of modern biology begin to appear. The most important advance was Charles Darwin's theories on the origin of species and on the evolution of the adaptations of plants and animals through natural selection. The theory of evolution, which today forms the main basis of ecology and ornithology, still rests to the greatest extent on Darwinian principles.

Direct field studies of migratory birds were not begun to any appreciable degree until the end of the 1700s and the 1800s. They consisted largely of registers of the time of the birds' yearly arrival and departure. During the late 1800s, the observations from different countries were co-ordinated so that the birds' migration paths could be mapped. In order to administer the exchange of information, several ornithological unions and journals were founded which are still in existence today. The first international ornithological congress assembled in Vienna in 1884, with the expressed purpose of being a co-ordinating forum for the mapping of the migration of birds. The ringing method was not introduced until the turn of the last century, by the Danish schoolmaster Hans Mortensen. Systematic observations and counts of passing birds at particular sites were a further decade or two in coming. Investigations into the orientation capacities of birds, with emphasis on the roles of the sun and the stars, did not gather momentum until about 1950. The latest two milestones in bird-migration studies – the use of radar to follow bird migration across vast areas and at the highest altitudes, and the exploring of the birds' sensory world and their inbuilt magnetic compass – are only about 35 and 25 years old, respectively.

My aim in writing this book is to provide a review of this 'newly awakened' knowledge of bird migration. I have attempted to introduce discussions on various issues so far as today's knowledge and speculation extend, to be used as starting-points for those who wish to try out their own ideas and carry out new studies.

In working on this book, I have not only tried to throw light on the question 'how?' for various aspects of bird migration. I have in addition attempted constantly to keep alive the question 'why?'. This is not so simple as it sounds. The question 'why?' is very easy to forget or to disregard. Why, then, is it so easy to forget? One reason is that it often leads to problems both difficult and far-reaching. Another reason is presumably that during our school education we have been trained to learn how things function, but seldom why. During our education, knowledge is presented in a way that is pedagogic but which at the same time inhibits the imagination, so that gaps in knowledge are covered up and a veil of oblivion is drawn over all the errors and misinterpretations that have been made during the meandering process when the knowledge was acquired. To deepen one's knowledge becomes mainly a question of having the stamina and a good memory, of swotting and writing carefully.

Before we have reached school age we constantly ask 'Why?'. Even though many questions of why are impossible to answer concisely and tangibly, they allow considerably greater scope for the imagination than do questions of how. Why does the Thrush Nightingale move all the way to tropical Africa in winter, but its close relative the Robin only travel to west and south Europe? And why does the Thrush Nightingale leave the tropics to move far to the north for a few months in summer? Why do some birds migrate by night and others by day? It gives us much useful food for thought if we try to revive the curiosity and the childlike mind by asking 'Why?'.

This book is, of course, also written extensively for those who do not have the time or ambition to make their own systematic studies of nature but who wish all the same to gain an insight into the lives of migratory birds. I find it difficult to believe that an increased knowledge of nature can lead to anything other than increased caring for our world, for life and for ourselves.

2 *The rotating world of migratory birds*

The earth above the moon's surface. That is a picture which, I believe, has engraved itself deep in the minds of many people and which is very important to our view of the world. Above a yellowish-white, sterile moon surface, unprotected by atmosphere and filled with craters, the earth is suspended in a sky of jet black emptiness. Across the earth's surface stretch winding trails and complex formations of clouds, the atmosphere and the seas alternate between sparkling blue and dull green, the landmasses are dimly visible in ochre, brown and dull violet. Had we been able to look at the earth for a while longer, then we should have seen how it was spinning around, how the trails of cloud all the time altered their state, became thinner and grew bigger.

In a thin crust around the earth – the land surface, the sea and the atmosphere – life is found. The plants trap the solar radiation and use the energy for growth; animals feed on the plants, and they in turn become food for other animals. The whole of life may be seen in this way: as a single great system, as a gigantic organism adapted to take advantage of solar radiation. Sunlight is difficult to capture. Despite their every fantastic adaptation, the plants succeed in trapping and utilising only $\frac{1}{2}$% (=five-thousandths) of the total radiation energy that strikes the earth. The plants use about a half of this quantity of energy for their respiration; the other half they use for growth. Only $\frac{1}{4}$% (=two- to three-thousandths) of the solar energy is turned into plant mass. This perhaps does not sound so impressive, but nevertheless it is the equivalent of an additional hundreds of thousands of millions of tonnes of plant mass per year on the earth. If we reflect on this, then we get a better understanding of how the plants' exploitation of sunlight can be the foundation of all life.

Now some may perhaps wonder what this view of the earth in such a broad perspective has to do with bird migration. Well, I shall explain this straightaway. It is of course the earth, the whole earth, that is the stage for these fantastic birds with their miraculous journeys. The travels of thousands upon thousands of millions of birds range across continents and oceans, on flight to richer food sources, fleeing from predators and competitors, for it is mobility and speed above all that are the hallmark of birds.

There are about 8700 species of birds in the world, and only a very small number, for example Ostriches, penguins, and rails on isolated oceanic islands, have lost the power of flight yet still succeeded in holding their own in the competitive world of nature. It has often been demonstrated how fragile the adaptation of flightless birds is: the species easily become extinct if other animals or man disturb their environment. Several species of flightless rails have become extinct on isolated groups such as the Hawaiian Islands, the Fiji Islands, the Auckland Islands and Tristan da Cunha. Among the best examples of birds lacking the power of flight that have died out in recent times are the three species of dodo (big birds in the 20-kg

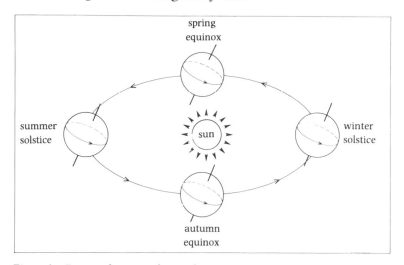

Figure 1   During the year, the earth moves around the sun on a faintly elliptical path. The diagram shows the seasons for the northern hemisphere. The seasons arise because the northern hemisphere is tilted, towards the sun in summer and away from the sun in winter. During spring and autumn equinoxes, the earth's axis is not tilted in either direction in relation to the sun. The earth spins on its axis from west to east, one revolution every 24 hours.

class, related to pigeons). The Grey Dodo on Mauritius, the Reunion Dodo on Reunion Island and the Solitaire on Rodriguez Island all disappeared during the 1600s and 1700s. The northern hemisphere counterpart of the penguins – the Great Auk – once had a wide distribution, but was driven from its final bastion, in Iceland, during the 1840s. Several other flightless species are today on the verge of extinction.

The great majority of birds depend on their wings – they flap, soar, hover, dive. We human beings cannot keep pace with the birds, but our thoughts and dreams are able to fly with them. Perhaps that is the reason why birds attract the interest, admiration and wonder of mankind to such a great degree.

Birds that migrate many thousands of miles are by no means unusual. Right outside our homes and houses, over the nearest wood, lake or shore, there are birds flying about which have just arrived from or will shortly be setting off for the Arctic Ocean, Siberia, Africa, India . . . But before I set about presenting in greater detail the actors in this drama on bird migration, I should like to outline what the stage looks like, and the stage is of course our and the migrating birds' rotating earth.

Why do the birds migrate? Many different kinds of answer can be given to this question, some of which I shall discuss in more depth in the section on the evolution of bird migration (3.10). Here I was going to give the answer that, well, birds migrate because we are tilted 23.5°. Let me at once explain in more detail. The axis around which the earth turns one revolution every 24 hours is not at right angles to the plane of the earth's path around the sun (if it had been, then there would probably not have been so many migratory birds), but it is tilted 23.5° (see figure 1). This results in the northern hemisphere being slanted away from the sun during the winter half of the year and towards the sun during the summer. The northern hemisphere accordingly receives much more solar energy during the summer than during the winter (summer and winter are of course transposed in the southern hemisphere).

This is the ultimate reason for the seasonal variations in climate and vegetation. The northern hemisphere is tilted most away from the sun, at exactly 23.5°, at the winter solstice; and most towards the sun, also at 23.5°, at the summer solstice. At the spring and autumn equinoxes, the earth's axis is not turned either towards or away from the sun. The earth moves on a very slightly elliptical, verging on circular, path around the sun. It is a little closer to the sun during the northern winter (147 million km) than during the summer (152 million km). The result of this difference in distances is that the solar radiation on the earth is a good 6% stronger in January than in July.

The earth's tilt, the shape of the earth's orbit, and the times when the distance between earth and sun are shortest and longest vary over the course of thousands of years. Thus, for example, the earth's tilt varies between approximately 21.8° and 24.4° over a period of about 40 000 years. These variations exert an influence on the earth's climate. A reduction in the tilt of the earth's axis leads to milder winters and cooler summers both in the northern and in the southern hemispheres, and can bring about extensive glaciation of the areas near the poles. By taking into account the variations in the movements of the earth and of the sun, we can analyse the origins of the ice ages and forecast that the warmer period in which we are living right now is temporary and that we are in all probability heading for a new ice age. I shall return to the developments that followed the last ice age and to their significance for bird migration in a later section.

The solar radiation that reaches the outer atmosphere has an energy flow of 2 calories per square centimetre per minute (= 1400 watts per square metre). The radiation becomes weaker as it passes through the atmosphere; only about half of the energy reaches the earth's surface. In the atmosphere, the ultra-violet radiation is absorbed in an ozone layer at an altitude of about 40 000 m. Much heat radiation (infra-red radiation) is absorbed by water vapour, and some of the light is diffused by molecules and dust particles in the atmosphere. Part of the diffused radiation is lost out in space and part of it reaches the earth and makes the atmosphere appear bright. Short-wave light (in the blue part of the spectrum) is more widely diffused than long-wave light (in the red part of the spectrum), and therefore the sky becomes blue. Well, then, since the absorption and the diffusion in the atmosphere are fairly similar at whatever latitude one happens to find oneself, I shall now proceed from the total radiation that reaches the outer atmosphere and see how the radiation energy is distributed at different latitudes during different times of the year. Let us look first at the situation at spring and autumn equinoxes, when the earth's axis is tilted neither towards nor away from the sun (see figure 2A). At these times, every spot on the earth has 12 hours of light and 12 hours of darkness during the day. Places on the equator receive most solar energy; when the sun is at its highest point in the sky, sunlight there falls at right angles to the earth's surface (the sun is at zenith). On every square centimetre 2 calories of radiation energy per minute fall at this time. In the northern and southern hemispheres the radiation is less intense, since here it falls at an angle to the surface; at latitudes 60° N or 60° S, for example, the radiation in the middle of the day is only 1 calorie per square centimetre per minute, and at the poles it is zero – there, no sunlight falls directly onto the earth's surface. When we look at the situation at the summer solstice, we get another picture (figure 2B). The earth's axis is now tilted 23.5° towards the sun. In the middle of the day the solar radiation is therefore at its most intense at latitude 23.5° N (the Tropic of Cancer), where the sun is at the zenith. In addition, the length of daytime is different at different latitudes: at the equator day and night are equal in length, while in the northern hemisphere day is longer than night (the reverse applies in the southern hemisphere). The day increases in length northwards, and it can be seen that north of the Arctic Circle, at latitude 66.5° N (90 − 23.5 = 66.5), it is light throughout the 24 hours,

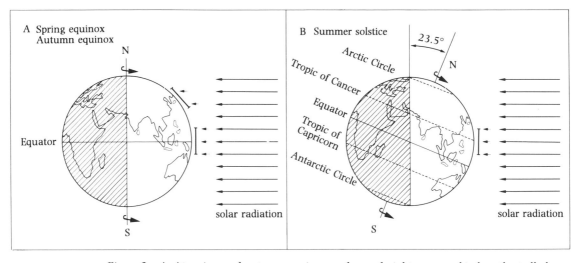

Figure 2   A. At spring and autumn equinoxes, day and night are equal in length at all places on the earth. Shaded areas in the figure show night, and unshaded day. At midday solar radiation is most intense at the equator, where the light falls at right angles to the ground (the sun is at zenith); solar radiation becomes less at other latitudes, where it falls at an angle to the earth's surface.

B. At the summer solstice, the northern hemisphere is tilted $23.5°$ towards the sun. At midday the sun is at the zenith at the Tropic of Cancer ($23.5°$N). Day is longer than night in the northern hemisphere, and night longer than day in the southern hemisphere. North of the Arctic Circle ($66.5°$N) it is light throughout the 24 hours of the day, while it is continuous night south of the Antarctic Circle.

whereas continuous night prevails south of the Antarctic Circle. At the winter solstice the situation is essentially the same, though of course with reversed roles for the northern and southern hemispheres.

When it comes to calculating how much energy, through sunlight, reaches different latitudes in one day at a certain time of year, we must accordingly take into consideration the day length and at what angle the solar radiation falls (=the sun's angle of elevation) during different times of the day. If we imagine that the earth is standing still and that the sun orbits around it at one revolution per 24 hours, we can construct the solar arcs during different seasons for different latitudes as shown in figure 3. I shall have occasion to come back to the difference in solar arcs between different places and times of year when I discuss the orientation of migratory birds (there are many theories on how migrants are guided by the sun).

The amount of total solar energy that reaches the earth during one day at different latitudes and times of the year is shown in figure 4. The situation between the northern and the southern hemispheres is not entirely symmetrical, owing to the fact that the sun is closest to the earth in January and farthest away from it in July. While there is a uniform supply of energy in areas near the equator, the amount of solar energy varies more and more substantially the closer one gets to the poles. For a short period of time, a month or two, there is more radiation energy streaming to the poles each day than to the equator, but for half of the year there is no sunlight at all at the poles. Over the year as a whole, the most energy reaches the equatorial areas and the least, not quite half as much (42%), gets to the poles.

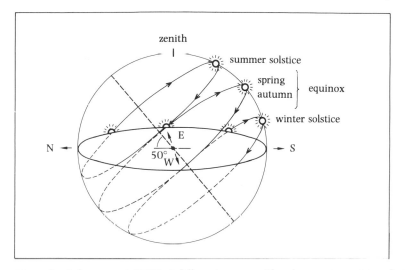

Figure 3   Solar arcs at 50° N at different seasons. The observer is positioned at the mid point on the horizontal plane, with the south to his right, the north to his left, east facing inwards into the diagram and west facing out of the diagram. The sun's apparent movement during one day traces circles on a celestial sphere around the observer. At spring and autumn equinoxes the sun's movement follows the sky's equator, at the summer solstice the celestial sphere's northern tropic (latitude 23.5°), and at the winter solstice the celestial sphere's southern tropic. The celestial sphere's axis is tilted the number of degrees of latitude where the observer is positioned on the earth from the horizontal plane, i.e. 50° for latitude 50° as in the diagram. One can easily discover, for example, that if one increases the angle (to 66.5° or more) the observer (who is then exactly on or north of the Arctic Circle) can see the midnight sun in the north at the summer solstice. If one reduces the angle to 0°, the observer (who is then at the equator) has the sun at the zenith in the middle of the day at spring and autumn equinoxes; the sun is always to the north during the summer months and to the south during the winter.

The diagram in figure 4 gives the clue to the climate and the vegetation on the earth. The plants are dependent on the sun for their growth. It can be seen that a steady plant production can take place throughout the year around the equator, while in the polar regions growth is possible only during a short period of the year. During these short periods, there is on the other hand solar energy enough to produce a vigorously intensive growth. The difference in energy supply between the equator and the poles leads to heat being transported on winds and ocean currents. The climate varies during different seasons as a result of the shifts in the supply of solar energy. Migratory birds follow the seasons' different climates and different plant communities.

Let us leave this external perspective on the earth and the sun and move down to the earth's surface to take a closer look at what is happening in the lower atmosphere, in what is known as the troposphere. There is every reason to make this restriction; the troposphere functions almost as a closed system. Very little exchange takes place with higher layers in the atmosphere. The upper limit of the troposphere is called the tropopause and lies on average at an altitude of not quite 10 000 m above the polar regions and at approximately 17 000 m altitude above the equator. In the troposphere all weather is enacted, clouds and precipitation are found there, the birds fly there.

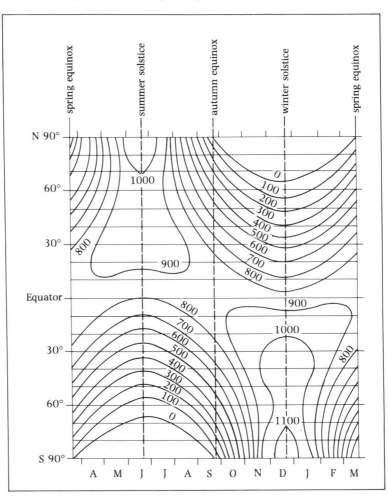

Figure 4   Daily solar energy, in number of calories per square centimetre, reaching the earth at different latitudes and times of the year. For those who prefer standard units: 100 calories per cm² is the equivalent of 4.2 million joules per m². Based on Lamb (1972).

It is the winds that shape the climate, that transport heat and moisture and that drive the ocean currents. I shall, therefore, dwell for a while on the passage of the winds across the surface of the earth. I do this not simply because the winds are the essential thing in the whole theory of climate, but also because the winds are of the very utmost importance for bird migration. Actually this is not that odd; when the birds, with their comparatively low flight speeds, launch themselves out into the open wind on long-distance journeys, it must be of vital importance that they are adapted to act in such a way that they do not waste valuable energy or risk flying dangerously off course and being wind-drifted. What is odd is that generations of ornithologists have taken most interest in the significance of the temperature and the landscape for bird migration and more often than not totally disregarded the wind. Perhaps man had first to gain a general experience of knowing how it feels to fly in a small aircraft in strong winds before he could properly imagine the situation of migrant birds in the air. In this book I shall frequently discuss the importance of the wind to migratory birds, to their routes, flight altitudes, speeds, direction of flight, and to their tendency to follow directional lines and to make use of thermals.

But now first to a global perspective on the winds.

The fact that the warming-up process is much greater at areas near the equator than at the poles leads to the warm air above the equator rising, creating an area of high pressure at high altitude and moving at high level towards lower pressure above the polar regions. This movement is compensated for by the fact that cold polar air at low altitude streams towards the equator and fills the low pressure near the ground beneath the rising, warmed air. The large-scale wind movement would presumably have been as simple as that had the earth not been spinning around its axis. The same sort of circulation arises in a room with a heating element on one wall. The warm air rises above the element, moves along the ceiling towards the opposite wall, is cooled, and falls, while cool air draws across close to the floor towards the element. At floor level there is, therefore, high pressure at the cold wall and low pressure at the element, while the situation is reversed beneath the ceiling. This corresponds to the situation in the troposphere. Just beneath the 'ceiling', that is the tropopause, there is an area of high pressure at the equator and low pressure over the North and South Poles. At the 'floor', that is the land and sea surfaces, there is a low-pressure area at the equator and a high-pressure area at the North and South Poles.

The same phenomenon also explains onshore and offshore breezes in coastal areas. During the day the land is warmed more than the sea, the warm air over the land rises and pushes out over the sea, while cooler air at low level blows in from the sea over the land (sea breeze). At night the circulation is the opposite: the land surface is then cooled down more than the sea, and at low altitude the offshore breeze blows out towards the sea.

I shall now set the earth spinning, one revolution per 24 hours, and something remarkable then happens. There is, then, at high altitude, beneath the tropopause, high pressure over the equator and low pressure at the poles. The air accordingly moves towards the poles, but through its movement is influenced by a force which pulls to the right in the northern hemisphere and to the left in the southern hemisphere. The result is that the winds are westerly at high altitude in both hemispheres, for an equilibrium is then set up where the intensity of pressure is balanced by the force turning to the right in the northern hemisphere and by the one turning to the left in the southern. This remarkable force that crops up when the earth rotates is the Coriolis force. This force is actually a kind of illusory force; it is due to the fact that we want to follow the wind movement in relation to a rotating system of reference. The earth spins one revolution every 24 hours, and, as the earth's radius is approximately 6370 km, we can easily work out how fast the earth's surface is rotating at different latitudes (see figure 5). If we are standing at the North Pole or the South Pole, then of course we do not revolve at all, but if we find ourselves on the equator we whiz around at 465 m every second = 1670 km per hour! It sounds almost unbelievable – surely we ought to shoot off the earth as a result of centrifugal force? No; if we work it out in a little more detail, we find that the centrifugal force at the equator is no greater than three-thousandths the force of gravity. Therefore, generally speaking, we have just as firm a foothold on the earth at the equator as at the North or South Pole (the earth is in fact not completely round but slightly squashed at the poles; this, combined with centrifugal force, makes the force of gravity a shade stronger at the poles than at the equator). But Coriolis force should not be confused with centrifugal force. Coriolis force is dependent on an object's *movement across the earth's surface*. If I stood at the North Pole and shot off a bullet towards the horizon, I should be able to see how the bullet gradually turned to the right (in reality I am not of course capable of perceiving this). The bullet is influenced by the right-handed turn of Coriolis force. What in

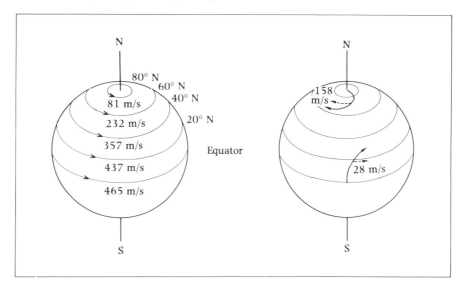

Figure 5   On the left is shown the speed of rotation at the earth's surface at different latitudes. When an object moves across the surface, it will be influenced by Coriolis force, which has a right-hand effect in the northern hemisphere and a left-hand one in the southern. An object that moves from the equator to latitude 20° N strives to maintain its speed, and will therefore take on an easterly-directed speed component in relation to the earth's surface of 28 m/s (the difference between the speed of rotation of the earth's surface at the equator and at latitude 20°). Near the North Pole the Coriolis effect is considerably greater – an equally long movement from the North Pole to latitude 70° N leads to a westerly-directed speed component in relation to the earth's surface of 158 m/s (= the speed of rotation of the earth's surface at latitude 70°). Coriolis force also has an effect when movement is in an east–west direction (see text).

fact happens is that the bullet maintains its rectilinear movement while the earth rotates beneath it. The earth spins from west to east and therefore the bullet moves to the west across the earth's surface.

Newton's second law of motion, the cornerstone of classical mechanics, states that a particle retains its direction of movement and its speed unaltered if it is not influenced by any force. If a particle moves from the North Pole to latitude 70° N, the earth will be spinning beneath it to the east at a speed of 158 m per second (figure 5).

If we take the earth's surface as our point of reference, the particle accelerates to the west and at 70° N reaches a westbound speed factor of 158 m/s. A force must be responsible for this acceleration: Coriolis force. If a particle moves northwards from the equator, it will maintain its rapid speed of rotation and drift to the east. When it reaches a latitude of 20° N, it has accelerated to an easterly-directed speed factor of 28 m/s.

Coriolis force also works with movement in an easterly or westerly direction. Owing to the earth's rotation, a particle has at any given moment a movement that points at a tangent obliquely outwards from the earth's surface in the direction of rotation. Coriolis force acts to maintain this movement when the particle moves across the earth's surface; with an eastward motion, therefore, Coriolis force acts obliquely outwards from the earth's surface, and with a westward motion obliquely inwards towards the earth. This force has a rightward-directed component along the earth's surface in the northern hemisphere and a leftward-directed one in the southern hemisphere. At the equator no force factor exists at all along the surface of the earth.

Now I shall not get further wrapped up in forces in rotating systems, but I can summarise them as follows: Coriolis force acts on objects in motion across the earth's surface and pulls straight to the right in the northern hemisphere and straight to the left in the southern. The force is directly proportional to the object's speed. It is zero at the equator, and increases as latitude increases.

The force has an effect on, for example, a flying bird, and it has been wondered whether migratory birds could be aware of it and so determine what latitude they are at. In all probability, however, Coriolis force is far too weak to be detected by flying birds. Coriolis force is on the other hand of the greatest import for wind movements. It is now easy to see why the wind which is on its way at high altitude from the equator to the poles is deflected to the right in the northern hemisphere and to the left in the southern one, so that stable west winds arise where Coriolis force counterbalances the pressure strength in both hemispheres.

The strongest winds originate in the middle latitudes, that is between latitudes 30° and 60° N and S, where the temperature extremes and therefore also the pressure extremes are at their greatest. Here, the high-level west wind is channelled into what are termed jetstreams just beneath the tropopause. There are two or three jetstreams in each hemisphere. Those nearest the poles are known as polar-front jetstreams. In the northern hemisphere a polar-front jetstream 200–500 km broad passes over Europe, farther north in summer than in winter. The winds in this stream as a rule blow at more than 25 m/second (= 90 km/hour), and in the winter, when the contrasts in temperature between northern and southern regions are at their greatest, often at over 50 m/s (180 km/h). Extreme speeds of over 100 m/s (360 km/h) have been measured over widely separated areas in Europe between the Mediterranean Sea and latitude 65°. The subtropical jetstream is as a rule both broader and stronger and extends just north of latitude 30° in the northern hemisphere, this stream too being more northerly in summer than in winter. On a few occasions when the severe winter cold over Siberia and northern North America has forced the polar-front jetstream exceptionally far to the south, it has overlapped and joined up with the subtropical jetstream, and then wind speeds of up to 150 m/s (540 km/h) have been measured, for example over Japan. Similar records for wind speeds in the jetstreams have been established in the southern hemisphere.

It was, incidentally, in Japan that the existence of the jetstreams was discovered, as recently as during the Second World War. The Japanese tried to exploit these powerful winds for a balloon offensive against the United States. Hydrogen balloons with a cargo of bombs were sent up to approximately 10 000 m altitude, on the calculation that they would rapidly be carried on the westerly jetstream across the Pacific Ocean to the North American continent. A good number of bombs in fact reached their target in this manner, but the attempt failed owing to the fact that the release mechanisms generally froze up in the severe cold of the jetstreams.

The high jetstreams and the west winds in the middle latitudes in fact blow so hard that Coriolis force prevails over pressure force (which acts towards the poles from the equator). The air is therefore pushed towards the equator across these areas, while the opposite situation obtains at other latitudes. The result is that air moving at high altitude from the equator towards the poles meets air travelling in the opposite direction at approximately latitude 30°; the air is then compressed downwards towards the surface of the earth and there creates the subtropical high-pressure belts (see figure 6).

Where the high-level west winds in the middle latitudes derive their extra energy from is

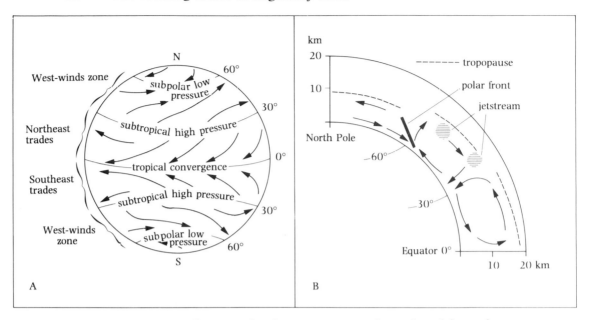

Figure 6   A. Different wind and pressure zones at the surface of the earth.
B. Cross-section through the atmosphere above the northern hemisphere to show the circulation in the troposphere. A mirror situation prevails over the southern hemisphere. The cross-section shows the transport of air in vertical and north–south directions which balances out heat. The most powerful wind movements in the atmosphere occur in an east–west direction, as for example in the westerly jetstreams, where the wind direction accordingly is pointing straight inwards into the diagram. Based on Lamb (1972) and Liljequist (1970).

not entirely clear. Many people believe that it comes from the cyclones that originate at the polar front and move across the middle latitudes.

The earth's various pressure and wind zones are shown in figure 6. The situation at high altitude is already clear: here, it is principally west winds that blow in both hemispheres, and they are strongest above the middle latitudes where the jetstreams are also blasting ahead. In total, a net transport of air takes place from the equator towards the poles on both sides of the middle latitudes, while the direction of net transport is the reverse at these latitudes (figure 6B).

At the earth's surface, the wind pattern is a different one – and it is the surface winds that are of the greatest interest when it comes to migrating birds. From the subtropical high-pressure belts winds blow towards the low pressure beneath the rising air at the equator. Owing to Coriolis force, these winds are deflected to the right in the northern hemisphere and to the left in the southern one; the result is the northeast trade-winds and the southeast trades. The power of friction for winds near the surface of the land and sea leads to the winds there blowing more obviously from high pressure towards low pressure than at higher altitudes. The trade-winds are stable in direction and strength. The wind strength is as a rule moderate, between 4 m/s and 10 m/s. The term 'trade-winds' indicates the value placed on them by merchants in olden times. One could always count on the winds filling the sails and enabling merchant vessels to make headway.

The meeting place of the trade-winds, the tropical convergence zone, is characterised by weak and variable winds or by dead calm weather. For sailing vessels, the risk of becoming becalmed there for long periods is high. Another belt with calm weather like this one is the subtropical high-pressure zone. These latitudes were termed 'horse latitudes' in the age of the sailing vessels. This odd term has its explanation. Quite a number of the vessels that sailed between the New World and the Old World carried horses. It sometimes happened that ships were held up for weeks on account of poor winds somewhere between latitudes 25° and 30°. Many horses died of thirst and starvation and had to be heaved overboard while the sailors waited disconsolately for the wind to rescue them.

The tropical convergence and the subtropical high-pressure zones differ radically, however, when it comes to other aspects of climate. The air rises at the tropical convergence, the water condenses on cooling, and cloud and precipitation is formed – more often than not heavy squalls and thundery weather. Within extensive areas around the equator there is thunder on more than 150 days of the year; indeed, at some places in the Amazon and Congo Basins, at Lake Victoria and on Java it thunders on more than 200 days of the year! In the subtropical high pressure zones, the weather is totally different. Here, dry air descends and there is no cloud but instead a completely clear sky with a relentless sun given free rein. At these latitudes we find the earth's desert regions: the southwest United States and north Mexico, the Sahara, the Arabian and central Asian deserts in the northern hemisphere, and the Kalahari desert and the deserts of Australia and beside the Andes in the southern hemisphere.

The air that moves from the subtropical high-pressure belt towards the subpolar low-pressure zone gives rise, because of Coriolis force, to west winds over the middle latitudes in both hemispheres. As a result of friction, the mean wind direction nearest to the earth's surface becomes southwesterly or west-southwest in the northern hemisphere. With the west winds come drifting low- and high-pressure systems in which the wind may vary in all different directions. With these drifting cyclones comes precipitation, too, which is due ultimately to the collision between the warmer southerly (in the northern hemisphere) air and the polar air. Indeed, we who live in these latitudes are only too aware of this. And we can also attest that the west winds can certainly become fresh! The west winds become even stronger in the southern hemisphere, where there is not so much landmass so the winds can sweep along across the open sea with minimal friction. Among seafarers, 'brave west winds' are talked of with great respect, and it is said to be the whalers who christened the southern latitudes, in view of the winds, 'the roaring forties, the howling fifties, and the screeching sixties'. We can easily imagine why the passage around the tip of South America, Cape Horn, just south of 50° S, has become so dreaded by and legendary to mariners. Now do not think that these west winds deter the birds. On the contrary, these sea areas above all others are the habitat of the albatrosses. As well as these, there are masses of other seabirds; the Arctic Tern, for example, spends the winter (that is to say the southern hemisphere's summer) in this region.

Over the polar regions, the prevailing winds are generally easterly ones. These penetrate south towards the subpolar low-pressure belt and there meet warmer air masses at the polar front.

What I have written above gives a rough summary of the wind pattern on the earth, an explanation for which is thus ultimately to be found in the differences between the amount of heat that solar radiation provides at different latitudes combined with the effect of the earth's rotation. This has been demonstrated in a convincing way in laboratory experiments. Through studying the circulation of water (the air of the atmosphere) in a rotating shell (the earth surface), which is heated at the edge (the equator) and cooled in the centre (the poles), it can be seen how the water circulation is modified from the simple 'heating element in room'

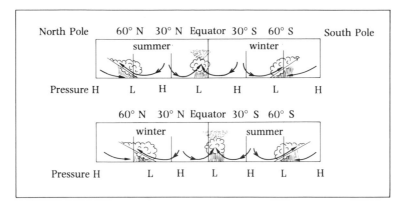

Figure 7 Outline of the earth's main climatic zones and their displacement at different seasons. Based on Liljequist (1970).

pattern at slow rotation to a more complicated pattern at faster rotation, when separate zones and frontal effects appear that are analogous with those that are found on the earth.

Wind and pressure zones are displaced northwards in the summer and southwards in the winter as a result of the varying amount of warming up from the sun. The redistribution of heat is a considerably sluggish process. The climatic zones are therefore not displaced all that much, on average 10–20 degrees of latitude during the year. In figure 7 it can be seen that this leads to two annual rainy seasons at the equator, in spring and in autumn when the tropical convergence zone passes. In the regions north of the equator there is only one rainy season, which falls during the northern hemisphere's summer. The rainy season south of the equator occurs during the northern hemisphere's winter. The farther from the equator and the nearer to latitude 30° one gets, the weaker and more uncertain the annual rainy seasons become. In the area where the subtropical high-pressure belt drifts backwards and forwards, a very small amount of precipitation falls during the course of the whole year. This is where the earth's desert areas are found. Nearer the poles, the polar front's precipitation is displaced with the seasons. Thus, the drifting areas of low pressure often reach south to the Mediterranean region in the winter and then bring rain. In the summer, the same area is covered by a desert climate with sun and heat.

This simple rough outline is fairly accurate in general, but there are important discrepancies. These discrepancies occur over large continents, especially in the zone where land and sea meet. There, the phenomenon can arise of sea and land breezes of continental dimensions. The summer sun's warming up of interior Asia brings rising air over land. A deep, low-pressure centre is formed over northern India. Moist southwest winds flow from the sea in over land towards the low pressure and lead to extensive monsoon rain over large parts of south and southeast Asia. In the winter the interior of Asia is very considerably cooled, with a gigantic area of high pressure over Siberia as a result. The winds now blow from the land out over the sea, and there is very little precipitation except in those parts of southeasternmost Asia – the Philippines, Indonesia and Sri Lanka – which are exposed to the northeast winds that pass over the sea.

An airstream coming from the sea and crossing over warmed-up land helps to keep the amount of precipitation plentiful and fairly evenly distributed throughout the year in some regions at the equator, for example on the River Amazon and the River Congo.

Mean temperatures at the earth's surface in January and July are shown in figure 8. While the temperature near the equator remains at a constant and high level all year, it obviously varies much more widely nearer the poles. The temperature difference between summer and winter is greatest in the northern hemisphere with its vast landmasses. The cooling of the northern and central parts of the North American and Asian continents during the winter is very considerable. The difference between the coldest and hottest months of the year is more than 40 °C here; indeed, in northern Siberia there are areas where the difference exceeds 60 °C! There are certainly not many birds that stay to winter in these parts! In northwest Europe, the contrast in temperature between summer and winter is considerably less drastic, between about 15 °C and 20 °C. This is due to the proximity of the sea and to the warming effect of the Gulf Stream.

There is every reason to study figure 8 in particular detail so as to imprint on one's mind how the 0° isotherm runs. Below-zero temperatures of course bring frost and snow, which prevent many birds from getting at their food. Birds that live on lakes or beside shallow water, or which take their food in and on the ground, are among the first to be hit. It is not, therefore, surprising that the 0° isotherm marks roughly the northern limit of winter distribution for

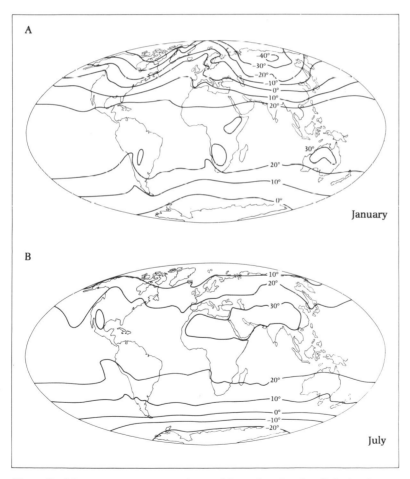

Figure 8 Mean temperatures on the earth's surface (sea level) during January (A) and July (B). What happens to the 0 °C isotherm is of particular interest. Based on Lamb (1972).

many such birds. Several of northern Europe's best-known birds are among those which only just manage to avoid frost and snow, for example the Lapwing, the Skylark, the Starling and the thrushes. Owing to the warmth from the sea in winter, the 0° isotherm runs almost directly north to south through Europe – from westernmost Norway across the southwest Baltic Sea (Falsterbo is the only Swedish weather station that during the 1900s has had mean temperatures above zero in all months of the year) and down to Switzerland and southeast France. Just north of the Mediterranean Sea, the 0° isotherm swings sharply eastwards towards the Black Sea and the Caspian Sea. West Europe, with its mild climate, is thus a favourable wintering area for masses of birds from north and east Europe; some species even come from Greenland and Siberia. The assertion that birds migrate south in autumn is therefore a greatly qualified truth in wide parts of Europe: here, the case is instead that many bird species fly west in the autumn.

The darkness and the cold place restrictions on life at the northernmost and southernmost regions of the world for large parts of the year. By contrast, the heat and the drought are limiting factors in many regions around the equator and the tropics. Figure 9A shows, for different areas on the earth, the length of the vegetation period during the year, that is the period when the plants are capable of effectively exploiting solar energy in order to germinate and grow. The vegetation period is a time when many new ecological niches are created for various organisms, a time when life can bubble over and expand. The migration of birds is an adaptation that makes it possible for them to get quickly to the unoccupied niches that are made available temporarily in various areas.

The vegetation period extends over the whole year in the rainforest areas in South America, Africa, Southeast Asia, Indonesia and New Guinea, where a stable, warm and humid climate prevails. There are few migrant birds in these forests; there is not even room for long-distance winter visitors from northern parts, despite the fact that life is more luxuriant there than anywhere else on earth. The rainforests provide their birds with a stable and enormously diversified food supply throughout the year. The birds therefore do best if they are resident within a smallish area. Only when man intrudes, chops down the forest and creates clearings and plantations, is the stability upset so that the migrant birds get the chance to force their way in and hold their own in competition with the resident species.

In southern Sweden one can reckon on a vegetation period of between 200 and 240 days, approximately the period of the year when the mean daily temperature is higher than $+3\,°C$. In the far north of Scandinavia the vegetation period is barely 130 days, half what it is in the south. In Fennoscandia the clear majority of birds are migratory, only a few percent remaining there during the winter.

The shortest and most intensive growth period occurs on the tundra and the dry savanna/steppe around the edges of the deserts, two widely differing types of country. The warmth from the sun and the winds melt the snow and the ice for a short period on the open tundra. The summer sunlight around the clock allows an explosion of growing mosses, lichens, grasses and shrubs. The ground is permanently frozen except in a thin surface layer of about half a metre to a metre, and the meltwater from the snow impregnates the ground, collects in depressions and forms lots of small lakes. What glorious country for such birds as waders! But those birds which nest here have only a short time. As soon as the young are ready to fly they must be off, before the waters and the ground freeze and the snow starts drifting.

For the greater part of the year, the savanna south of the Sahara is powder-dry beneath a fiery-hot sun. The grass is brown and burnt, bushes and trees are leafless. The ground is cracked in many places and the riverbeds are dried out. Only at the largest rivers and lakes

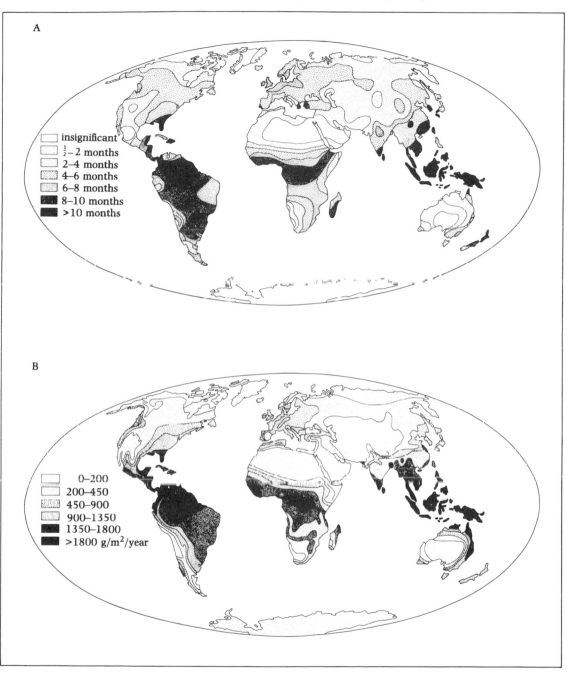

Figure 9   A. Length of vegetation period within different areas on the earth.
B. The average annual production of plant mass expressed in grams dry weight per square metre per year. The dry weight represents in round figures 20% of the wet weight, and 1 gram of plant mass (dry weight) corresponds to an energy amount of about 4 kilocalories. Based on Lieth & Whittaker (1975).

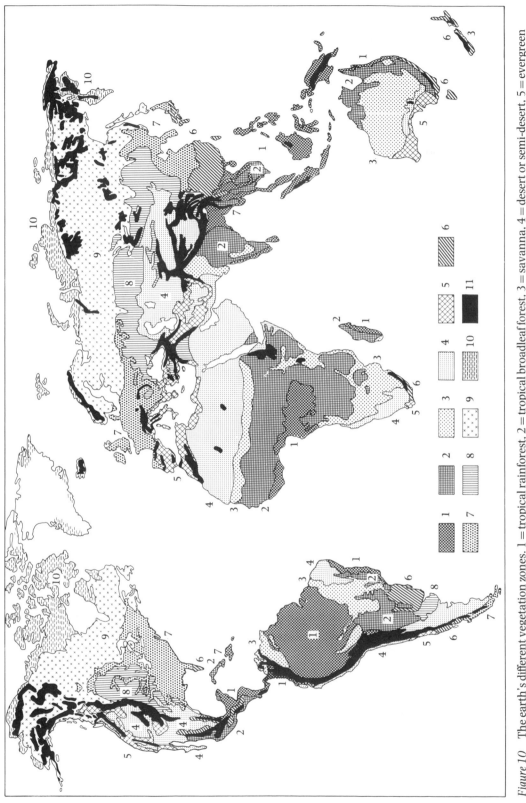

Figure 10 The earth's different vegetation zones. 1 = tropical rainforest, 2 = tropical broadleaf forest, 3 = savanna, 4 = desert or semi-desert, 5 = evergreen scrubland, 6 = temperate rainforest, 7 = temperate broadleaf forest, 8 = temperate steppe or grassland, 9 = coniferous forest, 10 = tundra, 11 = mountain regions. Note that the map projection is not accurate in terms of relative area. Based on Walter (1973).

can water and some green plants be found. Over large parts of the savanna fires rage every dry season, often the work of people who are preparing land for farming. The fires are not uncommonly allowed to rage freely over wide areas. Then the rainy season comes: the ground is flooded, the riverbeds and the lakes are filled, and plants grow at a furious pace. Bushes and trees burst into leaf; everywhere there is fresh, tall grass waving about. And at that time, of course, the birds are there – herons, storks, ducks, cuckoos, nightjars, bee-eaters, swallows, sunbirds and many others.

Bird migration is closely linked with the seasonal fluctuations – the tundra and the dry savanna present extreme instances. But the savanna in addition to this affords a real surprise. Many northern migrants winter during the dry season in the savanna regions south of the Sahara. Despite the heat and the drought, the warblers, flycatchers, shrikes and wheatears obviously find sufficient food. This may seem incredible when we take a cursory look at the dried-out landscape. We may wonder why in the world the birds do not fly a few thousand kilometres farther on to meet the rainy season, which is to be found south of the equator during the northern hemisphere's summer (see figure 7) and which should logically afford much better feeding conditions. This problem is usually known as 'Moreau's paradox', since attention was first drawn to it by the leading expert on African ornithology, the English ornithologist R. E. Moreau. There will be room for speculating on the solution to this paradox in a later section.

It is the sum total of climate and ground conditions that determines the distribution of various vegetation zones on the earth (see figure 10). The northernmost plant communities (tundra, coniferous forest, temperate broadleaf forest and steppe) are in the first place adapted to seasonal differences in temperature and amount of light, other plant communities to rainy/drought seasons, and the rainforests to more stable conditions throughout the year. The evergreen scrubland is an adaptation to a combination of temperature seasons and rainy/drought seasons, to a so-called Mediterranean climate with cool, rainy winters and warm, dry summers. This kind of scrubland is found in both the northern and the southern hemispheres around approximately latitude 40°, in California (where it is called chaparral), in the countries around the Mediterranean Sea (macchia in Italy, maquis in for example France and Israel), in Chile (matorral), in South Africa (fynbos) and in Australia.

Life in the seas, which together cover 70% of the earth's surface, is based on the production of marine plants. The migratory habits of the birds that are tied to the sea and coasts must be looked at in the light of the distribution of marine plants. In figure 11, the annual plant production in different sea areas is shown together with the most important ocean currents. Apart from some larger algae near the coasts, the plants are plankton that float free in the surface water where the sunlight can reach them. All kinds of small animals, mainly crustaceans, feed on phytoplankton; cephalopods, fishes and whales exploit the crustaceans, and so the cycle keeps turning. Plants and animals die and sink to the bottom, where an abundant community of organisms is standing by to receive this energy.

At first we are puzzled by the fact that, according to figure 11, the richest plant production in the seas occurs in northern and southern areas, mostly between latitudes 40° and 60°, both N and S. This is in stark contrast to the circumstances for land vegetation (compare figure 9). The difference is due to the fact that it is not the sunlight that is the controlling factor for the seas' phytoplankton but availability of nutrient salts, nitrogen and phosphorus compounds. These nutrient substances accompany the dead marine organisms to the bottom, where they are trapped in the cold, heavy bottom water beneath the warm surface water. An exchange of water takes place between the bottom and the surface only if the surface water is cooled down so that the temperature differences at different water depths

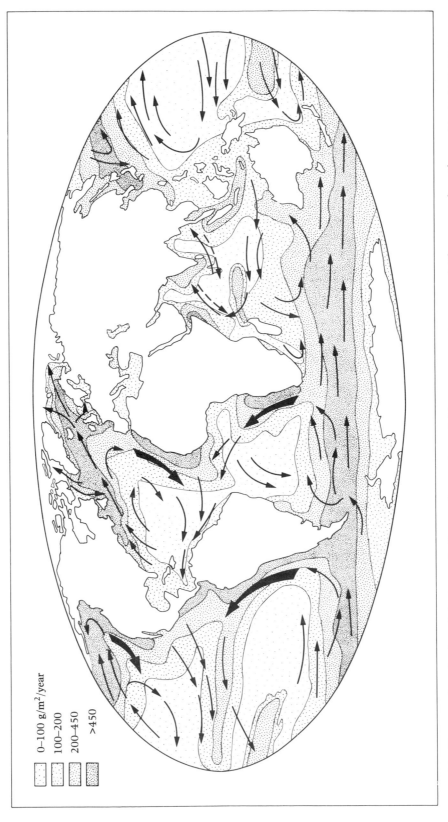

Figure 11 The annual production of plant mass in the seas (grams dry weight per square metre per year). The arrows show the most important ocean currents. The main directions of currents are similar all year around except in the northern Indian Ocean, where the currents are southwesterly during the summer (owing to the southwest monsoon) and northeasterly in the winter (during the northeast monsoon). The wider arrows show currents with upwelling water. Based on Lamb (1972) and Lieth & Whittaker (1975).

0–100 g/m²/year

100–200

200–450

>450

become minimal. This is exactly what happens in areas with rich plankton production. Around the equator, however, the surface water is warmed regularly; here, there is as a rule no chance for the nutrient-rich cold bottom water to reach the surface. The vast expanses of warm tropical ocean therefore do not harbour much life – they are the sea's deserts.

Rich plant production and an intensity of life also occur in some restricted areas with upwelling water, mainly on the west coasts of Africa and America. On these coasts the easterly trade-winds blow, driving surface-water currents away from land. When that happens, cold and nutrient-rich water then wells up from the bottom and provides good conditions for life in the sea. The Californian Current off the west coast of North America, the Canaries Current off North Africa, the Humboldt Current off South America and the Benguela Current off southwest Africa are the major currents with upwelling water.

The steering effect of the wind on the ocean currents can be clearly understood from figure 11; the effect of the trade-winds is seen near the equator and the influence of the west winds at the middle latitudes. Where the annual heat-balancing transportation of energy from the equator towards the poles is concerned, the wind is responsible for 90% and the ocean currents, among them the Gulf Stream which brings blessings for northwest Europe, for 10%.

A rough summary of the earth's various ecosystems, their surface area, plant production and plant mass is given in table 1. If we find the plants' low degree of exploitation of the solar energy that reaches the earth ($\frac{1}{2}$%) slightly disappointing, we must on the other hand be enormously impressed when we see how astronomically high the figures are for plant mass and annual production on a global scale. The total biomass of animals on the earth has been estimated at close on 1000 million tonnes (dry weight) for terrestrial animals and almost as much for marine animals. The plants' biomass is, therefore, approximately a thousand times greater than that of animals.

Perhaps the most surprising thing in table 1 is the big difference between the plant mass in the seas and that on land: the plant mass in the seas is only two-thousandths of the land plant mass, but the annual plant production on land is only twice as large as that in the seas. The yearly turnover is thus almost 100% in the seas, but only about 6% on land. Why are the organisms that consume plant mass on land not so efficient as the plant-eaters in the seas? Why do they not graze the land plants down to a negligible and sparse carpet? Indeed, why is the earth's surface so gloriously green, with the deepest forests? The reader can speculate as well as I can on this question, which certainly fires the imagination. We do, however, feel a sense of thankfulness that the land plants do so well against their consumers!

May I venture to make a conjectural calculation of how large a part of the total annual energy flow (= the annual production of plant matter) is exploited by the birds on the earth? I shall restrict myself to the situation on land. Recently, an estimate of the numbers of all bird species in Sweden was made (Ulfstrand and Högstedt 1976). The average density of breeding birds of all species was 200 pairs per square kilometre. If I assume that in addition there is on average throughout the year an equal number of immature birds, I can estimate the total density at 800 birds per square kilometre. If this density is a fairly representative average for the earth's total land area of 125 million square kilometres (desert and ice areas excluded, see table 1), then there should be approximately 100 000 million landbirds on the earth.

I would guess that the size of the average bird is roughly between that of a small warbler and that of a starling; let us say that it weighs about 50 g (= about 10 g dry weight). A bird such as this uses up about 50 kilocalories per day. I can now work out the total energy consumption during one year for 100 000 million birds (it is 1.8×10^{15} kcal). If I convert the annual plant production of 117 000 million tonnes (table 1) into energy (it is approximately 500×10^{15}), I come up with the result that birds use about *four-thousandths* of the total

Table 1. *The earth's different ecosystems. See also the map in figure 10*

Ecosystem	Surface area (millions of km²)	Plant production (dry weight)		Plant mass (dry weight)	
		Mean value (g/m², yr)	Total (1000 million tonnes/yr)	Mean (kg/m²)	Total (1000 million tonnes)
Land Tropical rainforest	17.0	2200	37.4	45	765
Tropical broadleaf forest	7.5	1600	12.0	35	260
Temperate rainforest	5.0	1300	6.5	35	175
Temperate broadleaf forest	7.0	1200	8.4	30	210
Coniferous forest	12.0	800	9.6	20	240
Scrub	8.5	700	6.0	6	50
Savanna	15.0	900	13.5	4	60
Temperate steppe/grassland	9.0	600	5.4	1.6	14
Tundra and mountain regions	8.0	140	1.1	0.6	5
Desert scrub	18.0	90	1.6	0.7	13
Extreme desert, rocky country, ice areas	24.0	3	0.07	0.02	0.5
Cultivated ground	14.0	650	9.1	1	14
Marshes and swamps	2.0	3000	6.0	15	30
Lakes/watercourses	2.0	400	0.8	0.02	0.05
Total land	149	780	117	12	1836
Sea River estuaries	1.4	1500	2.1	1	1.4
Alga belt	0.6	2500	1.6	2	1.2
Sea inside continental shelf	26.6	360	9.6	0.001	0.27
Upwelling waters	0.4	500	0.2	0.02	0.008
Open ocean	332.0	125	41.5	0.003	1.0
Total sea	361	150	55	0.01	4
Total earth	510	340	172	4	1840

energy flow from the plants, the rest being taken care of by other animals or exploited by bacteria.

The birds' total role as consumers of energy in nature is therefore very small. I should like to be more precise and say that the role of birds as primary consumers is insignificant, but that their importance as secondary consumers is a little greater. The organisms which live directly on what the plants produce are known as primary consumers, those which in turn feed on the primary consumers are described as secondary consumers, and so on. The primary consumers accordingly base their existence on the plant mass produced. Since the primary consumers breathe and lose heat, there is of course less energy left over for the next consumer in line. One can reckon on only one-tenth of the energy that is available for the primary consumers being available also for the secondary consumers.

Birds are to a large extent both primary and secondary consumers: their food comes both from the plant kingdom (grass, seeds, buds and fruit) and from the primary-consumer stage (insects, worms and small rodents). If we conjecture that the birds, seen in total, are half primary consumers and half secondary consumers (I am omitting the few birds that are found higher up the food chain), the birds' primary consumption can be estimated at only about two-thousandths and their secondary consumption at two-hundredths.

However we turn over the argument in our minds, birds are not a significant component in the total animal-and-plant community in terms of energy. It is scarcely possible to see bird migration as a material flow of energy across the earth's surface – it is a case rather of a small trickle of energy.

But we do not study migratory birds because they have an effect on the distribution of the world's energy, but because they awaken our interest and our imagination with their life and their journeys in different climates and in different environments, their fantastic adaptations in the pursuit of a place on the earth.

In this chapter, I have given an outline background to the varying climates and habitats that are important to migrant birds. It is on this magnificent revolving stage that the migratory birds appear, in light and darkness, in cold and warm, in rain and drought, among fanning winds and together with an indescribable multiplicity of plants and animals, on land and in the seas.

3 Summer and winter quarters

It is now time to introduce the actors in nature's great migration drama: the birds. My aim is to attempt to describe some of the environmental conditions that the birds come across and the resources that they exploit during their various stops on the earth. If we are to be able to understand bird migration, then we must really try to imagine how the life of birds is governed by the opportunities for finding food and avoiding predators, by the means for surviving and breeding. We must realise how drastically different conditions of life can be between summer quarters and winter quarters.

Avoid anthropomorphising the life of birds: so say some biologists who wish to caution against 'unscientificness'. This warning I think we may take lightly. In order to understand and to interpret the ecology of birds we must live their life in our imagination. I believe that we can learn much more about the birds' aerial journeys if we try to picture their situation as pilots than if we look at birds solely as small automatic flying machines. And in order to understand why they migrate as they do we must appreciate their more earth-bound daily life.

I have divided up the birds that occur regularly in north Europe into nine different ecological categories, according to habitat or type of food. This grouping into categories is far from definitive or clear-cut. On the contrary, a good many species are borderline cases and many birdwatchers will no doubt be of the opinion that I have sometimes chosen the wrong category. I hope, however, that the basic ecological characteristics in different bird types will be seen as correct. In the following nine chapter sections, I shall describe the general ways of life and migratory habits within these categories and select certain species or groups for more detailed treatment. The basics are of course where and how the birds live, but in the back of one's head there is also the question: Why? In a tenth section I attempt to provide a more general overall picture of bird migration as a natural phenomenon, its prerequisites and consequences, its occurrence in arctic, temperate and tropical regions, its development in the perspective of time.

3.1   Birds in wetlands

Birds that are specialists in seeking food in wet terrain are dealt with in this section. We are therefore looking mainly at birds which utilise the flooded summer tundra, marshes and swamps, lake- and sea-shores, and estuaries. These areas are often very productive and account for approximately 5% of the earth's total annual contribution of vegetable matter. In table 2 I have drawn up a list of birds which, according to my judgement, belong in this category and which are regular in north Europe.

Table 2. *Birds in wetlands (marshes, bogs, lake- and sea-shores etc) which occur regularly in north Europe. The data on wintering areas apply primarily to European birds or birds which pass through Europe on migration. Parentheses in this and later tables indicate scarcity in occurrence (not included in the total number of species)*

Species (NB = does not breed in Scandinavia)	Found in N Europe in winter	Important wintering areas exist in	
		West Europe & Mediterranean region	Africa south of Sahara
White Stork			+
Black Stork			+
Crane		+	(+)
Water Rail		+	
Spotted Crake		(+)	+
Little Crake			+
Moorhen		+	
Oystercatcher		+	
Lapwing		+	
Ringed Plover		+	+
Little Ringed Plover		(+)	+
Kentish Plover		(+)	+
Grey Plover NB		+	+
Golden Plover		+	
Turnstone		+	+
Snipe		+	+
Great Snipe			+
Jack Snipe		+	(+)
Curlew		+	(+)
Whimbrel			+
Black-tailed Godwit		+	+
Bar-tailed Godwit NB		+	+
Green Sandpiper		+	+
Wood Sandpiper			+
Common Sandpiper		(+)	+
Redshank		+	+
Spotted Redshank		(+)	+
Greenshank		(+)	+
Knot NB		+	+
Purple Sandpiper	+	+	
Little Stint			+
Temminck's Stint			+
Dunlin		+	(+)
Curlew Sandpiper NB			+
Sanderling NB		+	+
Broad-billed Sandpiper		?	?
Ruff			+
Avocet		+	
Rock Pipit	+	+	
Grey Wagtail		+	(+)
Total number of species	2	23	26

The list embraces such widely differing birds as storks, Crane, various crakes, waders and a couple of representatives of the passerines, Rock Pipit and Grey Wagtail. Quite the predominant group, however, is the waders. The term waders also gives an excellent illustration of that which is characteristic of all the birds in table 2: they feed on shallowly flooded ground or at the water's edge where they can find suitable aquatic insects, molluscs, crustaceans and worms. Sometimes plant material may also be included in the diet, and the larger birds such as storks and Crane readily take larger animal prey, for example frogs.

Several different types of habitat can be separated out for the birds in table 2. Species such as White Stork, Lapwing and Golden Plover show a rather low degree of attachment to water. Storks can catch grasshoppers in the most arid terrain in Africa. Lapwings and Golden Plovers often resort to open arable land. The Water Rail, the crakes and the Moorhen by contrast live in marshes with tall vegetation. Usually completely concealed from man, they wander about in the dark and the wet beneath the grass and sedge tussocks of the marsh, below the willow bushes and in the tall reedbeds. In the luxuriant marshlands of central and southern Europe, there are larger species which have a similar lifestyle: Purple Heron, Little Egret, Squacco Heron, Night Heron and Little Bittern (which all have their main wintering grounds in Africa south of the Sahara). In table 2, there are also species which are wholly tied to the arctic tundra for breeding and others which scamper along flat sandy beaches totally devoid of vegetation.

The migratory habits among the wetland birds in table 2 are varied, to say the least. Many species remain in west and south Europe during the winter, but just about as many winter in the tropics. The winter distribution of some species covers the most amazing range: that of the Knot, for example, extends from Scotland to South Africa! Only two species are found regularly in north Europe during the winter; this low number is of course due to the fact that the ice on the lakes and around the coasts makes it impossible for wading birds to find their food. The two species that occur in north Europe during the winter are associated particularly with boulder strewn and rocky skerries and shores with ice-free waters. Here, the Purple Sandpiper and the Rock Pipit run about on icy water-washed rocks and pick among the seaweed at the water's edge.

Winter waders are very few in the far north of Europe, but we do not need to move far to the southwest before the situation is completely different. As soon as the North Sea coast is reached in Denmark, Germany, the Netherlands and Britain, there are hundreds of thousands – indeed millions – of wintering waders. Widespread counts have been made in midwinter of coastal waders in west Europe and northwest Africa. Figure 12 shows those localities where the largest concentrations of waders are found; here, wintering waders can regularly be counted in five or six figures. The record is held by a large, shallow-water area on the coast of Mauretania known as the Banc d'Arguin, where the average number of waders in January is estimated at approximately 2 million. The number of wintering birds of the various species which have been counted in midwinter in west Europe and northwest Africa is shown in table 3. The numerical data in this table must be treated with some caution. In many cases, the figures are presumed to be too low; on the one hand there is a big risk of underestimating when counting birds in dense flocks of thousands of individuals, and on the other hand there are a great many localities with less sensational numbers of waders which have been insufficiently studied. But the reservation that we are presumably dealing with underestimates only makes the figures all the more impressive: almost 7 million waders in winter along the eastern coasts of the North Atlantic and of these about 3 million in west Europe. The reason why the numbers are so large is obviously that the coasts are ice-free and that the tidal water alternately washes over and uncovers large expanses of sand- and

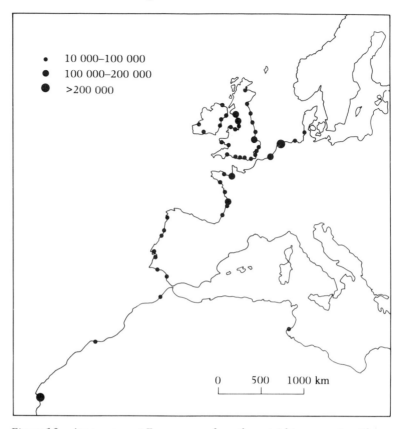

Figure 12   Areas on west European and northwest African coasts with an average of more than 10 000 wintering waders (see species in table 3). Based on Glutz von Blotzheim *et al.* (1975) and Prater (1976).

mudflats which are very rich in food. In the previous chapter (figure 11), I mentioned the intensive production of phytoplankton in the North Atlantic off west Europe and around the Canaries Current off northwest Africa. This rich plankton life supports, among other things, molluscs, bivalves and small crustaceans – in other words, the food of waders. Plankton from the sea and algae in the shallow water in this way form the basis for the whole animal community in the coastal zone. Life becomes exceptionally rich where nutrient substances from rivers are washed out into the sea. Often the very best conditions of all for waders are found in tidal areas in river mouths – estuaries.

Why do so many waders leave west Europe for other regions to breed when spring comes? If feeding conditions really are so favourable in west Europe, and if there is no earthly reason for believing that they would be so much worse in summer than in winter, should the waders then not stay there during that phase in the birds' life cycle – breeding and the growing period of the young – when food requirements are greatest? Certainly some waders breed on west European coasts, but they are in fact a rather small fragment of the large numbers that regularly use the area as a staging site during migration and to winter. The explanation is very probably that breeding is an exceedingly delicate business for the waders and that many demands, and not just that of an adequate food supply, must be met in order for it to turn out

Table 3. Numbers (in thousands) of wintering coastal waders in west Europe and northwest Africa. The figures are rounded to the nearest thousand, and + indicates that fewer than 500 individuals of the species are encountered in winter. In midwinter counts along the coasts of west Europe and northwest Africa, an average of 4000 wintering Black-winged Stilts. 2000 Spotted Redshanks. 9000 Greenshanks and 10 000 Common Sandpipers have been counted, on top of those species in the table. Waders which often occur in large numbers inland near the coast, such as Lapwing, Golden Plover and Snipe, are not shown in the table. Data from Smit & Piersma (1989)

	Ireland	England, Scotland	North Sea coast (Denmark–the Netherlands)	France	Portugal, Spain	Total west Europe	Morocco	Mauretania–Guinea-Bissau	Grand total
Oystercatcher	33	280	498	40	2	853	1	13	867
Ringed Plover	8	23	+	6	6	43	4	183	230
Kentish Plover	−	−	−	+	5	5	4	40	49
Grey Plover	1	21	12	17	10	61	4	77	142
Turnstone	5	45	7	6	1	64	1	29	94
Curlew	100	91	104	16	4	315	1	16	332
Whimbrel	−	−	−	−	+	+	1	63	64
Black-tailed Godwit	9	5	−	7	25	46	19	17	82
Bar-tailed Godwit	18	61	33	7	6	125	3	701	829
Redshank	15	75	19	4	9	122	7	135	264
Knot	30	223	73	20	2	348	3	509	860
Little Stint	−	−	−	+	1	1	2	169	172
Dunlin	115	433	279	240	68	1135	32	792	1959
Curlew Sandpiper	−	−	−	−	+	+	+	415	415
Sanderling	2	14	6	3	1	26	2	40	68
Avocet	−	+	1	16	17	34	3	15	52
Total	336	1271	1032	382	157	3178	87	3214	6479

well. Waders place their nests on the ground, and the predators which can profit from a situation with densely sited and poorly protected wader nests are many. The most critical factor is probably the exposed position in which the young find themselves during the period of their growing up. Wader young, like chickens, leave the nest immediately after hatching and then gather food themselves under their parents' supervision. During this time, the young cannot fly but squat motionless low on the ground when the parents warn of imminent danger. It is easy to realise that the young would be too exposed to all kinds of dangers if they ran up and down over the tidal shores of west Europe. It may be that there is plenty of food on the many sand- and mudflats, but the waders have no chance of taking advantage of it during the breeding season. Gulls and crows would quickly finish off the wader young.

During the summer, therefore, a considerable feeding niche for waders stands almost empty; it is at the disposal of only a relatively small number of summering waders not yet capable of breeding.

The birds' haste to return after breeding to the food-rich coasts is great. As early as July, migration is in full swing for most species. As a rule the adult females pass first, followed by the adult males and finally the young. For some species, the differences in the timetable are very striking and lead to different, clearly marked passage peaks. The female Curlew, for example, departs from the breeding site after egg-laying and leaves incubation and supervision of the young to the male. The female Curlews are already on their way south before midsummer (the same applies for example to Green Sandpiper and Spotted Redshank), the adult males follow at the end of July, and the juveniles then dominate the passage during August.

A number of northern wader species winter in tropical Africa, not only on the coasts but also at all sorts of lakes and watercourses inland. It is in fact the case that the wintering waders are a very prominent feature of the African watercourses. The Common Sandpiper is found in the most varied habitats in its African winter quarters: beside lakes and mountain streams up to an altitude of more than 2000 m, beside pools on the roads, on irrigated fields and along ditches. The Wood Sandpiper is also very common on shores of lakes and rivers, as too is the Green Sandpiper, which often prefers smaller waters and streams.

I have myself come across these species at their winter sites in Africa. When suddenly the Common Sandpiper's familiar piping whistle rang out beside the hippopotamus pool and the bird flew out low over the surface of the water with its rapid, stiff wingbeats, or when I flushed the Green Sandpiper in the yellow light of the eventide at a small pool in Zambia – it pitched away in flight like a swallow, black with pale gleaming underside, urgently shouting its alarm call – then I stopped, looked and listened with the delight of recognition. In the magnificent African landscape, abounding in strange and wonderful mammals and birds, the whistle of the Common Sandpiper sounds as natural as it does beside a summer tarn in northern Europe, and the Green Sandpiper could be one of those which always stop off in midsummer at the local river back home.

That large numbers of waders find good wintering sites at the food-rich African marshes, lakeshores and rivers is readily accepted. But what is noteworthy is that there are hardly any native sandpipers left behind to breed in tropical Africa when the winter visitors have departed to northern breeding sites. There are many African breeding birds among the herons and egrets, ibises, rails and plovers, but the sandpiper family (which includes calidrids, snipes, and curlews and godwits) contains only a single species which breeds in tropical Africa – the African Snipe (*Gallinago nigripennis*). In addition, the Common Sandpiper has been found on several occasions attempting to breed in Africa and has even succeeded in producing fledged young. As a comparison, it may be noted that 21 different species of sandpiper breed in Scandinavia.

It is amazing to stand beside an African swamp with fermenting mud and dazzling greenery, extensive fields of marsh-grass, papyrus and floating vegetation. Here one would really expect the sandpipers' song flight and display in exuberant chorus. But no, it is not so; other breeding birds occupy the stage. During the migration and wintering periods the northern waders, usually quiet and determined, are busy feeding and putting on fat in preparation for the long journey north for their next breeding effort.

A great many younger birds stay behind over the summer in Africa without breeding, and these summering birds seem to manage quite splendidly. This indicates that it cannot be food shortage that causes such an extremely small number of sandpipers to breed in Africa. Incidentally, it is not only in Africa that breeding sandpipers are rare – the situation applies pretty universally in tropical regions.

Can the explanation be that the sandpipers' young are far too exposed to the large numbers of predators which are found at the tropical waterbodies? The young of rails and crakes evidently manage well in the tropics, but these birds live in the very densest swamp vegetation. The sandpipers use the more open muddy shores and the water's edge just on the outside of the densest vegetation. Maybe the dangers there are too many for the vulnerable sandpiper young? If so, there is beside the waterbodies of the tropics, as well as on the open tidal shores, a large feeding niche for waders which cannot be used during the breeding season. Breeding adaptations may perhaps appear during the course of evolution so that the birds may in the future have an opportunity to take advantage of these great resources. There are sandpipers that nest high up in trees, for example the Green Sandpiper, which freely makes use of old thrush nests. It is therefore thought that good and reliable solutions to the siting of the nest could probably be found. It is more difficult to imagine how the safety of the young could be provided for.

Another possible explanation as to why so few sandpipers breed in the tropics is that the tropical sandpipers have been pushed out by the long-distance migratory species – there is more about this in the section on the evolution of bird migration.

Far and away the commonest coastal wader in west Europe is the Dunlin. The Dunlin has a circumpolar distribution during the breeding season and winters along many of the coasts of the northern hemisphere (figure 13). Wintering in west Europe and in northwest Africa are Dunlins from the shore meadows of Europe, from the coasts of Greenland and from the mountain and tundra regions of almost the whole of north Europe and Asia. Thanks to a sustained ringing effort in west Europe, our knowledge of the movements and staging sites of Dunlins is extensive. This applies also to most other species of wader. The major exception is the Broad-billed Sandpiper, for which I have been forced to leave a question-mark in table 2.

The Broad-billed Sandpiper breeds on the flat, soggy bogs of north Scandinavia, in the Kola Peninsula and eastwards to Siberia. The species is everywhere rare, and in Siberia there are so few observations that some ornithologists believe that the species is as good as extinct there. The Broad-billed Sandpiper is very shy and retiring and difficult to catch sight of. Furthermore, it usually inhabits what are known as quaking bogs, where the ground is often much too waterlogged for a human being to be able to make his way across. The Broad-billed Sandpiper resembles the Dunlin. Both species also use similar types of stopping-off places on migration. If the watcher has the patience to examine carefully the flocks of sandpipers that stop off in southern Scandinavia, then he is sometimes rewarded by finding a few occasional Broad-billed Sandpipers – if you see more than ten individuals you are really lucky. The end of July and the beginning of August are the best times; the Broad-billed Sandpipers have then

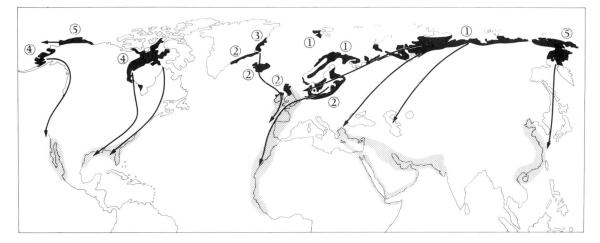

Figure 13 Breeding (black) and wintering (grey) areas of the Dunlin, west Europe's commonest coastal wader. The Dunlins in different breeding areas differ partly in size and appearance. Five different races are distinguished: (1) *Calidris alpina alpina* on the Soviet tundra and in the Scandinavian mountains (though in the latter area racial affiliation is uncertain). (2) *C. a. schinzii* around the Baltic Sea, in Britain, in Iceland and nearby parts of Greenland. This race has decreased considerably in recent years and is now very few in number; today, there are less than 300 breeding pairs in Continental Europe, in Finland about 200 pairs, in the Baltic countries a few thousand pairs at best, in Denmark 600 pairs, in Sweden 150 pairs and in Britain between 4000 and 6000 pairs. In Iceland, however, the Dunlin is very common. (3) *C. a. arctica* in northeast Greenland. (4) *C. a. pacifica* in western Alaska and Canada. (5) *C. a. sakhalina* on both sides of the Bering Strait. The first three races share wintering areas in west Europe and northwest Africa. Based on Glutz von Blotzheim *et al.* (1975).

just begun their journey to winter quarters. But where are the main wintering grounds situated? There are three recoveries of Broad-billed Sandpipers ringed in Sweden: one from the northwest German coast (August), one from Portugal (November) and one from France (February). This seems to indicate that west or southwest Europe is a wintering area, but in this well-covered region very few winter occurrences are known. Africa south of the Sahara does not appear to be a wintering area of any importance, either: only sporadic records of odd Broad-billed Sandpipers exist, from Chad, Uganda, Tanzania and South Africa. In the Mediterranean area, in particular in Tunisia, however, there seem, according to recent reports, to be regularly wintering Broad-billed Sandpipers, but the number is far too small to represent any significant proportion of the Broad-billed Sandpiper population. Not until we get to the Red Sea, the Persian Gulf and along the coasts of the Sea of Arabia do we find any important winter localities of the Broad-billed Sandpiper. Do most Scandinavian Broad-billed Sandpipers perhaps go there? If so, the Broad-billed Sandpiper is the only north European wader (apart from the Red-necked Phalarope) that seeks winter quarters in a southeasterly direction. We must, however, be cautious in drawing conclusions. The Broad-billed Sandpiper is easy to overlook; maybe there are, after all, substantial wintering populations in Europe or in west Africa, as the few ringing recoveries suggest. Interestingly, the very few found each year in Britain are recorded mainly on spring migration, in May–June. This presents a challenging problem for ringers and industrious birdwatchers with a desire to search for the Broad-billed Sandpiper's mystical winter quarters.

I shall give a more detailed picture of a number of wader species so as thus to show the many varieties of adaptations for coping with the winter – from the Purple Sandpiper, which winters in Greenland and at Norway's North Cape, to the Curlew Sandpiper and the Little Stint, which can often be seen running about on the shores of the Cape of Good Hope.

The Purple Sandpiper (*Calidris maritima*)

The Purple Sandpiper is *the* bird of the northern coasts (see breeding and winter distributions in figure 14). It breeds in a belt around the North Pole from the Taimyr Peninsula in the east to the arctic archipelago in north Canada in the west. The distribution shows a large gap on both sides of the Bering Sea, but in this gap its place is taken by a very closely related species, the Rock Sandpiper (*Calidris ptilocnemis*), and in all essentials this fills the same ecological niche as the Purple Sandpiper. The breeding habitat of the Purple Sandpiper is flat, stony moorland and tundra plateaux near the coast, around fiords or in country with many lakes and pools. In Scandinavia, it is found in the lichen region of the mountains and in the upper part of the willow zone. The Purple Sandpiper's most northerly breeding site is the islands of Franz Josef Land, just north of latitude 81° N.

It is obvious that the Purple Sandpiper must abandon its breeding sites in winter, when the landscape consists of nothing but snow and ice. The Purple Sandpiper exchanges its summer moorlands for a quite particular winter niche: stony, rocky shores, where the birds spend their time in the very zone which is washed over by the waves, often right near the foaming surf. Here, the Purple Sandpipers feed in particular on small molluscs and shore crabs. Many waders exchange their breeding habitat for a winter niche on shallow, sandy and muddy seashores; the migration flights must in this case be extended south to west Europe or beyond to more southerly coasts. The Purple Sandpiper's special rock habitat allows its wintering area to be much farther north. So long as there is open water on rocky shores, the Purple Sandpiper's requirements are satisfied.

In several places, such as in southwest Greenland, Iceland and the Faeroe Islands, the Purple Sandpiper is almost a resident species. The birds there move only the short distance from the moors to the rocks of the shores. Purple Sandpipers at more arctic breeding sites are compelled to make migratory journeys of greatly varying lengths. Large numbers of Purple Sandpipers have been ringed on the west coast of Greenland, and the many recoveries there show that the more northerly birds travel at least 500 to 1000 km south to winter on the ice-free coasts of southwest Greenland. There are two recoveries from Greenland of Purple Sandpipers that have been ringed in Iceland and in England. This indicates that some west Greenland birds winter in Europe. It may even be the case that some Canadian birds are present among the wintering flocks in Europe. That an exchange of Purple Sandpipers across the Atlantic takes place is shown by a couple of recoveries of sandpipers ringed in Iceland: one from the southeastern part of Baffin Island and a winter recovery from Newfoundland.

The Purple Sandpipers disappear altogether from east Greenland during the winter, and there is good reason to believe that north Iceland is their main winter refuge.

The most important wintering coasts in Europe can be seen in figure 14. Recent counts indicate that there are approximately 20 000 wintering Purple Sandpipers in Britain, mostly in the Shetlands and Orkneys (about one-third), around mainland Scotland (almost as many) and in northeast England. Between 1000 and 2000 Purple Sandpipers winter in France, on the rocky coasts of Brittany. The numbers in Belgium, the Netherlands and Denmark are low, a few hundred birds in each country; in these countries, the Purple Sandpipers are found near rocky breakwaters and harbours, which have to act as a substitute for the species' natural rocky-coast habitat. The main European wintering areas are in Norway. Everywhere along

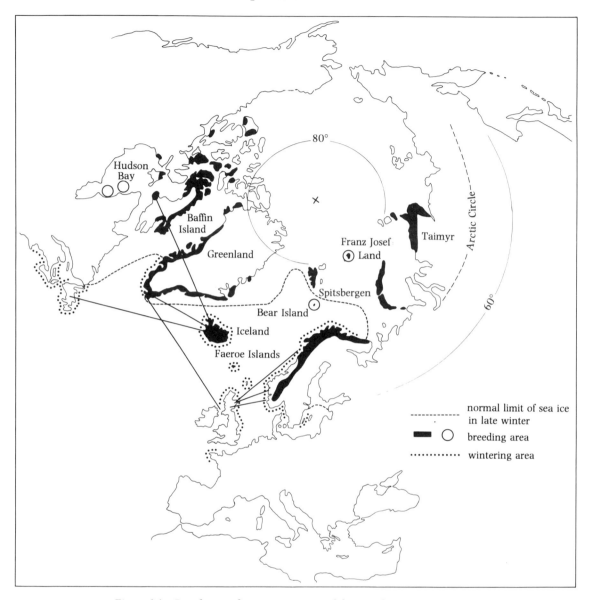

Figure 14   Breeding and wintering areas of the Purple Sandpiper. The unbroken lines show examples of movements of ringed birds.

the coast, deep into the fiords and on isolated outer skerries there are Purple Sandpipers, sometimes concentrated in flocks a thousand strong. No counts are available from Norway, but there must in all likelihood be hundreds of thousands of wintering birds involved. Along the coasts of Iceland, too, Purple Sandpipers can probably be counted in six figures. Not only the sandpipers from the Scandinavian mountains but no doubt also populations from the eastern coasts of the Arctic Ocean winter in Norway.

Purple Sandpipers winter regularly in Sweden, but only in modest numbers and then mainly on the west coast and on the rocky islets and shores of Blekinge, on Öland and on

Gotland. Purple Sandpipers are also found every winter in Scania, on the Kullen Peninsula and on Hallands Väderö. I remember a cold and sunny winter's day at the tip of Kullen. The sea was covered with broken ice as far as the eye could see; only here and there was a gap of clear water occasionally opened as the ice drifted slowly northwards. Right against the shoreline cliffs there was open water; crowds of different birds were resting there, preening, swimming and diving – Cormorants, Eiders, Red-breasted Mergansers and Goosanders, Common Scoters, Velvet Scoters and Great Crested Grebes, Tufted Ducks and Goldeneyes. But it was something else that chiefly captured my interest: half a dozen Purple Sandpipers were running around eagerly at the water's edge, picking among the seaweed, peeping into small rock crevices, preening in the sun, flying low over the water to other rocks. They really looked as if they were in splendid spirits. While I was watching them I thought of how impressive a migrant the Purple Sandpiper is, despite its wintering so far north. Perhaps the Purple Sandpipers I had right in front of me came all the way from the Taimyr Peninsula, 5000 km away!

The Lapwing (*Vanellus vanellus*)

How could anyone ignore the Lapwing when the subject is bird migration! Every year it is something special, the day when the Lapwing turns up again on its northern breeding grounds. There are only a few birds that are so eagerly awaited – the first lark, the first wagtail and the first Swallow. Delightful the day when the first Lapwings head northwards in the melting snow! What a graphic study in elegance it is, with the rich contrast of Lapwings over a mosaic of patches of snow and clear ground.

Lapwings are found on meadows and fields throughout Britain and Scandinavia apart from in the mountains and in the interior of the northern coniferous forests. In Britain, for example, there are very probably close on half a million breeding pairs, according to recent estimates, while in Sweden one can reckon on 100 000 pairs. The Lapwing is, therefore, a very common bird and its whole distribution extends over a large area (see figure 15).

Nearly half a million Lapwings have been ringed during the present century. The frequency of their being reported again is about 2%. This means that there are about 10 000 ringing recoveries of European Lapwings. An ornithologist from Switzerland, Christoph Imboden, has set to work on this large amount of material and, with the help of data

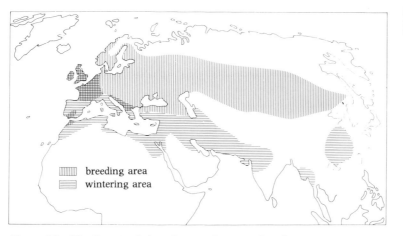

breeding area
wintering area

Figure 15   The Lapwing's breeding and winter distributions.

technology, has succeeded in putting it all together. To be exact, we are talking about a treatment of 9521 ringing recoveries in total of the Lapwing, of which 7252 birds had been ringed as chicks and the rest as fledged birds.

Recoveries of birds ringed as young are of course especially valuable, since we can be certain of the exact nesting site. Everything said henceforth about ringing data on the Lapwing is concerned with these recoveries. The large amount of ringing material provides us with unique opportunities to compare different wintering areas for Lapwings which breed within different areas. A series of maps (figure 16) shows recoveries in winter of Lapwings from England, the Netherlands, Denmark, Norway, Sweden, Finland and Hungary.

It is well worth the trouble to compare and reflect on the maps for a while. The first thing we notice is that some of England's and the Netherland's Lapwings remain at home during the winter. Of those Lapwings that stop in the country, the proportion of birds over one year old appears to be slightly greater than the proportion of one-year-olds. The data in figure 16 apply to first-year birds; corresponding frequency of winter recoveries for older English Lapwings within England is 20%, and for older Dutch Lapwings within the Netherlands 4%. Apart from this difference, the habits seem to be substantially the same for Lapwings on their first migration flight and for older Lapwings with experience of at least one previous winter.

It is striking that Lapwings from the different breeding areas have winter quarters which to a large extent coincide. Only minor disparities can be detected: (1) Hungarian Lapwings deviate most from the others and often spend the winter in Italy, where there are extremely few Lapwings from west and north Europe. (2) Many English Lapwings winter in the 'green isle' – Ireland. Particularly when cold wind sets in from the east, an intensive westward passage of English Lapwings fleeing from the cold and snow is triggered. This happens not only in autumn but even in the middle of winter or during early spring. This we term 'rush migration'. Similar 'hard-weather movements' also take place between England and the Netherlands – early in the spring Lapwings, Skylarks and Starlings try leaving the winter haunts in Britain and migrate eastwards to Continental Europe. If a set-back in the weather then occurs, such that winter makes a temporary advance from the east, the birds turn and make their way back to Britain, where they await milder westerly winds before making a fresh attempt to migrate east. In Scandinavia, too, it is fairly common for the early spring birds to be forced to make a temporary retreat, even though it is perhaps seldom so strikingly manifested as when birds migrate in hordes across and back over open waters in the southern North Sea or in the Irish Sea. These movements illustrate how sensitive the Lapwing is to frost and snow, which totally prevent it from getting food from the topmost layer of the ground (generally damp ground). As well as English Lapwings, even Norwegian and some Swedish Lapwings turn up in Ireland (where there are also a great many Irish Lapwings, birds which remain on the island during the winter). (3) The Danish, Norwegian and Swedish Lapwings have a wintering area which, when compared with that of the Finnish Lapwings, is all in all slightly displaced to the west. The most marked 'west-coast distribution' is shown by the Norwegian-ringed recoveries.

The coincident wintering areas naturally involve the migratory journeys being considerably longer for northerly populations of Lapwings than for southerly ones. Figure 17 shows a comparison of the mean distance between breeding site (place of ringing) and recovery site for English and Swedish Lapwings during different months of the year. The English Lapwings move on average only about 900 km; the Swedish Lapwings find themselves at this distance from the breeding site in November or in March–April, but in January and February they are on average 2000 km away. For the Finnish Lapwings, the migration flight is of course even longer; during midwinter, they are on average nearly 3000 km from their breeding sites. The

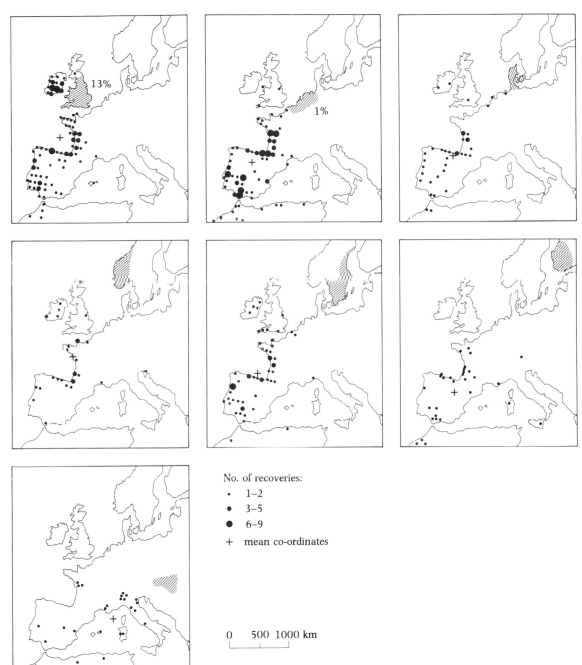

Figure 16   Map series of recoveries during the winter of Lapwings ringed as young within different breeding areas (hatched). The maps show recoveries during January (for English and Hungarian Lapwings, during January and February). The recoveries shown are of one-year-old birds; among the Hungarian Lapwings, however, are also recoveries of older birds. A certain percentage of English and Hungarian Lapwings have been found within the same country during the winter. The mean co-ordinates for ringing recoveries from the various regions are shown on each individual map. Based on Imboden (1974).

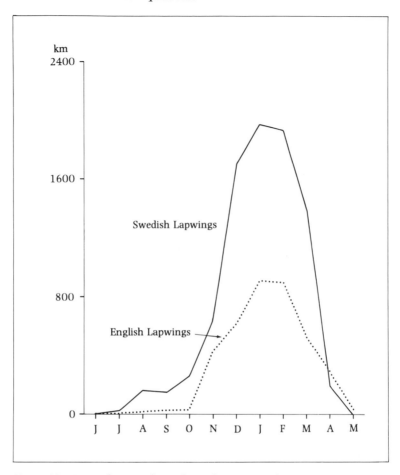

Figure 17   Mean distance from place of ringing to place of recovery during different months for English and Swedish Lapwings. Only recoveries of birds during their first year of life have been included. Based on Imboden (1974).

Lapwing which is responsible for the longest distance recorded is one which hatched at Oulu in west-central Finland and which was reported in winter from Morocco, 4330 km away.

The Lapwings' spring passage, which occurs in March and April, takes place quickly and directly. The Lapwings are clearly in a great hurry to reach their breeding areas. The southward passage is something quite different, a drawn-out course of events. It begins as early as June, but large flocks of Lapwings can still be seen leaving Scandinavia in November. The summer passage in June and July has been the subject of much discussion in the scientific literature. Which birds is it that move away so early? How far do they fly? What is the reason for this migration?

There is no completely sure answer to these questions, but some interesting points of view are worthy of mention. Much points to the adult males being the first to set off on the summer migration. At that time, the females and young are still behind on the breeding sites. Birds that have not bred or have failed in their breeding attempts probably also take part in the start of the summer migration. As the young of the year become capable of flight, so they set off, together with adults, on summer migration. In July, therefore, the migration is made up of

Table 4. *The site fidelity of the Lapwing during the breeding season. The table shows the proportion of ringing recoveries (between 1 May and 14 June) at various distances from place of ringing. The data include 368 Lapwings, ringed as juveniles at the breeding site in Europe and found again during later breeding seasons. Based on Imboden (1974)*

Distance from ringing site in kilometres						
0–20	20–60	60–200	200–500	500–1000	1000–2000	>2000
70%	10%	5%	4%	3%	5%	3%

widely different age classes. The summer passage is noted most in north and central Europe, and is almost completely unknown in Britain. The birds' course is, as later in the autumn, southwesterly, but the summer migration does not take them all that far towards the winter quarters but only a few hundred kilometres.

Another species which, like the Lapwing, has a concentrated summer migration is the Starling. With this species it is mostly juveniles that make up the summer flights.

A possible explanation for the summer migration of the Lapwing and the Starling is that the birds leave the breeding sites, where suitable food has been heavily exploited throughout the whole breeding period, in order to seek food at unexploited localities, places which are unsuitable for breeding but which are very good for off-passage birds to find food. That the birds take the opportunity at this time to set off in the direction of their winter quarters, and thus at an early stage put a small part of the whole migration journey behind them, appears to be an opportune adaptation. But, if this is correct, one wonders why only the Lapwing and the Starling among the terrestrial birds exhibit a summer migration. Should not the same thing apply for example to larks and pipits? That these small birds do not set off on summer passage may conceivably be due to the fact that breeding pairs are dispersed over all suitable summer terrain with the result that specific unexploited habitats for survival to which it is worth moving do not exist. This opens up a rewarding field for the interested birdwatcher: to map how the feeding niches of different species change during the summer and thereby to attempt to throw light on the reason for the summer passage.

Most birds display strong site fidelity and return to breed within the area in which they were hatched. The Lapwing's site fidelity is illustrated in table 4. Approximately eight out of ten Lapwings return to areas within 60 km of their hatching site – one can say with some justification that they are returning 'home'. But there are also Lapwings that certainly do not return home: one out of 12 move to a new breeding area more than 1000 km from their hatching site, and one Lapwing out of 30 breeds more than 2000 km away. In the map in figure 18, four examples of change of breeding area are shown. A Scottish Lapwing has been found breeding in Iceland: this is especially interesting since the Lapwing is only an occasional breeder there. The most impressive examples of change of breeding area are the three Lapwings which had hatched in Finland, England and the Netherlands and which were found again during the breeding season beyond the Ural Mountains by the Russian rivers of the Ob and Yenisei, at 3460 km, 4370 km and 5170 km from where they were hatched! An important basis for the Lapwing's change of breeding area is no doubt the large overlap of winter quarters for populations from widely separated breeding localities. We may guess that the Lapwings in part pair off in the wintering areas; if the parties concerned at that time come from different places, one bird must move with the other to a totally new breeding area.

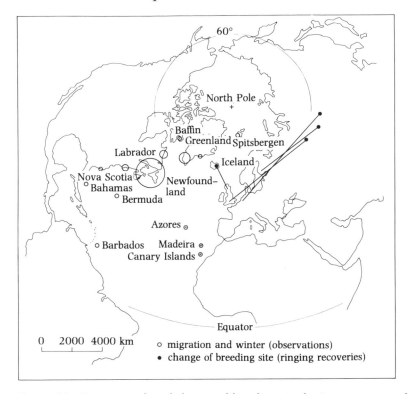

Figure 18 Four examples of change of breeding site by Lapwings according to ringing results. The open circles show places where Lapwings have been observed during migration and winter. Based on Bagg (1967), Imboden (1974) and others.

Perhaps there are some readers who, despite all that has been said above, do not think that the common Lapwing is a particularly exciting migrant. In that case, I shall tell of some of the Lapwing's wildest migratory escapades!

In the afternoon of 20 December 1927, the most remarkable thing happened on the island of Newfoundland in easternmost Canada. The inhabitants of Cape Bonavista saw birds completely unknown to them arriving from the east from the open North Atlantic. It was Lapwings which were flying in over the land in small flocks and in parties of two to three or 15–20 birds. The Lapwing is not normally found at all in America. When the people awoke the next morning, the surprise of their life awaited them – there were Lapwings in their hundreds, all over the cape.

One person counted approximately 500 Lapwings, another asserted that he had counted close on 2000 individuals. A hunter succeeded in shooting 60 Lapwings over the following days, of which one, shot on 27 December, had been ringed as a chick in May of the previous year in northwest England. It was not only at Cape Bonavista that the Lapwings turned up, but observations of hundreds of Lapwings were reported along the whole Newfoundland coast. Odd Lapwings were also reported from Nova Scotia on 20 December and on subsequent days. One Lapwing was observed too in Maine, the most northeasterly state of the USA. During the following weeks, the Lapwings remained and dispersed over Newfoundland and easternmost Canada, but the winter climate in these parts is harsh (mean temperature in February in eastern Newfoundland is $-6\,°C$) and before the winter had come to an end the

Lapwings had succumbed. The last Lapwing at Cape Bonavista was seen on 15 January, but there were still several left on a southern cape of Newfoundland in the middle of February.

When the winter was over and contact could be made with the people on the coast of the Labrador Peninsula, it was found that many Lapwings had also been seen in those arctic regions. Flocks of between ten and 50 birds had turned up during the last week before Christmas. A member of a scientific expedition had shot two Lapwings well to the north in Labrador. An American ornithologist received a live Lapwing which had been taken into care during the winter and kept in a cage. The same man received from an Eskimo woman a skin of a Lapwing and at the same time skins of three other European birds which, like the Lapwing, had been captured during the winter: a Snipe, a Jack Snipe and a Coot.

The explanation for the appearance of the Lapwings is not difficult to find. The winter had broken out with fearsome power from the east across Europe. In Paris, the daytime temperature at that time was about 6° below zero. An intense ridge of high pressure extended from Greenland across Iceland to north Scandinavia. South of the high pressure, cold easterly winds swept down over Europe and farther out over the Atlantic as far as Newfoundland. On 19 and 20 December, the east wind blew at storm force; Land's End in southwest England reported 20 m/s (72 km/h) on 19 December, and weather ships in the North Atlantic measured wind strengths between 20 and 24 m/s (86 km/h) on 20 December. The Lapwings were doubtless fleeing before the cold and frost. Many made off westwards, heading for Ireland (compare with what has been said above about hard-weather movements). The Lapwings were seized by the hard east wind, probably when they got to Ireland, and were driven westwards across the Atlantic. The normal flight speed of Lapwings in relation to the surrounding atmosphere is around 15 m/s (about 50 km/h). They cannot, therefore, control their flight when the wind blows still faster. Beyond or west of Ireland, the Lapwings were not able to fly back into the wind but were driven inexorably farther out over the sea.

If the Lapwings flew westwards across the Atlantic at 15 m/s and the east wind was blowing at 25 m/s, their total speed of travel would have been 40 m/s (approximately 150 km/h), and it should have taken them about 23 hours to fly the 3300 km across the Atlantic from Ireland to Labrador and Newfoundland. Since the first Lapwings reached Newfoundland in the afternoon of 20 December and most then arrived during the night, it is likely that they were swept away from the regions around Ireland in the evening and the early night of 19 December (the Lapwing often migrates at night).

A similar event, with many observations of Lapwings in Newfoundland, took place in the middle of January 1966. At least 30 Lapwings were reported at that time from various places. The wind situation was almost exactly the same as that which prevailed in December 1927, and the temperature in Paris was as low as −12 °C. Particularly interesting are the many reports from British birdwatchers of heavy westward movement of Lapwings at that time in 1966. One observer reported from Dublin Bay in Ireland that the Lapwings were streaming in from the Irish Sea throughout the day on 15 January; the birds were coming chiefly from Wales and continued westwards in over Ireland. There were approximately 600 Lapwings per hour coming in. The Lapwings which reached easternmost Canada in 1966 did not survive the winter there; the final observation in Nova Scotia on 26 February concerned a single Lapwing in a full blizzard.

A large number of observations of single Lapwings or of small Lapwing flocks have been made on the east coast of America at times other than in the winters of 1927/28 and 1966. Discoveries have been made from Baffin Island in the north to South Carolina in the south, but by far the majority of observations originate from the Newfoundland area.

Lapwings have even been found in the Bahamas and Bermuda, and, on two occasions, in

Barbados right down in the West Indies (figure 18). Practically all observations of Lapwings in the New World have occurred between the end of September and the middle of March, the exception being a single male Lapwing which was shot in Barbados on 25 July. The Lapwings have often been met with during periods of easterly winds, even though these have by no means always been of storm strength. Sometimes, they have turned up at the same time as other wind-driven vagrants from afar: for example, in November 1932, when Lapwings, frigatebirds and Little Auks galore were driven in over the east coast of North America.

Some further examples of the Lapwing's ventures while migrating are given below:

Greenland	There are about 20 reports of Lapwings during the winter months, mostly from the southwest coast where the climate is mildest and the sea is not completely ice-covered.
Iceland	In virtually every winter period, Lapwings turn up in Iceland. In some years only a few stray individuals are involved, in other years invasions of hundreds of birds occur. This happens in particular in fierce southeasterly winds, when the Lapwings are probably wind-driven from the regions around Ireland. It sometimes happens that Lapwings survive the winter in Iceland; occasional pairs have made breeding attempts, but no regular breeding stock has ever established itself.
Spitsbergen	A single Lapwing, presumably wind-driven, has been seen in August.
Madeira	This island, almost 1000 km from Portugal and 600 km from the nearest African coast, forms part of the Lapwing's normal winter range. A five-year-old bird, found in January, had been ringed as a juvenile on the south Swedish island of Öland.
Canary Islands	The Lapwing is a regular winter visitor.
The Azores	Here, too, 2000 km west of Portugal, wintering Lapwings turn up in most years. This may involve flocks of several hundred, especially following strong easterly winds. One Lapwing had been ringed in Denmark.

It is fairly common for migrants to be carried long distances over the open sea by the wind, even though they are rarely wind-driven in large flocks like the Lapwing. It is considerably more common for North American birds to be driven across the Atlantic to northwest Europe than the other way around. This is of course due to the fact that the North Atlantic lies in the zone of westerlies.

In more southerly latitudes, where the northeast trades blow, it is by contrast most common for birds to be driven by the wind from the Old World to the New. The most easterly island of the West Indies, Barbados, is a paradise for those birdwatchers who search for rarities. The northeast trade-winds have borne not only Lapwings but also many other European and African birds to this island.

It is natural to ask why the Lapwing has not started to breed and to spread in North America. There must have been many opportunities over the course of centuries for wind-blown Lapwings to colonise suitable nesting habitats in this land. Presumably, the case is that the Lapwing's niche is occupied by a bird species which exhibits optimal adaptations to particularly American conditions and which can therefore outcompete those Lapwings that try their hand at colonising new areas. The Lapwing's closest ecological counterpart in North America – and its superior competitor – is probably the Killdeer.

There are instances of wind-blown birds having found unoccupied niches to colonise. The easterly winter wind over the North Atlantic is responsible for one such case. Thousands of Fieldfares fleeing to Britain from the cold in Norway in January 1937 were caught up over the North Sea by storm-force southeasterly winds which carried them to southwest Greenland. There, they quickly dispersed all the way south to the southernmost cape. Even today a small breeding population survives in southwest Greenland.

The Cattle Egret's successful spread since the 1930s in South America, the West Indies and southernmost North America may be considered to be due to the effect of the northeast trade-winds. The Cattle Egret previously bred only in the Old World, and birds that had been driven by the wind from West Africa across the Atlantic laid the foundations for the species' expansion in America.

Birds have a magnificent capacity for extending their range. Wherever suitable niches exist, even if on the most remote oceanic islands, birds turn up and take advantage of them. When new areas, such as volcanic islands, are created, the birds are among the first colonists. The Lapwing is an example of a bird that is often forced out towards the unknown across thousands of kilometres of storm-whipped ocean.

The Redshank (*Tringa totanus*)

The Redshank is principally a bird of coastal meadows with a breeding distribution that coincides closely with that of the Lapwing. Unlike the Lapwing, however, the Redshank occurs as a breeding bird in the upland birch and willow zone. Redshanks winter in large numbers on western European coasts (table 3), but are also found along the west coast of Africa, south to the tropics.

The migration of the Redshank has been analysed in detail by Salomonsen (1955, 1972) and Hale (1973). The analysis has been based on the results of ringing, but also on studies of museum skins of Redshanks from different areas. The Redshank's appearance and size vary quite a lot between different areas; several different races can therefore be separated.

Hale made accurate measurements of wing, tail, tarsus and bill of Redshanks from different parts of the breeding range, and on the basis of these data worked out a programme for calculating, with the help of a computer, the most likely area of origin. Such calculations have then been made for birds found during the winter and whose skins were held in various museums. In figure 19, examples are shown of wintering Redshanks which, according to measurement data, most probably originate from an easterly breeding area in China and Mongolia.

With ringing recoveries we are absolutely assured of where the birds come from; when it is a matter of 'computer recoveries', based on measurements, we must be prepared for the fact that a lot of errors can occur. Birds from the same breeding locality naturally vary appreciably in shape and size; it is only the mean values from different areas and of different races that differ. Despite the fact that the individual computer recoveries must therefore be treated with a certain caution, in the absence of ringing data they can give a good pointer as to where the most important wintering areas are for birds from different breeding regions. Thus, there are, for example, several computer recoveries of Scandinavian Redshanks from the coasts around the Arabian Peninsula (figure 19). There are no ringing recoveries from this area, but then it should be noted that the reporting of ringed birds from this region is very rare (compare the Broad-billed Sandpiper). The people there do not have the great interest in birds that is found

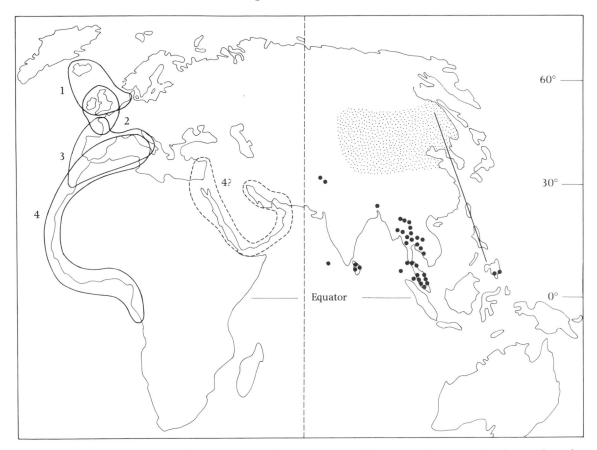

Figure 19   Winter quarters of Redshanks from different breeding areas. The dots on the right-hand side of the map show winter recoveries of Redshanks which, according to detailed studies of measurements, most probably come from the stippled part of the Redshank's breeding range. The unbroken line shows a winter recovery of a ringed Redshank from this area. The left part of the map shows the main winter quarters according to ringing results for Redshanks 1 = from Iceland, 2 = from Britain, 3 = from Denmark and the Netherlands, 4 = from Sweden, Norway and Finland. Biometric studies suggest the possibility that some Scandinavian Redshanks also winter in the Arabian Peninsula; there are, however, no ringing recoveries from there. Based on Hale (1973).

among the Europeans. Another major reason for ringing recoveries not being reported is no doubt that the words on the rings are in European characters, which are of course totally foreign to many people. The computer recoveries give us every reason to take note of the possibility that some of the Scandinavian Redshanks migrate in a southeasterly direction. Future studies and ringing recoveries may provide positive answers to how the situation stands with regard to this.

Ringing results show that Redshanks from different breeding areas in northwest Europe winter largely in separate areas. This is accordingly where the Redshank differs entirely from the Lapwing. Table 5 gives winter recoveries of birds that have been ringed in different parts of northwest Europe. The ringed birds include both those which have been caught at the breeding site and those which have been caught during autumn and spring. We can

Table 5. *Distribution of winter recoveries (from November to January inclusive) of Redshanks ringed within four different areas, in Iceland, Britain, Denmark + the Netherlands and Sweden + Norway + Finland. n = number of recoveries. Based on Hale (1973)*

	Area of ringing			
Winter recoveries from	Iceland $n=9$	Britain $n=148$	Denmark, the Netherlands $n=212$	Sweden, Norway, Finland $n=25$
Britain	89%	85%	1%	4%
Denmark, the Netherlands	11%	1%	7%	16%
France, Spain, Portugal, Italy	—	14%	85%	52%
Africa	—		7%	28%

therefore anticipate that some of the Redshanks which have been ringed, for example in Denmark and in the Netherlands, are passage migrants from more northerly areas.

The ringing results indicate that Redshanks winter principally within those areas which are shown in figure 19. Some Icelandic Redshanks (the race *Tringa totanus robusta*) winter in Iceland (unfortunately, any ringing recoveries there may be during the winter from Iceland are not accessible), but many birds set off for Britain and the North Sea coast. It is probable that the few Redshanks that sometimes winter in Norway and southernmost Sweden come from Iceland. This has recently been established as Redshanks have turned up in the extreme southwest of Scania in December, January or February: some of these late-arriving winter visitors have been caught, and measurement checks revealed them to belong to the race *robusta*.

British Redshanks as a rule remain throughout the winter in the British Isles; only a small proportion fly to France. Redshanks from Scandinavia have winter quarters farther to the south: Danish ones (and, according to Roos 1969, those from extreme south Sweden too) mainly in southernmost France, Spain and Italy; north Scandinavian ones mostly in Africa.

How can we be so bold as to conclude that the northernmost birds from Scandinavia move primarily to Africa despite the fact that almost three-quarters of winter recoveries of Swedish, Norwegian and Finnish birds fall within Europe (table 5)? The explanation is given in table 6. If we compare the number of recoveries during the winter with the number of recoveries early in the autumn, when the Redshanks have just begun their migration and are certainly still present within Europe, we find that there are, relatively speaking, very few winter recoveries reported of the northernmost Redshanks.

Where then do the northern Redshanks go to during the winter? The only plausible explanation is of course that they move farther on to places from where ringing recoveries are seldom reported. It is well known that considerable numbers of Redshanks winter along the west coast of Africa, and the supposition that north Scandinavian birds are involved here is a very likely one, all the more as there are in fact one or two ringing recoveries from there.

The example with the Redshank shows that we must be very cautious with interpretation of ringing results. The extent to which ringing recoveries are reported varies very widely in different areas, and the risk of getting a false picture of how the migration progresses is great.

The most important reason for bringing in the migration of the Redshank is that it is a fine example of what is known as leap-frog migration: northern populations move past more

Table 6. *Quotient for recoveries in winter (Nov–Jan) and during the early autumn (Aug–Sept) of Redshanks ringed within different areas. The figure in parentheses is for Redshanks ringed as fledglings on the breeding sites in Denmark and the Netherlands. Based on Hale (1973)*

	Area of ringing		
	Britain	Denmark, the Netherlands	Sweden, Norway, Finland
$\dfrac{\text{No. of recoveries Nov–Jan}}{\text{No. of recoveries Aug–Sept}} =$	3.1	0.54 (0.93)	0.13

southerly ones and end up in the southernmost winter quarters. If, for a moment, we disregard the Icelandic Redshanks, we get a perfect pattern of leap-frog migration for those Redshanks which breed in the area from Britain in the south to the Scandinavian mountains in the north. One can imagine at least two different possible explanations for leap-frog migration.

a *Competition and the timing of the migration.* Birds which breed in southerly, mild areas get started with their breeding early in the spring and also finish it early. Straight after breeding, they begin to occupy the winter quarters that lie as near as possible to the breeding sites so that the birds avoid wasting unnecessary energy and exposing themselves to risks on long migratory journeys. More northerly populations breeding later in the year move later; they then find the more northerly wintering areas already occupied and are forced to continue farther south in order to reach unoccupied wintering areas. Birds breeding still farther north are forced even farther south because of the competition, and so on in accordance with the principle shown in figure 20. There are at least two serious objections to this reasoning. One would think that the birds breeding farthest south should disperse over a wider area after breeding in order to avoid unnecessary competition, and not completely 'fill up' and crowd together on the most northerly wintering localities. It is also unclear what it is that prevents the late migrants from the north from trying to push their way into the northerly wintering areas and thereby compelling other birds to move farther south.

b *Competition and the predictability of spring.* We can assume that birds choose to live where it is most suitable for them, and that the suitability of a particular locality decreases with increasing density of competitors for the food. When birds come to unoccupied areas, the very optimal spots are accordingly occupied first (type A in figure 21). When the density of birds within these areas increases, the suitability diminishes; consequently, the birds soon occupy other types of areas (such as B and later C as per figure 21). In ideal instances, therefore, the birds are dispersed at different densities within different areas, but the different spots are on average equally favourable (or equally unfavourable) to the birds.

If the suitability of the wintering site is determined only by food supply, by the risk of attack from predators and perhaps by the cost of the extra flight from northern to southern wintering sites (the cost should be approximately the same for southern

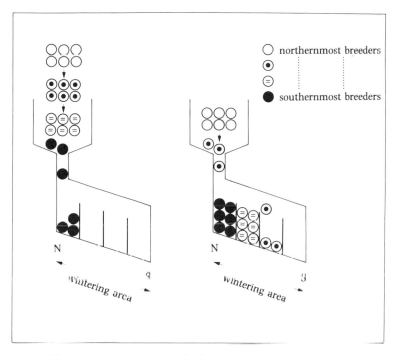

Figure 20 An imaginary principle for leap-frog migration. The southerly breeding birds migrate first and fill up the nearest wintering areas. More northerly breeders migrate later and are forced to carry on past the parts of the wintering area that are already occupied.

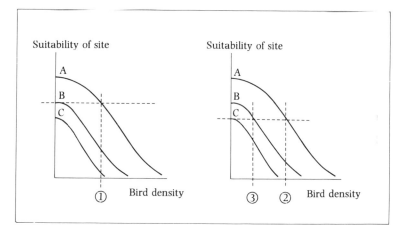

Figure 21 The effect of competition on dispersion of birds over different areas. The birds occupy the most suitable areas (type A) first. As a result of competition the suitability diminishes at a certain unknown density of birds, and when the density reaches more than (1) in A areas the birds will also occupy the type B areas (left-hand figure). When the bird density has increased to (2) at type A areas and to (3) at type B, it becomes profitable for the birds to occupy C-type sites too, as shown in the right-hand figure. Based on Fretwell (1972).

breeders as for northern ones), then we should be able to expect that the wintering sites for birds from different regions will overlap each other considerably. If on the other hand there is a direct adaptive value for the southern breeders but not for the northern ones in staying as close to their breeding sites as possible during the winter, then this will immediately result in separate winter quarters. The southern breeders are prepared to endure heavy competition in order to retain the advantage of wintering as near to the breeding sites as possible, and the northern breeders move farther south during the winter so as to avoid unnecessary competition.

It is important for the birds to occupy the breeding sites as early as the spring climate and food supply possibly allow. Studies of widely differing types of birds show that the earliest pairs are most successful in their breeding and produce most young. In mild and warm-temperate regions such as Britain, the spring is a long drawn-out process. The point of time at which the birds can proceed with nesting varies appreciably between different years.

It is of great adaptive value for birds which breed in warm-temperate climates to find themselves during the early spring as near to the breeding site as possible, within the same climatic region. This enables them to react immediately to the weather and to make their way to the breeding site as soon as the spring weather becomes favourable. The situation is totally different for arctic birds: once they have migrated out of the polar climatic zone, they are prevented from detecting spring progress at their breeding grounds. The birds of the north can frequently find themselves a long way away from the breeding sites and relying on their internal biological seasonal clock to take them to the breeding localities at the correct time every year.

Leap-frog migration is a fairly common phenomenon that has been described for widely differing types of migratory birds in a classic paper by the Danish ornithologist Finn Salomonsen (1955). Just as we might expect on the basis of the explanation above, it is chiefly in birds with a wide breeding distribution, from warm-temperate to arctic regions, that leap-frog migration occurs. Among the waders, the Ringed Plover and the Redshank are the most obvious examples. English Ringed Plovers are resident, south Scandinavian Ringed Plovers move to southwest Europe and northwest Africa, while Ringed Plovers from the mountains and the tundra in northernmost Scandinavia and Russia winter in tropical Africa and all the way south to the Cape of Good Hope. As examples of leap-frog migration among quite different groups of birds, we can give the migration of the Canada Goose and of the Fox Sparrow in North America and that of the Buzzard in Europe and Africa.

If two species are in close competition, we may expect them to exhibit leap-frog migration between themselves. This is probably the case with, for example, Whimbrel–Curlew, Great Snipe–Snipe and Curlew Sandpiper–Dunlin. The Whimbrel breeds on northern bogs and winters in Africa; the Curlew breeds on meadows and pastures in central and western Europe and winters mainly within Europe. The Great Snipe breeds in the Scandinavian mountains and migrates to Africa; the Snipe is found in marshes in large parts of Europe and in winter stays mostly in west Europe. The Curlew Sandpiper breeds on the very northernmost Siberian tundra and migrates to Africa; the Dunlin breeds somewhat farther south and winters in west Europe and in northwest Africa.

An interesting problem of detail is the migration of Icelandic waders. The Redshanks do not move far, and the same seems to apply to the Black-tailed Godwit: the Icelandic birds winter in Britain and west Europe, while Black-tailed Godwits from central and eastern Europe for example migrate to Africa. A totally different situation is found with the Icelandic

Ringed Plovers and Dunlins, which, according to indications from ringing results, migrate a long distance, to the west and northwest coasts of Africa.

Iceland is also a problem for climatologists. Some classify the climate as temperate, others again as arctic. Some draw the boundary between these climatic regions straight through the island. In Iceland there is never an intense winter cold as in the arctic regions, but on the other hand spring is late in arriving. One can therefore imagine that the Icelandic spring is very different for bird species with different ecological requirements.

Further studies are needed, however, for us to be able to explain leap-frog migration in more detail. This being so, Iceland is, then, a particularly important area, and the Redshank is a species of the greatest interest.

The Knot (*Calidris canutus*)

The Knot breeds on the arctic tundra, as close to the North Pole as it can possibly get (figure 22). It usually places its nest on flat, dry and stony tundra plateaux with low vegetation of mosses, lichens, low creeping willow and brilliant white mountain avens. After the eggs have hatched, the parents lead their young to nearby damp spots and along the edges of small pools, rivulets and rivers. During the breeding season the Knot feeds mainly on various insects, but also on some vegetable matter, tender leaves, buds, seeds and moss. The plant food is especially important when the birds arrive at the breeding sites at the beginning of June.

The stay at the breeding areas lasts no more than a month or two, June and July and, for the juveniles, also the beginning of August. As early as August the frost and snow return to

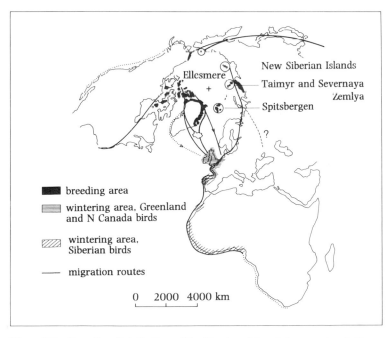

Figure 22   Breeding distribution of the Knot and its winter quarters in Europe and Africa. The lines show the main migration routes. Only relatively few birds follow the dotted-line routes. There are uncertain indications of a direct migration route overland between the Arctic Ocean and southern Africa. Based on Dick *et al.* (1976) and Glutz von Blotzheim *et al.* (1975).

the northernmost tundra. During the remaining nine or ten months of the year, the Knots live, frequently in large flocks, on extensive estuarine areas with sandy or muddy bottoms which are alternately flooded and drained in rhythm with the variations of the tide. Here the Knots find a wealth of small bivalves, molluscs and crustaceans on which to feed.

The Knot has a wintering range that is immensely large. The birds which breed in Alaska and on Wrangel Island in the far east of Siberia winter mainly on the coasts of southeast Asia, Australia and New Zealand. A smaller number move to the east coasts of the Pacific. Knots which breed in Hudson Bay and neighbouring regions of north Canada migrate to South America, where they are found during the winter right down to the southernmost parts.

The largest numbers of Knots winter on west European and west African coasts, where the Knot is the third commonest wader after the Dunlin and the Oystercatcher (see table 3). Knots come here from breeding sites on the islands of the far north of Canada (in particular the large island of Ellesmere) and in Greenland, Spitsbergen, the Taimyr Peninsula, Severnaya Zemlya and the New Siberian Islands (figure 22).

The greatest Knot concentrations of all are found in England. The two large shallow bays, The Wash on the east coast and Morecambe Bay on the west, regularly play host to over 100 000 wintering Knots between them. During the 1960s in England, a group was formed of ornithologists who specialised in ringing of waders in these bays. The birds are caught with what are known as rocket-nets which are laid out on the sandbanks and bars where the waders pack together in enormously dense flocks at high tide. By remote-controlled firing of the rockets, the nets can be shot over the startled waders, which do not have time to escape. A successful shot with big rocket-nets can produce a catch of hundreds of waders. After being measured, weighed and ringed, the birds are released again. As soon as the English ringers had learned to master this new catching technique to perfection, they made a series of expeditions during the early 1970s, to Iceland, Greenland, Ellesmere Island and northwest Africa, in order to study and catch waders.

Large numbers of recoveries of ringed waders have been reported. Our knowledge of their migration has been greatly increased. The Knot is a species that is a focus of interest for ringers. The brief description of the migration process which I present here is based largely on the work of the English wader-ringing group (Dick *et al.* 1976, Morrison 1977).

Knots from the far north of Canada and Greenland fly across the Atlantic to Europe in late summer (see figure 22). Many fly almost 3000 km over open ocean straight to the shallow North Sea shores in Denmark, Germany and the Netherlands (the area known as the Waddensee). Some of the Knots, however, make an interim stop on the coast of southwest Norway after 2000 km of flying across the sea. Other Knots fly from Greenland to Britain; some of them stop *en route* in Iceland.

At the beginning of autumn, when the Knots moult their flight feathers, most are therefore on the Waddensee. Almost half a million Knots have been counted there at this time. Later in the autumn, when the moult is completed, many birds move gradually westwards, from the Waddensee to England's east coast and from England's east coast to its west coast. This autumn and winter movement has been given the name 'the North Sea loop'.

During midwinter, quite the largest numbers of Canadian and Greenland Knots are found in Britain. The important wintering sites on the Waddensee and in the Bay of Biscay should not be forgotten, though (see table 3).

During the spring the Knots set off for Iceland, a very important staging area in May. The Knots whose home is in Canada and northwest Greenland fly direct from Iceland to the breeding area, first over sea and pack-ice between Iceland and Greenland and then straight across the inland ice of Greenland. The Greenland ice reaches to 3000 m above sea level, and the mean June temperature is $-17\,°C$. It has recently been discovered that several tens of thousands of Nearctic Knots do not migrate via Iceland but rather use alternative staging sites in north Norway (Davidson *et al.* 1986). Many Knots use the flightpath across the Greenland inland ice also on the southward migration in late summer; only a minority follow the west coast of Greenland south to Cape Farewell and then migrate from there out over the Atlantic towards Europe.

This latter migration route is used much more frequently by the Turnstone, another wader of the far north with a winter distribution that is similar to the Knot's. The rather stony and rocky coasts in southwest Greenland are excellently suited as stop-over places for the Canadian and Greenland Turnstones during the late-summer migration south towards Europe. During the spring, however, the Turnstones, like the Knots, take a short-cut across the inland ice of Greenland.

There are many ringing recoveries that support this picture of the migration of the Canadian and Greenland Knots. Some are more striking than others: a Knot was ringed in August 1968 on The Wash; its ring was found again in summer 1973 on Ellesmere Island – in a pellet from a Gyrfalcon! It happens now and again that bird rings are discovered in pellets of birds of prey. The Knot figures in a further two such cases, although the recoveries were made in Britain: in one case the Knot had become the victim of a Sparrowhawk, in the other of a Merlin.

Weighing has given us interesting information on how the Knot puts on fat as flight fuel for the long flight stages (figure 23). During the early spring, the mean weight of the birds in England is approximately 135 g. When the spring migration approaches, the birds rapidly put on large amounts of fat; the average weight rises to between 180 g and 200 g. The fat is used up on the long flight to Iceland, where the birds arrive at the beginning of May with an

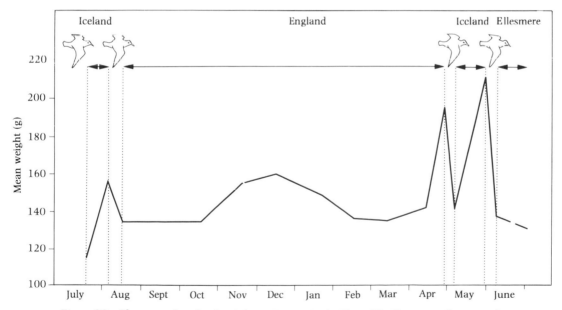

Figure 23 The annual cycle of weight variations in the Knot. The Knots rapidly put on large amounts of fat during late summer for the flight between Iceland and England, and during the spring for the return journey to Iceland and for the flight beyond from Iceland to the breeding grounds on Ellesmere Island in northernmost Canada. Based on Morrison (1977).

average weight of 135 g. There, the Knots put on weight again. At the end of May, the waders' mean weight is between 200 g and 220 g. At this time, the birds therefore consist of up to 40% fat. Those birds which arrive at the breeding grounds on Ellesmere Island once again weigh only around 135 g. All flight fuel has thus been burnt up during the 2500-km-long journey from Iceland.

Iceland serves as a 'refuelling station' during the late-summer migration, too, but the number of birds that stop off at this time is considerably smaller than during the spring and the weight data are more variable and unreliable. The Knots in England slowly increase their weight during the autumn, reaching a peak of around 160 g during the winter. The extra fat probably serves as a safety adaptation so that the birds can get by without food for a few days should heavy winter storms temporarily prevent their feeding on the shores.

Those Knots which breed in Siberia winter chiefly on the west coast of Africa, especially in Mauretania (table 3). There are occasional ringing recoveries showing that wintering occurs on a smaller scale also in west Europe. Off-passage Siberian Knots are by no means an uncommon sight in Sweden or in other Baltic countries during the period from the end of July to September and in early June. Some Knots winter as far south as the coasts of South Africa (see table 7). The question has been asked whether there might not be a direct migration route overland from Siberia to southern Africa. Sporadic observations of Knots around the Black Sea, in the eastern Mediterranean region and in east Africa could indicate that such a route exists.

Waders have been ringed on a large scale in South Africa. Several of the Knots have been recovered. All of the recoveries come from north and west European and African coasts, and this suggests that most Knots which winter in South Africa do after all travel along the Atlantic coast, as shown in figure 22. Ringing recoveries of Knots that have wintered in South Africa come from Mauretania in September and from England, Belgium, Denmark, the German Baltic coast and southeast Sweden in July and August.

Why do the Siberian and Greenland/Canadian Knots have separate wintering areas? Perhaps it is very worth while for the Greenland/Canadian Knots to fly to the staging sites in Iceland as soon as possible during the spring. It should therefore be of great adaptive value for them to winter in west Europe, as close to Iceland as possible. The Siberian Knots do not need to migrate through west Europe equally early in spring. They therefore extend the autumn migration to more southerly wintering areas in order to avoid competition with members of their own species from the New World.

The Knots demonstrate bird migration at its most magnificent: thousands of kilometres of inland ice, tundra and taiga, open oceans and tropical shores are traversed by the birds. Just leave it to our imagination to travel with them.

The Ruff (*Philomachus pugnax*)

The Ruff is best known for the males' remarkable ornamental neck collar (ruff) and ear tufts during the tournament displays (leks) in spring on damp meadows, on grassy bogs and on the northern tundra. The female (known as the Reeve), which keeps in the background during the display, looks more like a normal wader and has an unassuming plumage of brown and grey. The male moults into a similar plumage as early as summer and at that time differs from the female only in the fact that he is bigger.

The Ruff is most remarkable not only on account of its display, but also because of its migratory and wintering habits, for it changes its way of life radically when it leaves the breeding area and makes its way towards the wintering sites in Africa and, to a much lesser extent, southernmost Asia. From having fed almost exclusively on an animal diet – insects,

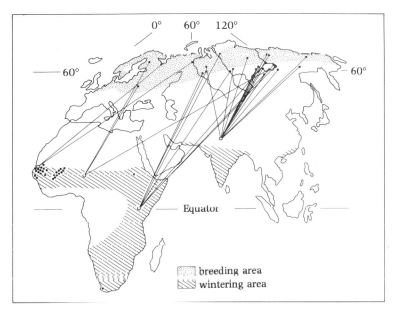

Figure 24 The breeding and wintering areas of the Ruff. The map also shows recoveries within or near the breeding area of Ruffs ringed on the Senegal River, at Lake Chad, at Lake Nakuru and Lake Naivasha in Kenya and at a staging site in northern India. The Indian birds are from wintering localities in southern Asia, probably also from eastern and southern Africa. The dots show recoveries within the wintering area of Ruffs ringed in northwest Europe, England and Scandinavia. Most recoveries have been made in the wintering areas around the Senegal and Niger Rivers. Based on Glutz von Blotzheim *et al.* (1975) and McClure (1974).

small bivalves, molluscs and crustaceans – during the breeding season and the greater part of the migration, the Ruff changes over to eating practically nothing but seeds on its African wintering grounds.

The Ruff is a very common winter visitor over wide areas of Africa (figure 24), where it generally lives in flocks on open grasslands associated with lakes or rivers. It does not show any particular attraction to the coast, but prefers areas beside fresh or brackish waters, both on migration and during the wintering period. In Africa the Ruff is especially common in the Sahel zone, just south of the Sahara, in the inundation areas around the Senegal and Niger Rivers and at Lake Chad. During the rainy season in summer, the rivers and lakes flood low-lying land. Immediately south of the town of Timbuktoo in Mali, for example, there is a large area of plain of 38 000 square kilometres (almost twice the size of Wales) around the Niger which is regularly flooded during summer and autumn. The first Ruffs arrive there at the end of July and the majority come in August. By this time, many different species of grasses have already had time to seed. Large flocks of Ruffs exploit this food source on the flooded savanna lands. Later in the winter and during the early spring, the flooded areas dry out. The grass then withers. The Ruffs remain behind, however, and pick the seeds from the dry, cracked ground. Flocks of thousands or tens of thousands of seed-eating Ruffs can be found, often together with large numbers of Turtle Doves and sandgrouse. Immense flocks fly to nearby waterholes to drink and bathe. There are altogether hundreds of thousands of wintering Ruffs in the inundation zone around the Niger.

The Ruffs can cause damage through attacking seeds of cultivated crops. In one area with new rice cultivations beside the River Senegal, 50 000 Ruffs (and 150 000 Turtle Doves) have been counted feeding on ricefields during the winter and early spring. The Ruffs have become so disposed towards eating rice that they even gather in flocks on those roads where spills of rice grain occur from leaking transport sacks. Only in the delta region of the Senegal are there normally more than half a million wintering Ruffs, and at times the number can be even higher. In February 1972, the number of Ruffs at a single roost site in this region was estimated at close on 1 million birds.

Other areas with immense concentrations of Ruffs are found at Lake Chad, where around 1 million have been counted within an area with a radius of only 20 km in a district with large wheat cultivations. In a neighbouring area, large numbers of Ruffs eat the seeds of cultivated millet, which ripen in February and March.

The Ruff has also been reported to feed for the most part on seeds during the winter from many other parts of Africa: for example, from Eritrea and Zambia, where the birds pick seeds of wild grass species. The Ruffs which winter in the southernmost parts of Africa, however, constitute an exception: they occur mostly on mudflats and lagoons, where no doubt they feed mainly on animal food.

Such a drastic change of living and feeding niche between the breeding and the wintering areas is unusual. Among the waders, there are only two other species which are known to alternate between animal and vegetable food on a fairly large scale. One is the Knot, which, as pointed out earlier, at times exploits plant matter at the breeding site. The other is the Black-tailed Godwit, which is a parallel case to the Ruff. Black-tailed Godwits winter in the inundation areas in Senegal (tens of thousands), beside the Niger (over 100 000) and at Lake Chad (probably over 100 000). The birds wade around in large flocks on the flooded plains and eat grass seeds. Like the Ruffs, the Black-tailed Godwits readily pick rice, though the godwits do not eat the ripened rice but instead the new-sown, which the birds are able to reach with their long bills on the waterlogged ricefields. When the drought comes during the late winter and early spring, many Black-tailed Godwits leave the Sahel zone (some go to the coast in Morocco). Those godwits which stay behind in the Sahel during this period keep to lands that have not yet dried out.

The large gatherings of Ruffs which winter in West Africa come not only from Europe but also from breeding grounds in the very easternmost parts of Siberia. This is shown by a whole series of easterly recoveries during the breeding season of Ruffs ringed at staging sites on the migration through northwest Europe (figure 25). The shortest route, following the Great Circle from far eastern Siberia to West Africa, passes through northwest Europe (see figure 134).

The most easterly recovery (see figure 25) of a Ruff that has passed through Europe comes from the lower reaches of the River Kolyma. This was made in June, two years after ringing in the Netherlands in August. The distance between ringing and recovery sites is nearly 8000 km. A bird was ringed in Uppland, Sweden, in August and recovered at the end of May in the River Lena area. Another Ruff was ringed off Copenhagen in September and recovered in May seven years later on the shore of the Sea of Okhotsk! Of particular interest is the Ruff that was ringed in the spring in East Germany and which was found again 32 days later more than 6000 km to the east; its average migration speed was therefore approximately 190 km per day. The record mean speed for a Ruff is, however, held by a bird which was ringed during the autumn migration at Ottenby in southeast Sweden and recovered four days later near Milan; this gives an average speed of 330 km per day.

The Ruff is reminiscent of the Lapwing when it comes to long-distance shifts to new

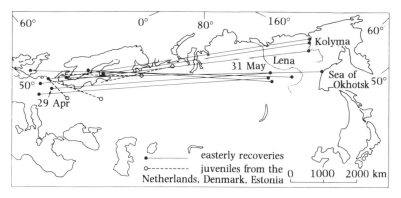

Figure 25   Recoveries in easternmost Siberia of Ruffs ringed during migration in northwest Europe. The broken lines show change of breeding site.

breeding sites. Figure 25 shows several cases in which Ruffs which have been ringed as young at breeding sites in the Netherlands, Denmark and Estonia have been found again later in completely different breeding areas. This widespread exchange of Ruffs between different parts of the breeding range leads to a considerable gene-flow between different Ruff populations and militates against the evolving of geographical differences in the birds' size and appearance. It is therefore not possible to distinguish different races either of Ruff or of Lapwing.

The Ruff also resembles the Lapwing in the respect that there are many chance observations of birds which have flown across the Atlantic and ended up in North and Central America. Well over a hundred different observations have been made of Ruffs along the American east coast, from Canada in the north to the West Indies in the south.

The Ruff is a wader of ecological contrasts. There is certainly a tremendous contrast between the magnificent display of the males on northern bogs and the eating of seeds in flocks on scorched-dry, almost desert-like country in Africa, where the temperature in the middle of the day approaches 40 °C.

Among Curlew Sandpipers (*Calidris ferruginea*) and Little Stints (*Calidris minuta*) at the southern tip of Africa

At the Cape of Good Hope in southernmost Africa, long, wide sandy shores alternate with sections of rocky cliffs. Here and there, there are river estuaries and lagoons with shallow sandy or muddy bottoms which are uncovered and flooded over by the tide. On the slopes behind the shore, a permanently green scrubland (*fynbos*) unfolds, the counterpart of the Mediterranean region's *maquis*. The coastal region in the very southernmost part of Africa has a typical Mediterranean climate, with cool and rainy winters from April to September (the southern African winter falls during the northern hemisphere's summer) and warm, sunny summers (October to March). During the height of summer in January, the mean temperature in Cape Province is 22 °C, the sun shines more often than not from a clear blue sky, and fresh southerly winds drive the sea against the coast in magnificent breakers. Jackass Penguins breed on islands off the coast. If you sit up on a cliff and scan the sea, you may catch sight of the magnificent Wandering Albatross passing by.

What remarkable and exciting species of wader one expects when wandering along these far-off shores! But a great surprise awaits the birdwatcher from northern Europe who is on a

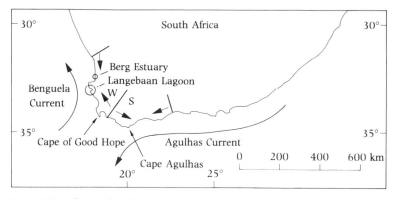

Figure 26   The waders have been censused along the coast of the southern tip of Africa. S and W indicate the stretches of coast that are designated as south and west coast respectively. The two localities which hold the largest concentrations of wintering waders are circled.

visit to the Cape Province in January. He will see many waders, but six birds out of seven are well known from the breeding or migration seasons back home, for that large a proportion of the waders on Africa's southernmost coasts actually come from northern breeding areas in Europe and Asia.

During December 1975 and January 1976, all waders were carefully counted along a 1090-km-long stretch of coast in the very south of Africa (as in the map in figure 26). The results are shown in table 7. It is clearly evident here how predominant the northern migrants are. The most numerous species are Curlew Sandpiper, Sanderling, Turnstone and Little Stint, all visitors from the northern hemisphere.

Most waders are found on coastal marshes, lagoons and river estuaries, especially along the stretch of coast west and north of the Cape of Good Hope. This is due to the fact that the cold and nutrient-rich Benguela Current has an effect on the western coastal regions; the warm easterly surface current off the south coast, which has been named after Africa's southernmost cape, Cape Agulhas, is by contrast poor in nutrients (figure 26). The largest wader concentrations of all are found at Langebaan Lagoon and the Berg Estuary. In January 1976, almost 40 000 waders were counted at Langebaan Lagoon and 20 000 at the Berg Estuary.

How great an importance the nutrient-rich Benguela Current has is also evident if one compares the number of waders per kilometre of sandy and rocky shore on the west coast with the number on the south coast (the delimitations are seen in figure 26). This comparison shows that waders are six to nine times more numerous on the west coast, where the proportion of northern migrants is also the highest (table 8). The Sanderling is the commonest species on the sandy beaches and the Turnstone the commonest on the rocky shores. The waders on the west coast often find their food in large banks of seaweed and seagrass which are dislodged from the luxuriant ocean bottoms and thrown up onto the shore by the waves. On the south coast, within the Agulhas Current's nutrient-impoverished waters, seaweed banks are present only to a small degree. The most favourable conditions of all for the waders are found on the mixed rocky and sandy shore along the west coast. Algae which cover the stony bottoms off the rocky headlands get torn away and are washed up onto the sandy shore in nearby bays. The circumstances on South Africa's coasts illustrate clearly how important the sea's production is to the feeding conditions of coastal waders.

Let me now describe in a little more detail the migration of the two waders in the Cape

Table 7. *The number of waders along the coast at the southern tip of Africa in December–January 1975/76. The area is shown in figure 26. The number of birds has been rounded off to the nearest hundred; + denotes that birds were observed but that the number was less than 50. As well as the northern migrants that are mentioned in the table, one Broad-billed Sandpiper, 15 Wood Sandpipers and four Red-necked Phalaropes were also counted. Based on Summers* et al. *(1977)*

	Lagoons and estuaries	Beach	Total
African waders			
Black Oystercatcher	100	2 600	2 700
White-fronted Sandplover	700	4 400	5 100
Chestnut-banded Sandplover	100		100
Kittlitz's Sandplover	4 200	300	4 500
Three-banded Plover	100	100	200
Blacksmith Plover	600	+	600
Avocet	2 100	100	2 200
Black-winged Stilt	800	+	800
Total African waders	8 700	7 500	16 200
Northern migrants			
Turnstone	1 100	8 000	9 100
Ringed Plover	1 200	300	1 500
Grey Plover	4 300	300	4 600
Curlew Sandpiper	50 800	3 900	54 700
Little Stint	5 800	200	6 000
Knot	3 800	+	3 800
Sanderling	2 600	12 100	14 700
Ruff	4 100	100	4 200
Terek Sandpiper	200	—	200
Common Sandpiper	100	300	400
Marsh Sandpiper	100	—	100
Greenshank	1 000	300	1 300
Bar-tailed Godwit	100	+	100
Curlew	400	+	400
Whimbrel	700	400	1 100
Total northern migrants	76 300	25 900	102 200
Grand total	85 000	33 400	118 400

region, the Curlew Sandpiper and the Little Stint, which have been studied most closely. More than 10 000 birds of each species have been ringed in South Africa; long-distance recoveries are shown in figure 27.

Both the Curlew Sandpiper and the Little Stint breed on the very northernmost arctic tundra. Ringing recoveries suggest that the birds take the shortest route between the Cape and the arctic coast, with important staging posts around the Black and Caspian Seas and at the Aral Sea. There is also a recovery of an off-passage Curlew Sandpiper in May in the Congo Basin.

The Curlew Sandpipers and Little Stints which pass through southern Scandinavia and Britain on migration no doubt winter in West Africa. Recoveries during the autumn of

Table 8. *Density of waders on different types of shores at the southern tip of Africa, December–January 1975/76. W = shores on the west coast and S = shores on the south coast (see figure 26). Northern migrants are strongly predominant, and commonest are Turnstone and Curlew Sandpiper, which are found mostly on rocky or mixed rocky/sandy shores, and Sanderling, which is found on all different types of shores. Based on Summers* et al. *(1977)*

Type of shore	No. of waders per km	Percentage of northern migrants
Sandy shore W	23	68
Sandy shore S	3	53
Rocky shore W	47	85
Rocky shore S	7	72
Mixed rocky/sandy shore W	79	79
Mixed rocky/sandy shore S	9	53

individuals, ringed mainly at Ottenby in south Sweden, show that the Curlew Sandpiper follows the Atlantic coast southwards; the Little Stint more often flies straight across the European interior (figure 27).

The Curlew Sandpiper has been studied with particular intensity at Langebaan. The birds arrive there in September, and disappear north again in April. The Curlew Sandpiper exhibits a high degree of site fidelity. Approximately half of the adult birds that winter at Langebaan return in following years; the remainder either shift wintering site or die.

Birds which fly long distances never or only extremely rarely moult their flight feathers during ongoing migration: the wings must be complete for the long flights to be as economic as possible. Adult Curlew Sandpipers therefore do not start to moult their flight feathers until after their arrival at Langebaan in September. The moult continues up to February; then the sandpipers can head northwards in the spring with entirely new wings. The whole course of the moult of the flight feathers accordingly takes between four and five months. The same case has been established for Curlew Sandpipers wintering in Tasmania.

The situation is, however, different for the Curlew Sandpipers which winter in West Africa. Many of them make an intermediate landing and stop off in Morocco for just under two months, from the end of July to the beginning of September, before continuing southwards to the final winter quarters. During the short stop in Morocco, the Curlew Sandpipers moult their wing feathers completely. The strategies for the annual moult of the wing feathers can, therefore, vary substantially between separate populations. The Curlew Sandpiper's close relative, the Dunlin, exhibits further different moulting habits. Dunlins breeding on the Alaskan tundra renew their wing feathers while still on the breeding grounds and when incubation is finished; the moult is completed within two months. The adult birds can begin to move from the breeding area with completely new wings. Those Dunlins which breed in northern Europe and Asia, however, have a far from complete moult (sometimes none at all) at the breeding sites, but interrupt the moult when the time comes to migrate south and set off with both old and new wing feathers; after their arrival in the wintering areas in west Europe and northwest Africa, the moult is completed.

The Curlew Sandpiper does not begin to breed until two years of age, and many immature birds stay behind and 'oversummer' at Langebaan during the southern African winter. The number of those immatures which oversummer in South Africa varies drastically between different years. In some years there are many thousands of immature Curlew Sandpipers at

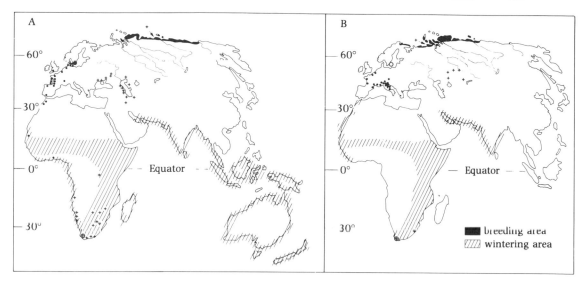

Figure 27 Migration of the Curlew Sandpiper (A) and the Little Stint (B). The crosses show recoveries of birds ringed or controlled at the southern tip of Africa. There is one recovery of each species from the breeding area – a Little Stint in June on the Jugor Peninsula 13 000 km from South Africa, and a Curlew Sandpiper in June on the lower reaches of the River Lena, over 15 000 km from the southern tip of Africa. The remaining recoveries are mainly from the autumn migration during the period August–October, with a few also from the spring migration in May and June. Most recoveries in southern Africa have been made during the period September–April.

The dots show recoveries of birds ringed in southern Sweden, mainly at Ottenby. The recoveries have been made during the period July–November in the same autumn as ringing. Exceptions are the two recoveries within the normal wintering area, both of birds ringed at Ottenby: one recovered after six years in November in Mauretania, the other after five years in May in Ghana. Based on Elliott et al. (1976) and Glutz von Blotzheim et al. (1975).

Langebaan (in 1973, for example, there were nearly 7000), but in other years there are practically none at all (in 1975, there were only four oversummering Curlew Sandpipers). Similar tremendous yearly fluctuations have also been reported from Tasmania. A probable explanation is that in some years the Curlew Sandpipers are so totally unsuccessful in their breeding efforts that almost no young at all reach the flying stage. Possible causes of this are heavy predation, mainly by arctic foxes (eating lemmings when available but switching to birds in years of low lemming abundance), and catastrophic spring and summer weather in the breeding range.

The immature Curlew Sandpipers moult only part of their flight feathers during their first year. They renew on average four or five of the ten primaries. This moult does not begin until February, when the adult Curlew Sandpipers are just finishing off their moult, and continues until July. By then, the adults have long since departed to the breeding grounds and left the young birds behind on their own at Langebaan. Between September and February, the $1\frac{1}{2}$-year-old Curlew Sandpipers then moult the wing feathers in full, at the same time as the adult birds which have returned in large droves to Langebaan after breeding. In April, by which time the young Curlew Sandpipers are two years old, it is time for them to migrate north for the first time in order to breed.

The Curlew Sandpipers at Langebaan have a mean weight from September to March of around 57 g. At the beginning of April, they begin to store up fat in order to use it as flight fuel during the long migration northwards. The mean weight increases to 80 g; some birds weigh up to 100 g. The birds therefore consist of between 30% and 40% fat when they set off from Langebaan. This quantity of flight fuel is enough for more than 2000 km flying. But the Curlew Sandpipers' entire journey to the breeding grounds is more than 15 000 km long. The birds must consequently stop off on several occasions in order to renew their energy reserves.

The Little Stint displays many similarities to the Curlew Sandpiper. The Little Stints arrive in southernmost Africa in October. Right up to March their mean weight is around 22 g. In April, the stints put on large amounts of fat before the spring migration. The mean weight increases to 32 g, some individuals reaching up to 40 g. Many Little Stints also winter at Lake Nakuru in Kenya. In May, approximately two weeks later than at the Cape, the mean weight of the stints there increases greatly. Then they disappear northwards. Among the Little Stints at Nakuru in May there are probably birds which have wintered at the southern tip of Africa and which are stopping off there in order to replenish their fat reserves.

The Little Stint is the smallest of the Old World waders. Its normal weight, 22 g, is the equivalent of the weight of slightly more than 2 centilitres of water. When the Little Stint has stocked up completely with fat, it still does not weigh more than about 4 centilitres of water. The Curlew Sandpiper's weight is not all that impressive, either. What is on the other hand impressive is the fantastic machinery that fits into these small avian bodies and which makes it possible for the birds to live for at least seven of the 12 months of the year on the blossoming shores in Cape Province, two months among melting snow on the far northern tundra, and the rest of the year on 10 000-km journeys between the two.

3.2   Birds which forage on lake and sea bottom

When winter releases its grip on the regions of the north, coasts, lakes and waterways are opened up to the birds which derive their sustenance from the bottom of lakes and seas. As seen in table 9, this category of birds includes ducks, swans, certain grebes, coots and even one representative of the passerines, namely the Dipper. The Dipper occupies an altogether special ecological niche: fast-flowing or rushing water in small rivers and streams, where it dives for bottom-dwelling insect larvae. The Harlequin Duck has a similar niche during the summer, when it lives on torrential rivers; during the winter it dives beside the breakers off rocky coasts. The nearest Harlequin Ducks to Britain are those in Iceland. The population moves only the short distance between the rivers of summer and the coasts of winter.

The smallest dabbling ducks wade or swim on shallow shores or at the water's edge to glean plant food and small animal matter (table 10). Larger dabbling ducks sometimes up-end, with the rear end and the tail visible above the water surface, in order to reach down to a depth of some tens of centimetres. The Pintail is the species that most often 'stands on its head' while searching for food. It has the longest neck among the dabbling ducks and reaches to a bottom depth of 50 cm. The long neck of swans is of course an adaptation: with it they can reach the bottom vegetation. Many ducks, as well as the grebes and the coots, dive for food. The heavy diving ducks (Velvet Scoter, Common Scoter, Steller's Eider, Eider, King Eider and Long-tailed Duck) are the most accomplished divers. In exceptional cases they have been discovered, when they have got caught up in fishing apparatus, at depths down to about 50 m, but the general rule is that they seldom forage at depths greater than 10–20 m. It is, of course, much more economical to seek food at moderate depths than to waste valuable

Table 9. *Birds which feed on lake and sea bottom and which occur regularly in north Europe. The data on wintering areas apply primarily to European birds or birds which pass through Europe on migration. A = non-diving species; B = diving species*

Species (NB = does not breed in Scandinavia)	Found in N Europe in winter	Important wintering areas	
		W Europe & Mediterranean region	Africa south of Sahara
A			
Mallard	+	+	
Teal		+	
Garganey			+
Gadwall				
Pintail		+	+
Shoveler		+	(+)
Shelduck		+	
Mute Swan	+	+	
Whooper Swan	+	+	
Bewick's Swan NB		+	
B			
Scaup	+	+	
Tufted Duck	+	+	
Pochard	+	+	
Long-tailed Duck	+	+	
Velvet Scoter	+	+	
Common Scoter	+	+	
Goldeneye	+	+	
Steller's Eider NB	+	+	
Eider	+	+	
King Eider NB			+	
Slavonian Grebe	(+)	+	
Black-necked Grebe		+	
Little Grebe	+	+	
Coot	+	+	
Dipper	+	+	
Total no. of species	16	24	2

energy on deep dives. The King Eider and the Long-tailed Duck are the two species that dive deepest.

The migratory habits of the birds in table 9 differ very markedly from those of the waders. A large proportion of the waders winter in Africa south of the Sahara, but of the wildfowl only the Garganey, the Pintail and to a certain extent the Shoveler cross the Sahara in large numbers. Particularly important wintering sites are the inundation zones around the Senegal, the Niger and Lake Chad, as well as certain lakes in the Ethiopian highlands.

Why do so few northern ducks penetrate south to African winter quarters? Many northern duck species throw out feelers southward to Africa south of the Sahara. Teal, Gadwall, Pochard and Tufted Duck are found there every winter, but the numbers are always small. Maybe the competition between the indigenous African wildfowl and the northern wildfowl

Table 10. *Normal and maximum bottom depths for foraging by various wildfowl and grebes (applies mainly to migration and wintering periods). Dabbling ducks, coots and lighter diving ducks exploit both vegetable and animal food, such as seeds, shoots, buds and roots from algae and aquatic plants, together with small animals such as insects, freshwater shrimps, molluscs etc. For dabbling ducks the emphasis is on a vegetarian diet, for the lighter diving ducks on an animal diet. The swans feed on aquatic plants (the Whooper Swan's diet also includes bivalves and terrestrial plants). Heavy diving ducks live on an animal diet, bivalves, molluscs, crustaceans, sea-urchins etc. A = non-diving species. B = diving species. Based on Bauer & Glutz von Blotzheim (1968, 1969) and Cramp & Simmons (1977)*

	Bottom depth (m)	
Species	Normal	Maximum
A		
Teal, Garganey	0–0.1	0.2
Shoveler	0–0.1	0.3
Gadwall	0.1	0.4
Mallard	0–0.4	0.48
Pintail	0.1–0.4	0.53
Shelduck	0–0.4	0.9
Whooper Swan	0–1	1
Mute Swan	0.2–1	1
B		
Pochard	0.5–2.5	5
Little Grebe, Coot	0–2	6
Scaup, Tufted Duck, Goldeneye	0.5–5	10
Black-necked & Slavonian Grebes	?	20
Velvet & Common Scoters	2–10	20
Eider, Steller's Eider	2–10	30
King Eider	5–20	50
Long-tailed Duck	3–25	55

creates a barrier at the Sahara, a barrier that only the Garganey and the Pintail effectively break through.

The number of indigenous African species of wildfowl is surprisingly low: only 18 in all. Only half a dozen species are found in West Africa, Sudan and the Congo area. This can be compared with the situation in north Europe. Sweden, for example, despite its comparatively small land area, harbours 23 species of regularly breeding ducks, swans and geese. A possible reason is that the food resources on the lake bottoms in Africa are used first by animals other than ducks: for example, by fish, which are present in abundance and with a great wealth of species in the tropics. As a result of the competition between the animal groups, the ecological niche for wildfowl is accordingly considerably smaller in Africa than in northern regions.

The migration of wildfowl and their most important stopping-off sites have been the subject of intensive mapping in Europe, in large parts of Asia and in North America. A major reason for these large-scale studies is the great importance of ducks and geese from the point of view of hunting. Other factors, such as human exploitation of important swamp and lake areas and habitat destruction as a result of, for example, oil spills, also give cause for monitoring of the most important staging sites of wildfowl. Since the late 1960s in Europe and the Middle East, international co-operation in the matter of counting wildfowl has been pursued under the direction of the International Waterfowl Research Bureau (IWRB), whose headquarters are in England. The counts, which have taken place mainly in midwinter, have been carried out partly from the shore, where thousands of birdwatchers have been engaged, and partly from boat and aeroplane. Aerial inventories have proved especially practicable when a good picture of wildfowl numbers over large areas is needed quickly.

The wintering areas of wildfowl are of course determined primarily by the extension of the ice. The northern limit for wintering dabbling ducks and the lighter diving ducks coincides well with the extent of the January 0 °C isotherm. Some of the heavier diving ducks winter farther north than this; they remain far out off the coasts, where the water does not freeze over until periods of cold weather become very severe and prolonged.

A close relative of the common Eider, the Spectacled Eider, breeds on the arctic coasts and on islands in the Bering Strait. It is estimated that the entire population comprises a quarter of a million birds. The big problem is: where do these birds spend the winter? Nobody has seen the birds in their winter quarters! There are only sporadic winter observations of occasional Spectacled Eiders from the coasts of the northernmost Pacific Ocean. In addition, a few winter observations have been made on quite different northern coasts, for example in north Norway and on the Kola Peninsula. We can, however, be fairly certain that the main wintering sites lie somewhere in the Pacific Ocean; this is because large numbers of Spectacled Eiders are seen on spring migration in May when they arrive from the area of the Pacific Ocean and migrate north through the Bering Strait.

On the northern coasts of the Pacific, along the edge of the pack-ice in the Bering Sea and off the Aleutian Islands, large numbers of heavy diving ducks winter, including King Eiders and Eiders, but nowhere can Spectacled Eiders be found. By a process of elimination, the conclusion has been reached that the Spectacled Eiders probably winter in large open channels in the pack-ice in northern parts of the Bering Sea. Satellite pictures show large openings of clear water in the pack-ice in sea areas with a depth of only some 10 m. The Spectacled Eiders could very well remain there and find plenty of food. The dispersion of the openings in the pack-ice is constantly changing. The birds must therefore be prepared the whole time to move to new gaps in the ice.

The most important wintering sites for dabbling ducks, lighter diving ducks and coots/gallinules according to the results from the international midwinter censuses are shown in figure 28. The shallow shores and bays around Denmark's coasts and islands constitute a first-rate area for wildfowl. I should think that Denmark, in sharp competition with the Netherlands, is the country with the highest density of wildfowl in Europe.

The tidal shores and estuaries in the North Sea area are central wintering sites for many wildfowl. In addition, there are quite a number of important wintering localities in the Mediterranean area and on the shores of the Black and Caspian Seas. It is striking how often the best sites are found at the mouth and delta areas of large rivers. Rivers with such deltas are the Guadalquivir in Spain, the Rhône in France (the Camargue), the Po in Italy and the Danube in Romania. The Danube Delta on the Black Sea, an area of about 5000 square kilometres of river arms, lakes and marshes, is a particularly fine staging site and wintering locality for wildfowl. In some winters, over 300 000 individuals of both Pochard and Mallard have been counted there, and more than 100 000 Coots and Tufted Ducks.

The Danube Delta is hit by frost in some winters. Many birds then leave the area, but the total number of wintering wildfowl never falls below 100 000. In favourable winters, the number may approach 1 million. Equally enormous numbers also stop off in the delta regularly during autumn and spring. On one occasion, in the period between the end of

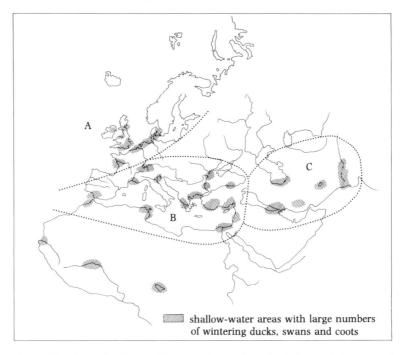

shallow-water areas with large numbers
of wintering ducks, swans and coots

Figure 28   Areas in Europe, the Near East and North Africa with large numbers of wintering dabbling ducks, lighter diving ducks, swans and coots. Bird presence has been mapped in international wildfowl counts, and the total populations have been estimated in various regions (see table 11). A = northwest Europe, B = Mediterranean area and Black Sea, C = Near East with Caspian Sea, Persian Gulf and Indus River. Based on Atkinson-Willes (1976).

November and the beginning of December, 1 million Pochards and more than half a million birds of other species, mostly Mallards, Teals and Coots, were counted.

The outlet of the Danube forms Europe's second largest delta area, the largest being where the Volga discharges into the Caspian Sea. As a rule this latter area is completely frozen over during the winter and therefore has no importance as a wintering site for wildfowl. At other seasons, however, it serves as a staging-post for gigantic hordes of birds. The Volga Delta is above all a moulting centre for ducks, not only from Europe but also from large parts of Siberia (see below regarding moult migration and the migration of the Garganey).

The midwinter wildfowl counts that have been carried out over many years make it possible to estimate the total number of wintering birds in separate regions (table 11).

Southerly regions contain many food-rich lagoons and delta areas, which are well suited for dabbling ducks and lighter diving ducks. Northwest Europe is alone in providing heavy diving ducks with large shallow sea areas with a rich life of bottom-dwelling animals (figure 29). The fiords along Norwegian and Icelandic coasts and the extensive shallow bottoms (with depths of less than 10–20 m) in the southern Baltic Sea, the North Sea and the Bay of Biscay are the major homes of the heavy diving ducks outside the breeding season. The situation for the heavy diving ducks is similar to that of many waders. The birds have ample access to suitable feeding environments which they cannot exploit when breeding. Such a disproportion between suitable feeding environments and suitable breeding environments influences the birds' lifestyle in a number of different ways. Breeding competition becomes

Table 11. *Numbers (in thousands) of wintering ducks, swans and coots in different regions. The areas A, B and C are shown in figure 28. Other species include Red-crested Pochard, Harlequin Duck and Barrow's Goldeneye in region A, and Red-crested Pochard, Ferruginous Duck, Ruddy Shelduck, Marbled Duck and White-headed Duck in regions B and C. The data are based for the most part on the international midwinter wildfowl counts according to Atkinson-Willes (1976) and Cramp & Simmons (1977), updated from Monval & Pirot (1989). The data from normal or mild winters in Denmark are from Joensen (1974). The data for Sweden are from Nilsson (1975). The figures are in many cases very rough estimates, particularly those concerning the diving ducks which live on open sea areas (Long-tailed Duck, Common Scoter, Velvet Scoter)*

Species	A	Denmark	Sweden	B	C
Mallard	5000	150	80	4000	1000
Teal	400	—	—	1000	1500
Garganey	—	—	—	—	?
Gadwall	12	—	—	75	?
Pintail	70			300	800
Shoveler	40	—	—	375	700
Shelduck	250	20	—	75	?
Mute Swan	180	70	8	20	?
Whooper Swan	42	10	2	17	?
Bewick's Swan	17	0.5	—	—	—
Scaup	150	100	1	50	40
Tufted Duck	750	200	80	600	600
Pochard	350	10	0.5	1250	350
Long-tailed Duck	1500?	50?	200?	—	—
Velvet Scoter	200?	20	3	—	—
Common Scoter	1500?	200?	1	—	—
Goldeneye	200	100	25	?	?
Steller's Eider	?	—	—	—	—
Eider	2000	750	5	—	—
King Eider	200?	—	—	—	—
Coot	1500	200	20	2500	2000
Other species	20	—	—	150	500
Totals (millions)	14.4	1.9	0.4	10.4	7.5

razor-sharp; old and experienced birds have much greater breeding success than young ones; the number of fledged and independent young per pair is low; the birds leave the breeding area as soon as possible; and they do not begin to breed until they have reached several years of age. Within the extensive feeding habitats, a long distance from the breeding sites, the competition is by contrast weaker. Here, the birds can live a fairly trouble-free existence so long as they have only their own survival to think about and do not need to cope with breeding. Young birds which are not yet ready to breed stay behind of course and 'oversummer' in the suitable survival environments.

The heavy diving ducks show several features that are in accordance with this picture. Competition seems to be slight in the winter quarters. Several different species occur there together and live on the same type of food. Sea mussels are often a key food for all of the species. The birds return to the mussel beds immediately after breeding. As early as June, many males of Eider, Velvet Scoter and Common Scoter are already back in the southern

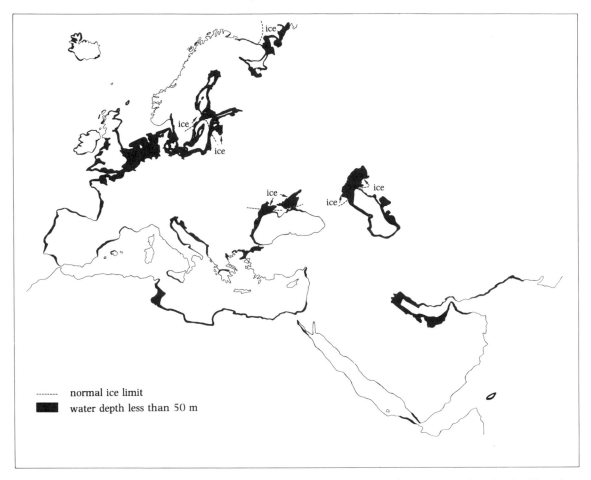

Figure 29   The map shows sea areas, ice-free during the winter, with a depth of less than 50 m. These areas, together with coastal waters, provide wintering opportunities for marine diving ducks.

Baltic Sea and the North Sea in order to moult. The young diving ducks do not begin to breed until two or three years of age but instead summer at the food-rich mussel beds (as a rule, the dabbling ducks and the lighter diving ducks breed as early as their first year).

For anybody who wishes to get a proper understanding of the migratory habits of wildfowl, it is important to be acquainted with their pair formation and moult. In contrast to most other birds, ducks pair in the winter quarters. Since different breeding populations often have common winter quarters, it is not such a rare event for two birds from widely separated breeding localities to pair off. The female generally returns to her breeding area and the drake accompanies her, even if he has grown up at an entirely different place or has previously bred somewhere else with a different female. This explains why recoveries of ringed males are often reported from far-off and unexpected breeding sites. The phenomenon is called 'abmigration'. This migration naturally leads to different genes being rapidly disseminated within the species' different breeding populations. As a logical consequence, the wildfowl species are only to a very limited degree divided into various races. The reason why it is the females and not the males that show breeding-site fidelity is probably that the former have full responsibility for nest-site selection and brood-rearing and therefore it is they who benefit from being in familiar terrain.

The birds' wing- and tail-feather moult is of course a sensitive operation since their powers of flight are inevitably affected. In most species, the feathers are changed gradually, step by step. Only a few new feathers are growing out at any one time. The birds can manage to fly at this time without any major difficulty. The gradual moult process is slow – it often takes several months before all the wing feathers are renewed – but necessary for birds which are dependent on keeping their powers of flight intact. During periods when good flight capability is of additional importance, such as during long-distance migrations, generally no moult takes place at all.

For all the species in table 9, however, it is the case that all the primaries of the wing are moulted simultaneously. The birds thereby lose the power of flight for the greater part of the moult period. A few species also in table 2, namely the Crane (which changes its flight feathers only every other year), the Water Rail and the Spotted and Little Crakes, shed all the primaries at the same time and lose the capacity to fly at that time. The advantage of renewing all the primaries in one go is that the moult can be completed quickly. In smaller ducks, new flight feathers have grown out and the birds have regained the power of flight after about three weeks; in the larger ducks this takes about five weeks, and in the swans six to seven weeks.

It requires special conditions for the birds to be able to get by without the capacity for flight during the moulting period. Rails and crakes live under cover of tall, dense vegetation in marshes and on lakeshores, and cope very well with catching food, even though they are flightless for several weeks. The same applies to the Crane, which, despite its size, leads a very secluded life in summer on extensive bogs and marshlands. The Dipper is one of the very few passerines that loses the power of flight completely during the moult. Since it makes use of its wings when diving and manoeuvring in rushing waters, it in fact loses not only the ability to fly but also partly the ability to dive during the moult. The moult takes place immediately after breeding has finished. At that time the Dipper keeps to the sides of streams, under overhanging roots, among tussocks, thicket and scrub, where it is virtually impossible to see or catch.

Several of the species in table 9 migrate impressive distances, sometimes thousands of kilometres, to moulting sites that are rich in food and safe. The Volga Delta serves as a moult centre during the summer and early autumn for enormous numbers of eastern Mallards, Teals, Garganeys, Pintails and Shovelers. The males of these species leave the incubating females on the breeding grounds and make off to a suitable moulting site. Females which for one reason or another are unsuccessful in their breeding attempts also take part in the summer migration, while the rest of the females remain on the breeding grounds and look after the young. The majority of the dabbling ducks that breed in Britain and northwest Europe, in contrast to the eastern populations, show no large-scale moult migration. The males gather together in small flocks on sheltered lakes and bays in the vicinity of the breeding site and renew their wing feathers there before the autumn migration to winter quarters.

The most striking examples of moult migration, in which hundreds of thousands of birds from wide regions concentrate in one restricted moult centre, are worth describing in more detail.

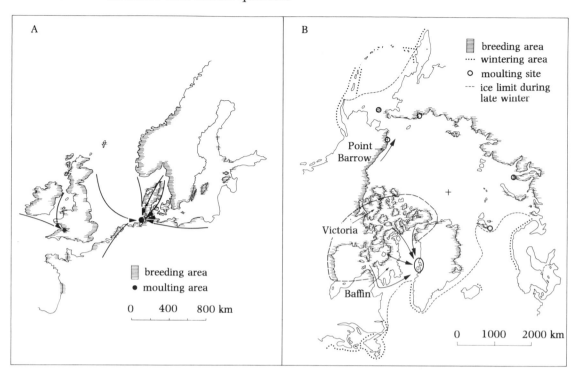

Figure 30   A. Shelducks from the whole of northwest Europe gather during the moulting period at shallow tidal shores along the German North Sea coast. In addition, a smaller moulting site, probably chiefly for Irish Shelducks, exists in the Bristol Channel in England. B. King Eider drakes migrate in summer from the breeding grounds in northernmost Canada and northwest Greenland to moult in Disko Bay on Greenland's west coast. The arrows indicate the main migration routes. The map also shows other sites with major gatherings (though not so large as in Greenland) of moulting males and non-breeding immatures.

Based on Bauer & Glutz von Blotzheim (1968, 1969), Cramp & Simmons (1977), Kistchinski (1973) and Salomonsen (1967).

The Shelduck (*Tadorna tadorna*) (figure 30A)

The majority of northwest Europe's Shelducks gather on the shallow tidal shores along the German North Sea coast in order to moult. At Grossen Knechtsand, which lies between the outlets of the rivers Weser and Elbe, around 100 000 moulting Shelducks have been counted. Over 2000 individuals have been ringed there and more than 200 recoveries have flooded in, from France, Ireland, England, Scotland, Belgium, the Netherlands, Denmark, Sweden and Norway. Apart from this large moulting site, there are only a few localities with concentrations of moulting Shelducks, in most cases only of a few hundred birds; the largest assembly is of 3000–4000 at Bridgwater Bay in the Bristol Channel, England.

In the last week of January first-year non-breeding Shelducks begin to gather at Knechtsand, but not until July does the actual moult migration of adults start. There are many reports from England, Scotland and southern Denmark of considerable overland passage of Shelducks. The flight heads straight towards the German North Sea coast. At the beginning of August Shelducks are at their highest numbers at Knechtsand, and at the end of the month many birds set off on new wings. Moulting birds can, however, still be found there right up to October. It is interesting that the British Shelducks, after moulting at Knechtsand, move at a leisurely pace 'homewards' to get back to southern England in November–December, where they then winter.

Scandinavian Shelducks which have failed in their breeding attempts, or which have left their young in the large 'crèches' which often occur with this species, fly to Knechtsand in July. The remaining adults set off at the beginning of August. After moulting, they all disperse over winter quarters in the countries bordering the North Sea.

The King Eider (*Somateria spectabilis*) (figure 30B)

The King Eider is a high-arctic duck which in winter often remains right at the edge of the pack-ice. Males from separate parts of the range gather at several different moult centres. The most magnificent migration, consisting of hundreds of thousands of birds, heads for Disko Bay on the western side of Greenland. The importance of this moult centre has been demonstrated by the Danish ornithologist Finn Salomonsen. The birds at the Greenland moult site consist of adult males and non-breeding immatures. In August there are at least 200 000 King Eiders present there. Close on 4000 have been ringed, and recoveries show that the birds come mainly from breeding areas in the Canadian Arctic archipelago. The ones that fly farthest are those which come from Victoria Island, about 2500 km to the west. Reports of migrating King Eiders tell of large flocks flying across the sea off Greenland's northwest coast and in the straits both south and north of Baffin Island towards the moulting site. The most intensive and impressive migration passes straight across the central parts of Baffin Island. At the end of July, many flocks of hundreds of drake King Eiders can be seen there daily flying east at altitudes of just over 1000 m through mountain passes and alpine valleys south of the glacier region of Baffin Island. What a fascinating sight that must be!

The female King Eiders do not participate in the moult migration but stay at the breeding site and moult there before they begin to move, together with the juveniles, to the winter quarters.

Other moult centres where thousands of King Eiders gather can be found off Alaska's northwest coast (a heavy westward passage can be observed at Point Barrow; in July–August this involves only drakes, later in August also large numbers of females), at St Lawrence Island in the north Bering Sea (King Eider males from east Siberia migrate southwards through the Bering Strait in July), and at the southern tip of Spitsbergen and off the mouth of the River Ob. In many places within the European and west Siberian part of the King Eider's range, long-distance moult migrations are not known. Scattered flocks of moulting males are encountered here and there in shallow sea areas off the breeding coasts.

The Eider (*Somateria mollissima*)

It is not, however, necessary to travel to Greenland or arctic Canada to witness moult migration of eiders. There is also a passage of hundreds of thousands of birds in the Baltic area, but in this case it is of course not King Eiders that are involved but common Eiders. When the females have started to incubate, the males assemble in large flocks on the open sea. Many of the Baltic Sea males then set off and move to Danish waters, where they moult largely within the same area that later serves as winter quarters.

Most flocks of migrating drakes are observed between about 10 June and 5 July in Kalmarsund (Kalmar Sound), where they are moving south (figure 31). Over the Danish islands the flocks fly westwards, often at high altitude. A particularly fine summer migration crosses southern Jutland: here, the Eiders fly overland in large flocks of hundreds or even

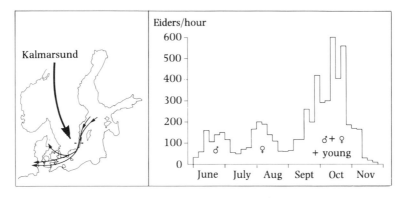

Figure 31   Numbers of individual Eiders per hour which migrate through Kalmarsund. The diagram is divided into five-day periods and shows the average for all days with more than four hours' observations. The data were collected over 15 years (1958–1972) by the Ottenby Bird Observatory.

   The first passage peak consists of males on route from the breeding grounds in the Swedish and Finnish archipelago to Danish moulting grounds; the second shows moult migration of females; and the final peak shows the passage of adults and birds of the year to winter quarters in Denmark.

thousands of birds, mostly at dawn or at dusk, and come down in the Danish Waddensee. On the basis of aerial surveys, it has been estimated that around 160 000 male Eiders moult in the Danish waters (the Baltic Sea, the North Sea and the Kattegat) during July and August. They spend most of their time in sea areas where underwater banks lie only a few metres below the surface, probably on account of their reduced diving capacity during the moult. The flocks resting on the water in these areas can become very dense and large. On one occasion, 40 000 drakes were counted in a single flock; the density was approximately one bird per square metre. It is particularly during the daytime that the flocks are as compact as this. At night, when feeding activity is at its most intense, the birds disperse out over the banks. When the month of August is at an end, almost all the male Eiders have completed the moult and regained their powers of flight.

   Some of the Baltic's females, too, move to Denmark to moult. A clear migration peak occurs at the beginning at August, when females which have probably lost eggs or young pass through Kalmarsund (figure 31). Several thousands of females reach Denmark and moult there from the end of August and during the following month. They do not assemble in large flocks like the males but are spread out in small groups. As well as adult males and females, immature non-breeding Eiders also moult in Danish waters. In total, approximately a quarter of a million Eiders have their moult centre in Denmark. In addition, Denmark serves as a moult centre for at least 60 000 Velvet Scoters, a species whose feeding and migratory habits are similar to the Eider's.

   Many male and female Eiders remain behind in the Baltic Sea during the moult. The late-autumn Eider migration (the third passage peak in Kalmarsund) therefore includes a mixture of adults of both sexes and young birds of the year.

The Common Scoter (*Melanitta nigra*)

The Common Scoter breeds at mountain and tundra lakes in northernmost Europe and in Siberia. It moults in the North Sea, such as off the west coast of Jutland, where, in aerial surveys at the end of July, between 100 000 and 150 000 Common Scoters, adult males together with summering immatures, have been counted. Aerial surveys in June off Jutland's west coast have informed us that over 30 000 immatures are to be found there. The problem with the Common Scoter is that it prefers deep waters, often 10–15 m deep. Because of this, the birds are so far out to sea that they are almost impossible to census – not even aircraft are a properly effective means. We cannot, therefore, know for certain how many Common Scoters moult in the North Sea.

There are many observations of heavy scoter passage during July and August. On the Estonian coast of the Gulf of Finland, for example, a total of 100 000 migrating Common Scoters has been counted during two weeks at the end of July and the beginning of August. The passage takes place at dusk and dawn. It is also well known, for example from radar observations in Estonia, that large numbers of Common Scoters pass through at night-time. Whistling calls from night-migrating flocks are heard in summer over Denmark and southernmost Sweden. At Kalmarsund, a migration peak occurs during the first week of August. Many flocks, sometimes a thousand strong, have been reported at dusk passing west overland in the southernmost part of Jutland or over Schleswig-Holstein. A particularly interesting report comes from one of the east Friesian Islands (Wanderooge) off the German North Sea coast. One observer there has made random observations over more than ten years of a heavy westward passage of Common Scoters over the sea off the island. The passage starts in the middle of June, increases in strength at the end of the month, and falls away in August. At the beginning of the migration period, the daily totals of passing Common Scoters vary between 500 and 2000. When the migration is at its most intensive of all, between 1000 and 2000 scoters per hour pass. In one year with a particularly high frequency of spot counts, the total number of Common Scoters passing (of which about 75% are males) was estimated at nearly 1 million.

The Danish ornithologists' method of studying the moult progression of the Common Scoter demonstrates how common and at the same time how difficult to study this species is in the North Sea. The ornithologists collect moulted wing feathers that are washed up on the west Jutland shores. Around 20 July, the black wing feathers can be found there, a sure sign that moult is underway. In September, the males and the non-breeding immatures are once more able to fly with new wings. At the same time the females begin to arrive from the breeding areas, and many of them moult immediately after their arrival in Denmark in September and October. A good many young of the year do not reach Denmark until November.

The Long-tailed Duck (*Clangula hyemalis*)

I shall not easily forget a February day in Scania, south Sweden, in the severe winter of 1979. The snow lay in huge drifts by the farmyards and roads of the flat landscape. On the south coast there was ice as far as the eye could see over the Baltic Sea, compressed into high banks at the edge of the water. Off Kåseberga, a fishing village in the southeast corner of Scania, however, the ice extended only some 100 m out from land. Looking down from the edge of the slope by the stones of the ship tumulus Ales on that cold winter's day, there was a fine view down over the edge of the ice and the open sea beyond. The sun only just penetrated through the mist that rose from the ice and the sea, and it hung suspended like a round disc in pale yellow pastel low in the south. A light, cold breeze draughted in from the sea over the snow-covered meadows. Long-tailed Duck song! A soughing, a cacophony of Long-tailed Duck voices rolled in over the shore. Occasionally the song rose like a bustling orchestral arrangement – bird voices were heard together in chorus from different directions and at different ranges. By slowly turning the head and listening in different directions, splendid effects could be experienced. Voices and chorus fluctuated, died away or increased in volume. The air was vibrant with the sound from hundreds or perhaps thousands of male Long-tailed Ducks all singing their nasal trisyllabic song.

Immediately at the foot of the hill slope, two Hooded Crows walked a short way out on the ice. They cawed a few times, the only sound that broke through the Long-tailed Duck orchestra.

As far as the eye could see in the haze there were Long-tailed Ducks on the sea, in aggregations large and small. There was constant restlessness, with many birds diving and others taking off in flocks and flying short distances so that the characteristic profile, with short wings and rounded belly, could be seen well. Small parties flew in playful pursuit: the females were chased by the males, which called continuously both in flight and while swimming on the surface of the water. Because of the birds' incessant diving and flying around it was difficult to count them accurately, but at a rough estimate there appeared to be almost 1000 Long-tailed Ducks. Farther along the coast there must have been more large gatherings, impossible to see on account of the haze: this was clearly evidenced by the distant hum of Long-tail song.

I had never before witnessed Long-tailed Duck display on such a magnificent scale as this. For me, there will always be an association between the music of Long-tails and the Ales stones.

The display of Long-tailed Ducks can be heard in late winter and in spring off many Baltic shores. I have myself heard it from the sea off the twisted pines of Gotland and the clapperstone fields of Öland. In the north Gotland and Öland areas, there are large sea areas with mussel-rich banks which are suitable wintering sites for tens of thousands of Long-tailed Ducks. Long-tails are also found in the ice-free archipelago areas, and Long-tailed Duck song is said to rise loud from many waters around rocky islets and skerries.

The Long-tailed Duck is the commonest duck in the Baltic during the winter months. The birds habitually dive to great depths and therefore stay so far off the coasts that it is difficult to estimate with any real accuracy how many they are. The organiser of wildfowl counts in Sweden, Leif Nilsson, has counted (from aircraft) about 110 000 in the Swedish waters of the Baltic, but aerial surveys have not provided complete coverage; the true number must be considerably greater. The Long-tailed Duck also winters in large numbers in ice-free parts of the Baltic waters of Denmark, East Germany, Poland, the Soviet Union and Finland. In the Baltic Sea, there is a total of at least half a million, perhaps even a million, wintering Long-tailed Ducks. Radar studies in Finland of seabird migration overland between the White Sea and the Gulf of Finland indicate that the number is greater than we might at first believe. During both spring and autumn, the most splendid bird passage moves through this area; of millions of birds, the Long-tailed Duck and the Common Scoter are the commonest species.

Is there sufficient food in the Baltic Sea for the many Long-tailed Ducks to manage to survive the winter without serious starvation? Leif Nilsson has made some interesting calculations for the situation in Hanöbukten, a large bay off the southeast coast of Scania. The clear majority of the 10 000 or so Long-tailed Ducks that normally winter there are found in areas where the water is less than 20 m deep. These ducks' main food consists of common mussels (*Mytilus edulis*) and Baltic clams (*Macoma baltica*). These bivalves are usually swallowed whole, and are not crushed until they get to the birds' gizzards. The Long-tailed Ducks in Hanöbukten are estimated to require a total of about 1700 tonnes of mussels in order to survive the entire winter period. According to extensive samplings of the sea bottom, however, within the Long-tails' area in Hanöbukten there are at least 20 000 tonnes of mussels! The mussels in Hanöbukten never reach more than about 2 cm in length and therefore make very good food for Long-tailed Ducks. Accordingly, over the whole winter, the Long-tailed Ducks take at most 8% of the food that is available. They would seem therefore to have quite a good living in winter, with a superabundance of food. (It is not, however, possible to draw definite conclusions: there can be a great difference between the quantity of mussel food that is theoretically possible and the quantity that is in practice accessible to the Long-tailed Ducks.)

The Long-tailed Duck winters as far north as the ice permits, in the northernmost coastal regions of the Atlantic as well as in the Pacific (figure 32). During the winter, the birds are also present in significant numbers on fresh water, namely on the Great Lakes in North America. The species has a circumpolar breeding distribution which reaches as far north as the birds can possibly get, to 83° N on Ellesmere Island and in Greenland. The Long-tailed Duck prefers small mountain and tundra lakes (if possible, waters where the char is not a competitor for food), but it also breeds at arctic archipelagos and fiords. The nests of coastal-breeding Long-tails are frequently sited in association with Arctic Tern colonies.

There are a lot of ringing results which shed light on the migratory habits of the Long-tailed Duck. Russian birds ringed on the Yamal Peninsula have provided many recoveries in the Baltic Sea area. These show that it is mainly Long-tails from the Russian tundra that winter in the Baltic. There are also two recoveries of Russian birds from the Norwegian coast, which suggests that some of the eastern Long-tailed Ducks do not migrate in over the White Sea prior to further passage overland to the Baltic Sea, but travel off the coast past the Kola Peninsula and the North Cape. Most of the Long-tails that breed in the Scandinavian mountains probably winter along the coast of Norway. Two summer recoveries of Russian-born males at breeding sites, one in northernmost Norway and the other in Sweden, demonstrate the results of abmigration.

Recoveries of Icelandic Long-tailed Ducks show that these do not always winter on the coasts of Iceland. Many move to southwest Greenland, to the same winter waters as most Greenland Long-tails. There are also, however, examples of long-distance recoveries of Greenland Long-tailed Ducks, including from Iceland and Newfoundland. Most astonishing is that one bird was found in January, when four years old, in the southern Baltic off the Danish island of Falster, 3400 km from where it was raised in Greenland. This recovery shows that the various regional populations of Long-tailed Ducks are not isolated. A favourable gene ought to be able to become widespread among Long-tailed Ducks right around the North Pole after moderately few generations. A further good example of the exchange of individuals between widely separated parts of the range is a splendid case of abmigration: a male, hatched in Greenland, was met with in June three years later at the Mackenzie River outlet on the Canadian tundra, 3000 km from its birthplace.

In west Alaska at the Yukon River outflow, there is a huge delta area where the tundra is studded with thousands of lakes. Naturally, enormous numbers of geese and ducks breed there. The number of breeding Long-tailed Ducks in the whole delta region, for example, is estimated at 300 000! Some lakes serve as moulting sites after breeding, an excellent example being Takslesluk, where around 10 000 ducks, many of them Long-tails, moult in July and August. This has long been known by the Eskimos. When the ducks have become flightless, these peoples take the opportunity to drive them with boats into a bay where they can club or shoot them, and in this way acquire a much-needed addition to their diet. The driving method has recently been practised by ornithologists in order to catch the birds in large net traps and ring them. With luck, the catch can be good: in one successful drive using both boats and aircraft, 3699 ducks were captured and ringed at one go. A total of nearly 2000 Long-tailed

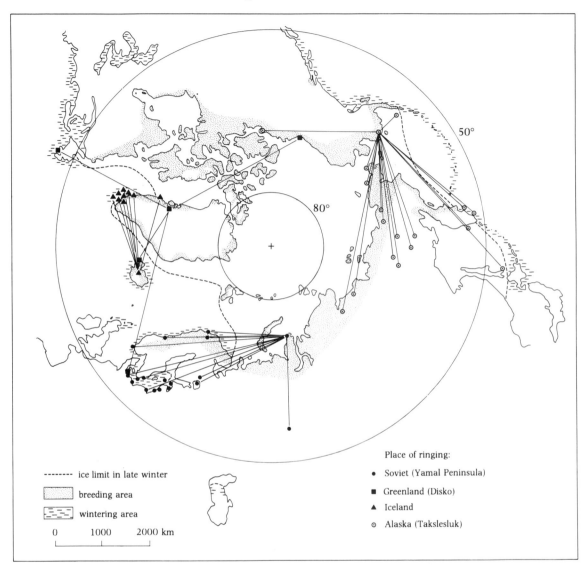

Figure 32   The summer and winter distributions of the Long-tailed Duck. Recoveries of ringed birds are discussed in greater detail in the text. Based on Bauer & Glutz von Blotzheim (1969), Cramp & Simmons (1977), King (1973) and Salomonsen (1967).

Ducks has been ringed at Takslesluk (the species that has been ringed in the largest numbers is the Scaup). The recoveries are quite surprising. Apart from reports of Long-tailed Ducks that have been caught by the Eskimos at Takslesluk several years after ringing, only a single recovery has been made within the American part of the Long-tailed Duck's breeding range. By contrast, there are several recoveries from the east Siberian breeding area. Since the recoveries from Siberia involve males only, a possible explanation is that abmigration by males which have hatched or previously bred in Alaska, not far from Takslesluk, has taken place. The winter recoveries provide some degree of support for this supposition, for these show that the Long-tailed Ducks from Alaska do not winter only on the American side of the

Bering Sea (one recovery) but probably also and on a larger scale on the Asiatic side, off Kamchatka and in the Sea of Okhotsk (four recoveries). The American Long-tails to a great extent therefore share winter quarters with birds from Siberia, a situation which naturally prepares the ground for many cases of abmigration across the Bering Sea. It would be best for us to ignore our artificial political boundaries and we should regard the Long-tailed Ducks in Alaska and east Siberia as members of one and the same population.

The recoveries in east Siberia need not necessarily be the result of abmigration. Maybe Takslesluk and the Yukon Delta are a moult centre for Long-tailed Ducks from large parts of Siberia (another important moulting site can be found at Wrangel Island).

An area of lakes within the deciduous belt immediately east of the southern Urals, more than 1000 km south of the Long-tailed Duck's breeding range, constitutes a notable moult centre of lesser magnitude. A Long-tailed Duck ringed on the Yamal Peninsula was recovered in that area (figure 32).

Let us hope that the Baltic will remain a reasonably unpolluted sea, without major oil spills, so that the hundreds of thousands of Long-tailed Ducks may continue to dive for bivalves there and to call in clamouring chorus every winter.

The Garganey (*Anas querquedula*)

The Garganey is the only species in table 9 whose entire winter distribution normally lies within the tropics, in Africa, India and southeast Asia. In Africa, the Garganey winters in the steppe zone immediately south of the Sahara. There, it is concentrated in areas which are flooded after the summer rains, mainly beside the Senegal and Niger Rivers and around Lake Chad. These are exactly the same regions that play a central role in winter for Ruffs and Black-tailed Godwits. On the whole the Garganey exhibits a striking number of parallels with these two wader species, not only as regards wintering but also in the matter of migratory and breeding habits.

The inundations of the Sahel zone reach their greatest proportions in October and November, immediately after the rainy season. During the course of the winter, the floods shift farther and farther downriver while the country around the upper reaches of the rivers dries out. At the same time the total extent of the flooding diminishes. Towards February and March, only deeper depressions and lakes are still filled with water. The Garganeys move with the floods. The central point of their occurrence therefore shifts gradually along the course of the rivers as the winter progresses. During the winter, the Garganeys feed mainly on seeds of wild rice and grass which they snatch up where the water is only a few centimetres deep. In late winter, when large areas are dried out, flocks of seed-eating Garganeys can be found walking about on dry riverside beds, sometimes in close association with large flocks of Ruffs.

Major environmental changes are currently taking place within the inundation zones of the Sahel: reservoirs are being built, and vast areas are being enclosed within embankments for rice cultivation on a large scale. In some places the waders and ducks are regarded as pests, and attempts are being made to reduce the numbers of birds with the help of visiting hunters from Europe. Fifteen ducks per gun per day is said to be the normal limit by licence during the six-month hunting season. How this development will affect the birds' future it is too early yet to say. Censuses have been made in order that events may be followed. The Senegal government has taken a commendable initiative in setting aside a part of the Senegal Delta as a national park in order to protect the large numbers of birds.

The wildfowl counts in the Senegal Delta have been carried out from boats and aircraft as well as from land and therefore no doubt give a reliable picture of the total number of wintering ducks. As can be seen in table 12, around 100 000 Garganeys winter there. The

Table 12. *Numbers (in thousands) of wildfowl in the flood regions around the Senegal and Niger Rivers. The data for Senegal are for the whole delta, but the counts on the Niger are incomplete (sample censuses from aircraft). Apart from Garganey, Pintail and Shoveler, small numbers of other northern winter visitors, including Teal, Pochard and Ferruginous Duck, also occur. Other African wildfowl comprise mostly Comb Duck, Egyptian Goose and Spur-winged Goose. Based on Roux (1973, 1976)*

Species	Senegal Delta		Sample counts from Niger	
	Jan 72	Jan 74	Jan 72	Jan 74
Garganey	135	85	94	68
Pintail	55	90	27	75
Shoveler	2	3	—	1
White-faced Whistling-duck	43	7	21	8
Other African wildfowl	3	1	6	1

number of Pintails is almost as large. Shovelers are present only in a few thousand. In addition, there are several hundred Teals, Pochards and Ferruginous Ducks, these, too, being winter visitors from the north. The northern migrants represent 80–95% of the total wildfowl population during the winter months in Senegal. The only indigenous African species that can to any extent compete in numbers with the Garganey and the Pintail is the White-faced Whistling-duck (*Dendrocygna viduata*). Between 1972 and 1974, the Sahel zone was hit by extreme drought (with devastating and tragic consequences for the inhabitants in the area). The effect of the drought is plain to see in the greatly reduced numbers of African wildfowl in 1974. Amazingly enough, the number of winter visitors from the north was not affected to any major extent.

The inundation region around the River Niger (just south of Timbuktoo) is much bigger than the Senegal Delta. It has not been possible to carry out complete counts there, but the studies have been confined to sample surveys from aircraft. The counts in January 1972 are based on approximately 2800 km of flying within the area, but despite this it is estimated that only a minor part of the total number of ducks in the whole region could be counted – clearly less than a third, perhaps not even a tenth. This means that the total number of Garganeys is perhaps something in the region of 1 million and that there are in addition several hundred thousand Pintails in the area.

Table 12 shows that the species composition of the wildfowl is of a very similar nature at both the Senegal and the Niger. The same applies to the effect of the extreme drought in 1972–1974. An interesting problem which has still not been investigated in detail is: in what way do the winter niches of the three predominant species, Garganey, Pintail and White-faced Whistling-duck, differ?

Much indicates that the Garganey lacks any advanced degree of site fidelity when it comes to choice of wintering site. The birds move considerable distances during a single winter, related to the shifting distribution of the floods. They turn up in widely varying numbers in different winters at different localities. The interchange of Garganeys between the inundation sites at the Senegal, the Niger and Chad is probably considerable, even though positive evidence of this from ringing is still lacking. For the Pintail, however, there are amazing ringing recoveries of individuals that have changed wintering site between different years. The Pintail does not winter only in Africa but also in Europe, north to the North Sea area (table 11). A Pintail ringed in winter in the Senegal Delta was reported again in a subsequent winter from the French coast on the English Channel; another found itself right up on the German North Sea coast (Schleswig). The same Pintail can therefore spend one winter at the northern edge of the species' winter range and another at the southern edge!

Ringing of Garganeys on the Senegal has shown that the birds do not come only from breeding sites in Europe but also from a long way east, from areas beyond longitude 80° E (figure 33). Recoveries within the breeding area of Garganeys ringed during the winter in India fall mainly between 70° E and 90° E. This means that in this eastern breeding area there is a mixture of birds that migrate west via Europe to West Africa and birds that migrate due

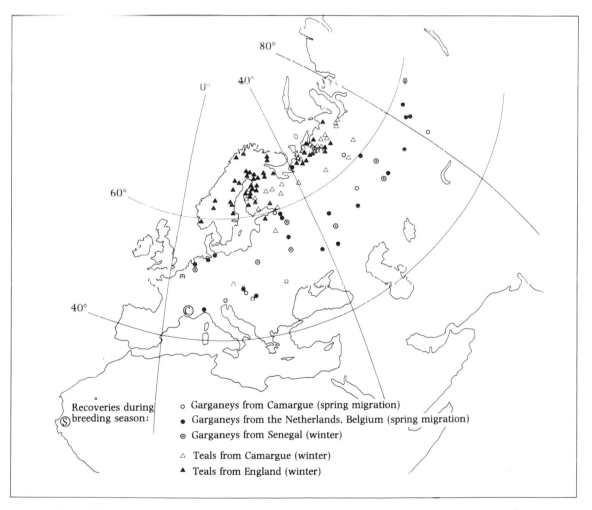

Figure 33 Recoveries in May and June (mostly the breeding period) of Garganeys ringed during the winter in Senegal (S) and during the spring migration (March–April) in Camargue (C) and in the Netherlands/Belgium. The map also shows for comparison recoveries during the breeding season (21 May–16 June) of Teals ringed during the winter in England and in Camargue. Based on Impekoven (1964), Roux *et al.* (1976) and Wolff (1966).

south to winter in India. Perhaps it sometimes happens that the same Garganey switches in different years between these two wintering areas, 10 000 km apart from each other?

The Garganey breeds within a belt between about 45° N and 60° N, from west Europe to the Asiatic coast of the Pacific Ocean. The centre of distribution lies in the steppe or deciduous forest steppe of the Soviet Union. The breeding habitat is eutrophic lakes with luxuriant flooded water-meadows. In Scandinavia and Britain, the Garganey occurs in sparse numbers in marshes and marshy meadows beside lowland lakes, where, incidentally, it breeds in company with its closest companions in the winter quarters, Ruffs and Black-tailed Godwits. (For that matter, I wonder whether the Pintail and the Ruff, which often meet each other in tropical winter quarters, also resort to a large extent to a common breeding habitat within the northern taiga.) The Garganey's breeding environment is susceptible to drying-out. Periodically, exceptional numbers of Garganeys turn up in west Europe during the breeding season; we could almost describe them as invasions. Nobody knows for sure what causes this, but presumably there are drought years on the Soviet steppe with the result that the Garganeys must look for wet swampy meadows elsewhere.

Garganeys pass through northwest Europe in considerably greater numbers in autumn than in spring. In the spring, the passage seems to be routed to a major extent over Italy, the Balkan Peninsula and other parts of the Mediterranean region, where winter rains create many suitable wetlands. A certain proportion of the population, however, migrates via west Europe in spring, too. A comparison between the distribution of recoveries during the breeding season of Garganeys ringed in spring in the Netherlands/Belgium and Garganeys ringed on the French Mediterranean coast (Camargue) reveals no obvious differences (figure 33).

In the southern Soviet Union there are enormous moulting concentrations of drake Garganeys, the largest being in the Volga Delta. Soviet ringing efforts there in July and August have provided recoveries from breeding localities 2000 km away, sometimes even farther (figure 34).

The Garganey has a sibling species: the Teal. Naturally, it is tempting to compare the Teal's migratory habits with the Garganey's. In the map in figure 33, I have included for comparison recoveries during the breeding season of Teals ringed in winter in England and in the Camargue. Two things are worth pointing out.

  a The distribution of recoveries from the two wintering localities is clearly different: Scandinavian Teals often winter in England, while more easterly populations are found to a greater extent in the Camargue. This tendency towards leap-frog migration in the Teal suggests that it has a greater degree of site fidelity, both during summer and during winter, than the Garganey. The Teal's environment is probably more stable than the Garganey's.
  b The distributions of the Garganey and the Teal border on each other both in summer and in winter. Where one species disappears, the other takes over. A question for future special studies: is there any direct state of competition between the two species that may have enforced this distinct geographical division?

The Dipper (*Cinclus cinclus*)

Amid the rushing torrents of streams and brooks the Dipper perches on a rock or perhaps at the edge of the winter ice and bobs up and down. Suddenly it dives into the water, where, by duly diverting the water current with its wings, it can walk around on the bottom, turn over small stones and search for insect larvae. Caddisfly larvae are a favourite food. Back up on a

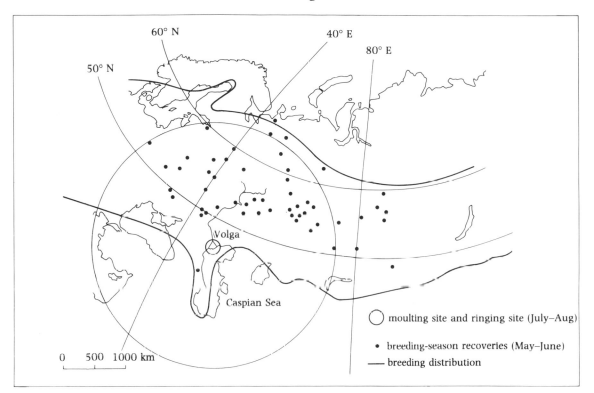

Figure 34   Recoveries during the breeding season (May–June) of Garganeys (mostly males) ringed while moulting in the Volga Delta. The large circle covers a 2000-km radius from the Volga Delta in the centre. Based on Impekoven (1964).

stone or the edge of the ice, the Dipper beats the caddisfly against the hard substrate to remove the casing so that it can get at the larva.

There are four different species of dipper in the world. They all have their core areas in the great mountain ranges, the Rockies, the Andes, Tibet, the Himalayas, the Urals, the Alps and so on, and exhibit a high degree of variation in appearance. At least 14 different races of the European Dipper can be distinguished. As well as the race which breeds in Britain, there is, for example, another in Ireland, a third in the Scandinavian countries, a fourth in the Pyrenees, a fifth in the Alps, and a sixth in the Atlas Mountains of North Africa.

This large number of races indicates that the interchange of individuals among populations in different regions is small and that the Dipper is to a high degree a resident bird. Access to open water, however, is of course a necessary requirement. In winter, the birds therefore abandon the coldest regions and the highest mountain peaks, where the watercourses freeze up. During the breeding season, the Dipper is found at up to 2000 m in the Alps and the Caucasus and at up to 5000 m in the Himalayas and Tibet. It is not the low temperatures as such that drive the birds out. In the mountains of arctic Scandinavia/the Kola Peninsula and in Siberia Dippers are known to be present during the winter beside open water torrents despite the fact that the temperature at times drops down to −40 °C!

In Scandinavia, the Dipper is found principally in the mountain regions during the breeding season; at times it is encountered right up at the snow limit in the high mountains.

The Dipper is particularly abundant on the watercourses in the strongly undulating terrain of Norway, where, under the agreeable name of *'fossekall'* (= old man of the waterfalls), it has been chosen as that country's national bird. The Dipper breeds here and there in low-lying country of north Scandinavia, in southeast Norway and in central and southern Sweden, but immeasurably more sparsely than in the mountains. In large parts of Finland and in Denmark only extremely small numbers of Dippers breed.

In many places in southern Finland, southern Sweden and Denmark, the Dipper is commoner during the winter than during the breeding season. Uncertainty has long reigned over where the wintering birds come from. Recently, two Swedish ornithologists and industrious ringers, Sune Andersson and Stig Wester, at Västervik in southeast Sweden, have gone some way towards explaining the Scandinavian Dippers' movements. Ringing of wintering Dippers in the Västervik district during the 1960s led to four recoveries during the breeding season from the Norwegian mountains between Bergen and Trondheim. Andersson and Wester lost no time in packing up their ringing gear and making for Norway in the summer. On the Norwegian breeding grounds they succeeded in catching several Dippers that had been ringed in winter in Denmark; one even came from West Germany. A Dipper was ringed initially in Jutland in November 1970 (it was later re-caught in the same winter in January, at the same site), controlled in summer 1971 at its breeding site in Norway, caught again at the Danish wintering locality as early as the end of October of the same year, and was finally seen once again at the Norwegian breeding site in 1972.

These exciting results stimulated interest in the Dipper's migration, and small groups specialising in ringing Dippers were formed in various places in the Nordic countries. Today, there are so many results (figure 35) that we are able to get a collective picture of the Dipper's migration in Scandinavia.

Dippers from the Swedish and Norwegian mountains migrate on a large scale to Finland, southern Sweden, Denmark and to a certain extent West Germany in winter. For many birds, the migratory flight covers 700 km or more. The birds arrive at the wintering sites in October and leave in March. They show a high degree of site fidelity, to wintering site as well as to breeding site. Some Scandinavian Dippers probably extend their southeastward migration to wintering localities on the other side of the Baltic Sea: there are two recoveries there, from Lithuania and East Germany respectively, of birds ringed in Västervik. Dippers also winter on the south Baltic islands of Bornholm and Gotland, and are sometimes seen resting on Gotska Sandön, a small isle northeast of Gotland. Many certainly migrate at night; this is indicated by reports of Dippers that have been killed at Danish lighthouses in the Kattegat, for example at Läsö, Anholt and Hesselö.

The Dipper's migration pattern is of particular interest bearing in mind that there is no other short-distance migrant in the north European avifauna that departs southeastwards in the winter. Why does the Dipper not migrate to the much milder Britain as, for example, do thrushes, larks and Starlings? There is no lack of suitable habitat, for in western England, Wales, Ireland and Scotland there are uplands (between 300 m and 1000 m above sea level) with the most choice rivers and streams. The reason is, if anything, that there are already so many Dippers in Britain that temporary winter visitors must face strong competition. The British and Irish Dippers are in the main residents and maintain their territories during both summer and winter. They frequently sing in the middle of winter, and are busy nest-building as early as February and March. Judging from the few British recoveries in winter of the northern race of the Dipper, Scandinavian Dippers turn up in eastern Britain only by chance. Northern Dippers are blackish-brown on the belly, where British ones are a warm rusty-brown. The 'guinea-pigs' obviously are not particularly successful in holding their own on

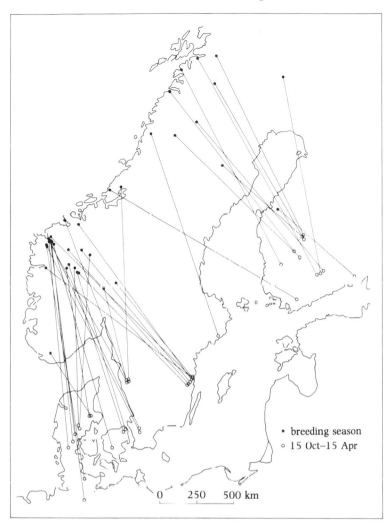

Figure 35 Breeding and wintering localities of ringed Dippers in Scandinavia. Only recoveries more than 250 km from place of ringing are shown on the map. Dippers breeding in the Swedish and Norwegian mountains migrate southeast to Denmark, southern Sweden and Finland. Based on Andersson & Wester (1976).

the British wintering grounds, since the habit of migrating to Britain has not become widespread among the Scandinavian Dippers.

So, we ask ourselves instead: if large numbers of Dippers manage to winter successfully in Denmark, in southern Sweden and in Finland, why then is the stock of breeding birds in these regions so sparse? There is a great difference between what is needed for an individual bird to survive and the conditions that are required for a pair of Dippers to succeed in feeding both themselves and five young. In southern Scandinavia, many streams and waterfalls are probably 'middling': there is easily enough food for an individual to survive, but hardly for a pair to achieve exceptionally good breeding success. The Dippers in the sparse breeding populations of Denmark, southern Sweden and Finland are perhaps prevented from increasing in numbers owing to competition from the many winter visitors from the mountain regions. The disadvantage the southerly Dippers have of attempting to breed at 'middling' sites is not compensated for by what may be the advantage for them in being already on the spot when the migrating birds arrive from the mountains.

Many Dippers winter in northerly areas where there is open water locally, and especially in the districts nearest the coast in Norway. Which birds stay behind in the north and which ones set off southeastwards on migration? Are the migratory habits different for adults and juveniles, for males and females? There are still sticky questions remaining for students of Dippers.

It is additionally enjoyable to go to one's Dipper habitats now that it has been 'discovered' in full that the Dipper is a migrant. I regularly go looking for wintering Dippers myself at several small waterfalls on rivers in west Scania, in the far south of my native Sweden. Dividing the year between a river in Scania and a Norwegian mountain stream – such may be the Dipper's life in Scandinavia.

3.3   Birds which feed on terrestrial plants

When we look around in a summer landscape of dazzling greenery and consider how enormous is the amount of energy that is converted annually into plant mass, it seems that there ought to be no end of birds that feed on the green plants. But this is by no means so! On the contrary, very few bird species obtain their principal nourishment through feeding on terrestrial plants. Only when it comes to exploiting the seeds and fruits of the plants do the birds really assert themselves. I shall deal with the seed-eating birds as a special category in section 3.8. In the present section I shall confine myself to birds that prefer to eat other parts of plants (table 13).

The herbivorous birds in particular eat the most nutritious and most tender plants and plant parts, the shoots, the buds and the tubers and sometimes also catkins, sprigs and thin twigs (grouse, Ptarmigan) and fresh pine needles (Capercaillie, Black Grouse). An important reason why the birds restrict themselves to eating the most tender parts of the plants is that their digestive systems are not particularly effective in breaking down plant food. The content of the plant cells is enclosed by cellulose, and the birds lack the capacity to break down the cellulose chemically. Herbivorous mammals harbour bacteria in the digestive system which break down cellulose, but these are not present in birds. Instead they have a special gizzard where the plants are reduced to pulp and the plant cell walls are torn into pieces purely mechanically by means of grit and small stones which the birds pick up and swallow.

The birds in table 13 comprise two main groups, geese and gallinaceous birds. To these are added the Woodpigeon and the Wigeon; the latter may, from an ecological point of view, very well be counted as a small goose. The geese are migratory birds; the gallinaceous birds are to a large extent residents.

It is, however, impossible to make a rigid distinction between migrants and residents. Ptarmigans, for example, set off on annual movements up and down the mountains. They breed mainly in the montane willow region. From there, they make their way at the height of summer to the boulder lands of the lichen region and stay there until the snows force them to descend to lower levels in the willow or upper birch-forest zone.

In arctic regions many animal populations fluctuate in regular cycles, so that peak numbers come about every fourth or tenth year. A four-year cycle occurs in, for example, the Norway lemming and a ten-year cycle in, among others, the snowshoe hare of arctic North America. The number of predators, such as arctic fox, Rough-legged Buzzard, Gyrfalcon, Snowy Owl and Goshawk, fluctuates concurrently with the availability of prey animals.

Table 13. *Wintering areas of birds which feed on terrestrial plants*

Species (NB=does not breed in Scandinavia	Present in N Europe in winter	Chiefly resident	Important wintering areas	
			NW & W Europe	SE Europe (Hungarian plain, around Black and Caspian Sea)
Wigeon			+	+
Greylag Goose			+	+
White-fronted Goose NB			+	+
Lesser White-fronted Goose				+
Bean Goose	+		+	+
Pink-footed Goose NB			+	
Brent Goose NB			+	
Barnacle Goose			+	
Canada Goose	+		+	
Willow Grouse	+	+		
Ptarmigan	+	+		
Black Grouse	+	+		
Capercaillie	+	+		
Hazel Hen	+	+		
Grey Partridge	+	+		
Pheasant	+	+		
Woodpigeon	(+)		+	+
Total no. of species	9	7	9	6

There is a very comprehensive literature on these cycles, and intensive research is being carried out in many parts of the world in order to provide a more detailed explanation of the causal connection behind these pronounced population fluctuations. Of interest in this connection is the fact that in many places the Ptarmigans, too, occur in regular cycles. (Perhaps the Ptarmigans' fluctuations are due to those predators which have become numerous when rodent availability has been good changing over to a Ptarmigan diet when the rodent stocks crash and at that time greatly reducing the Ptarmigan populations.) In Greenland, 'Ptarmigan years' come about every tenth year or so. The Ptarmigan then occurs in exceptionally large numbers. In normal years, Greenland Ptarmigans migrate to a minimal extent – only the very northernmost parts are vacated during midwinter. In Ptarmigan years the picture is totally different: masses of Ptarmigans, a large proportion of which are juveniles, migrate to south Greenland. Strikingly large flocks may be seen during this migration: we can with some justification talk of 'invasion migration'. Ringing has shown that the migration flights can extend over thousands of kilometres. After the winter, the Ptarmigans move back to the regions where they earlier bred or hatched. Invasions of Ptarmigans obviously occur when the production of young has been good and when competition for food in the breeding area has consequently become acute.

Grouse invasions also occur in the far north of the Soviet Union. Exceptionally severe winter conditions intensify the scale of the invasion. The Willow Grouse on the Kola Peninsula, the Kanin Peninsula and the nearby tundra island of Kolguyev (almost 100 km from the mainland) migrate each year to winter on more southerly shrub tundra and forest tundra. During invasion years, enormous numbers of Willow Grouse take part in the migration. Many follow the river valleys, where large areas of scrub provide suitable food and good cover for resting birds. Observers report flock after flock of between 100 and 300 grouse passing along the riversides at heights of 50–200 m. Such mass movement can continue uninterrupted for several days running. It must be a magnificent sight! Gyrfalcons and Snowy Owls follow in the wake of the grouse. The grouse's return journey in the spring often takes place without any lengthy stop-offs on the way. The birds frequently fly at rather high altitude, 'out of firing range' as those reporting the events put it.

The grouse's varying choice of winter quarters, which is in large part determined by the direction and strength of the wind, is of the greatest interest. Strong winds often blow across the tundra, and the grouse make use of energy-saving tail winds for as long a distance as possible. If, during the autumn and winter, there are strong northeasterly winds, the Willow Grouse from the Kanin tundra invade the areas to the southwest, in the direction of the White Sea regions. With strong westerlies or northwesterlies the migration is mostly in a southeasterly direction, to places such as the valleys of the Pechora River. It is unusual for the winds to have such a powerful influence over where migrant birds winter (though compare with the migrations of the Redwing, the Siskin and the Brambling, which are described in sections 3.7 and 3.8).

The grouse are not the only gallinaceous birds in table 13 that migrate considerable distances. Grey Partridges from central and eastern Russia, where the depth of snow in winter is around 0.5 m or more, are regular migrants, having their winter quarters around the Caspian Sea. In autumn and spring the passage of Grey Partridges is intense in many places, with family parties of four to 20 birds passing continually.

Wild geese are often seen as a symbol of migratory birds, and certainly a migrating goose flock in V formation is magnificent to behold. Not all geese, however, fly in V formation: a flock of hundreds of Brent Geese in a surging band is at least as captivating a sight as the most perfect V of other geese, swans and cranes. Like the Wigeon, the geese exploit mainly areas of low and tender vegetation. Their migratory habits should be looked at in relation to at least five different habitat types of central importance.

1 *Northern tundra and marshland.* These habitats are used by most goose species for breeding. Low-growing plants, which sprout up along lakeshores and in damp depressions during the short but intense growth period of the northern summer, are the staple food of the geese and the goslings. Presumably there is a surplus of plant food, for only locally does the vegetation become heavily grazed down by the geese. During the breeding season, the feeding areas of different goose species overlap each other to a great extent. Two factors are probably more important to the geese's life on the northern breeding grounds than immediate shortage of food: the short summer season and the danger from predators.

Let me give a good example of this with the Brent Goose, which is found on the most high-arctic tundra of all and also has the shortest breeding period. Brent Geese of the dark-bellied race *bernicla* pass through southern Sweden in autumn and spring. They breed mostly within the northernmost parts of the Taimyr Peninsula. After arriving at the breeding area, the geese need at least a day or two to locate a good nest site and to arrange their modest nest on the ground. To lay the eggs (four or five) takes approximately six days, and the incubation period is 25 days. The fledging period of the young lasts about 40 days, on top of which there is a further day or two before the young can fly well enough to be able to accompany their parents on migration. The entire breeding season for the Brent Goose accordingly lasts at least 75 days (or $2\frac{1}{2}$ months). The snow-free period in the high-arctic breeding area averages 75 days! The geese have no extra time margins. More often than not they arrive in the middle of the thaw, before suitable food is accessible. The birds must therefore lay their eggs while fasting and must rely on the fat reserves they have put on at the wintering and staging sites, on what is left after the spring migration flight. Studies of Barnacle Geese and Brent Geese on wintering grounds in the Netherlands show that it is quite possible in spring to predict which females will have the best breeding success and will come back in autumn with most young: they are of course those which are fattest immediately before the spring migration starts.

The northern geese place their nests on the first patches of snow-free ground, often beside a sunlit boulder on the top of one of the tundra's many hummocks. The Barnacle Goose, which is found in Greenland, Spitsbergen and Novaya Zemlya, breeds mainly in colonies high up on cliff ledges; we could describe these as a kind of bird cliff with geese. The cliff-face ledges become free of snow early on, and are difficult for predators to gain access to.

In some years, the spring and the thaw are so late that the geese cannot find any sites for their nests. They await the thaw, but only for one or, at most, two weeks. If breeding cannot get started during this time, then it is suspended for the year. The reason for this is the tight time schedule: should breeding begin too late, then the young do not have time to reach the flying stage before the cold and snow return to the tundra. A delayed thaw, or a reversal in the weather with blizzards just after the geese have laid their eggs, therefore leads to breeding being virtually completely ruined over wide areas. Figure 36 shows the annual number of wintering dark-bellied Brent Geese, both the number of first-years and the whole total, in west Europe between the mid 1950s and 1970s. In ten of the 23 years the Brents have suffered almost complete breeding failure. Another important cause of the large and partly cyclic variations in reproductive success of Brent Geese is probably the predators of lemmings (particularly the arctic fox) turning to the eggs and young of Brent Geese and waders in years when lemming populations have crashed (Summers & Underhill 1987). The annual survival of adult Brent Geese is as high as 83%; this means that, out of 100 Brents, an average of 47 individuals are still alive after four years and 16 after ten years. In general, therefore, there are many opportunities (on average six) for breeding attempts for the single individual Brent Geese. The fact that the Brent Goose is a strong competitive species despite the many years with non-production of young is clear, moreover, from the steep increase in the total population since the mid 1950s. The reason for this heartening development is probably that hunting has decreased and that the Brent Geese have changed their feeding habits in the winter quarters.

The geese which live in more southerly regions, on the taiga for example, show less variation in breeding success. In these species it is very rare for breeding to fail completely within large populations.

The danger from predators, such as fox or arctic fox, compels the geese to nest in places which are difficult of access, such as on islands or cliff precipices (Barnacle Goose). Some species assemble in colonies; the largest are those of the North American Snow Geese, in which up to 100 000 nests may be present in one colony.

An interesting special adaptation is shown by the Red-breasted Goose, which, like its close relative the Brent Goose, nests on the tundra of the Taimyr Peninsula. It winters around the Black Sea and the Caspian Sea, and occurs in west Europe only as

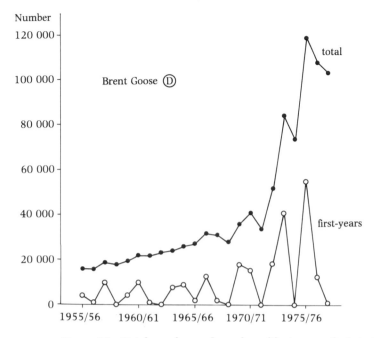

Figure 36   Total number and number of first-years of wintering dark-bellied Brent Geese in west Europe. The Brent Goose's breeding success varies dramatically from year to year: in some years hardly any young at all are produced, while in others first-winters may make up almost half of the total population. Based on Ogilvie (1978), with additional data from the Dutch journal *Watervogels* (1979). According to the most recent estimates, the population increase has levelled off at about 170 000 birds (IWRB 1989).

a rare vagrant. Red-breasted Geese breed in small colonies of three to 20 pairs, often right next to a pair of Peregrines or occasionally Rough-legged Buzzards. The Peregrine often places its nest on top of one of the low hummocks of the tundra. One observer found Red-breasted Goose nests around 19 of a total of 22 Peregrine nests. The geese usually position their nests between 10 m and 30 m from the falcon's nest, but in one case the distance was no more than 1.5 m. In this way the geese derive benefit from the falcon's aggressive defence of its nest against both winged and four-legged predators.

2 *Sandflats and mudflats* along shallow shores are an important passage and wintering habitat for the Brent Goose. On the northern coasts of the Atlantic and Pacific Oceans, there are shallow tidal shores where extensive bottom areas become accessible to the geese at low tide. A characteristic plant is the eel-grass (*Zostera*), of which there are several different species. The eel-grass grows in large beds and, when these are exposed, the shore can have the appearance of a fresh green mat. Here the Brent Goose finds its choicest feeding grounds. The Brent Geese are also very much at home in the lagoon areas which are not affected to any appreciable extent by tidal variations; while swimming, they feed on the eel-grass which reaches almost to the surface in the shallow water. Various kinds of seaweed which the Brent Goose grazes on the shore flats act as alternative and supplementary foods to the eel-grass.

During the 1930s, a disease caused by a parasitic micro-organism spread throughout the eel-grass community. The decline was catastrophic: the underwater meadows, formerly so immense, were suddenly gone. The eel-grass has since slowly but surely returned, but it is still not present in such abundance as at the beginning of the century.

The disappearance of the eel-grass combined with increasingly heavy hunting pressure seriously threatened the Brent Goose. In west Europe, only about 15 000 Siberian Brent Geese wintered in the mid 1950s; this race was endangered and was on the point of disappearing altogether. The Netherlands and Britain were the countries where the shooting of Brent Geese was first banned, and later the species was protected in the whole of the European wintering area. That a gratifying development has taken place since the 1950s is evident from figure 36. The disappearance of hunting and the gradual return of the eel-grass led to the Brent Goose population passing the 100 000 mark (and, in the 1980s, the 200 000 mark). A contributory reason for the steep increase may be that the Brent Goose is also in the process of broadening its winter niche. The habit of grazing grass on coastal meadows or autumn wheat on fields near the coast, particularly at high tide when the shore flats are inaccessible, is spreading among Brent Geese in Europe.

3 *Short-grass meadowland.* This habitat type is created and maintained by grazing cattle, but in addition is found in maritime regions, on coastal shore meadows where low-growing, salt-resistant grasses and plants are the natural plant community. On many shores the cattle help to keep the vegetation extra short. Barnacle Goose, White-fronted Goose, Pink-footed Goose and Wigeon are the principal species that feed on coastal grasslands during winter and on passage. Large numbers of Barnacle Geese stop off for example on headlands and islets with sheep pastures in the eastern and southern parts of the island of Gotland, while others stop off, mainly in autumn, at similar terrain on Öland to the west. Furthermore, a rapidly expanding breeding population has recently become established on Gotland and Öland. The Barnacle Geese's main wintering sites are on coastal meadows in the Netherlands (Russian breeders) and on treeless sheep-pasture islands on Irish and Scottish coasts (Greenland breeders).

Before man introduced stock-farming on a large scale, there was relatively little inland meadowland that was suitable habitat for the geese. It was presumably mainly on the North American prairies and on the central and eastern European steppe that the geese were able to find shorter grass vegetation, after grazing by wild buffalo, deer and antelope. In these areas there are still large grass plains which are suitable for grazing by geese, but today it is domestic cattle that fashion the landscape.

Large parts of west Europe and North America today have a landscape which is characterised by stock-farming and agriculture, and which provides resting and wintering geese with extensive areas of suitable meadowland. It is also important for the geese to have access to lakes and quiet sea bays where they can roost overnight safe from predators. This explains why the geese stay at meadowlands chiefly in coastal areas or in lakeland country.

4 *Marshland with rush and sedge meadows.* Some of the larger species of geese, such as the Snow Goose, the Canada Goose and the Greylag Goose, make use of marshlands with comparatively tall vegetation. The geese feed on the green parts of the plants, but also eat root parts under the water or in the mud. The Greylag readily breeds in this habitat, both at inland lakes and beside sea bays, and its breeding distribution

extends a long way down into south Europe. Of the goose species in table 13, the only one apart from the Greylag that breeds south of the northern coniferous belt is the Canada Goose, which has been introduced into Britain and Scandinavia.

Luxuriant marsh vegetation provides excellent cover. This is necessary not only for the goslings but also for the parent birds, which completely lose the power of flight during the summer moult. The Greylag's incredible ability to keep itself concealed and avoid detection during the breeding season is widely vouched for. Outside the breeding season, both the Greylag Goose and the Canada Goose graze on open meadowlands and arable land.

5 *Farmland.* The geese have adapted increasingly to feeding on arable lands during the migration and winter seasons. In early autumn there is plenty of spilled grain on the stubble fields. Another waste produce which comes in useful for the geese is potatoes. Despite the fact that potatoes have been cultivated in Britain and Ireland for 400 years, it was not until as recently as the 1920s that the geese really 'discovered' this food source in a big way. The habit of eating waste potatoes first became widespread in the Pink-footed Goose; during the 1940s, it spread further to British Greylags. Many geese frequent ploughed fields and pick up leftover rootsuckers from the harvested plants. Sprouting autumn crops are an important food source during late autumn and winter. In the early days of spring, the geese often resort to pasture cultivations.

In certain areas the geese's expansion has led to conflicts with farming interests and to proposals that more geese should be shot. There are also, however, splendid examples of how very well high farming efficiency can be reconciled with large numbers of geese. The Netherlands is an ideal country for geese. The habitat types, shallow shores, meadowlands, marshlands and arable lands, suit all different demands. There is no hunting here. The geese find themselves well at home in the country that has Europe's highest human population density. What else can be said of the goose numbers during for example the winter of 1977/78: 9000 Pinkfeet, 38 000 Greylags, 41 000 Brents, 45 000 Bean Geese, 53 000 Barnacle Geese and 183 000 Whitefronts?

This summary of the most important habitats of the geese sheds light on the ecological background to their migratory habits and provides some idea of how different species can operate between different feeding areas at different times of the year. Owing to their rather lengthy migratory movements, the geese are able to exploit extensive food resources and in this way survive during the winter months. It should not, therefore, in the first place be shortage of resources for survival that limits the goose populations (provided that man's persecution through hunting in the winter quarters is not extremely heavy), but rather the meagre availability of safe breeding sites in the geese's open living environments. In their struggle for breeding space most goose species are forced to retreat to inaccessible arctic regions. The situation is entirely different for the gallinaceous birds. Since they remain in northern parts all year round, in all probability they live under the chilly threat of food shortage during the winter. The sprouting vegetation of summer on the other hand provides food, cover and breeding opportunities for all age classes.

The migration of geese is comparatively well known; migration routes, staging sites and wintering sites are firmly bound by tradition. The geese are often given as typical examples of narrow-front migrants. Such birds move fixed distances within fairly narrow corridors. An important reason for the tradition-bound nature of sites is that the goose families keep

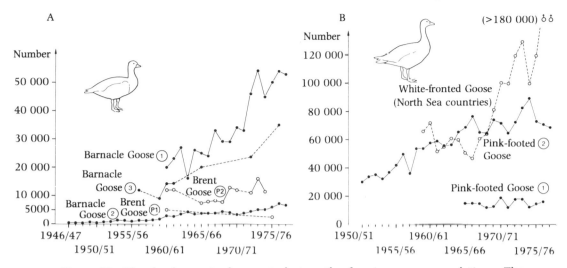

Figure 37 The development of numerical strength of various goose populations. The different populations of Brent Goose, Barnacle Goose and Pink-footed Goose are shown in figures 38 and 39. The data for White-fronted Geese refer to the population which breeds on the Russian tundra and winters in the coastal countries around the southern North Sea, in particular in the Netherlands. Based on Ogilvie (1978), updated from the journal *Watervogels* (1978 and 1979). The most recent population estimates are as follows: Barnacle Goose 1: 70 000, 2: 10 000, 3: 32 000; Brent Goose P1: 4000, P2: 20 000; Pink-footed Goose 1: 25 000, 2: 110 000; White-fronted Goose 300 000 (IWRB 1989).

together both on migration and during the winter, with the result that the young are able to learn the migratory habits of their parents. The parent geese as a rule remain together in pairs throughout their life.

In order that the reader may get a brief view of our extensive knowledge of the migration patterns of geese, I provide here a short and rhapsodic account of the migratory habits of the different species. I shall also say something about the development of the goose populations (figure 37), since many species are today undergoing a particularly dynamic phase of growth; they are expanding into new areas and are creating new passage routes. The migration of the Bean Goose is less well documented than those of the other goose species. Intensive studies are therefore currently being pursued in Scandinavia in order to remedy this deficiency. The Bean Goose's migration and the problems associated with it are discussed in a special section.

The Brent Goose (*Branta bernicla*) breeds on the tundra around the North Pole (figure 38). The species can be divided into six different populations with differing migratory habits.

1 The dark-bellied Brent Goose (D = the race *Branta bernicla bernicla*), which breeds in the regions of the Taimyr Peninsula, passes through southern Scandinavia and the Baltic on migration and winters in the southern North Sea area, on the coast of southern England and along the coasts of France. During spring and autumn, the majority of Brents migrate non-stop the 2000 km or so between the southern North Sea coast and the tundra areas of the White Sea; these latter serve as a halt (and as a moulting site for non-breeding immatures) on the way to and from the Taimyr.

2 The population of the pale-bellied Brent Goose (P = the race *Branta bernicla hrota*)

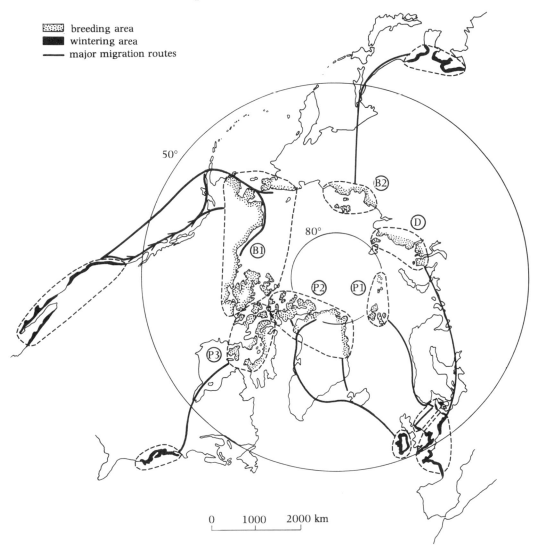

Figure 38 The Brent Goose's breeding and wintering areas and main migration routes. The Brent Goose is separated into three races: dark-bellied Brents = D, pale-bellied Brents = P, and 'Black Brant' = B. Based on Cramp & Simmons (1977) and Ogilvie (1978).

breeding in Spitsbergen and Franz Josef Land (P1) migrates along the Norwegian coast to Jutland. A minority of the population, approximately 500 of the mid 1970s total of 2500 geese (figure 37), continues on across the North Sea during the course of the winter to the coast of northeast England. The whole population re-assembles, however, in Jutland before the geese set off again in spring for the breeding grounds. A classic site is Nissum Fiord in northeast Jutland, where a large proportion gather in spring. The Brent Geese from Spitsbergen, in contrast to most other goose populations, decreased in number during the 1960s and 1970s (figure 37). The question was asked whether this might have been due to competition from the

Barnacle Goose, which was, and still is, increasing in Spitsbergen. This does not, however, seem particularly likely, at least not if we give credence to statements from the 1700s and 1800s that there were at that time 10 000 Brents in Spitsbergen and Franz Josef Land.

3 The 20 000 or so pale-bellied Brent Geese that originate from Greenland and the arctic archipelago in Canada (P2) winter around the coasts of Ireland. The geese stop off in Iceland for a month in both spring and autumn. Most birds fly about 800 km straight across the several-thousand-metres-high inland ice of Greenland: this is undeniably some achievement.

4 Pale-bellied Brent Geese from the northern Hudson Bay, Baffin Island and neighbouring parts of Canada (P3) stop off in spring and autumn in the very southernmost part of Hudson Bay, James Bay. The journey between James Bay and the winter quarters on the east coast of America around New York is managed in a 1500 km direct flight overland. The number of Brent Geese is between 100 000 and 200 000 individuals.

5 There is a third race of the Brent Goose, *Branta bernicla nigricans*, which in its native North America goes by the name of 'Brant' or 'Black Brant'. This name is a little misleading in so far as this race is not much darker than the dark-bellied Brent Goose (and it even has paler flanks). Brants from an enormous area in North America and easternmost Siberia (B1), with the Alaskan tundra lands around the Yukon River as a central point (up to 40 000 pairs breed there), gather in autumn at Izembek Bay on the Alaskan Peninsula. At this staging site, where the eel-grass grows in a dense green mat in extensive shallow-water areas, about 150 000 Brent Geese congregate. At the beginning of November, the geese set off and fly direct more than 3000 km over the sea to the wintering sites on the coast of the southwest United States and the Californian Peninsula in Mexico; the Brants' journey across the Pacific Ocean is a record for non-stop flying by geese. During the spring, the migration proceeds northwards by stages along the west coast of America, and some geese turn off towards the breeding sites without visiting Izembek Bay. This bay is, however, a central staging post for the majority of Brants in spring, too, especially for the very northernmost populations. The latter take the opportunity to feed and put on weight at Izembek during May, while waiting for the arctic tundra to become free of ice.

6 Our knowledge of the Brant population breeding on the New Siberian Islands (B2) is very meagre. The main population probably winters on both sides of the Korean Peninsula; only a few geese turn up in winter in Japan. Sporadic observations indicate that the geese migrate overland from the tundra to the Sea of Okhotsk and then follow the coast southwards.

The White-fronted Goose (*Anser albifrons*) is, so far as breeding is concerned, bound to the tundra, though not only to high-arctic tundra but also to more southerly tundra lands with shrub vegetation. It has near enough a circumpolar distribution, and its migration pattern in certain respects recalls that of the Brent Goose. Thus the Greenland White-fronted Geese (the population contains about 22 000 birds) winter in Ireland and in west Scotland. During the spring they use Iceland as a staging post, but in the autumn rather few resting Whitefronts are found there; the most likely thing is that at this season the Whitefronts fly the near 3000 km from west Greenland to north Britain in a single non-stop flight, straight over the inland ice of Greenland and the open North Atlantic south of Iceland.

The White-fronted Geese from the Russian tundra winter in west Europe, in particular in the Netherlands, Belgium and southern England and Wales. The flyway runs mainly south of the Baltic Sea via the Moscow area, Poland and East Germany. Smaller numbers of Whitefronts, however, pass southern Scandinavia each year and stop off in spring and autumn in Scania and in Denmark. In recent years the Whitefronts stopping off in south Scandinavia have increased in number to several thousand individuals. This increase must be connected with a greatly increasing population of wintering White-fronted Geese in west Europe (figure 37). The population curve climbed steeply upwards as early as about 1970, when the 100 000 mark was passed; an explosion took place in 1976, when the total number of wintering Whitefronts increased in one year from 120 000 to nearly 200 000. This increase is so great and so sudden that one is inclined to wonder whether it may have been due to a change in the species' migration pattern. Could a switch have taken place in the Siberian Whitefronts' winter quarters from southeast Europe to west Europe? Large numbers of eastern Whitefronts winter on the Hungarian plain, on the vast flats west of the Black Sea, in Turkey and on the south coast of the Caspian Sea; in all, perhaps half a million White-fronted Geese are involved, of which almost 100 000 are normally found in Austria, Hungary and Yugoslavia. Has some change adverse to the Whitefronts taken place in the landscape which has caused a large proportion of them to migrate instead to west Europe? The Whitefront population wintering in northwest Europe has continued to increase and has recently reached the 300 000 mark (IWRB 1989).

The Barnacle Goose (*Branta leucopsis*) can be divided into three well-delimited populations with separate migration paths (figure 39).

1   The geese which breed on the south island of Novaya Zemlya and the nearby island of Vaygach winter mainly in the Netherlands. The large islands in the Baltic Sea are important staging sites: for about one month in spring, from the end of April to the end of May, almost the entire population is found in the Baltic Sea area, on Gotland and on the Estonian islands of Hiiumaa and Saaremaa. The Russian population of Barnacle Geese has increased considerably and nowadays numbers more than 70 000 birds (figure 37). A remarkable event occurred in the 1980s when Barnacle Geese stayed behind in spring and established a breeding population in the traditional staging area, 2000 km from their normal arctic breeding sites; breeding occurs mainly on Gotland (with numbers rapidly approaching 1000 pairs), but has also started on Öland and at other parts of the Swedish and Estonian Baltic coasts (Larsson *et al.* 1988). The reason for this surprising colonisation is an open question: lack of human hunting pressure, or suitable food for breeders and their chicks becoming available through fertilising and intensive grazing of the shore meadows?

2   In the extreme southwest of Scotland, right on the border with England, lies the Solway Firth. This has long been known as a wintering site for Barnacle Geese. The island of Islay, less than 200 km to the northwest, is also renowned for its considerable numbers of Barnacle Geese. A total of well over 100 ringing recoveries of Greenland Barnacle Geese has been reported from Islay and from other islands and parts of the coast in west and northwest Scotland, and also from the west coast of Ireland. But not a single such recovery has been made on the nearby Solway Firth. The explanation for this was provided following a ringing campaign in Spitsbergen in the summer of 1962: in the following winter, no fewer than 90 recoveries of the ringed geese were reported from the Solway Firth. Spitsbergen's Barnacles are accordingly confined in winter to the Solway Firth, and here they are clearly segregated from the Greenland population despite the minimal differences in distance between the different populations' winter quarters. The migration between

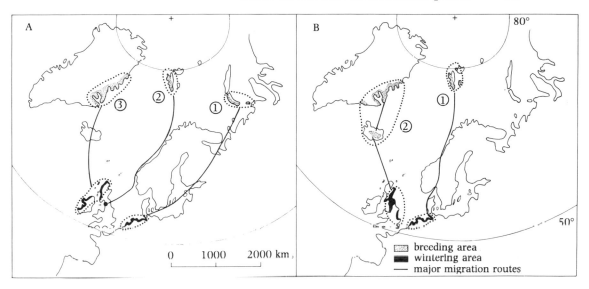

Figure 39   The migration pattern of the Barnacle Goose (A) and the Pink-footed Goose (B). Based on Cramp & Simmons (1977), Folkestad (1975), Haftorn (1971), Ogilvie (1978) and Salomonsen (1967).

Spitsbergen and the Solway, a total of 3200 km, takes place in two stages: the birds make a stop of several weeks at the Helgeland islands off Mo i Rana in Norway to replenish their fat reserves.

3 Greenland Barnacle Geese stop off in spring and autumn in Iceland and winter in coastal districts of western Scotland and Ireland. More than half of today's 35 000 or so birds (figure 37) are found in winter on Islay.

The Pink-footed Goose (*Anser brachyrhynchus*) breeds in Svalbard and also in Iceland and eastern Greenland (figure 39). Numbers are much larger in these last two areas than in Svalbard (figure 37).

1 The geese from Svalbard have their nearest major staging sites in Jutland. Of particular interest is the fact that, judging from recent field observations, they take an inland route across southern Norway by way of the inner Trondheim Fjord and the Oslo Fjord regions. In autumn and spring the majority are found in north Jutland; during the winter the Pinkfeet move southwards along the North Sea coast, a lot of them as far as the Netherlands.

2 The Pinkfeet from Iceland and Greenland winter in England and Scotland. Ringing results show that both of these breeding populations occur together at communal wintering grounds.

The Bean Goose (*Anser fabalis*) is tied primarily to the taiga and to the more southerly tundra within a broad belt extending right the way to the Soviet Pacific coast (figure 40). The eastern populations winter in Asia. In Europe, two distinct wintering centres exist: in west Europe, and on the Hungarian plain (i.e. in Hungary, eastern Austria and northern Yugoslavia). Further particulars on the migratory habits are discussed in a special section below.

The Lesser White-fronted Goose (*Anser erythropus*) breeds on upland bogs in the birch-forest region; its breeding range extends a long way east within the birch-forest tundra (figure 40).

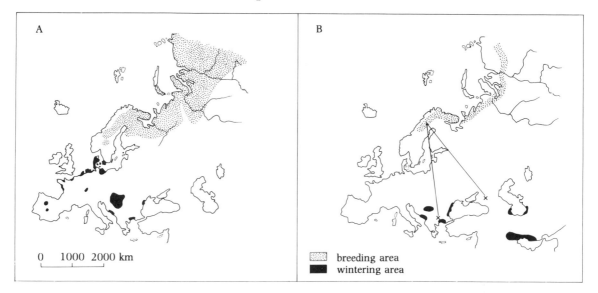

Figure 40   The breeding and wintering areas of the Bean Goose (A) and the Lesser White-fronted Goose (B). The Bean Goose is found in winter in west as well as in southeast Europe, a pattern which recurs in the Wigeon, the Greylag Goose and the White-fronted Goose. The Lesser Whitefront, on the other hand, like the Red-breasted Goose, winters only in southeast Europe. The map also shows two recoveries of Swedish-ringed Lesser Whitefronts, one from Macedonia (February) and the other from the north Caucasus (September). Based on Bauer & Glutz von Blotzheim (1968), Cramp & Simmons (1977) and Ogilvie (1978).

The main winter quarters are in southeast Europe (the Hungarian plain to the Black Sea) and in the Near East (around the Caspian Sea and the Persian Gulf). The total population in the Scandinavian mountains is today probably only a few hundred birds; on the coastland of upper Norrland, where a few decades ago resting flocks of Lesser Whitefronts were a regular sight in spring and autumn, the species has become a rarity. This heavy population decline leads one to suspect adverse changes in the countryside in the wintering areas. The Lesser Whitefront's status during the winter is very poorly known; this is in fact reflected in widely differing statements regarding the total number of Lesser White-fronted Geese in the winter quarters. The estimates or guesses vary between 5000 and 50 000 birds! What is clear in any case is that the Lesser Whitefront is in most places a comparatively rare species and that it is on the verge of disappearing altogether. It is to be hoped that these geese will change their habits and start to winter in west Europe. Occasional Lesser Whitefront families and individuals are seen every winter in the large goose flocks in west Europe, but no direct increase in its occurrence is detectable.

The Greylag Goose (*Anser anser*) has a wide breeding distribution in Scotland, in the Nordic countries and in large parts of central and eastern Europe. Large-scale ringing of non-breeding immature Greylags, which gather in thousands to moult on the Rone islets off Gotland in the southern Baltic, shows that the geese migrate both southwest and due south. Autumn and winter recoveries have been made in southern Spain (the Guadalquivir Delta) and in the Danube area in Austria, Hungary and Czechoslovakia. It is likely that the Greylags that moult at Gotland do not come only from the breeding population in Sweden but also from other Baltic countries, especially from Denmark, Germany and Poland.

The Canada Goose (*Branta canadensis*) was introduced into Britain as early as the 1600s, and the British and Irish population today amounts to about 40 000 individuals. The birds remain in these islands in the winter. In Sweden and Norway the Canada Goose was introduced during the first half of the 1900s, but not until the 1960s did it really start to spread. Numbers have since rapidly increased in Scandinavia, where the stock now numbers 50 000 or more birds. The Swedish geese have developed definite migratory habits. Several thousand Canada Geese winter on arable land in Scania, while the East German Baltic coast is another important wintering area for Swedish Canada Geese; some migrate even farther, to Denmark and the Netherlands, and during the harsh winter of 1979 Swedish-ringed Canada Geese turned up in England. The Scandinavian Canada Goose provides us with a unique opportunity to chart the way in which the migratory habits develop.

Like ducks and swans, the geese renew all their flight feathers at the same time and at that time lose the power of flight for a period of a month. Breeding birds, both male and female, shed the flight feathers when the young are about a month old, and they regain the power of flight at the same time as the young reach the flying stage.

Geese do not normally begin to breed until two or three years of age. The non-breeding immatures migrate during the summer to special moult centres. When the adults start breeding, the immatures leave and head for a suitable moulting site. Markedly often this movement takes them northwards. Figure 41 shows the largest and best-known moult

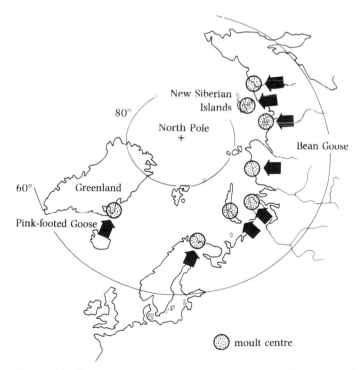

Figure 41 Important moult centres for Pink-footed Goose and Bean Goose to which immature non-breeding geese migrate in summer from adjacent parts of their species' breeding range. The moulting sites are situated on the northern edge of the breeding area. Based on Ogilvie (1978).

centres for Bean Goose and Pink-footed Goose; these sites all lie north of the normal breeding area or right on its northern edge. The moult migration sometimes heads over the open sea: immature Icelandic Pinkfeet moult in Greenland, and Siberian Bean Geese fly across the Arctic Ocean to the New Siberian Islands. There is a moult centre in Lapland and the Kola Peninsula, and here several thousand Bean Geese congregate.

Why are the moulting sites situated so far to the north? Two important factors may help to explain this.

1 The geese are additionally susceptible to disturbance and persecution during the moult period. They cannot fly but must stay on land in order to feed. The risk of persecution by predators and human beings is least in the very northernmost, most inaccessible, regions.
2 The summer on the northern tundras is too brief for the geese to have time to breed there, but there is easily enough time for a month's stay to moult. By migrating northwards, the moulting immature geese can therefore take advantage of an unexploited resource for survival and avoid competition for food with the breeding birds.

As early as the last century it was reported that Black Brants left the New Siberian Islands in summer and flew straight out over the Arctic Ocean basin! In olden times this was misinterpreted and it was considered that undiscovered land must exist somewhere in the Arctic Ocean (see figure 38). The correct explanation is that immature Brants on moult migration take the shortest route, almost 2000 km, over the pack-ice to Alaska. Just look at a globe. There we can best see the route 'as the crow flies' between the New Siberian Islands and Alaska. In Alaska, a heavy programme of goose-ringing is currently being carried out in order to clarify better the course of events involved in the moult migration.

The Bean Goose (*Anser fabalis*)

In Scandinavia, the Bean Goose breeds on boglands in the northern coniferous-forest country. In autumn big flocks of Bean Geese visit stop-off sites in south Sweden, in Scania and Halland and at Lake Tåkern. Most continue on before the winter, but significant numbers, at least 10 000 and in some years twice that number, stay on and winter, mainly in southwest Scania. During the spring migration, the Bean Geese once again stop off in south Sweden, but in smaller numbers than during the autumn.

Where does the north Scandinavian Bean Goose population migrate to? Where do the large flocks of Bean Geese that stop off in south Sweden in autumn come from, and where do they move to in the winter? What is the breeding area of the geese that winter in Scania? In order to answer these questions, a joint Scandinavian research project on the Bean Goose was recently instigated. The project leader in Sweden, Leif Nilsson, caught Bean Geese in Scania using rocket-nets. The birds were marked not only in the usual way, with a ring around the leg, but also with a brightly coloured numbered neck collar; the marked geese can be identified at long range by using a telescope. These efforts have yielded several interesting results. We now have, I believe, so many pieces of the jigsaw concerning the Bean Goose's migration in west Europe that it may be worth trying to see how far the puzzle fits together.

Two different races of Bean Geese occur in Europe. Those which breed in Scandinavia and eastwards within the coniferous-forest belt to the Ural Mountains belong to the race *fabalis*. The Bean Geese breeding on the Russian tundra, on the other hand, belong to the race *rossicus*. The tundra race is normally more compact than the forest race and also has a shorter and darker bill (figure 42). It is, however, seldom possible to separate the two races completely

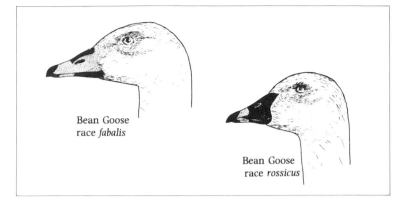

Figure 42   In Europe, two different races of the Bean Goose, *fabalis* and *rossicus*, occur. The race *fabalis* breeds in the northern forest belt and the race *rossicus* on the tundra. Typical individuals of the two races can be separated by, among other things, the bill. The forest Bean Goose has a long, narrow bill, suited to feeding on roots and leaves of bog plants. The tundra Bean Goose's deep and short bill is better adapted to feeding on leaves of low-growing plants on firmer ground. Based on Bauer & Glutz von Blotzheim (1968).

in field studies, since intermediate forms are much more common than racially typical individuals. To find wholly typical representatives of the race *rossicus* one must in fact go as far east as to the tundra north of the Urals. West of there the tundra geese have more or less obvious elements of *fabalis* characters. Despite the difficulties, however, it is often possible under good observation conditions to determine whether Bean Geese are principally of the forest type or of the tundra type. Such data provide important pointers to the migration patterns of different Bean Goose populations.

Ringing of wintering Bean Geese in the Netherlands during the 1950s and 1960s has resulted in many recoveries (figure 43A). The recoveries show that the Dutch winterers are mainly tundra Bean Geese and that they breed in the area around the Kanin Peninsula. Some recoveries have been reported from as far away as the regions of the River Ob east of the Urals. The migration route passes mainly south of the Baltic Sea. Recoveries during the winter following that of ringing show that the Dutch Bean Geese have an interchange link both with East German winterers and with the geese that winter in rather moderate numbers in France (normally up to 3000) and Spain (4000). A few recoveries from the Hungarian plain show, interestingly enough, that a certain interchange takes place between the two European wintering centres, the one in the northwest and the one in the southeast. We should bear in mind that ringing returns up to that point were made at a time when there were not particularly large numbers of wintering Bean Geese in the Netherlands (figure 44). Since then the number has steadily increased, and in 1976 it 'exploded' to over 50 000, ten times more than were present during the early 1960s. Despite this, the results from the early ringing efforts are probably still representative of the bulk of the Dutch wintering geese: with the exception of a small minority of forest Bean Geese in the extreme east of the Netherlands, they are considered to belong to the tundra type. The marking with yellow neck collars of 350 tundra Bean Geese in the Berlin district in East Germany in the autumn of 1977 produced reports in the following winter of a total of 50 birds from the Netherlands, two from France and eight from the Hungarian plain.

Let us now turn to the forest Bean Geese. Ringing of non-breeding immatures at moulting

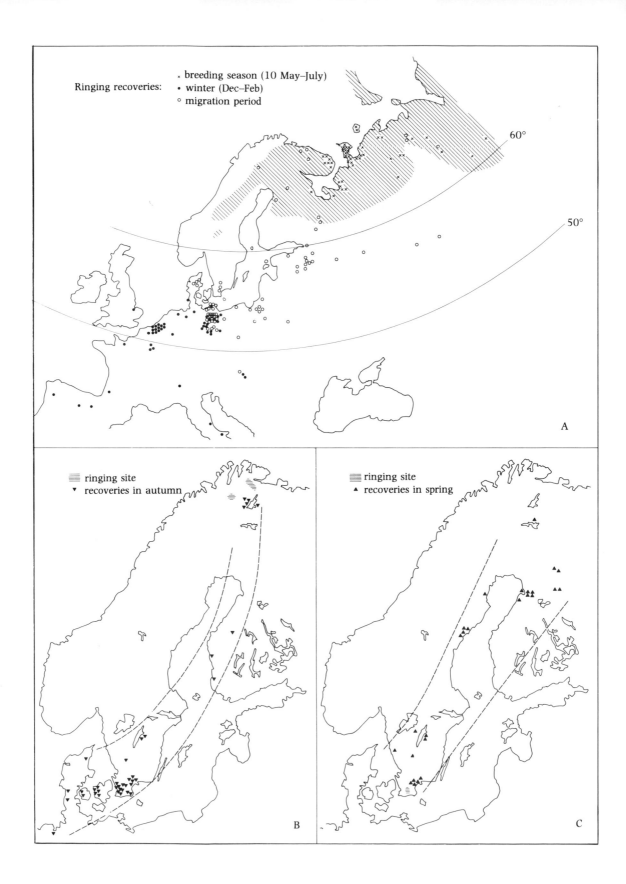

Ringing recoveries:
× breeding season (10 May–July)
• winter (Dec–Feb)
○ migration period

60°

50°

A

≣ ringing site
▼ recoveries in autumn

B

≣ ringing site
▲ recoveries in spring

C

sites in north Norway indicates that the autumn migration route is mainly via the eastern coast of the Gulf of Bothnia and straight across southern Sweden to Scania, Denmark and occasionally West Germany (figure 43B). Many of these geese belong to the north Scandinavian Bean Goose population, and in addition there are probably also birds from the adjoining areas of Russia on the Kola Peninsula and in north Karelia. The wintering area of these forest Bean Geese probably reaches from Scania, Denmark and West Germany to the Netherlands and Britain. Today the Bean Goose population wintering in Britain is negligible, only about 300 individuals, but around the turn of the century it was considerably greater. There is a recovery made during the winter in England of a Bean Goose ringed as a juvenile in Sweden.

The first results from the marking with neck rings of geese wintering in Scania in southernmost Sweden confirm that these originate largely from north Scandinavia and presumably also from the nearby areas of Russia (figure 43C). The fact that the Scanian winterers extend their migration flight southwestwards in unusually severe winter conditions emerged in 1979, when a dozen winter recoveries were reported from Denmark, West Germany and the Netherlands.

The number of Bean Geese that normally winter in Scania is 10 000 in round figures. If we add to this between 1000 and 2000 wintering forest Bean Geese in each one of the countries Denmark, West Germany and the Netherlands, then we get a total figure of up to 15 000 or possibly 20 000 geese. This is therefore the probable size of the population that breeds in Scandinavia and in the adjoining coniferous-forest areas of Russia. Considerably more forest Bean Geese than this, however, stop off in south Sweden during the autumn. The autumn geese in south Sweden were earlier estimated at up to 30 000 individuals, but in more recent years this figure has been surpassed by a long way. This is due largely to a stupendous increase in the number of Bean Geese that stop off in autumn around Lake Tåkern (figure 44). In October 1977, nearly 50 000 Bean Geese were counted in Sweden, of which those at Tåkern accounted for 17 000 and those in Scania for 25 000. In November 1978, this peak was exceeded: 37 000 Bean Geese were then present at Lake Tåkern at the same time as there were 20 000 in Scania. The rate of increase has since slowed down, and in the late 1980s the autumn population of Bean Geese in Sweden, comprising virtually the whole *fabalis* population, numbered about 75 000, of which close to 50 000 are found at Tåkern. Most of these geese therefore leave Sweden before the winter and do not return during the spring migration. Since they belong in the main to the forest type, they must in all likelihood come from the taiga country of Russia. But where do they get to once they have left south Sweden in late autumn?

In East Germany, three main regions holding large numbers of passage and wintering Bean Geese can be distinguished. The most northerly one encompasses the island of Rügen and the coastal districts on the Baltic. A good 100 km south of there in Neubrandenburg is a

Figure 43 Results from different ringing studies of the Bean Goose's migration. A. Recoveries of Bean Geese ringed in winter in the Netherlands indicate a migration route south of the Baltic Sea to breeding areas mainly on the tundra around the Kanin Peninsula. The cross-hatched area shows the Bean Goose's breeding distribution. Based on Bauer & Glutz von Blotzheim (1968). B. Immature Bean Geese ringed in July on the moulting grounds in north Norway are recovered mainly in Scania and Denmark. The study was carried out in 1967–1972 by G. Tveit and the map is based on the basic data hitherto available in Bulletins from *Norsk Viltforskning* series 2, no. 41 (1974), and series 3, no. 2 (1977). C. Observations during spring 1978 of Bean Geese marked with neck rings in late autumn in Scania. Based on Nilsson (1979, 1984).

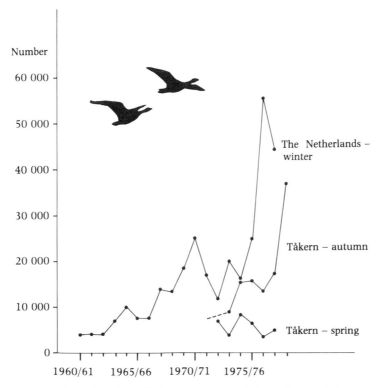

Figure 44   Growth in the number of Bean Geese wintering in the Netherlands and those which stop off in autumn and spring at Lake Tåkern in south Sweden. The Dutch geese are mainly of the tundra type; forest Bean Geese are in the majority at Tåkern. Based on Ogilvie (1978) for the Netherlands, and for Tåkern on Bulletins nos. 9–14 of the Tåkern field station (1973–1978). The increase has continued, and recent counts, from the late 1980s, total close to 50 000 at Tåkern during the autumn and exceed 150 000 individuals in the Netherlands during the winter.

lakeland area with good goose country (e.g. around Lake Müritz), and a further 100 km to the south, in Potsdam (the regions around Berlin), there is a similar area of lakes where Gülper Lake is the best-known goose site. Unfortunately, accurate and co-ordinated goose counts from East Germany are lacking, and it is therefore not known how many geese there really are in the country in midwinter. It should, however, be safe to say that several tens of thousands are involved. It is of interest that the tundra geese that were ringed in the Netherlands were recovered mainly in the interior of East Germany: 14 recoveries in Potsdam (nine in winter) and 19 recoveries in Neubrandenburg (five in winter). Only seven recoveries (one winter recovery) have been made on the Baltic coast. We may therefore guess that the wintering goose flocks in northernmost East Germany consist mainly of forest Bean Geese, probably those which have stopped off in autumn in south Sweden. That large numbers of geese are able to winter in northern East Germany during very harsh winters appears unlikely. Where the geese go to during such winters is totally unknown.

The sum of all various pieces of the puzzle could, then, be as follows. Tundra Bean Geese migrate mostly south of the Baltic Sea to winter quarters in interior East Germany and in the Netherlands; a smaller proportion continue on to France and Spain. This is similar to the migration pattern of the White-fronted Goose, and we may assume that tundra Bean Geese, like Whitefronts, pass through and stop off in southernmost Scandinavia during spring and autumn in minor numbers. Like White-fronted Geese, tundra Bean Geese also winter on the Hungarian plain.

Forest Bean Geese from Scandinavia and bordering parts of Russia winter mainly in Scania, Denmark, West Germany and eastern parts of the Netherlands. During the autumn in south Sweden, these geese are mixed with forest Bean Geese from more easterly areas of Russia. The latter birds, however, leave Scandinavia during the winter and then live on the Baltic coast of East Germany. Those which possibly continue a little farther south to the lakes country in East Germany come into contact there with wintering flocks of tundra Bean Geese. The eastern forest Bean Geese migrate in spring along the south coast of the Baltic to their breeding grounds. The increase in the number of geese in autumn at Lake Tåkern may possibly be due to the fact that eastern forest Bean Geese which formerly migrated direct to northern East Germany have diverted their migration routes.

These conclusions are of course only provisional and further studies are required, not least on the East German goose grounds, in order that we may be able to map definitively the Bean Goose's migration patterns. Further pieces of the puzzle, from ringers, goose counters, migration-watchers and from general scrupulous birdwatchers, are sought after. But we may of course also allow ourselves a day on the goose grounds simply delighting in the skeins of geese cackling as they fly to and from the roost sites or the gaggles grazing on fields and meadows. They set an inalienable stamp on the landscape, these splendid birds from taiga and tundra.

3.4   Birds which feed on fish

There are great demands for special adaptations in appearance and behaviour if a bird is to be an efficient catcher of fish. For this reason there are rather few examples of species that derive any major profit from fishing 'as a side line'. Some smaller gull species, the White-tailed Eagle and the Eagle Owl do, however, belong in this category.

In table 14 I have divided the fish-eating birds into three main categories, according to fishing technique and environment (figure 45). Surface feeders catch the fish right at the surface of the water. Birds which pursue the fish in the water are described as underwater feeders. These include many species which swim on the surface and from there dive under the water and hunt fish, sometimes at considerable depths, 10 m or more (divers, grebes, cormorants, sawbills, auks). The Gannet is also an underwater feeder, but it has a different technique: it plunges towards the water's surface from a height of 10–30 m and disappears under the water, where it catches the fish at a depth of several metres (though at times it dives to depths of as much as 12–15 m). The Gannet already has its sights fixed firmly on its prey when high up in the air: this can be seen when, during its dive towards the surface, it manoeuvres, changes angle or revolves its body, depending on the movements of the fish under the water. The shearwaters often dive from low-level flight diagonally towards the water's surface and disappear underwater. Their dives, however, are not deep; to a certain extent, therefore, the shearwaters occupy a position intermediate between surface feeders and typical underwater feeders and therefore, in table 14, they come at the beginning of the underwater feeders.

The Grey Heron and the Bittern fish by standing motionless in the shallows at the edge of the shore or reeds and suddenly stabbing a prey that comes within reach. The Kingfisher sits on the look-out on a branch suspended above the water; it swoops down like an arrow when

Table 14. *Birds which catch fish and which occur regularly in north Europe. The data on wintering areas apply primarily to European birds or birds which pass through Europe on migration*

Species (NB=does not breed in Scandinavia)	Present N Europe in winter	Important wintering areas				Breeds southern hemisphere. Found summer–autumn in N Atlantic
		W or NW Europe	Mediterranean, Black Sea, Caspian Sea	African coasts or inland S of Sahara	S American coasts	
Surface feeders						
Osprey			(+)	+		
Black Kite				+		
Lesser Black-backed Gull (Baltic Sea)			(+)	+		
Caspian Tern				+		
Common Tern				+		
Little Tern				+		
Sandwich Tern				+		
Underwater feeders						
Manx Shearwater (NB)			+	(+)	+	
Great Shearwater (NB)						+
Sooty Shearwater (NB)						+
Red-throated Diver	+	+	+			
Black-throated Diver	+	+	+			
White-billed Diver (NB)	+	+				
Great Crested Grebe	+	+	+			
Red-necked Grebe	+	+	+			
Gannet		+		(+)		
Cormorant	+	+	+			
Shag	+	+	+			
Smew	+	+	+			
Red-breasted Merganser	+	+	+			
Goosander	+	+				
Razorbill	+	+				
Guillemot	+	+				
Black Guillemot	+	+				
Puffin		+				
Feeders in rivers, by shore vegetation						
Bittern	(+)	+	+			
Grey Heron	+	+	+	(+)		
Kingfisher	+	+	+			
Total no. species	15	18	13	7	1	2

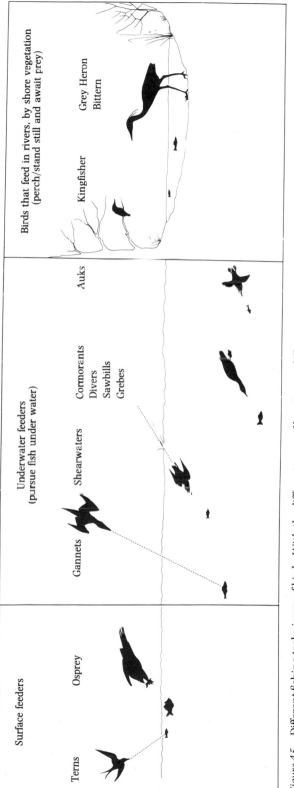

Figure 45 Different fishing techniques of birds. With the different ways of hunting go different migratory habits Surface feeders and shearwaters are **long-**distance migrants; the other species migrate shorter distances. Based on Ashmole (1971) and others.

it sights a suitable fish. These three species thus fish in shallow water where they can reach fish both at the surface and nearer the bottom.

Surface feeders as a rule migrate to distant tropical winter quarters. Underwater feeders, however, are short-distance migrants and many of them are present in northern Europe during the winter. The reason for this is that in northern latitudes the production of aquatic plants and phytoplankton ceases during the winter. The summer and autumn production of plankton sinks to the bottom. The animal life of sea and lakes leaves the surface waters and makes its way downwards, to greater depths. Fish are therefore still present in large numbers during the winter and at that time are often sluggish and easy to catch in the cold winter waters. Deep-diving birds consequently have no great problems with winter fishing so long as the water is not completely iced over, but species which catch fish at the surface must set off for warm waters where they can see the fish flashing in the sun at the water's surface.

The wintering of the Grey Heron, the Bittern and the Kingfisher in northern regions is wholly conditional on fishing in running water. In rivers and streams the birds can reach fish at the bottom.

In table 14, the Grey Heron is the species that shows the greatest variation in wintering habits. Although smaller numbers of herons regularly winter on open watercourses in southern Sweden, the majority of Scandinavian Grey Herons leave for Portugal, Spain and southern France, some continuing beyond the Mediterranean and the Sahara to tropical Africa. This shows that the Grey Heron has a great capacity for adapting to different environments and also that this shallow-water feeder has in its migratory habits certain features in common with the surface-feeding fish-eaters.

Some surface feeders spend the whole year chiefly at inland waters and on shallow coastal lagoons. This is the case, for example, with the two fishing species in the birds of prey family, the Osprey and the Black Kite. These two species differ from one another in several ways. The Osprey is a powerful and tenacious flyer, which is clearly demonstrated by its habit of hovering – a flight technique very wasteful in terms of energy – above the water before diving down towards a fish at the surface. On migration the Osprey soars and glides to a much lesser extent than other large birds of prey, and it flies long distances over open sea, for example over the Baltic Sea and across the Mediterranean Sea. The Ospreys counted annually at the main sites for regular fixed-route passage of soaring/gliding birds, such as Falsterbo and Gibraltar, are therefore not many in number, usually fewer than 100 birds.

The Black Kite is a neat and graceful bird which patrols along the shores of lakes and rivers, where it frequently soars on local upcurrents. It does not fish by diving powerfully like the Osprey. Instead it uses its talons to pick fish from the water's surface while in flight. This method of hunting often looks so easy and undramatic that one may think that the Black Kite feeds mainly on dead or diseased fish. But this is hardly the case; without rejecting easily accessible carrion and diseased fish, it just as often catches perfectly healthy fish. The Black Kite is common as a breeding bird throughout Africa, where in many places it is a bird of refuse tips and exploits all kinds of waste matter. The tropical kites also readily tackle carcases of small and large animals. Interestingly enough, the European kites stick to their fishing habits in the African winter quarters and seldom associate with the local Black Kites around villages and towns. At rice plantations in Senegal, wintering European kites are often seen picking fish out of the channels in the company of storks, herons and egrets.

The Black Kite makes great use of thermals for soaring during migration, and therefore travels as much as possible over the kind of country where thermal upwinds form. Thousands of Black Kites pass over the narrow sea passage-ways at the Bosporus and Gibraltar; at

Gibraltar, 39 000 Black Kites were counted in autumn 1972 and 25 000 in autumn 1974 on route to Africa.

The Caspian Tern, the world's largest tern, breeds in geographically isolated populations in all parts of the world except South America. The most important breeding area is on the Caspian Sea; in Europe there are in addition significant populations in Crimea on the Black Sea and in the Baltic, mainly in small colonies on treeless outer islands and skerries. The Caspian Terns prefer to fish in the inner islands and at inland lakes and therefore make fishing excursions of several tens of kilometres from the breeding islands. During migration and in winter, too, they prefer inland lakes and coastal lagoons. Ringing studies of the Baltic Caspian Terns show that to a large extent they migrate straight across the European interior via Italy and Tunisia and across the Sahara desert to winter quarters on the River Niger and in surrounding districts in West Africa. Here they come into contact with Caspian Terns from the Black Sea, for which, according to ringing recoveries, the Niger region is also a major wintering area. Caspian Terns from smaller breeding populations in Tunisia and on the West African coast winter there as well.

The Lesser Black-backed Gull is a common winter visitor to large lakes in tropical Africa and can often be seen out on the open waters of Lake Victoria. Lesser Black-backs which winter in tropical Africa belong to the race *Larus fuscus fuscus*, which has its breeding distribution around the Baltic Sea. Lesser Black-backed Gulls of other races breed in Norway and in Britain. Immatures from the North Sea migrate to the coasts of northwest Africa and southwest Europe. The same migratory habits were also demonstrated two or three decades ago by western Lesser Black-backs of breeding age, but many now stay behind to winter in west Europe. They have abandoned their winter fishing, and instead feed as omnivores and eat fish offal and refuse (see section 3.9).

The other three species of surface-fishing birds, the Common Tern, the Little Tern and the Sandwich Tern, winter in a marine environment, in tropical waters close to the shores.

What factors determine the migratory and wintering habits of marine fish-eating birds? Here there is occasion to remind ourselves of the distribution of nutrient production in the sea and of the most important wind zones and ocean currents (figures 6 and 11, pp. 14 and 22). Fish stocks are of course closely bound up with the rate of phytoproduction. We can therefore expect the richest fishing waters to be within the middle latitudes in both the northern and the southern hemispheres and within areas of upwelling water (figure 11). In the Atlantic, the most important currents carrying upwelling water are the Canaries Current off northwest Africa and the Benguela Current off South Africa and Angola. In addition, upwelling currents with a great abundance of phytoplankton, fish and other animals occur periodically in the Gulf of Guinea on the west coast of Africa and off the southeast coast of South America where the northward-moving Falklands Current advances. In warm tropical parts of the Atlantic, fish stocks are by contrast very sparse. Tropical birds which fish in these waters are specially adapted to move over enormous areas of water and to take advantage of the opportunity to catch flying fish or fish of other species which are forced up towards the water's surface in large shoals by attacking tunny fish. The birds must take the fish from the surface of the water or catch the flying fish in the air, for diving near the hunting tunnies is too dangerous by far.

It is thus easy to understand why so many birds which catch fish underwater winter in west European waters. Large numbers of herrings, sprats, sardines, anchovies, capelin, sand-eels, sticklebacks and other fish can be found there. Despite the fact that the fish shoals often keep quite a distance beneath the surface, they are within reach for the deep-diving bird species. Many underwater feeders do not migrate all that far, but are content with the nearest wintering waters that are rich in fish and free of ice. Thus the Shag, for example, is to a large extent a resident or migrates only short distances. In Norway, large parties of Shags gather in winter in fishing waters off Trondheim Fjord; birds come there from both northern and southern parts of the Norwegian coast. The guillemots leave their bird-cliffs in the winter and disperse over the immediate waters. Those guillemots which breed off southeast Sweden on the islands off west Gotland winter in the southernmost Baltic Sea. The Skagerrak and the Kattegat are visited in winter by Norwegian and British Guillemots. A part of the British Guillemot population migrates in winter to the Norwegian coast and there mingles with guillemots from nearby Norwegian breeding sites.

There are also examples of more spectacular migration patterns in the birds which fish in deeper water. The White-billed Diver is a bird of the tundra and Arctic Ocean which winters primarily along the coasts of arctic Scandinavia and the Kola Peninsula. As with the King Eider and Steller's Eider, however, a small proportion of the White-billed Diver population comes down into the Baltic Sea and occasional individuals are seen each year on passage through Kalmarsund; some even cross the North Sea.

Black-throated Divers from the far north of Scandinavia winter together with Russian and Siberian birds in the Black Sea and in the eastern Mediterranean Sea; divers from south Sweden and from Norway spend the winter in northwest Europe. Many Russian and Siberian Black-throated Divers take a long circuitous route back to the breeding sites in spring. They travel from the Black Sea to the Baltic Sea, continue northwards to the White Sea and the arctic coast and finally eastwards to the lakes on the tundra and in the taiga. By contrast, in the autumn these divers from eastern populations travel across the Soviet interior, a direct route to the winter quarters.

Cormorants breeding in the southern Baltic Sea area (a distinct subspecies) travel southwards straight across Europe to winter in the Mediterranean. The Cormorants that are common in winter on the south coasts of Scandinavia are visitors from Norway. In addition, some Cormorants from the Arctic arrive in the Baltic Sea from the White Sea.

Red-necked Grebes winter in amazingly large numbers along the Atlantic coast of Norway. They reach there by flying straight across the Scandinavian mountain chain from breeding sites in Finland and Russia.

The surface-feeding marine terns winter mainly along the west coast of Africa, including the Gulf of Guinea, where upwelling waters penetrate during the autumn. At this time large shoals of sardines are moving from deeper water in towards the Ivory Coast and Ghana in order to spawn. There are then considerable numbers of Common, Sandwich and Little Terns present there to help themselves to the fish. During the rainy season the rivers swell and transport large quantities of food out to the gulf, where the fish life, and thereby also the resting terns, flourish. Both the Little Tern and the Common Tern incidentally breed in local populations in the Gulf of Guinea. In winter, the Common Tern and to an even greater degree the Sandwich Tern are common southwards as far as the Cape of Good Hope, but east of there, where the nutrient-impoverished Agulhas Current flows, they are virtually absent.

The shearwaters are the real long-distance migrants. They are presented in greater detail in a special section below.

There are many environments which provide birds with the opportunity to fish, but the fishing areas that at the same time provide possibilities for breeding are comparatively limited. Fishing far out from the shore over the open sea is of course fine for a full-grown bird just so long as it has only its own survival to consider. When breeding, fishing must be restricted to waters near sheltered nest sites on land. For the birds that fish, the resources for survival are thus far more extensive than the breeding resources. In accordance with the arguments put forward regarding the marine diving ducks, this leads to a number of consequences:

1 Mortality in full-grown birds is very low. The annual mortality in adult Gannets and shearwaters is only about 5%, in terns 10–20%. With an annual mortality of 5%, out of 100 birds 50 individuals are left after 13 years, 25 individuals after 26 years and five after 58 years! Are there perhaps Gannets and shearwaters alive that are more than 50 years old? Or maybe the mortality increases at the top end of the age scale? This has not yet been clarified, since seabird studies have not been going on long enough. The oldest ringed Gannet is today 28 years old and the oldest Manx Shearwater at least 37 years.

2 The birds survive with fairly little trouble at good fishing waters, but competition becomes hard at suitable breeding sites. The production of young is low; many species lay only one egg. The birds are forced to make long and costly excursions out from dense breeding colonies in order to find food.

3 The immature birds do not start to breed until several years have passed. The terns and the Osprey as a rule breed for the first time at three or four years of age. The Gannet normally does not start to breed until five years and the Manx Shearwater not until six or seven.

4 The non-breeding first-year birds often summer in the winter quarters or in waters separate from the species' breeding areas. As they get older the immatures visit the breeding sites, but they arrive later in the spring and depart earlier in the autumn than the breeding adults. The period of their stay at the breeding site is gradually extended and eventually it comes into line with the adult birds' timetable in that year when their first breeding takes place.

Those fishing birds – divers, grebes, sawbills and auks – whose hunting method is not dependent on the ability to fly moult all their flight feathers simultaneously. The moult takes place immediately after breeding is completed, at which time the birds make for areas of open water where they can dive for fish and need not be troubled by the temporary incapacity for flight. The other fishing birds moult the flight feathers in stages and thus retain the power of flight. This is only barely the case for the shearwaters, however, for, during the moult, they are so heavily hampered in their flight that they are hardly capable of rising from the water in calm weather. In former times this was exploited by the Eskimos in Greenland, who caught moulting Great Shearwaters from their kayaks during the summer.

The Osprey, the Gannet and the terns have a continuous moult. They are dependent upon perfect flying ability for catching fish. Moreover, they use their wings for migratory journeys of thousands or tens of thousands of kilometres and more, and this imposes special demands on the moult of the flight feathers. The individual primaries are shed sequentially from the inner to the outer one. Not until a new primary is almost completely grown out is the next one in the sequence shed. The focus of the moult in this way moves slowly towards the tip of the wing, but before the focus has reached the wingtip the next moult wave begins from the inner primaries. Before the first moult wave has reached the outermost primary, the inner feathers can therefore have been renewed both twice and three times (the terns) or even four times (the Osprey). One individual wing feather is used by the Osprey for an average of $1\frac{1}{2}$ years before being exchanged for a new one. The moult of the secondaries also takes place according to a fairly complex continuously rolling schedule.

Owing to the gradual nature of the moult, which takes place symmetrically on both wings, the above species are able always to keep intact the requisite ability to fly and to hover. Even so, the Osprey's moult is suspended during the migration seasons. The terns break off the moult during the spring migration and the breeding period.

Following this general background to the life of birds that catch fish, I shall now turn to giving more detailed examples of the migratory habits of a surface-feeding bird (the Osprey), of an underwater feeder (the Gannet) and of the shearwaters.

The Osprey (*Pandion haliaetus*)

With a wingspan of over 1.5 m and a weight of nearly 2 kg, the Osprey is so big that it can easily take fish in the half-kilogram class: pike, roach, perch, bream, salmon etc. In fact the Osprey even manages to gather up fish weighing over 1 kg. Sometimes it attacks even larger prey by mistake. Such pursuits have on occasion come to a tragic end for the bird when the Osprey has not managed to release its talon grip on the fish but has been dragged under the water. Thus dead Ospreys or Osprey skeletons have for example been found stuck fast to 5-kg carp and bream and to pike weighing up to 9 kg (known as 'king pike')!

The Osprey's breeding distribution extends around the whole of the northern hemisphere. In many places, however, it occurs very sparsely, largely as a result of persecution through intensive hunting. In central Europe there are only some hundred breeding pairs, and in the north of Europe the Osprey is a common breeding bird only in Sweden, Finland and Russia. In more recent years the Osprey has, fortunately enough, returned as a breeding bird to Scotland, from where it had been absent since the turn of the century. One of the first colonisers was a ringed Osprey: even though it never proved possible to read the ring number, in all likelihood the bird involved had been ringed at a Swedish nest (an extensive ringing programme has been undertaken in that country). The first pair of Ospreys since 1916 bred in Scotland in 1959; the population had increased to at least 14 pairs in 1975, and to over 50 pairs by the late 1980s.

Up to 2000 pairs of Ospreys breed in Sweden, with the highest density in the central Swedish lakelands and in southern Småland. In Finland there are at least 500 pairs of Ospreys. In these two countries the species has been ringed on a large scale, despite the fact that it can be very difficult and at times hazardous to climb up to the Osprey's huge nest in the top of tall trees in order to ring the young. The ringing work has been most extensive in Sweden, where the project has been directed by the head of the central ringing office, Sten Österlöf. A total of around 13 500 Ospreys has so far been ringed in Sweden and Finland, and almost 1700 recoveries have been reported.

The Osprey usually lays three eggs. An average of two young survive to the flying stage. When the young have left the nest they often disperse in every possible direction, so it quite often happens that they set off in a northerly direction. In figure 46, several examples are shown of recoveries in early autumn (most often in August) which are farther north than the breeding site. As can be seen in the figure, the juveniles are dispersed over large areas; sometimes they may move up to 750 km away. This dispersal of juveniles is also well known for other species of fish-eating birds. The phenomenon is particularly well studied in the Grey Heron: the juvenile herons migrate in summer usually to areas 50–100 km from where they hatched, and not uncommonly they fly even longer distances, over 500 km. The young Ospreys and Grey Herons probably disperse in order to reduce the competition and to make use of unexploited rich fishing waters.

When the young Ospreys later depart southwards, they spread out on a strikingly broad-front migration. There are autumn recoveries of Swedish Ospreys from Portugal, France and England in the west and from the Ukraine, Georgia and the regions of the River Volga in the east (figure 46): a front 4500 km wide! We can guess that some of the easternmost birds end up in winter quarters around the Persian Gulf and the southern Caspian Sea. Juvenile

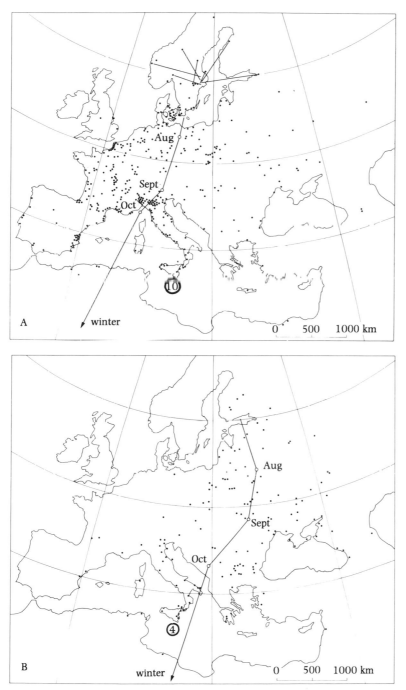

Figure 46 Recoveries in autumn (August–October) of juvenile Ospreys ringed as nestlings in Sweden (A) and Finland (B). The mean co-ordinates for recoveries during different months are also shown. The upper lines on map A show recoveries made at latitudes north of the ringing sites (which in all relevant cases are situated in central Sweden). There are ten recoveries of Swedish Ospreys from Malta. Two autumn recoveries in Morocco and four in West Africa fall outside the map. Four recoveries of Finnish Ospreys are from Malta, and one recovery in Chad is off the map. Based on Österlöf (1977).

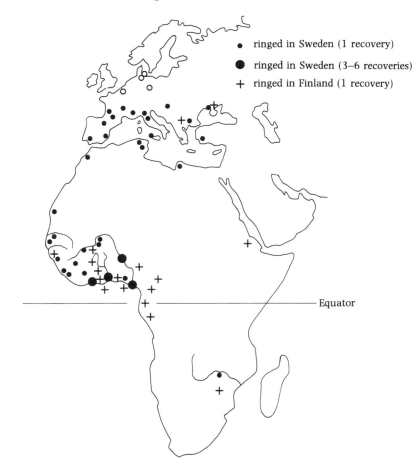

● ringed in Sweden (1 recovery)
⬤ ringed in Sweden (3–6 recoveries)
+ ringed in Finland (1 recovery)

Equator

Figure 47   Recoveries in winter (December–February inclusive) of Ospreys ringed in Sweden and Finland. Recoveries of adults as well as juveniles are included. Two recoveries south of the Zambezi River have been reported; whether these were made in winter or during other times of the year (possibly summering immatures) has not yet been published. The four most northerly recoveries are indicated by open circles since they probably do not show genuinely wintering birds (e.g. the rings have been found on birds that have lain dead for some time). Based on Österlöf (1977).

Ospreys from Finland also spread out on a broad front, but the central point of the migration is shifted to the east of that for Swedish birds. The Finnish Ospreys usually depart in a south to southeast direction on the eastern side of the Baltic Sea and turn off more towards the southwest in southern Europe. Amazingly few Finnish Ospreys seem to migrate via south Scandinavia: in fact there is not a single recovery from south Sweden, Denmark or West Germany. The first thought that comes to mind is that the Finnish Ospreys migrate over eastern areas so as to avoid competition with the large number of Swedish birds at the fishing waters at staging sites.

The young Ospreys break all ties with the family and fly south entirely on their own. The differences in migration directions between young from the same brood and between juveniles from different nests are of exactly the same magnitude.

Table 15. *Fidelity to home area in the Osprey. Percentage distribution of ringing recoveries of breeding Ospreys (three years or older) in relation to distance from site where hatched. Based on Österlöf (1977)*

0–50 km	50–100 km	100–1000 km	>1000 km
43%	16%	22%	19%

Figure 46 shows the recoveries of juveniles only, but if we put together the results for adult breeding birds we get the same pattern of broad-front migration. The only obvious difference is that older Ospreys are one to two weeks ahead of the juveniles in the average migration timetable. The temporal distribution of individual birds is, however, great: even in October there are still both young and adult Ospreys remaining north of latitude 50° N.

A few Ospreys stop and winter in the Mediterranean region, but the majority fly across the Sahara to tropical Africa (figure 47). The delta and inundation regions around the Senegal, Gambia, Volta and Niger Rivers are major wintering places. The winter recoveries of Finnish birds are on average farther east than those of Swedish-ringed birds; several recoveries have been made in the Congo Basin, for example.

The Osprey is found at coastal lagoons and inland waters virtually throughout tropical Africa. A large proportion of those in east and southeast Africa probably originate from Russian breeding sites. Recently recoveries from Zimbabwe have shown that Fennoscandian Ospreys also make their way there.

During the first summer the Ospreys remain in Africa. At two years of age most head northwards in the spring. The spring migration of second-years takes place about a month later than that of the adults; they do not therefore have time to breed that summer, but some of them locate a suitable breeding site and begin to build their large nest. Most Ospreys breed for the first time at three years of age. Approximately six out of ten individuals place their nests within 100 km of the site where they hatched (table 15); two out of ten breed more than 1000 km from their birthplace (for example the Swedish Ospreys that colonised Scotland). Once the Osprey has chosen a nesting site and built its nest, it returns to it year after year. This can be many years in succession – one ringed Osprey lived to 21 years.

Exceptionally, Ospreys may breed in tropical Africa. The continent abounds in lakes rich in fish and with apparently suitable nest-site facilities nearby. Both wintering and immature summering Ospreys obviously obtain a good living at the African fishing waters. It is therefore a mystery why the Osprey is not more numerous and better distributed as a breeding bird in Africa. The explanation possibly has something to do with the fact that it is in a situation of competition with the rather common African Fish Eagle.

One concluding question about the Osprey: What is the reason for the pronounced broad-front migration? The Ospreys are obviously dependent on a high success rate in their fishing at the sites where they stop off during the migration seasons. Maybe fishing success is affected by competition such that the Ospreys can gain much by spreading out and being all on their own at good fishing sites. Perhaps competition has a direct effect when Ospreys fish too close to each other? An interesting aspect for birdwatchers to investigate: How good is the Ospreys' fishing success (number of catches per unit of time) at a smallish lake or inlet when the birds are on their own and how good is it when there are several individuals present at the same time?

The Gannet (*Sula bassana*)

One can hardly fail to be impressed by the spectacle of Gannets fishing. When some Gannets have discovered a shoal of fish, the flashing white on the big birds as they dive headlong acts as a signal to the Gannets in the immediate waters. The birds quickly gather in large numbers and the scene becomes a chaotic mass of diving Gannets, Gannets disappearing beneath the waves and immediately floating up like corks, Gannets flying up only to plunge headlong once more. Gannets fish best in large aggregations. The massive attacks from the many birds probably upset the order in the shoals with the result that the fish become easily accessible.

I recall how engrossed I was myself by this spectacle when I was able for the first time to witness a party of fishing Gannets from a high cliff in the Shetlands. When the herring shoals are dense, so many Gannets can sometimes gather that even hardened fishermen are astonished. This is how the account from a herring fisherman at the banks between Scotland and Iceland goes:

> 'I could safely say that there wis a solid mass o' herrin' on these banks for aboot five miles. Ye could hardly see the sky wi' gannets. Ye wir almost afraid tae sail on it. No' an odd gannet hittin' here an' there – they wir comin' pourin' oot the sky lik' shrapnel the whole blessed day fae mornin' until night. A skyfu' o' gannets pourin' down . . .' (from *The Gannet* by Bryan Nelson, 1978, p. 226).

Gannets and fishermen are well acquainted with one another. The Gannets recognise those fishing boats which habitually move about in their waters, and approach in large flocks to inspect an unfamiliar boat. When fishing is good, the birds sometimes gorge themselves to such an extent that they are barely able to rise from the surface of the water. Fishermen in boats that arrive in waters where gorged Gannets are swimming around sometimes poke at a bird with an oar; the Gannet then regurgitates several fish, sometimes eight or nine, in order to relieve itself of some weight and be able to fly away. In this way it is possible to find out whether the Gannets have been catching herring or some other kind of fish.

In Europe, herring, sprats and sand-eels are important food for the Gannets. During the summer months, mackerel is the Gannets' favourite fish.

Gannets are found in both tropical and temperate seas and in both the northern and the southern hemispheres. Within the genus *Sula* there are nine different species; the biggest and heaviest, 3 kg in weight and with a wingspan of 2 m, is the North Atlantic species, the Gannet or Northern Gannet *Sula bassana* (the scientific specific name alludes to the classic colony on Bass Rock in the Firth of Forth, Scotland).

Today, Northern Gannets are distributed in 34 different breeding colonies with a total of approximately 213 000 pairs (figure 48). Most are in the eastern Atlantic; only about 50 000 pairs breed on the Canadian coast in the West Atlantic. Six colonies contain more than 10 000 breeding pairs; the largest of them all, where 60 000 pairs were counted in 1973, is on the remote island of St Kilda in the Outer Hebrides. Apart from the 400 000 plus breeding birds, there are, at a rough estimate, close on 100 000 non-breeding immatures. The sum total of Northern Gannets is therefore around half a million. Fortunately enough, the number of Northern Gannets has increased steadily throughout the 1900s (at the turn of the century, there were only about 60 000 pairs), and the population is well on the way to recovering from man's ravages of the breeding colonies during past centuries.

The migration often shows marked differences between immature and adult Northern Gannets. The immatures are responsible for the feat of making a fairly long journey from the breeding sites, 3000 km to 5000 km in distance, while a large proportion of the adults

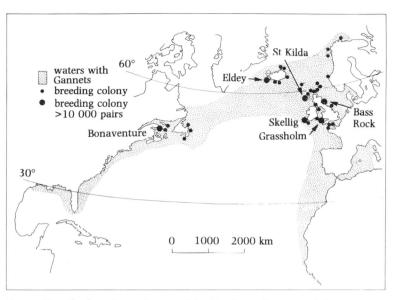

Figure 48   The breeding colonies of the Northern Gannet and the approximate limits of the species' distribution throughout the year. The traffic of Gannets across the open sea between the East and West Atlantic is probably very slight. Based on Nelson (1978).

disperse over wintering waters within a considerably closer range. The life cycle and migration of a European Gannet proceeds in the main according to the following pattern:

The Gannet pair's single young has a long fledging period, approximately three months. When it is ready to fly, at the end of August/beginning of September, it has become really fat and weighs half a kilogram more than its parents. In actual fact, when the young takes the big step of leaving the nest site on its own wings its powers of flight are not up to much. Having launched itself out from the nest on an elevated cliff ledge and got air beneath its wings, it is unable to maintain height but glides and alights on the sea only a few hundred metres from the nesting colony. Occasionally, however, the young do manage to get as far as 3 km on their first flight excursion. Young which have left the nesting ledge are chased off by adult birds in the vicinity, both in the breeding colony and when the young are resting on the water after their first flight. The young Gannets immediately begin to swim away from the breeding colony. The parents do not bother about them at all once they have left the nesting ledge, but stay behind at the breeding site right up to October. The young's extra fat is sufficient for it to get by for a week or two without eating. It swims for two or three days and covers up to 100 km. Then it must take to the air and learn to fish at the very soonest. The mortality during the first year of life is high, approximately 65%.

Juvenile Northern Gannets migrate to the waters off northwest Africa. There are many recoveries of ringed birds which show the migration routes (figure 49). The migration passes on both sides of Britain and sometimes over the Irish Sea, passes through the Bay of Biscay and rounds the Iberian Peninsula. A minority of young Gannets fly into the western Mediterranean, but the destination of the great mass is the northwest African Atlantic coast; some reach as far south as tropical coasts, to the coast of Senegal for example. Many Gannets make this migratory journey very quickly. A juvenile ringed in Scotland was reported again only 14 days later in Morocco. The young birds remain in African waters not only during the winter but also during the summer following it.

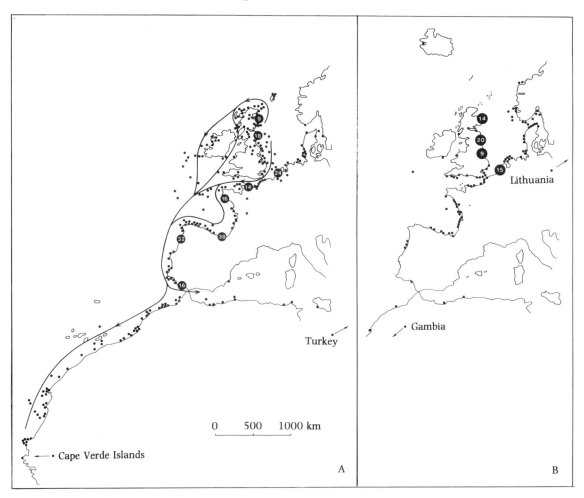

Figure 49   Recoveries of Northern Gannets ringed as juveniles in the breeding colony at Bass Rock (see figure 48). Each dot denotes one recovery and larger circles several recoveries as per the figures shown. A shows recoveries and main passage routes during the Gannet's first year of life. B shows recoveries during the fourth year of life, the year in which the birds start to breed. It is clear that young Gannets generally move farther south than older birds. The juveniles usually migrate towards the southern waters immediately after leaving the breeding colonies, and very few Northern Gannets visit the Skagerrak and Kattegat during their first year. These waters are visited regularly by Gannets of two years and older; one recovery shows that occasional individuals get as far as the Baltic Sea. Based on Nelson (1978).

The Northern Gannets return in their second year to the home colonies where, between fishing excursions, they rest at a general loafing site for non-breeding younger birds, two-, three- and four-year-olds. The time of the year when these younger Gannets are not at the home colonies they spend mostly in European waters: in the North Sea, in the Bay of Biscay and off the coast of Portugal. At four years of age the Gannets generally manage to secure a nesting site in the actual breeding colony. In the fifth year they are ready to follow the routine of the adults: the breeding site is occupied as early as January, egg-laying takes place in April, the breeding Gannets see the young of the year disappear in August–September, and themselves leave the colony in October.

Recoveries of older ringed Northern Gannets show that they are found during the winter mainly in the North Sea, the English Channel and the Bay of Biscay. It is more a case of dispersal from the breeding colonies rather than one of migration to a special wintering area. Adult Northern Gannets are, however, also seen on passage off northwest Spain and at Gibraltar. The English authority on the Northern Gannet, Bryan Nelson, is of the opinion that a fairly large proportion of the adults move south in October and November, to the waters off Portugal and northwest Africa. The movements and winter distribution of both the adult and the 'intermediate-age' Gannets in European waters are determined by the position of the large fish shoals. Unfortunately, studies and analyses of this association are lacking. At some headlands there are birdwatchers who keep a careful watch and count passing seabirds. How much more interesting and enlightening such data would be if at the same time we were to learn about the migrations of the fish!

The best observation points for those who wish to see large numbers of passing Gannets are Cape Clear at the southwest tip of Ireland and the northwest headland of Spain. Here, hundreds of Gannets per hour pass at times. The record at Cape Clear is as high as 4000 per hour. If one is content with smaller numbers, then the Danish and Swedish waters are also suitable for Gannet studies. Northern Gannets appear there all year around, but only in September and October are significant numbers present in the south Scandinavian waters (figure 50). At the Danish bird observatory of Blåvandshuk, more than two-thirds of Northern Gannets observed pass southwards. During early autumn, a large proportion of them are of 'intermediate' age. Not until the end of October do adult Northern Gannets in full white plumage (five years and older) become clearly predominant.

A surprising discovery made by Swedish migration watchers during the last 20 years is that there is a significant autumn passage of seabirds in the Kattegat. Particularly when low-pressure areas pass and associated strong westerly winds, often gale- or storm-force, are blowing, the seabirds migrate along the west coast of Sweden. Most follow a characteristic pattern of travel – the 'Kattegat loop'. This loop runs southward along the Swedish west coast, turns off at Kullen and veers northward over the western Kattegat. This flight pattern is probably accounted for by the fact that the strong winds make it advantageous for the birds to fly south along the west coast in oblique tail winds or cross winds and then to return with a certain amount of shelter from the wind behind Jutland. Some of the seabirds which follow the Kattegat loop have probably been wind-driven all the way from the North Sea during the autumn storms. In addition, there are certainly many that also visit the open Skagerrak and Kattegat in autumn when calm weather prevails and which are driven by fresh westerlies inshore and within sight of observers on land. Seabirds are well adapted to strong winds and are in full control of the situation during the passage through the Kattegat, where they have absolutely no trouble at all in avoiding being driven in overland by the westerly storms.

The Northern Gannet often follows the Kattegat loop. On windy days in September and October, from a good viewing spot on the west coast, one can in prime cases count several hundred passing Northern Gannets. Now this is not to say that all the Gannets follow the Kattegat loop; some birds also move northwards along the west coast of Sweden (in 1977, four out of ten Gannets were passing north off Gothenburg and one out of ten north off Kullen). Sometimes the Gannets also stop to fish. As at Blåvandshuk, 'intermediate-age' birds are in the majority in the Kattegat during September and at the beginning of October; adult Gannets are seen mainly during the latter half of October.

What is the reason for the Gannets' autumn visits to these waters? I should guess that they follow the migrations of the large herring shoals. There are two important populations of autumn-spawning herrings in the North Sea. During the early summer both are at the Great Fisher Bank. One population moves from there to the Scottish coast of the North Sea, where

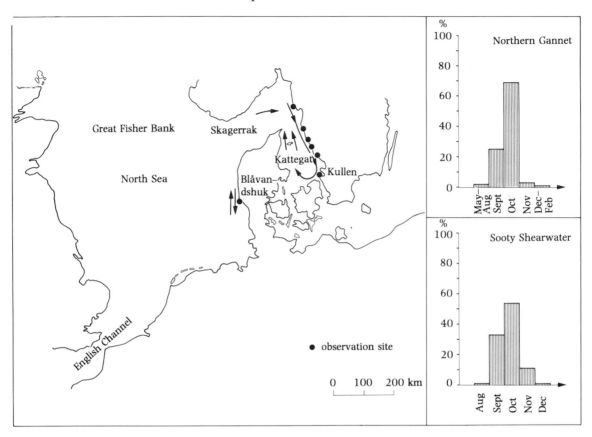

Figure 50   Counts of migrating seabirds in southern Scandinavia have been carried out at Blåvandshuk and at several sites on the Swedish west coast. In strong westerly winds, many seabirds are seen to follow the Kattegat loop: they pass south along the west coast of Sweden (often far offshore) and turn west at Kullen and North Sjaelland. The histograms show the monthly distributions for the two commonest fish-eating seabirds here, the Northern Gannet and the Sooty Shearwater. That for the Northern Gannet is based on birds observed at Kullen during 1970–1978 (a total of 1459 individuals) according to the journal *Fåglar i Skåne*; and that for the Sooty Shearwater is based on observations along the west coast of Sweden during 1960–1969 according to Pettersson & Unger (1972), and at Kullen in 1970–1978 according to the sources above (a total of 297 Sooty Shearwaters).

spawning takes place during late summer and early autumn. After spawning, this population migrates in autumn to the Skagerrak to spend the winter there. It is very likely that the Gannets follow the herring's autumn migration to the Skagerrak. In the Kattegat, there is in addition a separate population of herring which spawns in September and October. It may therefore very well be profitable for Gannets and other fish-eating seabirds to swing into the Kattegat.

The other large herring population migrates from Great Fisher to the southern North Sea to spawn during late autumn and early winter. After spawning, these herring migrate on to wintering waters in the English Channel. The Gannets at Blåvandshuk are perhaps mainly after these herring.

The interchange of Northern Gannets between European and Canadian breeding colonies is minimal. This is because the Canadian Gannets have their own migration routes. On the American side, too, the juveniles migrate farthest, occasionally 5000 km to Florida and the Gulf of Mexico. The adults do not move so far on average but, owing to the hard winter climate around the breeding colonies, they are forced to migrate several thousand kilometres to the southeast coasts of the United States. For this reason they do not return to the nesting sites until April; there is sometimes still a lot of snow remaining when the Gannets begin to incubate in May.

The 10 000-km cross-sea journeys of shearwaters.
The Manx Shearwater (*Puffinus puffinus*), **the Great Shearwater** (*Puffinus gravis*)
and the Sooty Shearwater (*Puffinus griseus*)

The shearwaters' method of feeding consists mainly in diving for fish immediately beneath the surface. Occasionally pelagic crustaceans and cephalopods are also caught. The oceans are the shearwaters' natural environment. In their characteristic gliding flight, which takes them down into the troughs of the waves and up over the crests, they are able to exploit fully the upwinds and the wind gradients over the sea. This method of gliding, which is practised by most oceanic birds, conserves a lot of energy and makes long-distance migratory journeys possible. It is part of the shearwaters' normal life to set off on journeys of tens of thousands of kilometres across the seas to good fishing waters. A shearwater which migrates 30 000 km annually has, when 30 years of age, covered 900 000 km over the seas. This is roughly the equivalent of 23 circuits around the globe! Our own human perception of the vastness of the oceans has no validity for the shearwaters.

The Atlantic race of the Manx Shearwater breeds mainly in Britain and Ireland, in the Faeroe Islands and in Iceland. The birds breed in large colonies on small, undisturbed islands. The best-known colonies, those on the islands of Skokholm and Skomer off the Welsh coast, hold approximately 40 000 and 100 000 breeding pairs, respectively. The total population of Manx Shearwaters in west Europe probably exceeds 300 000 pairs. The shearwaters nest in underground burrows which they either excavate themselves or take over from rabbits; they lay only one egg. They fly to and from the nesting burrows mainly at night. The female and the male take turns to incubate and each spell of incubation lasts about six days without a break. During the latter part of the development period of the young, the parent shearwaters visit the nesting burrow every night in order to feed their single offspring.

Those shearwaters which are sexually mature, birds of six years or older, return to the home colony after migration at the end of February or the beginning of March. Immediately after arrival an astonishing thing happens. While the males stay around the colony and remain in the burrows at night, the females set off again on a migratory journey. Females from the colonies in the Irish Sea fly 800 km south to the Bay of Biscay. There they remain for at least two weeks to fatten themselves up with food and to store nourishment for the egg which will soon be laid. The female shearwaters exploit the abundant presence of sardines during the spring in the Bay of Biscay. Shearwaters sometimes get caught in the sardine nets of fishing boats, and many ringing recoveries have therefore been reported during this period.

The sardine shoals migrate north in spring across the Bay of Biscay and the shearwaters of course accompany them. Towards the summer considerable numbers of sardines reach the waters off southern England and Ireland and are then within feeding distance of the shearwaters' breeding colonies. In May the female shearwaters return from the Bay of Biscay to lay their eggs. During the incubation period and the fledging period of the young, the adults, judging from ringing recoveries, forage in fishing waters within a distance of

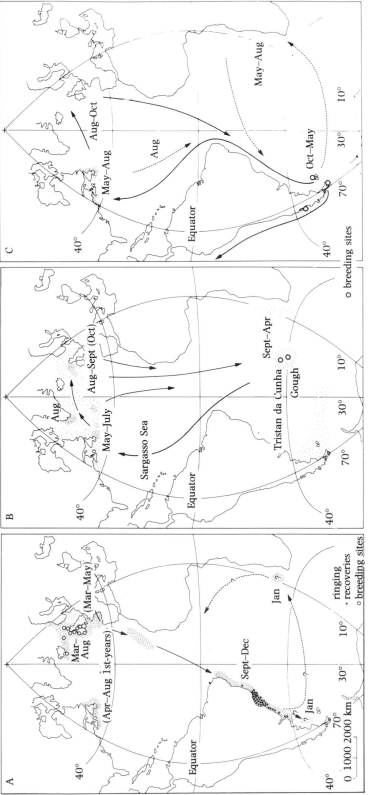

Figure 51 Migration routes and distribution of shearwaters at different times of the year. A. Manx Shearwater. B. Great Shearwater. C. Sooty Shearwater. The shearwaters' migration patterns are determined partly by the availability of fish in the surface layer of the sea and partly by the global wind pattern. The birds migrate in such a way that they can derive the maximum possible benefit from favourable tail winds. Based on Cramp & Simmons (1977), Perrins *et al.* (1973), Phillips (1963), Thomson (1965) and Voous & Wattel (1963) among others.

300–400 km from their nest sites. This is certainly no small feeding area: it takes about six hours for the birds to fly directly to fishing waters 300 km away. Imagine if a landbird with a nest in London were sometimes to fly to Paris, Lands End, North Yorkshire or Amsterdam in order to gather food for the young!

At the end of August the shearwaters have fed their young so much that it weighs appreciably more than the parents themselves. The latter at this point abandon the young in the nest burrow and quickly head south. Their journey takes them through the waters around the Canary Islands and the Cape Verde Islands; then the northeast trade-winds provide the shearwaters with a tail wind so that they can rapidly cross the Atlantic and the equator to the east coast of South America (figure 51). Between September and December the Manx Shearwaters spend their time mainly along the coast from Rio de Janeiro in the north to the Rio de la Plata in the south. Along these coasts, nutrient-rich water advances during this period with the Falklands Current, and the supply of fish in the surface water is good.

What happens to the abandoned young shearwaters in the nest burrows? Well, when the parent birds have departed, they still have another week or so before they are capable of flying. After several days they peep out from the nest burrows at night; then they come out each night during the following three or four days before they finally depart. Then off they go at high speed fast on the heels of the adults. A ringed juvenile was recovered at the Canaries, nearly 3000 km away, after only six days; another was found 9600 km away, in Brazil, after 16 days, and then it had probably been lying dead for two or three days before it was discovered. The Manx Shearwater really does make a tearing start in this world: immediately after it has fledged it flies 10 000 km away, and this in only a fortnight!

When the young leaves the breeding colony, it has on average 200 g of extra fat (out of a total weight of around 550 g). For a bird which flies using active wingbeats the whole time, this amount of 'fuel reserves' is sufficient for a maximum of 3000 km flying, but the shearwaters, with their energy-saving gliding flight, can probably get as far as South America without taking in food on the way. Some of the young leave the breeding site with much less surplus fat than the average. They must start fishing when still in European waters in order to get by. The average survival during the shearwaters' first year is about 30%, but for those which weigh less than 400 g when they leave the colony it is only around 10%. (Survival for adult shearwaters exceeds 90%.)

Both adult and young Manx Shearwaters thus spend the autumn and early winter off the coast of Brazil. In December, however, observations and ringing recoveries begin to decrease, and in January the shearwaters appear to have disappeared from there. Where they go to is still not entirely clear, although there are a number of January observations of Manx Shearwaters off the coasts of Argentina and South Africa which suggest that the following occurs. The shearwaters move south during the autumn along the South American coast in association with the fish-rich waters of the Falklands Current. When they get to south of about 40°S they arrive in the zone of westerlies. This may explain the lack of ringing recoveries. In the zone of easterly trade-winds dead shearwaters are washed ashore in South America, but within the westerlies belt dead birds drift out away from the coast. The shearwaters probably cross the Atlantic within the southern belt of westerlies and pass back into the trades zone at the nutrient-rich Benguela Current off southwest Africa. By flying in this way across the Atlantic, the shearwaters derive maximum benefit from the tail winds within the different zones.

Manx Shearwaters do not return to Europe during their first year of life but spend the summer mainly at the fertile fishing banks off Newfoundland, from where several ringing recoveries have been reported. A ringed first-year has also been found in south Australia. The

Manx Shearwaters of intermediate age, those of two to five years and not yet mature enough to breed, visit the natal colonies each summer.

For the sake of completeness, it should be mentioned that two races of the Manx Shearwater also breed in the Mediterranean Sea area (these are considered by some to constitute a separate species). The birds from the west Mediterranean migrate between July and October to waters off Morocco and Portugal and further to Biscay and the English Channel and sometimes also to the North Sea (there are even occasional observations from the Kattegat). The shearwaters first of all follow the migrations of the large sardine shoals; where they move to after October is unknown. The Manx Shearwaters in the east Mediterranean remain within that region throughout the year; large flocks assemble in the Black Sea during the early winter to fish for anchovies.

Now take the map of the Atlantic Manx Shearwater's migration pattern (figure 51A), and turn it upside-down so that the northern and southern hemispheres swap places. Look at it as if reversed such that the wind systems are correct (just get a mirror, or try to look at the map from the reverse side of the page). What you will see is the migration pattern of the Great Shearwater (cf. figure 51B). Here we are reminded of the whole chain of reasons behind it: of the earth's passage around the sun, of the energy flow from the sun which alternates symmetrically between the northern and southern hemispheres, of how temperature gains and temperature losses determine the winds, of how the winds determine the ocean currents and upwelling zones, of the importance of the temperature and the currents for the production in the sea, and of the associated variation in time and space in the occurrence of fish. The symmetry between the two hemispheres is maintained through all these physical and biological steps to the next link in the chain, the shearwaters' migrations on favourable winds to good fishing waters.

The Great Shearwater has three main breeding colonies, on Nightingale and Inaccessible, two islands in the Tristan da Cunha group, and on the nearby Gough Island. In spite of the small number of colonies, the Great Shearwater is a common bird: the number of breeding pairs exceeds 3 million. On the little island of Nightingale, which is no more than 1 km long, an estimated 2 million pairs exist.

The Great Shearwaters breed during the southern hemisphere's summer. In April the adults abandon their young in the nest burrows. They quickly migrate about 10 000 km to the fishing banks off Newfoundland, and after a while are followed by the young. The journey follows a quite well-defined migration route, past the horn of Brazil and across the Sargasso Sea. An observer on board a vessel travelling across the Sargasso Sea one day at the beginning of June estimated that, on that one day alone, 25 000 Great Shearwaters passed northwards on a 30-km-wide front.

During the early summer, the Great Shearwater is common at Newfoundland and Grand Bank, and also at the banks thousands of kilometres out in the Atlantic, Outer Bank and Midway Bank. The shearwaters are attracted in the first place by the large shoals of spawning capelin, a small salmonid fish 10–20 cm long which is found in great abundance in northern waters. In August, the Great Shearwater is common at the fishing banks off Greenland. A single flock of moulting shearwaters resting on the water immediately off the outermost skerries in southwest Greenland was estimated to contain at least 30 000 birds.

The Great Shearwaters make use of the westerly winds over the northern waters and during the course of the summer travel farther eastwards. Many reach Rockall Bank, northwest of Scotland, in August. There are still Great Shearwaters remaining in European waters right up to October. Here, the herring is an important food source. The adult birds leave the North Atlantic as early as August so as to arrive at the breeding localities at the end of the month or at the beginning of September. Great Shearwaters which are seen off west European coasts in September and October are therefore mainly intermediate-age birds which migrate at a more leisurely pace and do not arrive at their home colonies until the late spring in the southern hemisphere. During the period from December to April inclusive, the North Atlantic is practically empty of Great Shearwaters.

The migration of Great Shearwaters northeastward and eastward over the North Atlantic during the course of the summer should be seen as a shift in the central point of their occurrence. Great Shearwaters can still be found off Newfoundland, though not in any large aggregations, right up to October and November.

During the southern summer, intermediate-aged shearwaters often visit the Falklands Current and the Benguela Current. The whereabouts of the first-year birds during this period is not clear.

The Great Shearwater is a bird of the high seas which does not go near the coasts even during fierce storms. The Sooty Shearwater occupies a different niche. It often keeps near the coast and in very strong winds seeks shelter near the shores, sometimes even entering harbours.

The Sooty Shearwater breeds in two main areas: one in New Zealand, Tasmania and adjacent parts of Australia, and the other in southernmost South America. The population in the first of these areas migrates north in the Pacific Ocean to waters around Japan. Some of these Sooty Shearwaters continue eastwards to the west coast of North America and make a 'clockwise circuit' in the Pacific.

The majority of the South American shearwaters migrate along the Humboldt Current northwards, following the west side of the continent. They continue along the North American west coast to the nutrient-rich waters of the California Current off the coast of Canada, where they live during the northern summer. A smaller number of South American Sooty Shearwaters instead migrate up in the Atlantic. This migration, which is of greatest interest from a European angle, is shown in figure 51C.

Enormous numbers of Sooty Shearwaters are found in the colonies on the Falkland Islands, in the Cape Horn area and on the Chilean coast. Charles Darwin, who was not a man given to exaggeration, noted in January 1835 during his voyage with the *Beagle* off Chile:

> 'The Sooty Shearwater generally frequents the inland sounds in very large flocks: I do not think I ever saw so many birds of any other sort together, as I once saw of these behind the island of Chiloe. Hundreds of thousands flew in an irregular line for several hours in one direction. When part of the flock settled on the water the surface was blackened, and a noise proceeded from them as of human beings talking in the distance.' (from *The Voyage of the Beagle*).

In recent decades, off the Californian coast, immense flocks of Sooty Shearwaters have been encountered in the vicinity of unusually large shoals of sprats; the number of birds has been estimated at around 2 million.

In the Atlantic, no enormous gatherings of this kind are found. Despite the fact that the Sooty Shearwater is quite a common sight during the early summer at the Newfoundland fishing banks, the ratio between the number of Great Shearwaters and the number of Sooty Shearwaters is 100 to 1. There should at all events be several tens of thousands of Sooty Shearwaters migrating to the North Atlantic. So far as the geographical and temporal progress of the migration is concerned, the Sooty Shearwater is very reminiscent of the Great Shearwater. The former does not, however, visit the northernmost waters around Greenland and it generally keeps well in sight of the coast, in Europe often around the Faeroes, Ireland

and Scotland. Sooty Shearwaters also occur in the North Sea in considerable numbers. In the West Atlantic the Sooty Shearwater feeds largely on capelin and to some extent on cephalopods; the herring is the main food in Europe. The species also readily takes fish offal cast overboard from boats.

Adult Sooty Shearwaters return south as early as August. Birds of intermediate age remain behind in the North Atlantic, in particular in west Europe, right up to late autumn.

At the same time as the clockwise migration in the North Atlantic is taking place, an anti-clockwise circulation of Sooty Shearwaters is going on south of the equator. Sooty Shearwaters are regularly observed 'overwintering' off the southwest coast of Africa. The shearwaters probably fly there in tail winds within the southern belt of westerlies and then migrate back farther to the north assisted by easterly trade-winds.

In the autumn, shearwaters in the eastern North Sea follow the Kattegat loop (figure 50). The Great Shearwater is the least common here, having been seen on only a few occasions. This is of course due to its habit of keeping far offshore. For the Great Shearwater, the Kattegat is only a minor sea bay and is visited to a minimal extent. The Manx Shearwater is not seen particularly often in the Kattegat either, its main migration routes not passing by there. The Sooty Shearwater, however, turns up regularly in the Kattegat loop, in particular between September and November (figure 50). This means that most of the birds are of intermediate age.

Seeing the shearwaters in the North Sea and the Kattegat fires the imagination. These birds unite the familiar waters of northwest Europe with Newfoundland, the Cape of Good Hope, Tristan da Cunha, the Rio de la Plata and Cape Horn.

3.5   Birds which obtain food at the water's surface

The earth's surface is more than two-thirds covered by water. In the surface water large quantities of phyto- and zooplankton are produced. The seabirds in table 16 gather their food at the water's surface. They spend the greater part of their life on the vast expanses of the open sea and feed on various organisms in the surface water, on small crustaceans, cephalopods and jellyfish and on larval stages of various oceanic organisms. A particularly important food source is euphausiids (shrimp-like crustaceans), which occur in enormous shoals, mainly in the cold seas. They also, incidentally, form the principal food for many species of baleen whale, and together go under the name given them by the Norwegian whalers: krill.

Seabirds often follow ships and eat fish offal floating on the water's surface. Many species also take small fish. No clear dividing line exists between the species in table 16 and those in the previous category of 'birds which feed on fish'. In particular the Arctic Tern and the Long-tailed Skua are accomplished fish-catching birds which hover and plummet for fish in best 'tern fashion'. On a cursory acquaintance one may perhaps think that the Common Tern and the Arctic Tern are each other's ecological counterparts in different environments – the Common Tern mainly at lakes and the inner parts of coastal bays and the Arctic Tern in the uplands and in the outer bays. Closer study, however, reveals that the Common Tern really does live up to the name it has in some languages of 'Fishing Tern', and that the Arctic Tern is very flexible in its choice of food and readily picks insect larvae and small crustaceans from the water's surface. The Arctic Tern's marked seabird character is reflected most clearly in its migration habits. In contrast to the Common Tern, which is tied to inshore fishing waters, the Arctic Tern generally speaking lives out over the wide expanses of the open ocean. The Arctic Tern's graceful buoyant flight, its slender wings and its long tail reveal the adaptations of a seabird.

Table 16. *Birds which take food at the water's surface and which occur regularly in north Europe. Major wintering waters have been specified primarily for the Atlantic area. Many of the long-distance migrants in addition occur in very significant numbers in association with the Humboldt Current off the west coast of South America*

Species (NB = does not breed in Scandinavia)	Occurs in N Europe in winter	Important wintering areas: sea and coastal regions		Major wintering waters include
		North of 30°N	South of 30°N	
Storm Petrel NB			+	Benguela Current
Leach's Storm Petrel			+	Equatorial waters in the Atlantic
Fulmar	+	+		N Atlantic
Red-necked Phalarope			+	Arabian Sea
Grey Phalarope NB		(+)	+	Canaries, Benguela & Falklands Currents
Long-tailed Skua			+	?
Arctic Skua			+	Benguela & Falklands Currents
Pomarine Skua NB	(+)		+	Canaries, Benguela & Falklands Currents
Great Skua	(+)	+		N Atlantic
Sabine's Gull NB			+	Benguela Current
Little Gull		+		N Atlantic–Mediterranean
Kittiwake	+	+		N Atlantic
Black Tern			+	West coast of Africa
Arctic Tern			+	Antarctic Ocean
Little Auk	+	+		N Atlantic
Total no. species	3	5	10	

Gulls, too, sometimes take food at the surface, mainly in inshore areas. These gulls, however, feed on totally different types of food as well, largely on such kinds as they find on dry land; they are therefore placed in the category of omnivores (section 3.9).

The various techniques which the birds use to obtain food at the water's surface are shown in figure 52. Terns and the smaller sea gulls manoeuvre smartly in flight and snatch up the food without alighting on the water. The storm petrels have a singular ability to 'walk on the water' while in flight and to gather up food particles at a fast rate. In windy weather the dangling feet serve as an 'anchorage' in the water's surface so that the storm petrels, using their wings, are able to glide upwind and at the same time pick up food. The storm petrels thus work like miniature kites over the water's surface, with the legs and the feet as line and anchorage. To practise this feeding technique over a sea surface of breaking waves and ripples demands flying skills of a high order.

Fulmars readily alight on the water at fishing banks where a good deal of offal is thrown overboard from fishing vessels. They also seek out concentrations of shrimps or cephalopods, which they reach while swimming on the water or, when necessary, by diving, at the most to about 5 m depth. The Little Auk is specially adapted to dive a few metres for krill in the arctic waters near the ice limit. The feathers on the phalaropes' underparts are designed in such a way that they trap air; the birds are therefore floating on an air cushion as they swim in circles and pick up food.

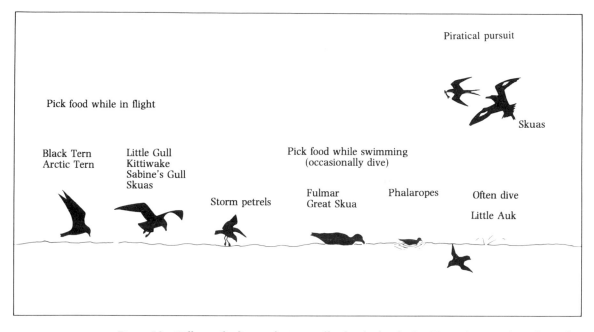

Figure 52 Different feeding techniques of birds which take food from the water's surface. The skuas often chase other seabirds to induce them to release their food. Sometimes the skuas also behave like raptors and catch, for example, phalaropes. Based on, among others, Ashmole (1971).

The skuas' feeding habits on the open sea are very imperfectly known. To some extent they probably obtain their food themselves from the water's surface. They also feed, however, by pursuing gulls, terns and shearwaters to induce them to release, or regurgitate, recently caught food. This habit is most pronounced in the Arctic Skua, which on migration often stays near Arctic (and other) Terns and subjects them to piratical attacks.

All of the species in table 16 live at sea during the winter and so are faced with the same situation as fish-eating birds and marine diving ducks: they must forsake large parts of their basic living environment in order to breed. Many of them obtain their food from the sea during the breeding season, too, and at this time they concentrate in immense colonies. These can contain thousands of nesting storm petrels (which, like shearwaters, nest in burrows), Fulmars, Kittiwakes or Little Auks. There are Kittiwake colonies numbering several hundred thousand pairs, and in the high-arctic bird-cliffs in Greenland, Spitsbergen or Novaya Zemlya there are millions of Little Auks.

Several of the species in table 16 change their feeding niche and during the summer live in entirely different surroundings from those during the winter. The Little Gull and the Black Tern change from a basic marine living environment to a freshwater breeding habitat. From its winter home in the North Atlantic, the Little Gull sets off perhaps for a marsh on a small island in the southern Baltic. The Black Tern leaves the waters off the west coast of Africa to breed at a lowland lake in south Scandinavia or the Netherlands. At inland lakes in southern and eastern Europe, White-winged Black and Whiskered Terns, close relatives of the Black Tern, breed. Both these species forage in a manner similar to that of the Black Tern. Interestingly enough, these two terns winter in freshwater environments, where they pick food from the water's surface on African rivers, lakes and marshes. What determines the

Black Tern's characteristic of being principally a marine bird during the winter, when the other species utilise a freshwater survival niche, is not known.

The phalaropes also make a radical change of environment following the winter. They leave the open sea to become mosquito-hunters at all the many small pools of the northern uplands and tundra. Swimming on an upland lake, the little phalaropes now and then spin around in a tight circle in order to stir up food and, while doing so, intensively pick after mosquito larvae.

When it comes to drastic changes of niche, the skuas are in a class of their own. The ecological roles of the Long-tailed Skua and the Pomarine Skua during the winter differ so completely from their ecological roles during the summer that there is probably no example to match them in the rest of the avian world. For these two skua species are transformed from seabirds into pure landbirds specialising in capturing lemmings and voles on the tundra and upland moorland. When hunting lemmings, the Long-tailed Skua hovers and then plunges down to peck its quarry to death with its bill. In contrast to owls, falcons and buzzards, the skuas do not therefore take their prey with the feet but use the bill as a weapon. Sometimes the lemming or vole that is being hunted manages to escape into an underground small-rodent passage: the Pomarine Skua may then spend a long time opening up the passage with its bill in order eventually to get at the prey. The skuas do not only eat small rodents, but also take food such as beetles, cranefly larvae and crowberries. Breeding success, however, is wholly dependent on the availability of small rodents, and in years when small-rodent populations have crashed the skuas do not breed at all but leave their breeding territories early in the summer having achieved nothing.

The Arctic Skua and the Great Skua breed in coastal situations, often in smallish colonies. During the breeding season they are opportunists in their choice of food. They pirate terns, gulls, auks and Gannets, they steal eggs and young from Kittiwakes and auks, and they take insects and berries and whatever else they can get hold of.

The above-mentioned changes of niche are of course a result of the considerable selection pressure on seabirds with extensive resources for survival: they must win breeding space. The species in table 16 exhibit the typical characteristics of birds with extensive opportunities for survival and limited breeding resources: high age of first breeding (often four years or older; in the Fulmar six to nine years or more), low breeding productivity, juveniles and immatures summering in the basic survival environments, and low mortality in full-grown adult birds.

The two phalarope species appear to be an exception. They are said to start breeding at only one year of age, something which is probably not universally valid judging from recurring reports of summering non-breeding phalaropes. The clutch consists of four eggs. One might perhaps be inclined to think that this has something to do with the birds' small size: the Red-necked Phalarope weighs around 30 g and the Grey Phalarope about 60 g. Yet the Storm Petrel weighs only 25 g and Leach's Storm Petrel only about 40 g, and these birds lay only one egg and do not start breeding until four or five years of age. No, the explanation has scarcely anything to do with size. A deeper insight into the phalaropes' life is required for us to be able to explain why their population dynamics differ from those of other seabirds. One important factor is possibly that the phalaropes, unlike other seabirds, have nidifugous young.

The species in table 16 exhibit widely differing migration habits. The Little Auk, which dives for food, remains in the North Atlantic. So, too, do the Fulmar and the Great Skua, which readily take advantage of fish offal. The fact that the other three skua species move far to the south appears natural as they are partly dependent on pirating surface-fishing birds, which in winter of course are found mainly in southerly waters. If we disregard these aspects,

I cannot find any clues to an explanation of the ecological factors that determine whether a seabird which takes food from the surface of the water is a long-distance or a short-distance migrant.

Why does the Arctic Tern depart for the Antarctic Ocean while the Kittiwake stays in the North Atlantic? And why do the Sabine's Gull and the Black Tern travel to the waters off Africa while the Little Gull is present in the European and North American parts of the Atlantic? This is difficult to explain. Sabine's Gull, a long-distance migrant, is a high-arctic species which often places its nest in a colony of Arctic Terns. Another high-arctic gull, Ross's Gull, also takes food from the surface and nests in association with Arctic Tern colonies. Its main breeding area is on the east Siberian tundra. During the winter, Ross's Gull is presumed to stay at open areas in the ice within the northern drift-ice and pack-ice region in the Bering Sea (compare the wintering habits of the Spectacled Eider, section 3.2). It is impossible to find any simple and straightforward reasons even for this drastic difference between the migration habits of Sabine's Gull and those of Ross's Gull. The species' feeding ecology during their time at sea has been very poorly researched.

The Long-tailed Skua's main wintering quarters are a mystery. Significant numbers of Arctic Skuas have been observed at the Falklands Current and at the Benguela Current. Also found at these currents is the Pomarine Skua, which occurs in large numbers at the Canaries Current as well. The Long-tailed Skua, on the other hand, has been reported only in small numbers from these areas; only from the Humboldt Current are there reports of any significant strength of numbers. Do Long-tailed Skuas migrate from the Atlantic to winter quarters in the Pacific Ocean, or are there after all some winter haunts thus far overlooked in the Atlantic area?

Among those birds which take food from the sea surface, there are examples of species which, like the Great and Sooty Shearwaters, migrate from breeding sites in the southern hemisphere to 'winter quarters' in the northern hemisphere. These species are not included in table 16 since they have not been encountered in the northern part of Europe. The best-known is a species in the storm petrel family: Wilson's Petrel. Its migration in the Atlantic is similar to that of the Sooty Shearwater. Wilson's Petrel breeds at the southern tip of South America and along antarctic coasts and between April and September migrates in a clockwise direction over the Atlantic. During the late summer it is fairly common in Europe north to the Bay of Biscay and off the Iberian peninsula. It prefers to stay well out from the coast. A smaller number of Wilson's Petrels regularly visit the waters of northwest Europe, in the Western Approaches and off western Ireland, but these have been detected only recently, during pelagic trips organised by seabird enthusiasts (although as long ago as 1839 John Gould recorded them 'in abundance' off Land's End). Another southern species that migrates to the northern hemisphere is the South Polar Skua; I shall return to this below.

When it comes to fitting in the moult in the migration timetable, the species in table 16 have three different basic strategies.

1 Five species, which are 'short-distance migrants' and winter in the North Atlantic, moult the flight feathers during a limited period immediately after breeding. Typical months for moulting are August, September and October. The fastest moult is that of the Little Auk, which sheds all the flight feathers simultaneously and at this time loses the power of flight. The process takes longest in the Great Skua, which often has not completed the moult of the wing feathers until towards January. The Little Gull normally completes the moult around September/October time. This is particularly interesting since there is a double peak in the Little Gull's autumn passage in

Scandinavia and the Baltic Sea area: Little Gull migration is already in progress in July and August, that is at the beginning of the moult, and a new migration wave occurs in October and November, after the adults have completed the moult (compare the similar situation in the Black-headed and Common Gulls, section 3.9).

2 The Black Tern and the storm petrels moult the flight feathers at a slow rate. The moult begins after breeding and continues through to the winter or, in the storm petrels, right up to March and April. The Black Tern follows the pattern of continuous waves of moult which was described in the previous section. The storm petrels change their feathers once a year in straightforward descendent sequence from the innermost to the outermost primaries. The slow rate of the moult results in a minimal effect on the birds' powers of flight.

3 The remaining long-distance migrants moult in the winter quarters. Typical months for moulting are December, January and February. The strategy is the same as that of the shearwaters, which was described in the previous section. As in the shearwaters, the moult in some species, the Arctic Tern among others, proceeds so quickly that the birds' flight is impeded. During this time, it is of course vital for the birds to have ample access to easily caught food.

The Kittiwake (*Rissa tridactyla*)

If we look up the Kittiwake in a bird book, then we find that it looks practically identical to the Common Gull. But my word, what a difference there is when we see the bird in real life! Certainly the Common Gull is an accomplished flyer; not least it is skilled at soaring in thermals over land. But the Kittiwake above the waves of the sea is something entirely different: its flight is light, graceful and airy, and with great elegance it picks up food from the surface of a stormy sea in flight as easily as blinking.

The Kittiwake has a circumpolar breeding range in the northern hemisphere; the distribution in the Atlantic area is shown in figure 53. In Sweden there is a minor breeding colony off the southwest coast, and on the Danish side of the Kattegat there are further colonies on small islands in the northwest; in Britain and Ireland there are many colonies, some containing tens of thousands of pairs. Some of the largest bird-cliffs, however, are in Norway, especially on Rundö and in the North Cape region, with colonies of hundreds of thousands of breeding Kittiwakes.

The Kittiwake is present all year in the Kattegat. It is especially numerous from September to November, when the food supply in the Kattegat is presumably at its richest. When the birds follow the Kattegat loop (see figure 50), thousands of Kittiwakes, both adults and immatures, can be counted on stormy days by birdwatchers on the Swedish west coast. Record figures for the number passing on one day approach 14 000.

During the winter the Kittiwake lives a pelagic life and is widespread over the whole of the North Atlantic between about 40° N and the ice limit. Ringing studies have shown that birds from different breeding sites are dispersed over vast areas. Thus many European Kittiwakes, in particular juveniles, migrate across the Atlantic to the New World (in Newfoundland, recoveries have been reported of Kittiwakes from, for example, Iceland, the Faeroes, Britain, Denmark, Norway and Russia). In the West Atlantic the European Kittiwakes mix with those from Canada and Greenland. There are also ringing recoveries which show that West Atlantic Kittiwakes, during their wanderings over the sea, sometimes reach European waters and there mingle with local birds (recoveries of west Greenland Kittiwakes have been made in Britain, the Netherlands and France). According to observations from ships, Kittiwakes spend the winter not only on the seas over the European and North American continental shelves

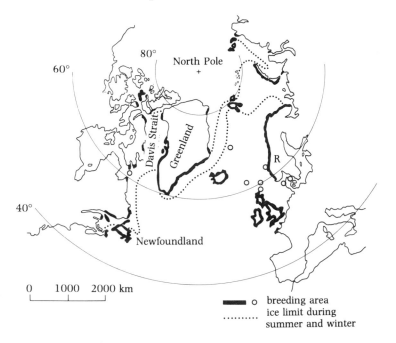

Figure 53   The breeding distribution of the Kittiwake in the North Atlantic. The breeding range extends right up to the drift-ice zone. During the winter the birds live on the open sea between 40°N and the ice limit. Particularly important wintering waters exist off Newfoundland. Many birds of intermediate age (one to three years) visit the Davis Strait off west Greenland during the summer, May–October. Large numbers of Kittiwakes have been ringed as young at the breeding sites in Britain, Norway (mainly on Rundö=R) and in northern Russia (mainly the Kola Peninsula).

but also right out in the middle of the open North Atlantic. Recoveries of ringed birds, which are of course made mainly on the coasts, do not therefore reflect the Kittiwake's distribution particularly well. If we bear this reservation in mind, however, the ringing material provides a great deal of valuable information on the course of events involved in the migration.

A calculation of the mean distance of ringing recoveries for Kittiwakes of different ages (figure 54) shows clearly that birds in their first three years of life, when they are not yet breeding, generally find themselves much farther from the home colony than the adults (four years and older). A lot of the second- and third-year birds visit the waters of their home colony for a short period during the summer; this results in the mean distance for these age classes in figure 54 being less in summer than in winter. This difference between seasons of course applies to a marked extent to adult birds, which in summer are present at the breeding colony and during the winter months lead a pelagic existence in surrounding waters. The gulls demonstrate great site fidelity and usually breed within or very close to the colony where they were born. The mean distance from breeding site to winter recovery of adult British birds, about 700 km, is, however, probably misleadingly short since adult Kittiwakes, to a greater extent than immatures, keep to open waters far offshore, from where ringing recoveries are lacking.

Recoveries of European Kittiwakes in Newfoundland and the Davis Strait off Greenland confirm that by far the majority consist of non-breeding immatures. Only 6% of recoveries involve birds that are four years or older (table 17). It is above all during summer and autumn, from May to October inclusive, that the gulls are found in Greenland waters. In Newfoundland, on the other hand, they appear mainly during late autumn and winter, mostly on the island's ice-free south coast and at the large fishing banks, the Grand Banks. In these waters large numbers of Greenland-breeding Kittiwakes also winter during their first year of life; older birds from Greenland are more oceanic during the winter, living in the northernmost Atlantic.

The young Kittiwakes usually fledge in July. It is a regular event for some juveniles from Europe to reach Greenland as early as September and October (6% of all ringing recoveries according to table 17). These of course involve mainly young coming from west Europe (especially from Iceland) and to a lesser extent those coming from the remote areas of north Russia (less than 2% of Greenland recoveries in the first autumn). From the middle of October and onwards the young birds arrive in large numbers in Newfoundland. Almost half of all European Kittiwakes recovered there are birds which are experiencing their first winter. In Greenland recoveries of first-year birds predominate during the following summer. The age

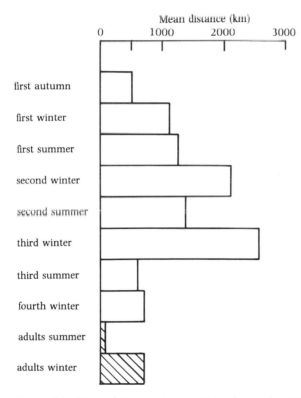

Figure 54 Mean distances from natal colonies for Kittiwakes of different ages based on recoveries of birds ringed in Britain. The first autumn covers the period July–September, the winter October–March and the summer April–September. The Kittiwakes are largely pelagic in winter in the North Atlantic; during the second and third years of life they are farthest away from the home colonies. As a rule they start to breed at four years of age. The scatter around the mean distance is usually very wide, with accumulations of recoveries in the East Atlantic, most often within 1000 km, and on the coasts in the West Atlantic, 3000–4000 km away. Based on Coulson (1966).

Table 17. *Age distribution of ringed European Kittiwakes (mainly from Russia, Norway and Britain) recovered in Greenland and Newfoundland. Recoveries in Greenland are principally from the summer months, May–October; Newfoundland recoveries mostly derive from late autumn and winter, October February. Some Kittiwakes reach Greenland in September and October of the same year that they hatched. Based on Salomonsen (1967) and Tuck (1971)*

Average age (years)	Proportion Greenland ($n = 206$)	Proportion Newfoundland ($n = 87$)
0.25	6%	
0.5		47%
1	67%	
1.5		28%
2	18%	
2.5		16%
3	3%	
3.5		3%
4 and older	6%	6%

distribution among those Kittiwakes which have been recovered in the New World is by and large the same whether the birds have been ringed in Britain, in Norway or in Russia. If, on the other hand, we compare the proportion of recoveries from Greenland with the proportion from Newfoundland for the various populations, then we find clear differences (table 18). For British birds there are approximately as many recoveries from Greenland as from Newfoundland, but 81% of recoveries of Russian birds are from Greenland. The Norwegian population (the ringed birds originate from Rundö), which breeds roughly midway between the British population and the Russian Kola Peninsula population, ends up in a neat in-between situation: 65% of the recoveries are from Greenland.

The Kittiwakes' journeys of 3000–4000 km westward across the Atlantic are impressive, not least if we bear in mind that the migration takes place within the westerlies belt with its constant passages of depressions and strong head winds. Notwithstanding this, however, in the seabirds tribe the Kittiwake is regarded as a 'short-distance migrant'.

The phalaropes (Red-necked Phalarope *Phalaropus lobatus* and Grey Phalarope *P. fulicarius*)

In winter the phalaropes lead a pelagic existence, and at this time they display great versatility in exploiting various feeding situations. They forage in drifting masses of seaweed, and are assisted by turbulence in the water; this is because the turbulence brings plankton up within their reach as they swim around like cork floats on the water's surface. They are therefore attracted to large shoals of fish where the fishes' movements make the surface water circulate. Sometimes the phalaropes take the opportunity to pick up food in the wake of boats. The Grey Phalarope occasionally follows the paths of the large baleen whales; from the northern hemisphere there are observations of Grey Phalaropes associating with white whales and walruses. The phalaropes are sometimes seen sitting and pecking at the backs of whales. The record for initiative was set by the Grey Phalarope which was seen on the back of

Table 18. *Number of recoveries in the New World of Kittiwakes ringed in Britain, Norway (Rundö) and the Soviet Union (Kola Peninsula), from Tuck (1971) and Salomonsen (1967). Birds from northern European breeding sites on average keep to more northerly waters than birds from southern European colonies during their stay on the western side of the Atlantic*

	Recoveries in the New World		
Country where ringed	Newfoundland	Greenland	Total
Britain	39	38 (49%)	77
Norway	18	34 (65%)	52
Soviet Union	30	126 (81%)	156

a killer whale; the bird was shot, and its stomach was found to be full of different lice of a rare species which lives just in the skin of the killer whale.

The Red-necked Phalarope, which often stops off at shallow shores during migration, sometimes feeds right next to ducks and grebes. Here the phalarope avails itself of the chance to take food that is stirred up by the larger birds' movements.

During the breeding season, the Red-necked Phalarope is a first-rate specialist in taking mosquito larvae on small waters in the birch and willow region of the mountains and tundra. Owing to its fearlessness, it is probably well known to all hill-walkers within its breeding range. It is no doubt best known for the fact that the female is much more brightly coloured in summer plumage than the male. The Red-necked Phalarope has three major wintering areas: the northern parts of the Humboldt Current and adjacent waters around the Tropical Convergence, the Arabian Sea, and also the southwest Pacific Ocean between Japan and New Guinea. The extent of this last area is poorly known (figure 55). There are also reports of Red-necked Phalaropes from the Canaries Current off West Africa, but it is most likely that these are due to confusion with the Grey Phalarope, which has an important and well-documented wintering area there; the situation has not, however, been investigated in any detail, and the possibility that some Red-necked Phalaropes may also winter in these waters should be borne in mind.

Ringing of Red-necked Phalaropes in Scandinavia has provided 18 recoveries during the migration period (all but one during the autumn). Without exception they show that the birds depart southeastwards straight across east Europe, heading for the Arabian Sea. Most birds have been ringed at the Varanger peninsula in the extreme north of Norway. Two of the phalaropes recovered (found on the Caspian Sea coast) were ringed during the migration period in Halland in southwest Sweden. Red-necked Phalaropes are quite scarce in south Sweden and Denmark on migration; the birds involved are probably from the north Scottish islands, the Faeroes and Iceland and pass through there on route to and from the Arabian Sea.

Female Red-necked Phalaropes leave the breeding grounds early and pass through the Black Sea and the Caspian Sea as early as August. In the same month they also begin to arrive in the Arabian Sea. Males and juveniles follow later, in September. The return passage takes place mostly in May, at which time resting flocks of hundreds of birds can be seen in the coastland of north Sweden. Wintering numbers in the Arabian Sea can reach fantastic proportions. A Dutch sea captain has described a voyage between the Gulf of Aden and the Gulf of Oman: on 21 January, throughout the day from dawn to dusk, the vessel passed flock

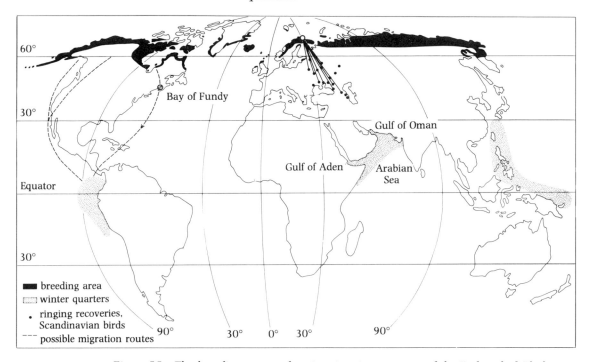

Figure 55   The breeding area and main wintering quarters of the Red-necked Phalarope. Recoveries of birds ringed in Norway, Finland and Sweden are also shown. Eleven of the birds recovered were ringed at the Varanger Peninsula (this site is connected to recoveries by straight lines on the map). The Scandinavian birds migrate straight across east Europe and the Middle East to winter quarters in the Arabian Sea. Based on Glutz von Blotzheim *et al.* (1977) and Schiemann (1972).

upon flock of Red-necked Phalaropes; altogether there must have been at the very least 100 000.

A similar observation from another corner of the world – off the Pacific coast of north Mexico – is recounted by Kai Curry-Lindahl in his book *Fåglar över Land och Hav* ('Birds over Land and Sea') (1975):

> 'On the lee side beneath the 200-m plus cliffs of the Coronado Islands, the sea was calm but not flat, for here it was alive with Red-necked Phalaropes on migration from arctic tundras to warm oceans. Immense flocks of tens of thousands of phalaropes flew up like huge clouds as our boat literally ploughed its way through the living mass. Through the binoculars even bigger flocks were detected, in many places covering the sea like a mat. It was impossible even to attempt to make an estimate of the birds' numbers; they must have involved a hundred thousand.'

I have still not exhausted reports of large flocks of phalaropes. Here is the best of all. Red-necked Phalaropes from large parts of northeast Canada gather on migration in August in the waters around the Nova Scotia peninsula. The concentrations can become enormous, especially in the Bay of Fundy (figure 55). From the 1930s and 1940s there are estimates of 250 000 phalaropes. In more recent years there have been attempts to make reliable estimates of phalarope numbers in the Bay of Fundy: on 21 and 22 August 1974 about

500 000 were recorded, and on 21 August 1977 the number was estimated at over 2 million! It is not really possible to imagine what a fantastic spectacle that must be.

Amazingly enough, the phalaropes' onward migration routes from the Bay of Fundy are unknown. Some ornithologists have put forward the hypothesis that they would fly straight across the Atlantic to presumed winter quarters in the Canaries Current. On the map in figure 55 I have indicated another suggestion. I assume that the birds derive a certain advantage from the frequently occurring strong northwesterly winds around Nova Scotia and fly out over the Atlantic, and then, when they reach the trade-winds belt of easterlies, turn off in a sweep which takes them across the West Indies and Central America to winter quarters in the Pacific Ocean. This suggestion is not based entirely on pure imaginings – the arc-shaped migration route over the West Atlantic is taken by many waders and even some small passerines in the autumn. They fly direct from the Nova Scotia area to South America. I shall give more details about this incredible long-distance flight, which has been mapped through extensive radar studies, in section 4.6.

In inland western North America Red-necked Phalaropes are seen regularly during the migration season. Obviously a considerable passage takes place there thousands of kilometres overland, exactly as in east Europe (see figure 55).

The Grey Phalarope is a more markedly arctic-breeding bird than the Red-necked Phalarope and is found in the tundra regions on the Arctic Ocean coast. The breeding area is circumpolar. The nearest the species occurs to northern Europe is in Greenland, Iceland, Spitsbergen and Novaya Zemlya. After hatching, Grey Phalaropes gather in the drift-ice zone of the Arctic Ocean, where flocks of thousands of birds can be seen in July and August. During this period the birds exchange their handsome red summer plumage for a winter plumage of gull- or tern-like grey and white. The majority migrate south in September, but there are still birds remaining in the Arctic Ocean in October. The migration takes place almost exclusively over the sea. The main wintering seas are the Humboldt Current off Chile, the Falklands Current as far south as Tierra del Fuego, and the Benguela Current. An important wintering area exists also in the northern hemisphere at the Cape Verde Islands in the waters of the Canaries Current. Occasional individuals winter here and there in even more northerly areas, including the coasts of west Europe, and wintering Grey Phalaropes have even been recorded a few times in Sweden. The likelihood of being able to see this bird in Britain and northern Europe, however, is greatest during the autumn migration in October, when westerly winds sometimes drive some of the phalaropes passing over the northeast Atlantic on to the shores of northwest Europe. On rare occasions, such as during the violent storm that hit southern Britain in October 1987, exceptional numbers may be blown ashore and even well inland.

From the North Pole to the South Pole with the Arctic Tern (*Sterna paradisaea*)
Let us transport ourselves to springtime on the southernmost tip of Africa and share in Gustaf Rudebeck's observations of the terns off the coast at Cape Town (Rudebeck 1957).

'1950. Oct. 20–30. Sea Point, Cape Town. Most days, large flocks of *Sterna hirundo* and/or *macrura* [= *paradisaea*] were observed from the shore, though rarely at close quarters. Often several hundred specimens were noted, but just as often it was obvious that still larger numbers were present farther out, at such a distance that they were barely visible from land. – The terns frequently congregated in very dense flocks of 50–100 specimens, which showed a remarkable behaviour, similar to that of packs of small waders such as *Calidris* spp.: They moved very swiftly, dashing and dodging now to one side, now to the other, sometimes breaking up into smaller units,

&c. On some occasions I also noted that such flocks suddenly changed their behaviour and flew off in their usual quiet manner towards the south. It seems probable that a considerable migration was going on, many flocks continuing their journey, while others arrived from the north; but the terns were usually much too far out to allow of detailed observations.'

Gustaf Rudebeck's caution in the matter of identifying the species is due to the fact that Common Terns and Arctic Terns in winter plumage are extremely difficult to tell apart in the field. Common Terns 'winter' commonly off the coast of South Africa. Those terns which migrate past and continue southwards are Arctic Terns. The passage past the Cape can carry on right in to early December. In March the Arctic Terns reappear there; this time they are passing on northward migration towards the Arctic.

What an amazing experience to find oneself as far south as Africa's southern tip and to see one of the northernmost breeding birds in the world migrating past! When, between October and December, the Arctic Terns leave the southern tip of Africa behind them, they enter the southern belt of westerlies where depressions pass daily and very strong winds blow. Their migration destination is the edge of the pack-ice in the Southern Ocean.

The Arctic Tern regularly makes the longest migratory journey of all bird species. It has a circumpolar breeding distribution in the northern hemisphere as far northward as there is land at all (figure 56). Breeding discoveries have in fact been made at the world's most northerly headland, Cape Morris Jesup in Greenland, almost 84° N. At the other end of the world, in the Antarctic, the Arctic Tern has been encountered south to 78° S. From the breeding colonies on northern shores, islands, tundra moorlands and fiords, Arctic Terns assemble for a grand migration over the seas. The main route of the autumn passage runs south off the eastern coasts of the Atlantic. This route is taken not only by birds from Europe and west Siberia, but also by those from Greenland and large parts of North America; Arctic Terns from the New World migrate across the Atlantic within the northern belt of westerlies, and off Britain and France join up with the birds migrating along the eastern shores of the Atlantic. A minority of the terns leave Africa's northwest coast, cross the Atlantic within the trade-winds zone, and continue south along the east coast of South America. Migrating Arctic Terns have been observed off Argentina.

A less important migration route runs down the Pacific Ocean off the west coasts of the two American continents. The waters of the Humboldt Current are a well-known 'summering centre' for first-year, non-breeding Arctic Terns. It is therefore possible that the at times intense passage southward off South America's west coast in Peru and Chile consists in the main of immature terns leaving their summering sites.

The Arctic Tern's migration pattern is unclear in several respects, owing to the terns' habit of travelling over the sea far out of sight from land. There is of course uncertainty in particular about what happens when the birds lose contact with southernmost South America and

Figure 56 Distribution and migration of the Arctic Tern. A. General picture, where migration paths that have not been definitively proven have been indicated by question marks. B. The migration routes to wintering areas along the pack-ice limit in the Antarctic Ocean. Most Arctic Terns arrive via Africa's west coast, but significant southward migration also takes place off South American coasts. A large proportion, perhaps most, of those terns which come via the waters of the Humboldt Current are first-years leaving their summering quarters. The terns arrive at the pack-ice mainly in November and December. A small proportion winter in waters off southernmost Africa and South America. C. The migration

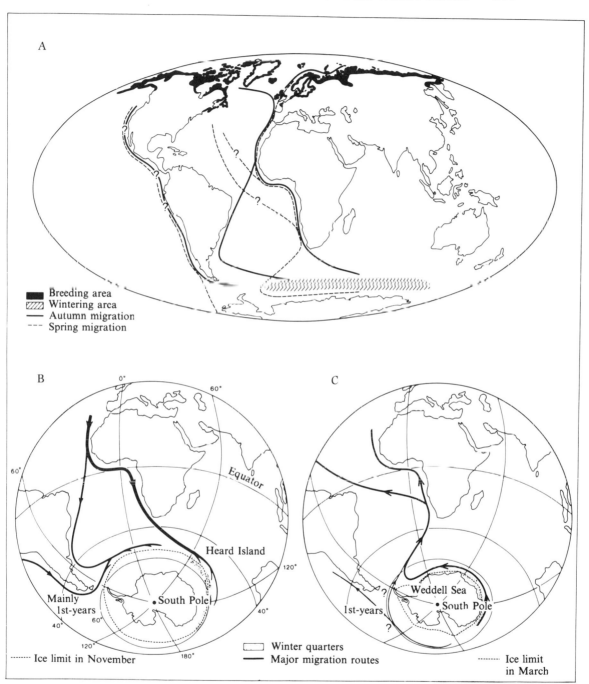

routes from the wintering areas in March. Adult and intermediate-age terns migrate mainly westwards off the Antarctic coast before continuing north in the Atlantic. There are indications that first-years in particular migrate around the South Pole to summer on the Humboldt Current. A theoretical possibility is also that those terns which winter farthest east of all do not migrate west along the edge of the ice but travel around the South Pole before turning northwards in the Atlantic. Based on Salomonsen (1967).

Africa and continue south across the open sea. Finn Salomonsen (1967) has attempted to put together available observations to form a total picture:

In November, the edge of the pack-ice in the Antarctic Ocean is roughly 3000 km south of the southern tip of Africa. The large numbers of Arctic Terns which pass southernmost Africa do not, however, fly a straight course in that direction but make use of the fresh westerlies in the southern middle latitudes to reach the edge of the ice between longitudes 30° and 150° E; the main area lies between 60° and 120° (see figure 56). From Heard Island, in November less than 1000 km from the edge of the ice, accurate observations have been reported of considerable Arctic Tern passage: the earliest date for Arctic Terns passing is 22 October; during November and December the terns pass daily; and the passage peters out in the first week of January. As a rule the terns do not stop for any length of time around Heard Island, perhaps just the odd day, before the flocks move on southeastwards. The passage probably takes place within a fairly narrow corridor between the island and the edge of the ice, for at Kerguelen Island, which lies only 500 km north of Heard Island, there is no significant migration of Arctic Terns.

The mean wind direction within the southern belt of westerlies is northwest to west at surface level, and the wind strength averages 10 m/s. The Arctic Terns accordingly have very favourable tail winds during their southeastward migration. The biggest problem for them is no doubt to avoid being drifted too far eastwards by the wind. Violent storms occasionally advance across the 'roaring forties' and the 'howling fifties'. There are many instances of Arctic Terns having been wind-blown to Australia, Tasmania and New Zealand.

The Arctic Terns which migrate via South America exploit the west winds south of Cape Horn and reach the ice-limit at approximately 30° W and east thereof. To all appearances it is very rare for Arctic Terns to winter west of the Antarctic Peninsula (longitude 60° W).

At the limit of the pack-ice in the Antarctic Ocean there are large inlets, bays and holes in the ice at the 'ice-coast' with lots of ice-floes. Vessels which attempt to penetrate farther south give up after only a kilometre or two as the ice becomes too solid. Scattered channels with open water exist within the pack-ice zone, but these do not seem to be of any great importance to the Arctic Terns. The narrow zone at the edge of the pack-ice is the terns' chief wintering environment. Various research vessels have reported on their behaviour. The terns keep together in flocks of varying size, from ten individuals to many thousands. Often the birds sit and rest on the ice; particularly in rough storms, they may gather in enormous flocks which at a distance look like big dark patches of shadow on the ice. During a two-month period in January and February the terns moult their flight feathers, and at this time their flight is clearly impeded. In this they resemble the Great Shearwater, though that species' moult takes place in arctic waters in the northern hemisphere. It is of course vital for the terns to be able to find easily obtainable food during the moult period. In the Southern Ocean there are large amounts of krill which can be fished from the water between the ice-floes. Those terns which are experiencing their first year of life do not have such a tight timetable as the adult birds; most of them will not return to the northern hemisphere during the following summer. The immature terns therefore do not begin to moult until February, and the moult is not completed until April.

During the terns' stay at the edge of the pack-ice one important thing happens: the ice melts! The extent of the pack-ice is at its greatest in November, roughly at the same time as the first Arctic Terns arrive. As shown in figure 56, the ice-limit at that time is on average between 1000 km and 2000 km off the Antarctic continent. During the summer thaw the pack-ice contracts, reaching its smallest extent in March, when in many places it is right up against the continental coast of Antarctica. In November the ice-limit was out in the windy westerlies belt; from there it shifts to the considerably calmer polar belt of easterlies. Throughout this period the Arctic Terns are found at the melting ice edge. One recovery, incidentally, has been made in February at the pack-ice just off Wilkes Land (112° E); the tern had been ringed on the Danish island of Saltholm off Copenhagen, a long-distance recovery worthy of the appellation.

The adult Arctic Terns begin their return migration in the first days of March. They then make use of the polar east winds and fly westwards along the edge of the pack-ice and the coast of Antarctica, where they find good sites for stopping off and fishing. A real problem now manifests itself: the birds must soon cross the westerlies belt, several thousand kilometres broad, over the open sea. The terns' solution to this problem is that they make a detour along the edge of the ice, all the way to the Weddell Sea, and from there head off northeastwards through the westerlies belt and in this way once more derive advantage from tail winds. Everything indicates that this flight route is used by the clear majority of adult terns, which thus migrate north to the breeding sites via the Atlantic. The same probably applies to intermediate-age birds (second- and third-years), which are not yet breeding but which often visit the northern breeding area during the summer.

First-year birds, however, migrate in large flocks to the waters of the Humboldt Current in the South Pacific, where they summer. How do the young terns find their way there from the edge of the pack-ice? To begin with, it would seem very likely that they continue the migration westwards along the edge of the ice to the Antarctic Peninsula and that on route from there they cross the Drake Passage to the southern tip of South America and then travel farther along the continent's west coast. The Drake Passage is, however, one of the world's most notorious thoroughfares; whining westerly gales blow there all the time. It must be perilous for the terns to try to cross the Drake Passage in head winds or cross winds.

In that case it is more plausible to imagine that the young terns do not migrate west at all from the wintering areas as the adults do, but that instead they move eastwards around the South Pole (!) so as to be able to cross the westerlies belt and proceed to the coast of Chile in the direction of tail winds. This interpretation is supported by a number of observations in February and March of Arctic Terns west of the Antarctic Peninsula (between roughly 70° and 130° W), in other words in areas where wintering Arctic Terns are otherwise virtually completely absent.

Some immatures get caught out by wind-drift and end up in odd summering waters, as for instance the tern which was ringed as a juvenile in the Stockholm archipelago and was recovered in June the next year in southwest Australia.

Much still remains to be confirmed and discovered regarding the Arctic Tern's fantastic migration. It would be especially interesting were we to acquire a better knowledge of the species' ecological role in the Antarctic. There is a local tern, the Antarctic Tern (*Sterna vittata*), which is extraordinarily similar to the Arctic Tern. The Antarctic Tern occurs mainly at islands and relatively snow-free and ice-free coasts of Antarctica, e.g. at the Antarctic Peninsula, where it fishes for krill over wide areas of open water; the Arctic Tern, on the other hand, is found at ice-floes along the limit of the pack-ice. There are clear signs of competition between the two species. The Antarctic Tern breeds mainly where wintering Arctic Terns are scarce. Or is it the case that the Arctic Terns prefer to winter where there are few Antarctic Terns? At Heard Island the species seem to come into conflict: the Antarctic Terns do not start to breed there before January, by which time the Arctic Terns have migrated past the island.

North of the Arctic Circle, it is possible to stand and watch the Arctic Tern in summer, in the light of the midnight sun, flying around and picking mosquitoes from the water's surface, and plunging for fish and small crustaceans. And during that time of the year when we in the north have our darkest and coldest days, the Arctic Terns are flying in the light of the midnight sun, at the other pole of the globe. No migratory bird follows the sun as far as the Arctic Tern does.

The Great Skua (*Catharacta skua*) **and the South Polar Skua** (*C. maccormicki*)
The Great Skua and the South Polar Skua are sibling species. They are so similar in general size, shape and appearance that they were formerly considered to be different races of the same species. More recently it has been discovered that the two are both met with and breed within the same region in the Antarctic Peninsula without interbreeding to any great extent; there is, therefore, every reason to regard them as distinct species. It has gradually been revealed that the two species are more dissimilar than was at first believed. A more recent discovery is that even their migratory habits differ markedly.

The Great Skua has a bipolar distribution in that it breeds within the westerlies zones in both the northern and the southern hemispheres, in other words on either side of a broad distribution gap around the equator. In the southern hemisphere it has a circumpolar distribution and breeds on smaller islands within the westerlies drift and in southernmost New Zealand, on the coast of southern Chile (recently considered a separate species, *C. chilensis*) and on Tierra del Fuego and the Falkland Islands. The southernmost breeding site is in the South Shetland and South Orkney Islands and on other islands on the north tip of the Antarctic Peninsula. Birds from different parts of the range vary in appearance and size; at least three different races of southern Great Skuas can be distinguished (Furness 1987).

In the northern hemisphere the Great Skua breeds only in the North Atlantic; strongholds exist in the 'proper' Orkney and Shetland Islands. In addition to these, breeding colonies are found in the Hebrides and on some other islands and headlands in northernmost Scotland and in the Faeroes, Iceland, Svalbard and Norway. The total population of northern Great Skuas amounts to about 12 500 pairs. The North Atlantic Great Skuas are very like their kin species in the southern hemisphere, resembling in particular those that nest on Tristan da Cunha and Gough Islands in the South Atlantic. Their occurrence in the north is very likely the result of colonisation by southern Great Skuas. The Great Skua, however, was already present in the North Atlantic in Viking times, this being revealed in its name in Shetland, 'Bonxie', which originates from Old Norse.

On the ground the Great Skua perhaps appears stocky and ungainly, like a well-fed dark brown Herring Gull, but in the air it is imposing. The first day that I was, as a young lad, in the Shetland Islands, I walked along the front at Lerwick in the evening. The promenade skirts a small golf course above steep cliffs facing the sea. Twites were flying up from under one's feet, the song of a Wren echoed among the shore cliffs, Fulmars were gliding on motionless wings in the upwinds of the precipices. But there, almost the first thing I caught sight of, was a Great Skua. With powerful and purposeful flight it passed, manoeuvring elegantly in a graceful dive, low over the waves and up over the shore cliffs, a sweep off the outermost headland, and so on away over the sea. All the time the white flashes on the broad wings contrasted with the dark chocolate-brown plumage. What power and dynamics its flight displayed!

When hunting for food, the Great Skua uses its magnificent manoeuvring skills to the full. It is a marauder at cliffs with breeding seabirds, and steals young of Kittiwakes and others from the rock ledges. Occasionally it succeeds in seizing Puffins in flight. It often chases gulls, both large and small, terns and Puffins in order to induce them to release or disgorge their recently caught fish. It keeps a particular eye on the Gannet. Sometimes it has only to begin to chase a Gannet for the latter to disgorge its fish. Often, however, the skua has to drive home the attack, and in this case it seizes the Gannet's tail or wing with its bill in flight. Sometimes it attacks from above and forces the Gannet down by positioning itself with its feet on the Gannet's back while in flight. Despite this, the Gannet at times refuses to give up its food and lands on the water, where it uses its bill to defend itself against the Great Skua's attacks. This is not always effective, however: the skua sometimes succeeds in clutching hold of the Gannet's neck and bill and pushing its head under the water until it finally disgorges its catch of fish beneath the surface. The flying skill and the power in the Great Skua's piratical attacks are best appreciated if we bear in mind that a Gannet weighs more than twice as much as a Great Skua.

The Great Skua is a first-rate opportunist in its choice of food: it readily takes young and eggs of birds, offal, fish carcases, whale carrion, dead birds etc. It sometimes fishes for itself, but how this is done and to what extent it occurs are largely unknown.

The Great Skua is a regular visitor to the Kattegat during the autumn, mainly in September and October, at which time it follows the Kattegat loop when there are strong westerlies blowing. Its migration in the North Atlantic can be described briefly as follows:

Adult, breeding-age Great Skuas (four to six years and older) do not migrate far. They winter mainly in west European waters, around Britain and Ireland, in the North Sea and off France.

Juveniles and immatures make much longer migratory journeys. The juveniles leave the breeding colonies in August, disperse over the waters of west Europe and move slowly southwards. Variations in the timetable for the journey south are very great. A ringed juvenile was reported from the Canary Islands as early as 30 August, but a number of recoveries show that young birds can still be found in the North Sea area from September to November. The juvenile Great Skuas seem to be particularly susceptible to wind-drift during periods of repeated gales. In such 'disaster autumns' they have been found windblown far in over the land, in Germany, Austria, Poland and the Soviet Union. Towards midwinter, almost all juvenile Great Skuas have reached south of 50° N, to Biscay and the areas south of there to northwest Africa, where they remain for the next year to come. Occasional Great Skuas head south of the equator, and chance recoveries of ringed birds have been reported for example from Brazil and Guyana.

Some juveniles do not move south during their first autumn but instead make a transatlantic migration to the food-rich banks off Newfoundland, where the cold Labrador Current collides with offshoots from the Gulf Stream.

In the spring following their second winter, many of the Great Skuas which have up to then remained at the Canaries Current also make a transatlantic migration to summering waters off Labrador and Greenland; others migrate north to summer near Norway and Spitsbergen. In winter most second-years return southwards to the eastern Atlantic off Portugal and northwest Africa. At this age the Great Skuas make their longest migratory flights. The nearer they get to breeding age, the more often they visit the waters around the breeding sites and the shorter the annual migratory journeys become.

The North Atlantic Great Skuas therefore stay in the northern hemisphere at all times of the year, mostly within the sea areas shown in figure 57. Only in quite exceptional cases do odd birds wander south to the equator. What, then, is the situation with the southern Great Skuas? Do they at any time move north of the equator?

It has long been known that Great Skuas visit the North Pacific in summer. They are sighted mainly between May and July off Japan, and later during the summer and autumn, from July to October, they appear off the west coast of North America, from British Columbia to south California. In the latter area they have occasionally been seen as late as the end of October. No Great Skuas breed in the North Pacific. Many ornithologists have therefore made

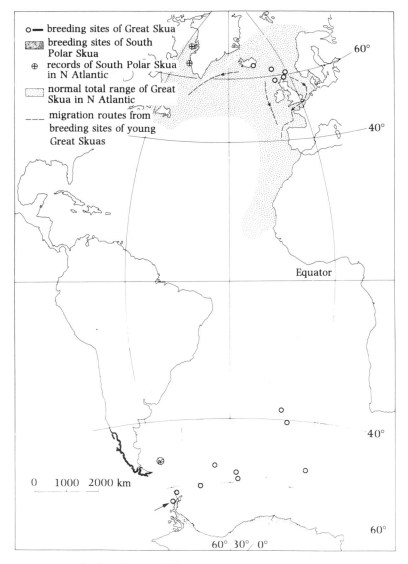

Figure 57   The distribution of the Great Skua and the South Polar Skua in the Atlantic area. Both species have a circumpolar breeding distribution in the southern hemisphere, the Great Skua within the westerlies belt and the South Polar Skua on the Antarctic continent, occasionally even in its interior several hundred kilometres from the coast. In the northern hemisphere the Great Skua breeds only within the Atlantic. Based on, among others, Furness (1978), Salomonsen (1976) and Thomson (1966).

the plausible assumption that southern Great Skuas wander northwards across the equator. Recently the Belgian ornithologist Pierre Devillers (1977) has compiled a critical register of the Great Skua's occurrence in the North Pacific, an initiative which became the start of an intriguing story.

Pierre Devillers inspected museum skins of Great Skuas from the North Pacific, examined in detail photographs of some of the birds observed and searchingly read available descriptions of observations. It became apparent that in all those cases which could be determined with certainty it was not the southern Great Skua that we were dealing with but the South Polar Skua instead. This was surprising since the South Polar Skua breeds south of the Great Skua, namely along the coast of the Antarctic mainland. The South Polar Skua is distinguished from the Great Skua mainly by the pale neck and by the fact that the underparts contrast with the dark brown upperparts and with the dark wings. In addition there is a rather uncommon all-dark phase of the South Polar Skua; expert knowledge is required to identify it correctly.

Pierre Devillers pieced together the reports of South Polar Skuas in the Pacific Ocean and was able to show that a regular clockwise migration takes place there during the summer months. It is probably mainly first-year birds, perhaps also those of intermediate age, that perform this migration. The South Polar Skuas moult their flight feathers during the time they are north of the equator, in particular in the Japanese waters. In their moulting habits they are thus reminiscent of the Great Shearwater, which also moults after migrating to the northern hemisphere.

Summer observations of what have been considered to be southern Great Skuas have also been reported from the North Indian Ocean. Pierre Devillers scrutinised available reports from India and Sri Lanka and found that all but one corresponded better with South Polar Skua than with southern Great Skua. The first recovery of a ringed South Polar Skua in the northern hemisphere was reported from India: the bird had been ringed in the Antarctic Peninsula and was found in August 1964 on the west coast of India.

Pierre Devillers then went further and became interested in what was happening in the North Atlantic. It is of course exceptionally difficult to track down skuas from the southern hemisphere in the North Atlantic since northern Great Skuas occur there. Large skin collections of Great Skuas from the North Atlantic can, however, be found in the zoological museums of London and Copenhagen. Pierre Devillers went carefully through all the specimens. In Copenhagen things came to a head. In the museum's collection he discovered a South Polar Skua, wrongly labelled as a Great Skua. It had been shot in mid July 1902 by an Eskimo hunter off Disko Island in west Greenland (almost 70° N, see figure 57). The bird had been killed in the middle of the moult period: the three innermost primaries were new and fully grown out, primaries 4 and 5 were still growing, and primary 6 was missing altogether; the three outermost primaries were old and very worn.

The story does not, however, end there. In the same year as Pierre Devillers completed his research with the visit to the museum in Copenhagen, a new discovery was made of a South Polar Skua in Greenland. The bird was shot on 31 July 1975 near 65° N on the west coast of Greenland. It had been ringed by an American research team on an island off the Antarctic Peninsula at latitude 65° S. The bird was ringed as a juvenile, yet to fledge, on 20 January, six months before it was shot in Greenland. The distance between the place of ringing and the recovery site is more than 15 000 km. This South Polar Skua must have made the longest migratory journey ever established for a ringed bird, in sharp competition with the Arctic Tern which was ringed on Saltholm in Denmark and recovered in the Antarctic.

The American research team incidentally had exceptional luck with its ringing programme. A further juvenile South Polar Skua, a bird which was ringed in the same month and within the same breeding area as the one just mentioned, was recovered in the northern hemisphere: at the Baja California peninsula in the Pacific Ocean!

The South Polar Skua is in fact the bird species that has been sighted the farthest south in the world, right at the South Pole itself. It will be most interesting to follow the future mapping of its migration habits. Perhaps it will be seen that the South Polar Skua is a perfectly regular summer visitor not only to the North Pacific Ocean but also to the North Atlantic; there have already been several reports of this species in west European waters in the 1980s. It is worthwhile for seawatchers in southwest Britain and on North Sea coasts to clean their binoculars and thoroughly examine every Great Skua in detail.

Why do South Polar Skuas migrate such a long way? Why are the Great Skua's movements much more restricted? We should be able to find clues by comparing the two species' feeding niches. Some South Polar Skuas breed near penguin colonies and often steal eggs and small young from the penguins. This food source does not, however, play such a large part for the skuas as we might perhaps believe. The penguins are usually a long way ahead of the skuas in the breeding timetable, and the penguin young soon grow too big to serve as prey for the skuas. Even those skua pairs which have many penguins within their breeding territory must therefore change over to other food during the latter part of the young skuas' growing period. This alternative food consists in the main of fish and krill, which the skuas catch in open coastal waters. For most South Polar Skuas fish and krill are their staple food throughout the breeding period and they rarely or never visit the penguin colonies.

At one colony of South Polar Skuas birds flew in a continuous stream across the pack-ice to and from open water which was dimly visible on the horizon several tens of kilometres away. Where there was plenty of food several tens of skuas, sometimes up to 100 individuals, gathered. The birds flew only a metre above the water surface and plunged their head and breast beneath the water when they had sighted prey; they often caught fish 10–20 cm in length. The harsh south polar climate left its mark on the birds returning to feed their young after fishing. At times the skuas could have difficulty in seeing for all the ice which formed when the water froze around their bill and on their head and breast.

These observations indicate that the South Polar Skua catches live fish and krill in the surface water to a greater extent than the Great Skua does. What sort of influence this difference in food choice has had on the evolution of the two species' different migration patterns, however, is difficult to determine. Perhaps the widely different migratory habits are a result, paradoxically enough, of the great similarities between the two sibling species. Maybe it is the competition with the southern Great Skua that compels young South Polar Skuas to make the leap-frog migration that takes them all the way to the food-rich cold waters of the northern hemisphere?

3.6   Birds of prey

Owing to their dramatic pursuits of quarry, the birds of prey attract special interest among appreciators of nature. This great interest and the associated efforts to protect the birds of prey is something we can be thankful for after the long periods of time during which they were heavily persecuted by hunters, who looked upon them as undesirable competitors for their game animals. The bird-of-prey populations in Europe are still suffering from the after-effects of this persecution, and in many places the acts of persecution are still going on full blast. The situation is made worse by the fact that birds of prey are the top consumers in the food chains and therefore are additionally susceptible to poisons in the environment. The great work that is being done today from the nature conservation point of view to protect the birds of prey – in Britain, for example, there are special projects for the Peregrine, the Golden Eagle, the Red Kite and the Osprey among others, and in Sweden for the White-tailed Eagle, the Eagle Owl and the Red Kite – provides hope that the assault on bird-of-prey populations need go no further than it already has and that the development of populations can gradually be turned in a positive direction.

By birds of prey in this section I mean birds which feed mainly on terrestrial vertebrates (table 19). This ecological definition does not entirely accord with the taxonomic delimitation of diurnal and nocturnal bird-of-prey orders. Within the order of diurnal birds of prey, or raptors, there are species which are first and foremost fish-eaters, such as the Osprey and the Black Kite, or insectivores, such as the Honey Buzzard, the Hobby and the Red-footed Falcon. These species are therefore not included in table 19. In this table there is, on the other hand, a passerine, the Great Grey Shrike, which is an expert vole-hunter and which also on occasion catches small birds.

In table 19 I have divided the birds of prey into groups based on the extent to which the different species take mammals, birds, reptiles, amphibians etc. This grouping is very rough: many of those species which feed mainly on mammals do not reject birds when these are easily accessible; the reverse applies for species which are by preference bird-catchers. The diet of birds of prey varies with the availability of easily captured prey. Thus Buzzards may be seen walking on wet meadowlands and picking up earthworms, feeding on small game killed by traffic on winter roads, hunting frogs in marshlands in spring and catching recently fledged birds in summer. Despite this variation the total picture shows that the Buzzard is a pronounced hunter of small rodents. The comprehensive studies of the Buzzard's diet which have been made in different parts of Europe show that small rodents generally make up more than 70%, in many cases over 90%, of the prey animals caught. The most extreme generalist in diet among the birds of prey is the Red Kite. In a study of its food selection in northeast France, remains of 469 mammals (mostly common vole, mole and water vole), 220 birds, 207 fishes, 60 reptiles and amphibians, and 255 insects (mostly beetles) were collected. Furthermore, the Red Kite is well known for its habit of readily taking carrion of various smaller animals. A varied diet indeed!

There is only one reptile specialist in table 19. This is the Short-toed Eagle, which to a great extent lives on snakes. This food specialisation explains why the species is a long-distance migrant and winters in the steppe and savanna zone immediately south of the Sahara. The nearest breeding areas of the Short-toed Eagle are in central France, Poland and the Baltic countries. Wandering Short-toed Eagles are encountered in southernmost Scandinavia mainly during the early autumn. Probably many of them are juveniles which have set off on nomadic travels when they became independent and before the autumn migration southward really gets underway. The dispersal of juveniles in various directions over longer or shorter distances from the nest site in late summer and early autumn is a common phenomenon in many birds of prey (cf. the migrations of the Grey Heron and the Osprey, section 3.4).

What is striking is that such a large proportion of the birds of prey are residents or short-distance migrants. In Sweden, for example, all of 23 species are present in the winter and only eight have important wintering areas in Africa south of the Sahara. The mammal prey are still present in the north during the winter, the problem for the birds being above all to catch them in spite of snow and the harsh winter climate. The owls have special adaptations in the form of large and asymmetrically designed ears and can locate and strike small rodents through a complete blanket of snow. Several bird-of-prey species, however, get by only if the snow cover is thin or if there is no snow on the ground. The Buzzard, the Kestrel and the Hen Harrier therefore are present in winter only in the very southernmost parts of Scandinavia, mainly in Denmark and the extreme south of Sweden (Scania, Halland).

An important factor which favours the birds of prey during the winter is that the prey animals, especially the birds, become more easily accessible, on the one hand because the winter environment provides poor cover in general and on the other because the prey

Table 19. *Birds of prey which occur regularly in north Europe. The data refer primarily to northwest European birds or birds which pass through northwest Europe on migration*

Food	Species (NB = does not breed in Scandinavia)	Largely resident	Found in N Europe in winter	Important wintering areas — W & N Europe	Mediterranean region	Africa S of Sahara	Invasion migration
Mostly reptiles/amphibians	Short-toed Eagle (NB)					+	
Mostly mammals	Buzzard		+	+	+	+	
	Rough-legged Buzzard		+	+			
	Lesser Spotted Eagle (NB)					+	
	Kestrel		+	+	+	+	
	Barn Owl	+	+	+			
	Snowy Owl	+	+	+			++
	Hawk Owl	+	+	+			++
	Tawny Owl	+	+	+			
	Ural Owl	+	+	+			
	Great Grey Owl	+	+	+			
	Long-eared Owl		+	+	(+)		
	Short-eared Owl		+	+	+	+	
	Tengmalm's Owl	+	+	+			+
	Great Grey Shrike		+	+			
Mostly birds	Goshawk	+	+	+			
	Sparrowhawk		+	+	++	(+)	
	Merlin		+	+	++		
	Gyrfalcon	+	+	+			
	Peregrine		(+)	+	+	(+)	
Mammals + birds	Hen Harrier		+	+	++		
	Spotted Eagle (NB)		(+)	(+)	++		
	Golden Eagle	+	+	+			
	Pygmy Owl	+	+	+			
Mammals + birds + fish	White-tailed Eagle	+	+	+			
	Eagle Owl	+	+	+			
Mammals + birds + reptiles/ amphibians etc.	Marsh Harrier			(+)	(+)	+	
	Pallid Harrier (NB)				(+)	++	+
	Montagu's Harrier					+	
	Red Kite		+	(+)	+		
Total no. species		13	23	23	9	8	4

animals are forced to cast aside their own safety in order to obtain sufficient food themselves to survive the bitter winter period. At a rough count there are about 200 million breeding birds in Sweden, of which over 90% are passerines. After the breeding season's addition of young, the number in summer will probably rise to nearer 500 million individuals. The total number is very high during the autumn, too, when great multitudes of birds pass through the country on migration from northern and eastern areas. Thereafter the crowds of birds thin out quite incredibly, and in midwinter there are probably 30 million birds at the most in Sweden. Despite the fact that the number of wintering birds is thus not much more than a tenth of the number of those breeding and no more than one-twentieth of the number during the summer, many birds of prey that hunt birds overwinter in Sweden. For the Sparrowhawk it is evidently a fairly easy task to set about the winter's unprotected flocks of small birds around birdtables and in leafless deciduous trees and bushes.

One factor of the greatest import for those birds of prey which feed on small mammals is the regular population fluctuations of their prey. These fluctuations are not particularly great in south and central Europe, nor in Denmark or the very south of Sweden, but in northern areas, such as central Sweden, Norway and Finland, they are all the more drastic. Every fourth or sometimes every third year extreme peaks occur there in the vole numbers. Between these 'vole years' the small-rodent populations crash to an exceedingly low level. Several rodent species vary in parallel in these three- or four-year cycles. During the 1970s the vole population was carefully monitored in Västerbotten in northern Sweden. The three commonest species, the bank vole, the grey-sided vole and the field vole, exhibit almost identical variation patterns in this region, with marked peak years in 1974 and 1978. The vole population in Svealand followed a similar pattern during this period (figure 58).

The small-rodent cycles in different geographical regions may vary asynchronously with one another. The fluctuation pattern shown in figure 58 is therefore valid only for central and northern Sweden and for Norway. By contrast, in Finland 1972 and 1975 were big vole years; in Sweden and Norway the populations then were at rock bottom.

The Norway lemming population also varies in three- or four-year cycles following a similar pattern. In years when the lemming numbers have reached a maximum and are heading for a crash, 'lemming marches' take place down into the woodlands, far from the animals' normal habitat in the mountains. Lemming migrations have been the subject of much myth-creation. The scale of the invasions has been exaggerated and the migration has been alleged to lead to certain death for the lemmings. This of course is not true, and it is a highly interesting exercise for students of small rodents to demonstrate which factors trigger the migrations and how the lemmings improve their future prospects by setting off on migration.

In favourable conditions small-rodent populations increase very rapidly; the females can produce new litters, with four to eight young in each, at intervals of only three weeks. Young females become pregnant at no more than a few weeks of age, and breeding can carry on even in the winter, under the snow. The populations probably increase so fast that the rodents overexploit their food resources: a population crash becomes unavoidable. The plants which make up the rodents' food have time to recover while the rodent populations are at their lowest, and when the animals' food situation has again become favourable the whole course of events can be repeated. Predators (foxes, weasels and birds of prey) which feed on rodents also contribute to the shaping of the rodents' population cycles. In vole and lemming years, the birds of prey reproduce on a large scale; they lay big clutches of eggs and many young fledge. When the rodent stocks begin to fall off, the predators are therefore at their most numerous. Consequently, predator pressure on the diminishing rodent populations becomes

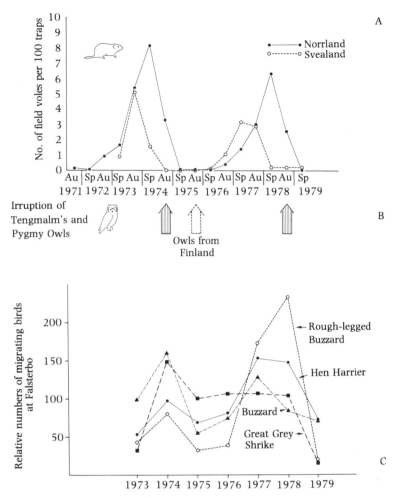

Figure 58   Variations in the vole population and in numbers of migratory birds of prey for which small rodents are an important food. A. Graph showing the fluctuations in the field vole population from studies in Västerbotten (in Norrland, northern Sweden) and Västmanland (in Svealand, southern Sweden). Several vole species in northern Sweden, such as grey-sided vole and bank vole, fluctuate in parallel with the field vole. Based on B. Hörnfeldt (1978, and Report 12 of Base Inventory of Rodents, Umeå University). B. In the autumns of 1974 and 1978 irruptions occurred in southern Sweden of Pygmy Owls and Tengmalm's Owls. The owls' invasion migration happens when the vole population crashes. In autumn 1975 an invasion of Finnish Tengmalm's Owls occurred on Åland, in Södermanland (Hartsö-Enskär bird observatory) and on Öland (Ottenby bird observatory), as well as elsewhere. Based on annual bird observatory reports in *Vår Fågelvärld* 1972–1979. C. Numbers of migrating birds of prey (species which often take voles) at Falsterbo during the autumns of 1973–1979. The 100 index corresponds to the average annual totals during the 1970s: for Buzzard = 10 719 individuals, Rough-legged Buzzard = 696, Hen Harrier = 144, and Great Grey Shrike = 31. Based on Gunnar Roos's reports on migration counts at Falsterbo (*Vår Fågelvärld* 1974, 1977; *Anser* 1977–1979). Compare also Lundgren (1979) and Roos (1979).

exceedingly great. It can easily be imagined that this would lead to a protraction of the period between rodent peaks: the rodent stocks cannot begin to increase until the food situation is good *and* the predators have decreased in number. In regions where not only voles or lemmings but also other food is available to the predators, the latter's populations may be maintained in such a healthy state that they prevent the small-rodent populations from 'disengaging themselves' from the predator pressure and rocketing up. A research group at Lund in south Sweden, under the leadership of Sam Erlinge, has pursued studies over many years which suggest that the rabbit in particular is an alternative food for the predators in southernmost Sweden; the main reason why the vole populations there do not show cyclical fluctuations would therefore be the presence of the rabbit.

In northern regions, where to a large extent predators have no food other than voles or lemmings, a difficult situation arises for the birds of prey when the vole supply starts to diminish. As we can see in figure 58, the crash in small rodents happens very rapidly. During both 1974 and 1978, the vole population in Norrland decreased in just over half a year from a maximum during spring–summer to only one-hundredth of that level in the following winter. In this situation, those birds of prey which are normally for the most part residents set off on invasion migration. In most years typical invasion birds are not seen at all on migration, but during invasion years they wander far outside the normal range. Hawk Owls and Snowy Owls leave the taiga and the mountains and winter in central and southern Scandinavia and other places to the south. By trapping birds at night, hundreds of Pygmy and Tengmalm's Owls can be ringed at a single bird observatory during a good invasion autumn. The most recent owl invasions took place in the autumns/winters of 1964, 1967, 1971, 1974, 1978, 1982 and 1985. The invasions thus occur about every third or fourth year, and the link with the collapses in the small-rodent populations is clearly evident from figure 58. Irruptions of Tengmalm's Owls occurred in Sweden, however, not only in autumn 1974 but also in the following autumn. The invasion on that occasion was felt mainly on the east coast, where the largest number of Tengmalm's Owls was caught at the Hartsö Enskär bird observatory not far from Stockholm. No doubt this was a case of Finnish Tengmalm's Owls having set off on invasion migration with the collapse of small-rodent numbers in Finland in 1975; this supposition is borne out by the large catches of Tengmalm's Owls that autumn at the bird observatories on Åland, in the central Baltic midway between Finland and Sweden.

Most invasion owls are doubtless juveniles. Many of the adults brave the scarcity of small rodents and stay behind to defend their territory and their nest site. During poor rodent years they often do not breed at all; the males then remain silent and do not call in the spring. It has been conjectured (Lundberg 1979) that females would take part in the invasion migration to a greater extent than males, which have the primary task of maintaining and guarding a good breeding site. This idea is supported by a check on Tengmalm's Owls found dead during the migration period in spring and autumn: of a total of 78, 51 (65%) were females. This has recently been corroborated by ringing results. Adult females of Tengmalm's Owl regularly depart on migration between vole peaks, and they also show breeding nomadism, i.e. they readily change to new breeding sites, sometimes hundreds of kilometres away (Löfgren *et al.* 1986).

The invasion migration does not as a rule take the birds particularly far. In southernmost Scandinavia the population fluctuations of small rodents are comparatively undramatic and with no sharp trough years, so the invading birds can generally find tolerable feeding conditions there. A ringing recovery in northern East Germany of a Tengmalm's Owl ringed on autumn passage at Falsterbo shows that a minor proportion of the owls get as far south as northern Continental Europe.

The availability of small rodents influences breeding success in a number of those bird-of-prey species which are regular migrants. In years when rodents are present in abundance in spring and summer, many young birds of prey reach the flying stage and strengthen the species' populations. This can be seen quite well from the annual autumn migration counts at Falsterbo (figure 58).

Natural selection results in every species becoming specially adapted to its ecological niche. In a stable situation the competition between closely related species is reduced in various ways. Species differ in geographical distribution or are adapted to different habitat types. Species which live in the same habitat exploit different foods to a large extent. These three factors, different geographical distribution, habitat segregation and food segregation, are the key mechanisms behind the ecological isolation of different animal species.

In many environments a whole succession of bird-of-prey species, both mammal-hunters and bird-hunters, co-exist. A good indication of differences in food choice between these species may be got by comparing their sizes. Different-sized birds of prey often use different hunting techniques, they strike prey with different force and they are able to manoeuvre in relation to the vegetation in different ways. Even though both Pygmy Owl and Ural Owl take field voles, the Pygmy Owl gets to its voles in circumstances and in a particular setting where these prey are inaccessible to the Ural Owl, and vice versa. In figure 59 the weights of competing birds of prey are compared. It is striking how species in the same environment differ in size, where in many cases one species is roughly twice as heavy as the next biggest one. This weight quotient of about 2 (corresponding to a length quotient of about 1.25) between closely related species in the same environment is a frequently recurring 'ecological constant' in animal communities of widely differing kinds. Some researchers have attempted theoretically to deduce this quotient to the limits of equality between co-existing closely related species in a stable ecological system. The theory surrounding the ecological equality quotient, however, is unclear, and some researchers have warned against exaggerated theorising by pointing out that the quotient also works quite well for, among other things, flutes (from high treble to deep bass), bicycles (child models up to adult cycles), kitchen saucepans, and plates and glasses in service sets!

Food segregation between competing species is not always manifested best in differences in the birds' body sizes. In many cases, differences in length and design of the bill, an instrument which is of primary importance in determining choice of food in many bird types, reflects different feeding habits in a more reliable way. Thus co-existing wader species, for example, differ most obviously in bill length. Body size mirrors food differences best in birds whose choice of food is determined largely by the accumulated power or by the manoeuvring precision in the bird's performance. Apart from birds of prey, this applies to such birds as fish-catching terns (in the Baltic there is a series of size-models: Little Tern, Common Tern, Sandwich Tern, Caspian Tern), woodpeckers (Lesser Spotted, Middle Spotted, Great Spotted, Green and Black Woodpeckers) and omnivorous birds (section 3.9).

Among the birds of prey there are departures from the 'doubling-of-weight rule of thumb' that attract particular interest (figure 59). The Ural Owl and the Great Grey Owl have closely similar body weights and are both primarily vole-hunters in the northern coniferous forests. In Scandinavia these species are largely geographically separated; the Great Grey Owl's distribution lies to the north of that of the Ural Owl. In the Russian taiga, however, there are wide areas where both species occur commonly. Do the species differ there in their demands on the environment? The Hawk Owl, the Ural Owl and the Great Grey Owl are absent from the owl community in the forests of south Sweden, where their places are taken by the Long-eared Owl, with a weight of around 300 g, and the Tawny Owl, with a weight of around 500 g.

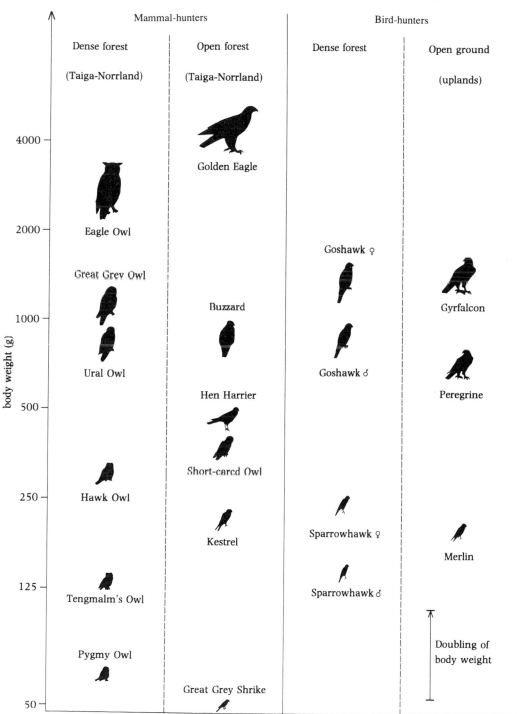

Figure 59   Body weights of the species in different bird-of-prey communities. An empiric rule of thumb in ecology states that competing species within the same habitat differ by a weight quotient of around 2. This rule works quite well within bird-of-prey communities, even though clashes in competition sometimes appear to come about. The body weight is shown on a logarithmic scale, where a given distance on the scale always corresponds to a fixed multiplicative or percentage ratio.

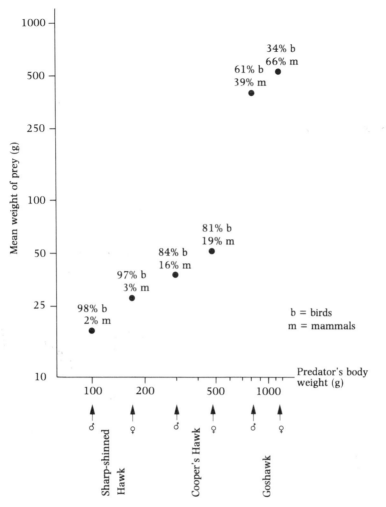

Figure 60   Birds of prey of different weights largely take different prey, which is demonstrated in a study of three species which exist together in North America. The different sexes of the same species are different in weight, and also show differences in diet. The mean weight of the prey increases with increasing size of the hawks, and at the same time the number of mammals caught in proportion to the number of birds increases. The scales on the figure are logarithmic. Based on R. W. Storer in Lack (1971).

Among mammal-hunting birds of prey in open woodlands a clash of competition seems to occur between the Hen Harrier and the Short-eared Owl. Both species have very similar ranges, migratory habits, habitat requirements and general choice of food. That they are, besides, generally speaking equal in size breaks all ecological rules. The only really obvious difference between them that I can see is that the Short-eared Owl hunts largely at dusk and in darkness and that the Hen Harrier is markedly diurnal. This difference may well be crucial for the co-existence of these species. The threat to the prey species from the nocturnal predator will tend to make the prey more active during daytime and thus more readily available to the diurnal predator, and vice versa.

In the bird-of-prey community in open woodlands there is a striking weight gap between the Buzzard and the Golden Eagle. Perhaps there is an absence of suitable prey animals for a bird of prey with a weight of around 2 kg, that is twice the Buzzard's weight and half that of the Golden Eagle? Or is there a niche here that is available for occupation? In that case a suitable candidate to fill the niche would be the Spotted Eagle, whose breeding range includes the Baltic countries and large parts of Russia. This species breeds now and then in Finland, and very sporadically also in the forest land of Norrbotten in the extreme northeast of Sweden.

There are rather few species that are typical bird-hunters. In almost all of them the female is considerably bigger than the male; female hawks are more than 50% heavier than males. It is therefore almost a case of the different sexes of the same hawk species filling different niches as per the ecological rule mentioned above (figure 59). The fact that the weights of the different sexes and species really are linked up with material differences in food composition has been neatly demonstrated for co-existing species of hawk in North America (figure 60).

An interesting question, which we may note in passing, is: why do only two species of *Accipiter* co-exist in north Europe when there is room for three in North America? Perhaps there is an unoccupied niche in Europe such that a species could squeeze itself in between the Sparrowhawk and the Goshawk? The space appears, however, to be rather too tight according to figure 59. That this niche really is narrow is suggested by the fact that the 'intermediate hawk', Cooper's Hawk, is quite an uncommon bird in North America.

The above discussion on the competition between different species and on the shaping of bird communities illuminates some significant ecological rules which are important for an understanding of the migration patterns of different species. A given species must fit into a stable niche in the total community of organisms both in summering and in wintering areas.

Northeastern and eastern populations of several bird-of-prey species migrate farther than the populations of northwest Europe. This applies to a large degree to the Buzzard, the Peregrine and the Kestrel. The northeastern populations are long-distance migrants and fly to areas south of the Sahara, whereas for example British and Scandinavian populations winter in west Europe. The Kestrel's migration pattern is given as an example in figure 61, where ringing recoveries during the winter of Swedish Kestrels are compared with summer recoveries of those birds which migrate in spring via Cape Bon in North Africa. A large proportion of the Kestrels which pass Cape Bon no doubt come from winter quarters south of the Sahara, within the steppe and savanna zones south to Tanzania. There the Kestrels are often attracted to large grass fires which drive out easily caught prey – insects, especially grasshoppers, and small rodents.

In Scandinavia the Red Kite occurs as a breeding bird mainly in Scania, where the population is currently at least 150 pairs. Over the last 30 years or so a certain proportion of the kites, at times over half, have wintered each year in south Scania; others migrate, often via Falsterbo, where the proportion of juveniles among the passing kites is about 80%, to winter quarters in the Mediterranean area. About 90% of the Scanian wintering kites are adults, which are concentrated in a small area where slaughterhouse refuse is available. Occasional Red Kites are sometimes seen in winter at other places, too, at localities where they are able to find food on dunghills or refuse-tips.

Similarly, the small Welsh population of about 60 pairs is largely resident, though some juveniles do wander away from the breeding range, mostly to other parts of southern Britain. Their feeding habits in winter are much as those of the Scanian wintering birds.

The harriers are adapted to open plains, moors, bogs and marshes, where they hunt by

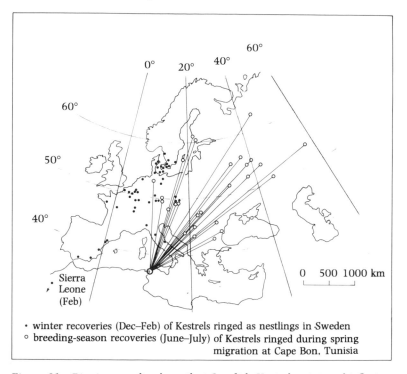

Figure 61 Ringing results show that Swedish Kestrels winter chiefly in west and central Europe, from the Mediterranean in the south to south Scandinavia in the north. Kestrels which pass Cape Bon in Tunisia during the spring migration, most of which come from wintering areas in Africa south of the Sahara, originate from breeding areas in east Europe. Kestrels from northern Sweden generally migrate farther south than south Swedish Kestrels; some from Norrland reach as far as West Africa, as shown by the recovery from Sierra Leone. Kestrels breeding in west Europe are mainly resident. Based on Ringing Office records, Riksmuseum, Stockholm (recoveries of Swedish Kestrels up to 1975), and Glutz von Blotzheim *et al.* (1971).

patrolling low over the ground, constantly turning, pitching and dropping down in flight. Where are the most extensive and suitable hunting environments for the harriers if not on Africa's open steppes and savannas and floodplains. Yet virtually no harriers breed in Africa's vast savanna and steppe areas; just one rare species, the Black Harrier, lives on scrub steppe, meadowlands and lagoon areas within a relatively small range in South Africa. The species is found chiefly within the zone that has a Mediterranean climate and is therefore compelled, like the north European Hen Harrier, regularly to hunt over frosty winter terrain.

Are there, then, no breeding harriers on the great African swamplands either? Yes, there is in fact one species, the African Marsh Harrier, which is a very close relative of the Marsh Harrier of Europe and Asia. The African Marsh Harrier, however, does not occur in the northern tropics. The huge inundation areas within the savanna zone between the equator and the Sahara are thus totally devoid of breeding harriers.

In autumn, however, a great upheaval takes place. Pallid Harriers and Montagu's Harriers arrive from breeding areas in Europe and Asia and take possession of the African savanna lands. Their winter distribution extends from the Sahara in the north all the way to South

Africa in the south. In addition, Marsh Harriers invade African floodlands and swamplands southwards as far as Zambia. In East Africa the Marsh Harrier and the closely related African Marsh Harrier occur side by side. How the competition situation between these two species works out is totally unknown. As a rule the sedentary African Marsh Harriers are very much fewer in number than the wintering northern birds.

The three harrier species which invade Africa in winter have a very varying diet in their northern breeding areas. Montagu's Harrier, as well as catching small rodents, takes a great many insects (crickets, grasshoppers, dragonflies and beetles) and lizards and small birds; it frequently robs small birds' nests of eggs and young. The Pallid Harrier also catches birds, especially young birds, and insects; its basic food in years when small rodents are abundant is the steppe lemming. The Marsh Harrier is tied to food-rich marshes and lakes, where it exploits what the summer has to offer in the way of water voles, young of all manner of birds, and frogs, fishes and larger insects.

The harriers retain their flexible feeding habits in their African winter quarters. They are attracted to areas where there is a temporary surplus of rodents, grasshoppers or the like. During migration and in winter the roaming harriers sometimes appear in parties and frequently roost together at sites with long grass where they can perch in sheltered spots on the ground. Such communal roost sites may on occasion be used for many weeks by several tens of harriers; at times individuals of all three long-distance migrants are included in the assembly of roosting birds.

I can very well imagine that the tropical savanna, steppe and floodlands are in actual fact the harriers' original home. Warm periods during the earth's climatic evolution have opened up northern regions to a magnificent summer blooming, a temporary explosion of greenery and animal life. The harriers which have migrated north during the summer in order to exploit this temporary food surplus have been so advantaged that they have driven out the harrier populations that were resident in Africa. An important reason why the harriers which have been resident in the tropical grasslands have failed to hold their own in competition with migratory populations is doubtless the uneven and erratic distribution of food. This of course compels the harriers to a nomadic winter existence in the tropics. The resident birds have accordingly been deprived of their main trump card: the advantage normally implicit for a resident bird of claiming and really getting to know a permanent territory, the 'home grounds'.

Earlier this century, Pallid and Montagu's Harriers were a strikingly common sight in the African wintering grounds. In more recent years the situation has changed radically. Observers who previously saw 20 or 30 harriers in a single day in the savanna regions nowadays record roughly that total number throughout a whole winter season. The reasons for this catastrophic decline are not known, and an investigation is very urgently needed. For the Marsh Harrier the situation is more reassuring; in Britain and in Scandinavia the breeding populations have increased markedly during recent decades. The commonest harrier in Scandinavia and Britain, the Hen Harrier, does not figure at all in the above discussion: it migrates only as far as west Europe and the Mediterranean region.

Many birds of prey begin to moult their flight feathers while breeding. The change of feathers is completed during the summer so that the migrating birds can depart on new wings. This does not, however, apply to the very northernmost of the birds of prey which are long-distance migrants. Arctic Peregrines have time to change only a few wing feathers in the northern breeding area; the moult then ceases during the migration and is not completed until in the tropical winter quarters. The same applies to the northernmost Buzzards, which migrate to winter quarters in Africa. Detailed studies in South Africa have shown that one wing feather or other may already be new on the adult Buzzards which arrive in October and early November; from November to February, however, full-scale moult takes place. In March the birds leave for the north on new wings.

West and south Scandinavian Buzzards are thus summer moulters, but Finnish and Russian Buzzards renew their flight feathers during the winter. This is tied up with an amazing difference in migration patterns among the different Buzzard populations, something that calls for more detailed presentation.

The Buzzard (*Buteo buteo*) – resident and long-distance traveller

Along with the Sparrowhawk, the Buzzard is the commonest bird of prey in north Europe. It is often seen in magnificent soaring flight above the forests or perched on fences in wooded pastureland on the look-out for voles. In Denmark and southernmost Sweden it is also a common winter bird. It frequently resorts to open terrain along roads and to the outskirts of towns. The fact that the Buzzard is so common certainly does not mean that it is unworthy of consideration. Quite the reverse – of the bird-of-prey species in table 19 it is the one that exhibits the most adventurous migration pattern. Between different Buzzard populations in the north there exist the most mind-boggling differences in migratory habits. Thus those Buzzards which breed in the far south of Sweden are generally, like those in Britain, sedentary the year round within their territories. However, we need move no farther than 1000 km away, to east Finland, to find Buzzards which migrate to southernmost Africa in winter.

In the Cape Province of South Africa, 'wintering' Buzzards are widespread in occurrence. There they perch on the telegraph poles along the busiest vehicle-roads and keep a look-out for small rodents; along the roads around Cape Town, South African birdwatchers have counted on average one Buzzard every 6 km. Back at my Scanian home in south Sweden, I often have cause to drive along the Öresund coastal motorway in winter, when I sometimes entertain myself by counting the Buzzards perched along the road on telegraph poles, fence posts or in trees; one Buzzard every 4 km is a fairly good midwinter figure. The Buzzards are an equally familiar and characteristic feature along the sunny summer roads at the Cape of Good Hope as they are on the snowy and slushy winter roads in Scania.

Let us look more closely at the migratory habits of various Buzzard populations (figure 62). Of those Buzzards recovered in winter which had been ringed as nestlings in Denmark, as many as 56% have turned up within 100 km of the nesting site; winter recoveries outside Denmark, more than 100 km from the nest sites, are shown in figure 62A. Of the Danish Buzzards, it is mainly the juveniles that migrate, to northernmost France at the farthest. When, at two to three years of age, Danish (and south Scanian) Buzzards have established territories and begun to breed, more often than not they remain in the territory throughout

Figure 62 Migration patterns according to ringing studies of Buzzards which breed in Denmark (A), in southern Sweden (B), in northern Sweden and Finland (C); and Buzzards which winter in Cape Province in South Africa (D).

A. Recoveries in winter (15 Nov–1 Mar) of Buzzards ringed as nestlings in Denmark. Of the Danish birds 56% can be regarded as residents, in that they have been recovered within 100 km of the nesting site. Based on Nielsen (1977).

B. Winter recoveries (15 Nov–1 Mar) of Buzzards from Sweden north to 60° N are distributed within a well-defined zone to the southwest. Of winter recoveries 13% are from south Sweden (all but two from Scania). Based on Ringing Office annual reports for 1960–1969 (S. Österlöf, Riksmuseum, Stockholm).

C. Recoveries of Buzzards ringed in Sweden north of 63° N (crosses) and in Finland. The

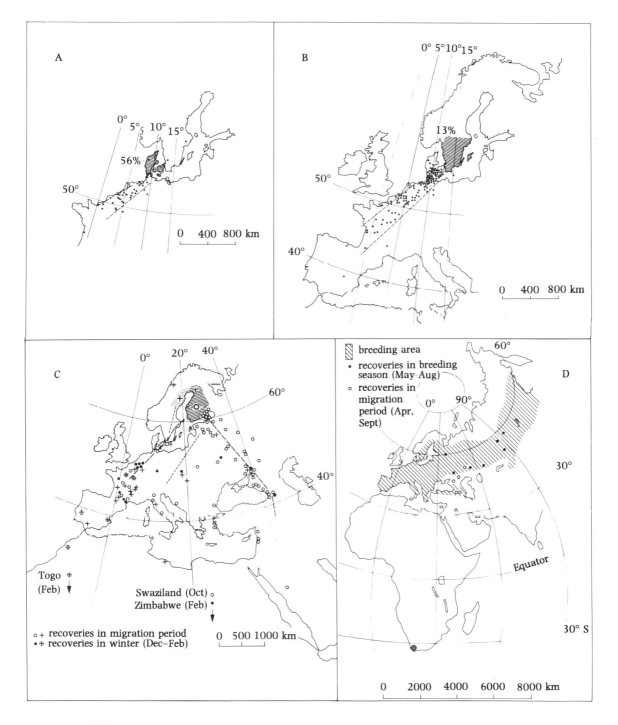

probable migration routes are indicated by broken lines. Data for Sweden based on Ringing Office annual reports for 1960–1969 (Stockholm), and for Finland on Nordström (1963) together with Finnish ringing reports in *Mem. Soc. Fauna et Flora Fennica* for the years up to and including 1967 (cf. Saurola 1977).

D. Recoveries of Buzzards ringed between October and March in South Africa. Based on Glutz von Blotzheim *et al.* (1971) and on ringing reports in *Ostrich* 1970–1974.

the year. The Buzzards which breed in westernmost Europe, from the Netherlands and Britain in the north to Spain in the south, can all be described as residents.

When we move from Denmark to south Sweden, we come to Buzzard populations which, with the exception of the birds in south Scania, consist entirely of migrants. The winter quarters of south Swedish Buzzards extend from Scania in the north (where 13% of winter recoveries of those ringed in south Sweden have been made) to southwest France in the south. The Buzzards migrate on a narrow front, within a neat corridor (figure 62B), and thereby differ dramatically from broad-front migrants such as the Osprey (cf. figure 46). They migrate only in the very best thermal conditions and so far as possible avoid sea passages where there are no thermals. The Swedish Buzzards concentrate on passage on a large scale in the Öresund area, at Falsterbo and on the Danish islands so that the sea crossing to Continental Europe is as short as possible.

In the wintering area Swedish Buzzards mix with local west European Buzzards and with some migrant Danish Buzzards, but to a surprisingly small extent with migrant German and Swiss birds. German and Swiss Buzzards migrate down through France on a narrow front on southwesterly courses which run parallel with the main route of Swedish birds. The result is that they winter mainly in regions to the east of the Swedish birds' main winter quarters. In France, Swedish Buzzards winter in a zone centred around a line drawn from Paris to Bordeaux, German Buzzards around a line from Strasbourg to Andorra and Swiss ones farther east still, along a line from Geneva to the Golfe du Lion in the Mediterranean Sea. What causes these different populations to winter in geographically parallelly displaced corridors and not to be mixed together within a wider wintering area, as for example the Lapwings (cf. figure 16), is a tricky and unsolved problem. It can probably be taken for granted that an effect of competition is in some way involved.

The importance of France as winter quarters for south Swedish Buzzards seems to have diminished in recent decades. According to a compilation of ringing results up to 1958 (carried out by Viking Olsson), around 40% of winter recoveries were from France. Since then the central point in the winter distribution has shifted towards more northerly regions of west Europe: fewer than 20% of winter recoveries during the 1960s, which are shown in figure 62B, come from France. This change is probably connected with the general decrease in the Swedish Buzzard population that has taken place over the last 20–30 years judging from migration counts at Falsterbo (where, during the 1940s and early 1950s, twice as many Buzzards on average passed as in recent years). The reasons for the changes in population size and winter occurrence are not known.

Ringing recoveries of Buzzards from the northern half of Sweden (north of 63° N) show that these birds migrate predominantly southwestwards and that broadly they follow the same route as the south Swedish Buzzards. Following the principle of leap-frog migration, the northern Buzzards on average travel to more southerly wintering areas than the south Swedish ones. They winter mainly in France; Spain and Portugal; in addition, there are two instances of birds having continued on to Africa, in one case as far as Togo in tropical West Africa. It is probably a regular occurrence for northern Buzzards to migrate to West Africa, at least judging by counts of migrants at Gibraltar, where between 2000 and 3000 Buzzards have been seen during the autumn on route to Africa. Our knowledge of the northern Buzzards' way of life in the exceedingly sparse winter populations in West Africa is practically non-existent.

Some recoveries of north Swedish Buzzards during their autumn migration are of particular interest. A bird from Västerbotten (around 65° N) was found in Russia southeast of the Gulf of Finland, and one from Norrbotten (Arctic Circle) was recovered at the Sea of Azov.

These interesting recoveries are seen in their proper light when we study the Finnish ringing material on the Buzzard. As can be seen in figure 62C, most of the Finnish Buzzards migrate southeast, round the Black Sea on its eastern side and then pass the Middle East and Arabia on their journey down to Africa. The Buzzard is present in winter in the whole of eastern and southern Africa, from the Sudan in the north to the Cape of Good Hope in the south. This is where the great majority of the southeastward-migrating Finnish (and north Swedish) Buzzards have their winter quarters. In large parts of eastern Africa the midwinter populations of Buzzard are rather sparse, but in South Africa the Buzzards are considerably more numerous. During the movements of the many Buzzards to and from southernmost Africa impressive passage figures have been reported from several places. Thus, during the month of March one observer counted more than 2000 Buzzards on passage northwards along the cliff faces of the Rift Valley following the Luangwa River in Zambia, and in the small mountainous country of Rwanda between Lake Victoria and Lake Tanganyika 3300 southward-moving (in October) and 4700 northward-moving Buzzards (most in March) have been counted recently. Altogether there must be hundreds of thousands of Buzzards that migrate to Africa. The British ornithologist and African authority R. E. Moreau even speculates on total figures of up to 6 million Buzzards wintering in Africa. The Finnish birds are of course only a small percentage of these.

As is apparent from figure 62, not all Finnish Buzzards migrate southeast but a certain proportion travel to west Europe. Most of these southwestward-moving migrants come from westernmost Finland and migrate via the island of Åland to Sweden and then onwards along the same general route as followed by the Swedish birds. There are also ringing recoveries of Finnish Buzzards in central Europe, in Italy and in Yugoslavia; these suggest that some Buzzards break off to the southwest after having rounded the Gulf of Finland in company with those flocks that migrate southeast.

In Finland and northernmost Sweden there are thus breeding Buzzards with completely different migratory habits: either they migrate southwest primarily to west and southwest Europe, or else southeast to Africa, quite often as far as South Africa. Buzzards that migrate southwest predominate in north Sweden, those that migrate southeast predominate in Finland. No sharp geographical dividing line can be drawn between the different Buzzard populations. Of Buzzards ringed as nestlings in the same region around Tampere in southwest Finland, some have been recovered in the southwest and others in the southeast. It is not only the migration pattern that separates the different populations but great dissimilarities exist also in the moulting habits. In parallel with the diffuse migratory divide that runs through Lapland and west Finland, an equally obscure borderline separates two different races of the Buzzard, the nominate race *Buteo b. buteo* and the so-called Steppe Buzzard *B. b. vulpinus* (the nominate race is greyer in colour and larger than *vulpinus*). The differences in appearance and size between the two races are the result of adaptation to different winter quarters: *buteo* Buzzards are adapted to a winter residence in the climate of west Europe, and *vulpinus* Buzzards to living in Africa. One wonders what happens in those regions where breeding Buzzards of different races overlap. Do the various migrant populations occupy breeding territories at separate points of time? Do the different Buzzards have separate habitat requirements? Do mixed pairs occur, and if so how do the migratory and moulting habits of the young turn out?

Recoveries of *vulpinus* Buzzards ringed during the 'winter' (the southern African summer) at the southern tip of Africa (figure 62D) show that they originate from a breeding area which encompasses large parts of the Soviet Union as far eastwards as central Siberia around Novosibirsk. The most easterly recovery comes from 93° E, more than 12 000 km from the site of ringing. Buzzards which breed even farther east in Siberia, Central Asia and Japan migrate in winter to India, Burma and Indo-China.

When they arrive in South Africa in October–November, the Buzzards weigh on average only about 650 g. They put on weight there and in February the mean weight has increased to 740 g. The fat reserves, at roughly 12% of the body weight, stand them in good stead during the coming long-distance migration north. As flight fuel such reserves last only 500–1000 km in active flight, but as the Buzzard conserves energy by soaring and gliding they may perhaps at the best be sufficient for three or four times that distance. The Buzzards from South Africa must 'top up' their fat reserves repeatedly during the 10 000-km or so migratory journey.

Many readers have now no doubt had time to ponder about the reasons for the Buzzard's diversified migration patterns. That the principle of leap-frog migration plays an important part is inescapable. Further illumination of possible causal connections can be gained by studying the distribution in summer and winter of the various species in the buzzard genus (figure 63). In the Old World there are eight different buzzards (the genus *Buteo*), with mean weights of about 0.5 kg (Madagascar Buzzard and Mountain Buzzard in Africa) up to 1.5 kg (Upland Buzzard in Tibet). Some races within the same species differ noticeably from each other in size. I have already mentioned that birds of the race *vulpinus* of the Buzzard, with a mean weight of around 700 g, are lighter than the western Buzzards; the mean weight of the race *buteo* is about 850 g. Even more extreme is the difference between the North African race of the Long-legged Buzzard (*Buteo rufinus cirtensis*, mean weight 800 g) and the nominate eastern race (*B. r. rufinus*, mean weight nearly 1300 g).

The similarities in the various buzzards' hunting methods and habitat requirements make stable co-existence difficult, and we can therefore expect the buzzard species to be to a great extent geographically isolated from each other. This is very clearly the case during the summer (figure 63A), when only within a few areas is there more than one buzzard species. A case of stable co-existence between two species can be seen in Africa, where the distribution of the Augur Buzzard completely overlaps that of the Mountain Buzzard. This is an 'ecologically permissible' combination in that the Augur Buzzard weighs almost twice as much as the Mountain Buzzard. The ecological separation is facilitated because the two species largely have separate habitat requirements: the Mountain Buzzard occurs locally in enclosed mountain forests 2000–3000 m above sea level, and the Augur Buzzard is widespread in savanna habitat and in open forest habitat.

How is the balance between the different species' distributions affected when north Europe and Siberia are shut off to the buzzards by the winter? Figure 63B shows that the Buzzard is affected most drastically: the Rough-legged Buzzard retreats south, while the Long-legged Buzzard and the Upland Buzzard to a large extent maintain their stations. The Buzzard gets squeezed out and, except in west and south Europe, moves southwards to the tropics. The friction between the buzzard species increases during the winter and the areas of overlap in the distribution become more numerous. In parts of Africa there are three species occurring together; the medium-sized Buzzard wedges itself in between the Augur Buzzard and the Mountain Buzzard. The fact that this combination can work out without any species being forced out by competition is remarkable. The Mountain Buzzard seems to cope with the

Figure 63   The distribution of various species of buzzard in the genus *Buteo* during summer and winter. In winter the total distribution is compressed and the degree of overlap and friction between the competing species increases. Based on, among others, Brown (1970), Dementev & Gladkov (1966), Glutz von Blotzheim *et al.* (1971) and Voous (1960).

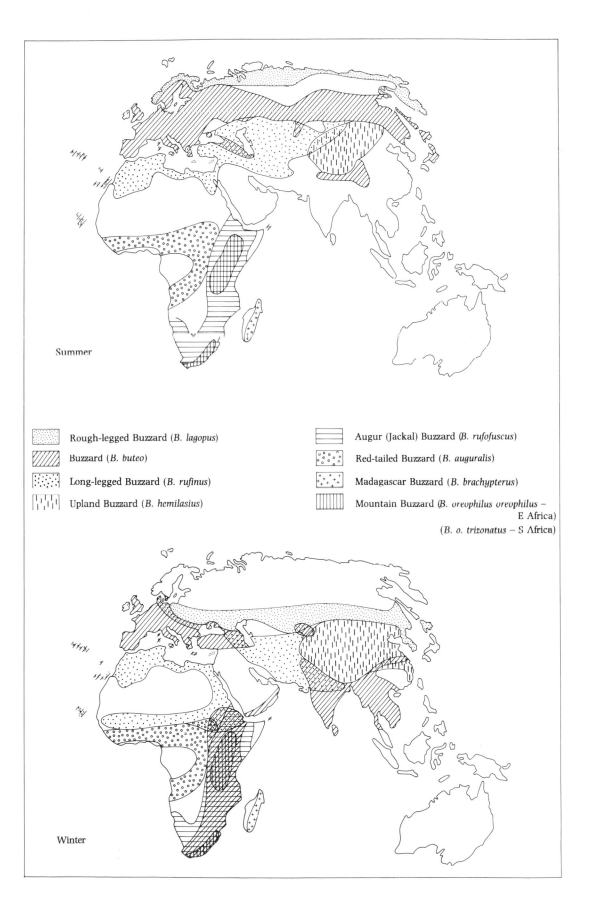

Summer

Winter

Rough-legged Buzzard (*B. lagopus*)

Buzzard (*B. buteo*)

Long-legged Buzzard (*B. rufinus*)

Upland Buzzard (*B. hemilasius*)

Augur (Jackal) Buzzard (*B. rufofuscus*)

Red-tailed Buzzard (*B. auguralis*)

Madagascar Buzzard (*B. brachypterus*)

Mountain Buzzard (*B. oreophilus oreophilus* –
E Africa)
(*B. o. trizonatus* – S Africa)

competition only by the skin of its teeth, thanks mainly to its particular habitat requirements; it is always very uncommon and occurs only locally. When, in 1963, Gustaf Rudebeck totted up his bird-of-prey observations from various journeys in southernmost Africa, he found that he had recorded in round figures 400 Buzzards but only five Mountain Buzzards. These figures do not give any precise measure of the relative numbers of the two species, for he looked specially for the Mountain Buzzards in dense forests and the Buzzards he saw pretty well everywhere during his journeys in the open South African landscape. Compared with the Buzzard, the Augur Buzzard occurs sparsely in South Africa, where it prefers mountainous terrain; by contrast, it is common on the East African high plateau, where the winter population of Buzzards is sparse. Apparently a delicate balance prevails on the buzzard grounds of Africa.

During the winter, friction arises between different buzzard species in other areas also. Thus, for example, the Long-legged Buzzard and the Buzzard come together in large parts of northern India. In south Scandinavia and in eastern Europe the Buzzard and the Rough-legged Buzzard clash. Magnus Sylvén (1978) has studied competition between Buzzard and Rough-legged Buzzard on wintering grounds in south Sweden (Scania). The two species

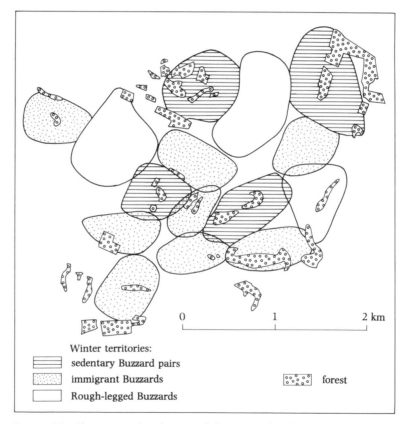

Figure 64 The winter distribution of the Buzzard and the Rough-legged Buzzard overlap each other in such places as southernmost Sweden (cf. figure 63). The map shows that suitable hunting grounds in a closely studied area in south Scania are divided into exclusive territories both of the sedentary pairs of Buzzards and of immigrant Buzzards and Rough-legged Buzzards. Based on Sylvén (1978).

exhibit only slight differences in habitat choice, hunting technique and food choice. The Rough-legged Buzzard is of course known for its habit of hovering while hunting, a good adaptation for open upland moors and tundra moor where perches suitable for looking out for voles are lacking, but in the Scanian winter terrain it spends less than two-hundredths of the time hovering. Magnus Sylvén carefully entered in his record books the two species' choice of perch when vole-hunting and found an overwhelming unanimity between them. The conclusion must be that the competition between individuals of the two different buzzard species is very nearly as great as that between the individuals within the same species. That the state of competition is acute is confirmed by the fact that the buzzards defend winter territories; intruders both of their own species and of the opposing species are chased out.

Figure 64 shows a mapping of the buzzards' winter territories in one area in southern Scania. Here there are sedentary pairs of Buzzards which retain their breeding territories, although reduced in extent, migrant Buzzards which arrive in the autumn and set up territories until the spring migration, and finally wintering Rough-legged Buzzards. The wintering Buzzards and Rough-legged Buzzards show strong site fidelity, with the same individuals returning year after year to their particular territories (such site fidelity, incidentally, occurs also in those Buzzards which winter in South Africa, where ringing has shown that the same individuals return each year to the same area).

Bearing in mind the Buzzard's multifarious travels and living environments, we may ask the question whether it does not take the prize of most fascinating migrant bird of prey, notwithstanding all the rare falcons and eagles. Just consider this when you see Buzzards soaring on motionless wings beneath the cumulus clouds.

3.7   Insect-eaters

At the present time there exist on the earth approximately $1\frac{1}{2}$ million different species of living organisms known to science. (The true number of species is in reality considerably greater, since specific knowledge of certain groups of micro-organisms and insects is very imperfect. One recent guess is that there are at least 20 million insect species, living mainly in the rainforest, which remain to be discovered and scientifically described.) Around 300 000 of the known species are plants. All the rest are therefore animals. As much as three-quarters of the animals, 900 000 species, are insects. The beetles, with 350 000 known species, form the largest group of insects, though this group is still far from completely explored. In this perspective, 8700 bird species or 3500 mammal species seem like a drop in the ocean.

The secret behind the insects' enormous profusion of diversity and successful dispersion on the earth presumably lies in a combination of three key factors: small size, metamorphosis and wings. Small size opens up a rich world of different living niches even in the smallest of passages in nature's complex architecture. Metamorphosis makes it possible to combine different niche fragments into a complete foundation for existence; in typical cases the insects undergo one phase of life as larvae in a temporary environment, and then, through pupation and metamorphosis, totally change both their shape and their ecological niche. The ability to fly makes insects efficient colonists; they can move rapidly between the widely differing subsistence bases of the various stages of life.

When sun and warmth return in spring to the northerly regions, not only an effusive blooming of chlorophyll but also an explosion of insect life takes place. Bearing in mind the insects' amazing multiplicity of forms, it is therefore not that remarkable that of the various 'ecological' bird groups the insect-eating birds constitute the category with the greatest wealth of species (table 20).

Table 20. *Insectivorous birds which occur regularly in north Europe. The data on wintering areas apply primarily to the species' populations in northwest Europe, particularly in Scandinavia. The Lesser Whitethroat forages not only in deciduous foliage but also very often in coniferous vegetation. Scandinavian Chiffchaffs prefer coniferous greenery for foraging; in more southerly parts of Europe the species occurs to a large extent in pure deciduous woodland*

Feeding habits	Species (NB=does not breed in Scandinavia)	Largely resident	Occurs in N Europe in winter	Important wintering areas				
				W or N Europe	Med. region	Africa S of Sahara	Southern Asia	Invasion migration
Catch prey in air	Hobby					+		
	Red-footed Falcon (NB)					+		
	Bee-eater (NB)					+		
	Nightjar					+		
	Swift					+		
	Sand Martin					+		
	Swallow					+		
	House Martin					+		
	Waxwing		+	+				+
	Spotted Flycatcher					+		
	Red-breasted Flycatcher						+	
	Collared Flycatcher					+		
	Pied Flycatcher					+		
	No. of species		1	1		11	1	1
Catch exposed prey on ground or other surface	Honey Buzzard					+		
	Roller					+		
	Tawny Pipit					+		
	Yellow Wagtail					+		
	White Wagtail				+	+		
	Redstart					+		
	Black Redstart			(+)	+	(+)		
	Whinchat					+		
	Stonechat			(+)	+			
	Wheatear					+		
	Red-backed Shrike					+		
	No. of species			1	3	9		

Foraging method	Species	1	2	3	4	5	6	7
Pick prey from herbaceous vegetation	Corncrake					+		
	Grasshopper Warbler					+		
	River Warbler					+		
	Savi's Warbler					+		
	Sedge Warbler					+		
	Blyth's Reed Warbler							
	Marsh Warbler					+		
	Reed Warbler					+		
	Great Reed Warbler					+		
	Bearded Tit	+	+				+	?
	No. of species	1	1			8	1	?
Pick prey from deciduous foliage (bushes, trees)	Cuckoo					+		
	Golden Oriole					+		
	Icterine Warbler					+		
	Barred Warbler					+		
	Lesser Whitethroat					+		
	Whitethroat					+		
	Garden Warbler					+		
	Blackcap		(+)			+		
	Greenish Warbler						+	
	Arctic Warbler						+	
	Wood Warbler					+		
	Willow Warbler					+		
	Penduline Tit	(+)		?	–			
	No. of species	1	1	1	–	10	2	
Pick prey from coniferous greenery	Chiffchaff			+				
	Goldcrest			+		+	+	+
	Firecrest			+				+
	Coal Tit			+	–			
	No. of species			3	–	1	1	2
Pick prey from woody parts of plants (twigs, branches, trunks, stumps)	Grey-headed Woodpecker		+	+				(+)
	Green Woodpecker		+	+				+
	Black Woodpecker		+	+				+
	Great Spotted Woodpecker		+	+	+			
	Middle Spotted Woodpecker		+	+				
	Lesser Spotted Woodpecker		+	+				(+)
	White-backed Woodpecker		+	+				(+)
	Three-toed Woodpecker		+	+				(+)
	Nuthatch		+	+				(+)
	No. of species		2	3	1			1

Table 20 (cont.)

				Important wintering areas				
Feeding habits	Species (NB=does not breed in Scandinavia)	Largely resident	Occurs in N Europe in winter	W or N Europe	Med. region	Africa S of Sahara	Southern Asia	Invasion migration
	Treecreeper	+	+	+				(+)
	Long-tailed Tit	+	+	+				
	Marsh Tit	+	+	+				(+)
	Willow Tit	+	+	+				
	Crested Tit	+	+	+				
	Blue Tit	+	+	+				+
	Great Tit	+	+	+				+
	Siberian Tit	+	+	+				
	Siberian Jay	+	+	+				
	No. of species	18	18	18				3
Pick prey concealed in ground layer	Woodcock	(+)		+	+			
	Dotterel				+			
	Hoopoe					+		
	Wryneck					+		
	Tree Pipit					+		
	Red-throated Pipit					+		
	Meadow Pipit			+	+			
	Wren		+	+	+			
	Dunnock			+	+			
	Robin		(+)	+	+			
	Thrush Nightingale					+		
	Bluethroat					+	+	
	Ring Ouzel				+			
	Blackbird		+	+				
	Fieldfare		+	+	+			(+)
	Song Thrush			+	+			
	Redwing			+	+			
	Mistle Thrush			+	+			
	Starling		(+)	+				
	No. of species		3	11	11	6	1	
All categories	Grand total of species	22	25	35	19	45	5	5

I have called the birds in this category insect-eaters despite the fact that most do not eat only insects in the strict sense but also all kinds of other small invertebrate 'creepy-crawlies' such as spiders, centipedes, earthworms or small crustaceans. The dividing line between this category and that of seed-eating birds, which is dealt with in the next section, is not sharp. Several insect-eating birds often eat a lot of seeds, berries and fruits, especially those species which winter in northern latitudes, and many seed-eaters change over to an insectivorous diet during the breeding season. In some cases, therefore, it is almost a matter of personal preference whether one describes the species as primarily an insect-eater or a seed-eater.

Table 20 shows that birds of widely different genera have adapted to a life as insect-hunters, from the large Honey Buzzard to the little Goldcrest. If we look beyond the confines of northern Europe, there are further surprising examples of insect-eaters. The Gull-billed Tern, whose nearest breeding area is in Denmark/West Germany, has largely given up fishing and instead occupies a niche on land. At its breeding sites it feeds mainly on beetles and grasshoppers, with frogs, lizards, voles and young of birds as important additional items. This species winters in the interior of Africa south of the Sahara, where it has been observed flying back and forth through the dense smoke of grass fires in pursuit of insects driven out by the fire. In southern Europe there are a couple of wader species of the genus *Glareola* (pratincoles) which specialise in catching flying insects in the manner of swallows; the Black-winged Pratincole moves in winter, sometimes in flocks a thousand strong, to grass plains and steppes in southern Africa to hunt swarming locusts and grasshoppers, termites, beetles etc. There is even an owl, the little Scops Owl, which is an insect specialist: from its summer home in southern and central Europe it leaves on a long migration across the Mediterranean Sea and the Sahara in order to continue catching insects at tropical forest edges and in small woodlands.

I have subdivided the species in table 20 into different groups according to hunting technique and feeding habitat. The first two groups comprise species which are adapted to pursuing a mobile and exposed prey, either in the air or on the ground. This method of catching food is of course impossible during the winter in northern regions, and all the species but one, the Waxwing, migrate far to the south. In order to survive wintering in the north the Waxwing must make a drastic change of niche: during the summer months it is a mosquito-catcher in the northern coniferous forest, but at other times of the year it lives by eating fruit and berries, especially rowanberries. In years when the breeding population is large and at the same time the rowanberries fail, large parties of Waxwings are forced to head south to areas with richer crops of fruit and berries; in these years, ringers who catch Waxwings passing through Scandinavia can get recoveries from the whole of Europe, from Scotland in the west to Turkey in the southeast.

Birds which take insects among tall herbs or in deciduous foliage naturally enough generally migrate to the tropics in the winter. There are, however, one or two exceptions. Like the Waxwing, these species make a radical change of niche. In summer the Bearded Tit eats insects among the lush vegetation at the edge of food-rich lakes, but during the winter months it specialises on the seeds of reeds.

A pronounced granivorous diet places different demands on the digestive system from those imposed by insect food. The Bearded Tit's stomach and intestines weigh roughly twice as much during the winter months as during the summer months. This weight increase is due to a massive strengthening of the muscle tissue around the alimentary canal and takes place simultaneously with the change-over to the difficult-to-digest diet of reed seeds. Even though the changes are not so radical as they are in the Bearded Tit, it is probably not particularly unusual among birds for the anatomy of the digestive system to be altered concurrently with varying dietary habits at different seasons. The Bearded Tit's dependence on reed seeds as a key food during the winter gives us reason to suspect that it is adapted for invasion migration in years when the reeds' seed production is exceptionally poor. Owing to the species' sparse occurrence and patchy distribution, what we know of the scale and scope of the irruptions and how they come about is thus far very blurred.

The Penduline Tit's winter ecology is also poorly known and calls for more detailed future studies. Like the Bearded Tit, the Penduline Tit resorts to large reedbeds during the winter, but to what extent it feeds on insects and to what extent it feeds on seeds is not known. Its migratory habits appear confusing. Wintering Penduline Tits have been found in reedbeds in Scania in south Sweden, but breeding-season recoveries have also been made in Scania of Penduline Tits ringed in winter outside Lisbon in Portugal. Furthermore, it is known that many Polish Penduline Tits, the south Swedish population's nearest neighbours, migrate south to the vast reedbeds at Lake Neusiedl in Austria. What exactly are we to make of the species' migration pattern on the basis of this?

Insect-eaters which catch their food in coniferous greenery or from the woody parts of trees and bushes have entirely different migratory habits from the above-mentioned groups. The conifer needles, the tree trunks, the branches and the stumps are of course still to be found in winter in northern regions, and so too are the birds that belong there: woodpeckers, tits, Nuthatch, Treecreeper, Goldcrest and Siberian Jay (ecologically speaking, the Siberian Jay is nothing other than a large tit). Overwintering is far from trouble-free for these birds, which often make use of special adaptations to enable them to survive the winter's bottleneck of limited supply of insects, difficult to get at and immobile.

Several species change over partly to plant food, to berries, seeds and nuts. In this way the Great Spotted Woodpecker is in large part dependent during the winter months on the availability of pine-cone seeds, the Coal Tit on the availability of spruce seeds, and the Great Tit and Blue Tit on the availability of beechmast. When the seed or nut production fails, large numbers of these species set off on migration towards milder winter quarters. The Great Tit's migratory habits are set out in more detail later in this chapter. The ecological interplay and counterplay between fruit-bearing and seed-bearing perennial plants and the birds dependent on them are dealt with in the next section.

Goldcrest, Treecreeper and Long-tailed Tit are what are known as partial migrants. Part of the population moves south in winter while another part stays behind. The intensity of the migration fluctuates greatly: in certain autumns it is of invasion proportions, while in other years it is far less conspicuous. These fluctuations occur without the species being, so far as we know, dependent as the 'classic' invasion birds are on any special key food. The partial migration pattern is probably maintained through competition for the limited winter supply of insect food. During mild winters a large proportion of the birds survive and give rise to a large breeding population with many young during the following summer and autumn. The normal winter supply of insects is not sufficient for all these birds. When bird densities are high, an innate instinct triggers migration to milder regions. In the first place it is the young of the year, those which are lowest in rank competition-wise, that migrate. In autumns following a mild winter or after several mild winters the migration takes on invasion proportions. After harsh winters, on the other hand, the population density is greatly thinned out; there is room for the young of the year to overwinter in the vicinity of the breeding localities. Flexibility is an important adaptation for the birds to be able to survive the winter as northern insect-eaters. To migrate or not – it depends on the density of food competitors, on the individual's own status in the competition, and on the expected supply of winter food. So, the final result in life for an adult individual is that it migrated most often as an immature but more infrequently when it became older.

Some ornithologists do not believe the birds to have such flexible migratory preparedness. They argue that partial migration is instead explained by the fact that one part of the population consists of strictly migratory birds and another part of rather pronounced residents; the genes govern the category to which a certain individual will belong. Both the alternatives are, according to this explanation, equally successful in the long run; it leads to a balance between the categories, even though constantly changing as one or the other category is temporarily favoured by the capricious games of the forces of the weather. That the balance would rest on a knife-edge, so that winter survival would in the long-term perspective be almost exactly the same for migrants and for residents, is most improbable. In that case, it is more likely that the balance rests on interdependent advantages and disadvantages, so that, for example, migrants are favoured in comparison with residents where winter survival is concerned but disadvantaged in the competition for the best breeding sites. It has recently been demonstrated for partially migrating Blackcaps that there is a definite element of genetic influence on the tendency of individuals to migrate or not (Berthold 1988). In bird species that are partial migrants, however, the age composition among migrating birds, as well as ringing results (cf. the Great Tit below), bears out the fact that the birds also possess the individual migratory flexibility that was discussed above. Hence, the relative importance of genetic and environmental factors in partial migration remains to be clarified – an extremely interesting problem from an evolutionary point of view.

Alongside change of niche, invasion migration and partial migration (these two migration strategies are for the most part based on common ecological principles), there is also the habit of caching a winter store of food during summer and autumn. The most zealous hoarders, those which conceal within their territories insects, berries and seeds in bark crevices and in slits in branches, beneath lichen and among conifer needles, are the most extreme examples of resident birds, the Marsh Tit, the Willow Tit, the Siberian Tit, the Crested Tit, the Nuthatch and the Siberian Jay. If we ignore the late-summer dispersal of juveniles, which is of widespread occurrence, in most of these species no significant migration is observed as a rule. To a certain extent the Willow Tit and the Siberian Tit are exceptions; in some autumns they migrate invasion-style. Swedish Willow Tits have been recorded wandering west to Norway, while Siberian Tit irruptions periodically reach from the northern taiga to southern Finland and central Sweden. It is probably juvenile tits that migrate instead of staking out territories for hoarding. The Coal Tit is also well known as a hoarder. At the same time it is an outright irruptive species which in certain autumns of poor spruce-cone production undertakes mass migrations far south into Europe. According to Swiss studies, 97–99% of the invading Coal Tits are juveniles. The Coal Tit, which thus has not only a flexible preparedness for long-distance migration but also the flexibility either to look after its winter fortunes by preparatory caching or to flock to areas with temporary food surpluses, is a tempting species for ecological specialist study.

The species in the last group in table 20, those which take small insects on the ground, present the most discordant picture of migration. Some winter as far to the north as the ground is free of snow and frost; others are tropical migrants. Some of the thrushes, foremost among them the Fieldfare, have made themselves partially independent of the snow by changing their feeding niche and becoming berry- and fruit-eaters in the winter. The Fieldfare shows marked invasion tendencies in relation to the supply of rowanberries. What special food requirements compel the Thrush Nightingale to migrate south of the equator in Africa when the Robin as well as the smaller thrushes are able to overwinter in west Europe? Besides, it is not at all certain that the Thrush Nightingale does have individual food requirements; the competition for west Europe's limited winter resources may be decisive without any conspicuous differences between the species' food niches being prevalent.

Remember that Swedish Buzzards migrate to west Europe while Finnish and Russian ones winter in South Africa!

In commenting on the species in table 20, I have so far dwelt mostly upon the strategies and special adaptations that allow certain species to succeed in defying the northern winter. Yet there is every reason not to forget the achievements of the long-distance migrants, which are at least as impressive. To underline this I have given a few examples of the most striking movements in figure 65. The little Arctic Warbler of the montane birch forest, with a weight of less than 10 g, winters in the East Indies and surrounding islands; for the north Scandinavian population this involves a migratory journey of more than 9000 km diagonally across the European and Asian continents. At the same time the Wheatears that breed in easternmost Siberia and Alaska migrate across Asia in the opposite direction, towards winter quarters in the African savanna zone south of the Sahara. The question, though, is whether Wheatears from Greenland and adjacent parts of Canada (*Oenanthe oenanthe leucorhoa*), despite the fact that they fly a somewhat shorter route, are responsible for an even more hazardous migratory achievement. Their flight route stretches across the inland ice of Greenland and over the open North Atlantic and then continues on across west Europe and the Sahara to the normal winter quarters in westernmost Africa. In autumn these Wheatears make use of the fresh northwest winds following the passages of depressions and fly direct, in one stage of about 2500 km, from the southern tip of Greenland, Cape Farewell, to west Europe. In the spring, on the other hand, their journey takes them across Britain and

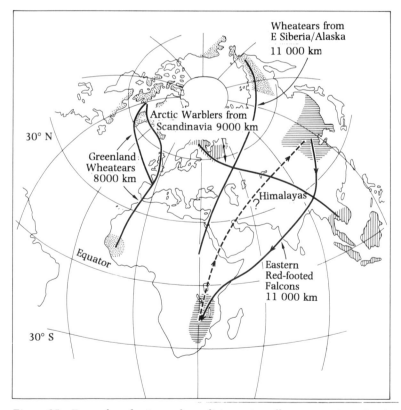

Figure 65   Examples of extreme long-distance travellers among insect-eaters. Based mainly on Moreau (1972).

Iceland, which are used as intermediate staging posts. Over ice, sea and desert, the Wheatears commute between summer quarters where the mean temperature is only a few degrees centigrade and winter quarters where it is 30 °C.

The most striking migration pattern of all among the insect-eaters is perhaps found in the Eastern or Manchurian Red-footed Falcon, a close relative of the Red-footed Falcon. From summer quarters between the Amur River in the north and the Yellow River, the Hwang Ho, in the south, the falcons migrate in the autumn on the south side of the Himalayan massif to India, where they arrive in November; they stop there for several weeks in order to put on large amounts of fat before a 3000-km flight across the Indian Ocean. This demanding flight stage across the ocean takes place at just the right time after the northeast monsoon of late autumn has stabilised so that it can provide the falcons with tail winds for migration. Not until December do the falcons reach the heart of their winter quarters, around Zimbabwe. In March–April it is time for the return journey northwards, but now the trail of the migrating falcons comes to a halt; we do not know what route they follow. Spring observations from India for example are lacking, even though the southwesterly monsoon stretches out over the Indian Ocean and offers the birds such good winds that they would be able to return the same way as they came. An observation of a flock of Eastern Red-footed Falcons passing Mecca in April suggests that they travel instead via Arabia and from there continue on the north side of the Himalayas and the highlands of Tibet.

How much of the insect food available do the birds exploit? With few exceptions, studies during summer in temperate and arctic regions have shown that they consume only a few per cent of the insects, sometimes so little that it can be counted in thousandths of the whole. A typical example is provided by the larvae of the winter moth, the staple food of the tits of deciduous woodland in England during the breeding season. On average, only two out of every 100 larvae fall prey to the tits; and in years when there are mass flushes of insects, often with devastating damage to the woodlands, the birds, despite increased breeding density, eat only an exceedingly small proportion of the myriads of insects that are available. The conclusion must be that in summer insectivorous birds as a rule exist amid a liberal superabundance of food. The clutch size and the number of young reared by a pair of adults are only to an insignificant extent limited by the absolute quantity of food.

Rather, the number of young is limited by the rate at which the birds can collect food for them combined with the risk of poor weather (when efficient insect-catching becomes impossible) and the risk that predators will discover the nest and plunder it. Breeding is therefore a battle against various risk factors. It is important for the birds to choose a clutch of just the right size so that the risks that come with increasing clutch size do not impair the total breeding success.

During the winter the situation is entirely different. The birds' offtake of available insects then regularly amounts to several tens of percent, in certain cases up to 98%. The British tits take around one-fifth of all adult females of the winter moth, thus a far greater proportion than of the larvae during the early summer. A classic ecological study carried out in England concerns the winter survival of the pupae of the pine-cone roller (Gibb 1958). This tortricid moth lays its eggs on pine cones. After the egg hatches the larva burrows into the cone, where it eats about seven pine seeds before gnawing a small chamber just beneath the cone's surface; pupation takes place in this chamber. After the final adult moths have formed and have left the pine cones, the fallen cones can be gathered up and the number of adults that have developed can be counted from the number of small round exit holes that the tortricids leave behind when they abandon the pupal chamber in the pine-cone scales. All holes to the pupal chambers, however, are not made by the moths. Some very small holes are the result of a parasitic wasp which has been at work and attacked the moth pupae. Yet other holes are extra large and have broken edges. These show that Blue Tits or Coal Tits have hammered through and eaten the pupae: by pecking with the bill at the cone scales, the tits can work out whether there is a pupal chamber in the cone. By gathering large numbers of cones and counting the different kinds of holes in the scales, it was possible to establish that in normal winters the two tit species take between 50% and 60% of all tortricid pupae. The tits' offtake is exceptionally high, around 80%, in those tree stands which have been subject to the heaviest tortricid attacks. Despite these high offtake figures, the tortricid pupae account for only a minor part of the tits' total food consumption during the winter.

Another method of estimating food offtake of insectivorous birds during the winter is to use nets to shut off parts of trees or bushes to the birds. At the end of the winter the supply of small insects in these shielded areas can be measured and compared with equivalent areas where the birds have had free access. Using this method, it has been estimated that during the course of the winter the coniferous-forest tits and the Goldcrests in a spruce forest in west Sweden eat more than half of the large nutritious spiders which live on spruce twigs. A similar technique using protective netting on tree trunks has been used in North America in order to calculate how many larvae and pupae of bark beetles the woodpeckers consume during the winter. Estimates of the offtake from different areas and in different years generally vary between 20% and 98%; in several cases the estimate has been around 80%. The woodpeckers collect in forest regions with a high density of beetles, for example in districts which have been ravaged by forest fire. Here, of course, the dead trees form dispersion centres for mass explosions of pest insects. In a further valuable study from North America, it was estimated that wintering American Downy Woodpeckers (*Dendrocopos pubescens*) and Black-capped Chickadees (*Parus atricapillus*) consumed 95% of the pine-wasp cocoons that were accessible in bushes and trees above the snow-line.

That insect-eaters are hard pressed by food scarcity during the northern winter is also evident from feeding experiments. By putting out extra food for Willow Tits and Crested Tits in the coniferous forests, the number of birds surviving the winter is increased significantly. The same has been shown for Great Tits in deciduous woodlands in south Sweden.

Northern insect-eaters thus live close to their 'food ceiling' during the winter; the summer months, on the other hand, allow of a life amid food abundance. A similar surplus of breeding resources during the summer compared with limited resources for survival during the winter probably also holds true for the insect-eaters that are long-distance migrants. This conclusion we may venture to draw despite the fact that no studies have been made of food offtake in the tropical winter quarters. The northern migrants do not winter in the most fertile tropical environments, but live mainly in arid savanna areas, in country that has been influenced by cultivation, burnt woodland or new young tree saplings, where they occasionally lead a nomadic life in search of irregular and temporary concentrations of food. Obviously, native tropical birds for the most part drive out visiting northern migrants from the most stable and most fertile tropical habitats, from the rainforests and the luxuriant deciduous forests south of the equator.

The surplus of breeding resources in comparison with survival space results in bird populations being restricted by the winter's 'bottle-neck', but when the summer comes there are resources sufficient for all age categories to breed and to rear a large brood, sometimes even two or more broods per season. Most species complete the exacting moult of the flight feathers immediately after breeding, while they can still take advantage of the northern summer's surplus of insect food. Some tropical migrants, however, have other moulting habits.

A small number of species have a less marked surplus of breeding space than the bountiful food supply during the summer would suggest. These species live their life in exposed, open terrain. The inevitable flight traffic to and from the nest, especially during the period when young are being fed, easily reveals the nest site to various predators. In order to prevent nest predation, therefore, these species must build their nests in protected sites, in tree holes, in rock crevices, beneath roof tiles, in buildings and in similar sites where the nests are out of reach of predators. High demands for secure nest sites naturally cut down the size of the breeding niche, in some cases so drastically that the first-year birds are excluded from breeding and have to content themselves with a good living on the summer abundance of food. This applies, for example, to the northern populations of Starling and Swift.

By comparison, attention may be drawn to the fact that tropical insect-eaters do not experience the same pronounced abundance of breeding resources relative to survival resources as their northern counterparts. Mean clutch size in tropical passerines is generally only two or three eggs; in their relatives in Europe it is five or six eggs (the difference is greatest in the tits: the mean clutch size in tropical species is 3.5, in European ones 9.2). The start of breeding in the tropics by no means always takes place at one year of age. One of the important factors limiting the birds' breeding niches in the tropics is the widespread nest predation; snakes are often the birds' most dangerous enemies. In certain tropical environments, seasonal variations, with their attendant variations in food supply, are subdued. This results in there being a consistently rich food base for a good stock of insectivorous birds, with high survival, but on the other hand a narrow scope being left for breeding and for rearing of young, which demand extra resources.

To give an example of the migratory habits of insectivores, I shall look more closely at the Great Tit, which often overwinters far to the north, and the Redwing, which is a middle-distance migrant. I shall also provide some insight into the life of tropical migrants in their winter quarters.

The Great Tit *(Parus major)*

The Great Tit in a book on migratory birds? I can understand if some readers react with incredulity. Just a glance at the birdtable in the depths of winter is enough to verify that the Great Tit is still in northern Europe. How does this tally with migration? This is a problem that has proved to be surprisingly complicated and which has still not been finally resolved. Let me put together some pieces of the puzzle.

Ringing recoveries of Great Tits which have been caught, chiefly in irruption years, in Switzerland (mainly at the Col de Bretolet pass in the Alps), at the Baltic bird observatory of Rybachi and at Falsterbo in southernmost Sweden are shown in figure 66. The Great Tits that migrate through Switzerland winter largely on the French Riviera and in the Rhône area between the Cevennes hills in the west and the Alps in the east. The majority are juveniles, and only a few return to France in subsequent winters. Instead they then stay closely associated with the breeding sites, which are sometimes as far away as central Russia, 2000 km from Switzerland.

The Great Tits from northwest Russia and the Baltic States also set off on amazingly long-distance movements. Some move southwest towards southern France, and in so doing a few touch Switzerland. Most maintain a more westerly course and winter in north Germany and occasionally as far west as Ireland. The element of individuals over one year old in these movements appears to be somewhat greater than in Switzerland.

At Falsterbo the migration picture is totally different. From here the Great Tits continue no further than to Copenhagen and the adjacent Danish islands. Some do not even cross the

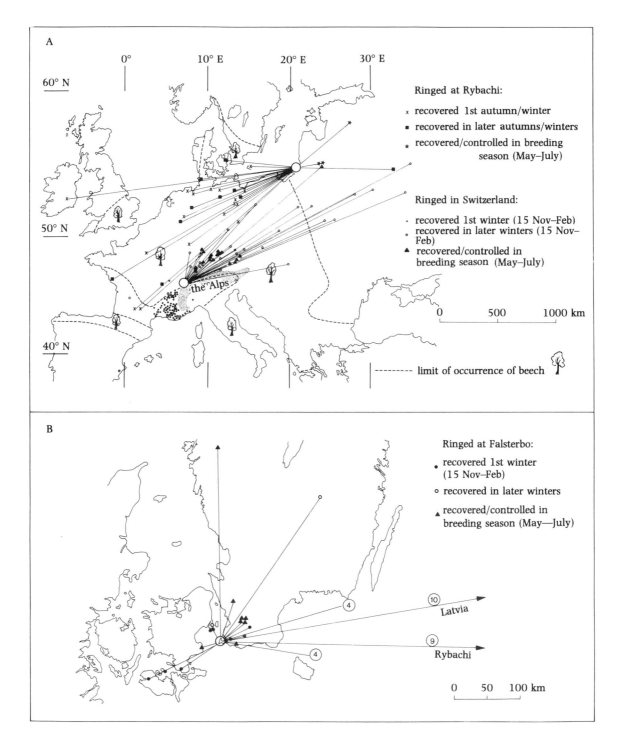

A

60° N

0°   10° E   20° E   30° E

50° N

40° N

the Alps

Ringed at Rybachi:

x recovered 1st autumn/winter

■ recovered in later autumns/winters

★ recovered/controlled in breeding
season (May–July)

Ringed in Switzerland:

· recovered 1st winter (15 Nov–Feb)

○ recovered in later winters (15 Nov–
Feb)

▲ recovered/controlled in
breeding season (May–July)

0   500   1000 km

------- limit of occurrence of beech

B

Ringed at Falsterbo:

· recovered 1st winter
(15 Nov–Feb)

○ recovered in later winters

▲ recovered/controlled in
breeding season (May—July)

⑩ Latvia

⑨ Rybachi

④

④

0   50   100 km

Figure 66 The Great Tit's migration pattern according to ringing results from the Russian bird observatory at Rybachi and from Switzerland (A) and also from Falsterbo bird observatory (B). In the upper map the distribution of the beech in Europe is also shown. Basic data from Switzerland include recoveries 100 km or more from place of ringing and are taken from the annual ringing reports for 1955–1976 in *Ornithologische Beobachter*. Data for

narrow strait between Sweden and Denmark but turn back and overwinter in southwest Scania. The Great Tits at Falsterbo come partly from breeding areas in southernmost Sweden and partly from regions east of the Baltic Sea. The latter represent the northernmost spur in the migration system that has been mapped through ringing at Rybachi.

Even though the mainstream of migrating Great Tits from northwest Russia is generally diverted along the eastern and southern shores of the Baltic, sometimes, as in the autumns of 1975 and 1976, the birds may cross the Baltic Sea in large numbers (figure 67). In both those autumns the tits were greatly assisted in their crossing of the sea by southeast winds – on some days the flocks came in to the Swedish coast in a tail wind of 15 m/s. Radar observations showed that the birds were travelling west–northwest on a broad front over the whole of the southern Baltic Sea, at a speed which, on days when the tail wind was strongest, reached 84 km per hour. Ringing recoveries show that some of the Great Tits came ashore only briefly in south Sweden and rapidly turned southwest or southwards back to Continental Europe. One bird was recovered after a month as far away as Belgium. Other Great Tits continued their migration through southern Sweden and in one autumn (1975) reached Falsterbo, where almost 2000 were logged moving out over the Öresund Strait. In the other autumn, 1976, however, not a single Great Tit was recorded passing through Falsterbo, despite the fact that the influx from the east to Öland was heavier than usual. The most likely explanation for this is that the tits' migration was checked by the ample supply of beechmast that autumn in the south Swedish beech woods. The previous autumn was by contrast a poor one for beechmast, and the tits then extended their migration beyond the beech woods to the mild districts of the Öresund and to the Danish islands.

The first person to point out the great importance of the beechmast supply in the migration of the Great Tit was the Swedish ornithologist Staffan Ulfstrand. He observed that Great (and Blue) Tits concentrated in the Scanian beech woods during autumns and winters of 'mast-years' in order to feed on the fallen mast. In poor years for mast only a few Great Tits are present in the beech woods. During most of these autumns there is instead an invasion of migrants recorded at Falsterbo. The annual fruit production of the beech is determined by, among other things, wide-scale variations in climate; peaks and lows in the availability of mast therefore often, but not always, occur simultaneously over large parts of the beech's range of distribution in Europe (see figure 66A). A rare exception of this sort occurred for example in 1954, a good mast-year in south Scandinavia and a very poor one in southern Germany. It was revealing that in that autumn the Great Tits were conspicuous by their absence at Falsterbo and that a heavy invasion was recorded in Switzerland. During mast-years the beeches invest so much energy in fruit production that they do not form flower buds the next year; this becomes a 'rest-year'. A mast-year is consequently followed almost without exception by a low year. Mast-years therefore usually occur every other year; on average, however, there are almost four years between the really highest of peak years. Staffan Ulfstrand's conclusion was that the Great Tits' migration from northern areas where beech woods are absent stops as soon as the birds reach the beech-wood region during years of good mast supply. At the same time the Great Tits within this latter region show no signs of migration but, like the others, exploit the beechmast as winter food. In those years when the beechmast fails completely, however, massive emigrations of Great Tits take place.

Rybachi, where a number of short-distance recoveries in the Baltic States and on the Polish coast have not been marked on the map, are from Paevskii (1973). From Falsterbo bird observatory material has been made available by Gunnar Roos (cf. Roos 1984). The eastern recoveries of Great Tits that have been found at Falsterbo were made during the migration periods in spring and autumn as per the months shown in figures in (B).

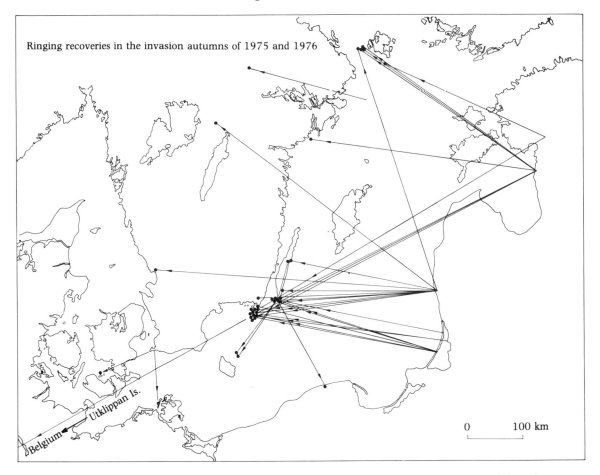

Ringing recoveries in the invasion autumns of 1975 and 1976

Utklippan Is.

Belgium

0   100 km

Figure 67   In the autumns of 1975 and 1976, migrating Great Tits crossed the Baltic Sea on an exceptionally large scale in association with moderate to strong southeast winds. Ringing in autumn at various sites along the Baltic coast yielded many recoveries on the west side of the Baltic Sea. Two recoveries are of Great Tits ringed as nestlings in Estonia. Also shown on the map are recoveries of Great Tits ringed in Sweden in the above autumns. Based on Lindholm (1978).

This is not, however, the whole truth of the matter. The story contains several additional facts. An important piece of the jigsaw is that the invasions consistently occur in those years when the breeding populations of Great Tits are at their most dense. The number of breeding tits is in turn determined by the food supply during the previous winter. Here the beechmast again comes in as an important factor, for a rich beech crop leads to high winter survival and thereby to a large breeding population the following summer. As a rule this happens simultaneously in, for example, England and other parts of Europe and is the reason why peak years for breeding Great Tits in England coincide with irruption autumns in quite different parts of Europe despite the fact that the English tits never take part in any migration. It has recently been shown that the winter climate, too, has an important bearing on breeding populations of Great Tits in north Europe. A mild winter results in high survival and a good breeding population in the following summer; a harsh winter has the opposite effect.

If we put together these pieces of the puzzle, we get a picture where the winter climate in combination with the presence of beechmast determines whether the Great Tit's population density in the following summer will be high enough for a widespread migratory preparedness to be triggered among primarily the juvenile birds. Even if this does happen, it does not necessarily lead to irruptive migration. If the beechmast supply during the autumn is sufficiently good, the tits' migration fervour in the beech-wood region is of course dampened. If the beech crop is poor at the same time as the bird population is extra large, then the massive movements take place. The Great Tits seek winter refuges in such places as the mild western Europe, in Denmark and in the very southernmost part of Sweden. A food source that probably attracts many invasion tits is what human beings in northwest Europe put out on birdtables, especially in town parks and gardens. The studies in Switzerland and at Falsterbo unanimously affirm that a clear majority, around 80%, of the irrupting Great Tits are juveniles and that two-thirds are females. The preponderance of females is hardly surprising when we consider that they are somewhat smaller in size than the males and thus lower-ranking in the competition for the winter resources in the north.

In figure 68 I have made a diagrammatic comparison between this chain of causes and the Great Tit invasions which have been recorded at Falsterbo during a 35-year period. The concordance, even though not 100%, is good. The heaviest invasions should occur in those autumns when the beechmast crop is very poor and when the preceding winter has been both abounding in mast and mild (to an even greater extent if two preceding winters have been mild). These circumstances have occurred during five of the years of migration-watching at Falsterbo, and in all of these years the irrupting flocks of Great Tits have been exceptionally dense. Only in one winter has a rich mast crop been counteracted by unusually harsh weather; in the following autumn no Great Tit invasion was recorded.

The fundamental similarity between the migratory habits of the Great Tit and those of what we call partial migrants, the Goldcrest, Treecreeper or Wren, is striking. The only difference really is that the Great Tit's winter survival is dependent not only on the climate and on the difficulties in finding sufficient insect food conditional on this, but also on a particular and highly variable food source – on beechmast.

The migration of the Blue Tit provides a parallel case to that of the Great Tit. It is susceptible to harsh winters and frequently chooses beechmast as a winter food. The Blue Tit's migration pattern is very like the Great Tit's. The two species generally appear in the same autumns on irruptive migration. The Coal Tit on the other hand is sometimes out of step with these two species in its irruptions. This is explained by the fact that it is dependent for its survival on spruce-seed production and not on the beechmast crop. The winter climate otherwise plays the same important role in the Coal Tit's population dynamics as it does for other tits. The Willow Tit's irruptive movements generally occur in those autumns when the Coal Tits are on the move.

Even if the reasons outlined above explain the fundamental features of the Great Tit's migration, several unsolved problems remain. Why do the north Scandinavian Great Tits not show such clear irruptive tendencies (it has not even been established that they reach south to Falsterbo, see figure 66) as the Russian ones? Perhaps the winter food commonly provided on Scandinavian birdtables checks these tendencies? Or do the invasions pass off fairly unnoticed because they do not run so much southwards but instead westwards (there are in fact ringing recoveries which indicate emigration of Great Tits from north Sweden to Norway)?

A second puzzling factor is the Great Tit's lack of fidelity to birthplace in certain situations. Normally the invading Great Tits return in March and April towards their homelands in the

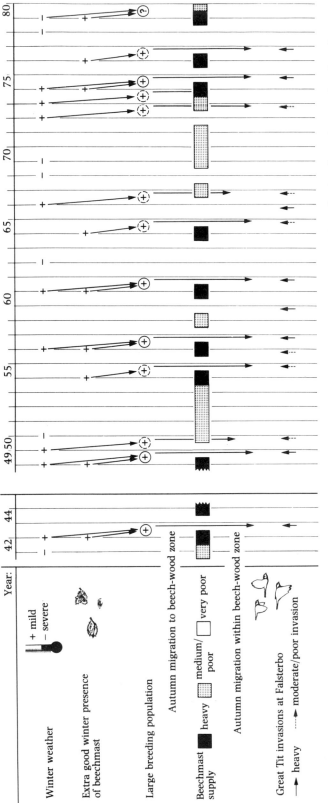

Figure 68 The reasons behind the Great Tit's irruptive migration. Mild winters with a rich supply of beechmast are beneficial to the tits' survival and result in a large breeding population. This gives rise to an increased migratory preparedness that autumn, and an irruption, chiefly of juveniles, materialises if the migration is not slowed down or arrested by a large crop of mast in the beech woods. The chain of events is illustrated diagrammatically for 35 years, between 1945 and 1980, whereby the expected invasions are compared with the observations at Falsterbo (from annual reports and various compilations of trapping and passage figures; Falsterbo data for 1956 are conflicting, and whether an invasion occurred or not in that autumn is unclear). Data on winter weather are based on mean temperatures during December to February in England and Scandinavia, which give an indication, even though imperfect, of the winter climate each year over extensive parts of north Europe. The data on beechmast supply come from Swedish and Danish forest research institutes.

north and east. What sometimes happens, however, is that tits which have hatched in, for example, central Germany stay behind in Belgium to breed, almost 500 km from 'home'. This is perhaps not that startling. But what are we to make of those cases where tits which have hatched and been ringed in, for example, Switzerland have been rediscovered later 500 and even 1000 km away, in Czechoslovakia, Poland and White Russia?

Another unexplained peculiarity of the Great Tit's life is that it does not store winter food and so try to avoid long-distance emigrations.

Where breeding biology is concerned, the Great Tit is, as a result of its general commonness and its habit of taking readily to nestboxes, probably the world's best-studied bird species. Its life as a migratory bird it has, on the other hand, managed to lead with considerably more freedom from the inquisitiveness of man.

The Redwing (*Turdus iliacus*)

How many dark autumn evenings I have listened to and enjoyed the calls of migrating Redwings as I have walked through my home town of Lund in south Sweden. Redwings are one of our commonest nocturnal migrants; they call frequently and clearly, especially when flying through the diffuse dome of weak light that hangs above cities at night. So I carry with me memories from many different towns and cities in autumn of the way the migration calls from Redwings suddenly reach down through the roar of the traffic to the pavement or to the parkway, always just as welcome to the ears. One evening at the end of October I was standing inside the Eiffel Tower and saw the great city in the same luminous perspective as the Redwings which, invisible in the dark of night, were flying past and uttering their calls.

The most intensive Redwing night I can remember, however, was in the Danish town of Frederikshavn near the north tip of Jutland. The continual calls bounced off the narrow streets, penetrated into the houses through window slits and air vents. I heard them unceasingly in the October darkness for several hours before I fell asleep at midnight and missed the chance to find out how the migration unfolded during the early hours of the morning. On that occasion I did not reflect in any detail upon the reason for this mass migration. I assumed that the Redwings had set off from Sweden southwest over the Skagerrak and the Kattegat towards winter quarters in southwest Europe. Today I suspect that an entirely different explanation is the most likely. This is because I have since then helped to reveal, through radar studies, that Scandinavian Redwings do not migrate only to the southwest. A heavily used migration route leads also to the southeast: the main route runs from southernmost Norway, over the Skagerrak, north Jutland, the Kattegat and on across south Sweden (Scania) and the Baltic Sea and down through east Europe to winter quarters between the Black and Caspian Seas and on the eastern shores of the Mediterranean. Let me tell of the discovery of this remarkable migration route.

According to comprehensive ringing results, the winter quarters of the Nordic thrushes are largely in southwest and west Europe. The central point for the Blackbird is the most westerly: Britain plays host in winter to most of the Scandinavian Blackbirds. The main wintering areas of the other species are more to the southwest, in France or in the Iberian Peninsula. So far as the Redwing is concerned, this is evident from table 21. Only 2% of Redwing recoveries fall outside the frame in that they are far to the southeast – hardly enough to instil any particular expectations when I began the mapping of bird migration over southernmost Sweden using radar.

The radar observations, however, soon revealed amazing things concerning the migratory movements. Judging from the type and speed of the radar echoes, during the late autumn they consisted mostly of nocturnal thrush migration. Certainly the radar showed migration

Table 21. *Recoveries in winter (December to Feb-*
ruary) of Redwings ringed at breeding sites in
Sweden and Norway. By far the majority of birds
were ringed as nestlings. The basic data comprise
Norwegian ringing results up to and including 1971
(Mork 1974) and Swedish data for 1960–1971 ac-
cording to annual reports of the Central Ringing Off-
ice (1964–1980, Sten Österlöf, Riksmuseet,
Stockholm)

Country	No.	%
Scotland	1	0.7
Ireland	1	0.7
England	1	0.7
The Netherlands	1	0.7
Belgium	3	2.0
France	87	58.4
Spain	18	12.1
Portugal	21	14.1
Italy	13	8.7
Lebanon	1	0.7
Georgia (USSR)	2	1.3
	149	100

towards winter quarters in the southwest on nights with northerly and easterly winds. But nights of southeasterly migration occurred even more frequently, and the intensity of migration then reached even higher levels. On some days southeasterly passage was still going on after dawn. Radar echoes then changed in character and showed that in the early hours of daylight the birds carried on flying further in well-demarcated flocks. The flocks that arrived in Scania across the Kattegat often reduced altitude when they passed Scanian country and climbed again when they flew out over the Baltic heading southeast.

The typical weather situation for southeastward migration occurs immediately after a depression with associated cold front has passed. The wind is then fresh and from the northwest. Owing to the tail wind the birds' mean speed, according to the radar measurements, was all of 80 km per hour; on some nights and mornings they were moving at 100 km per hour.

I remember every feeling of astonishment and wonder over this large-scale passage, revealed so emphatically by the radar. Experienced migration-observers and ringers were almost entirely anticipating that the passage in south Sweden would be in a southwesterly direction. Then it was discovered that millions of migrants were streaming off towards the southeast! Could it really be thrushes that were behaving in this way? If so, which one or which ones of the various species?

Field observations on the east coast of Scania made at the same time as the radar surveillance provided an early-morning answer to the above question. In a westerly gale the observers counted more than 100 000 Redwings moving past; at the same time the radar showed intensive southeasterly migration. Further field observations show that we cannot entirely exclude the possibility that Fieldfares also occasionally take part in southeasterly movements, even though the Redwings easily predominate.

Immediately we are faced with a new problem. Do the large flocks of Redwings continue on to winter quarters in the southeast or do they turn off towards the southwest when they reach Continental Europe? Today there are, from within the suspected region in the southeast, so many winter recoveries in total of Redwings which have been ringed above all in Britain that there can hardly be any doubt that there is an important wintering area there (figure 69A). The ringing in Britain had been carried out during the autumn and winter of the previous year and in consequence showed that one and the same Redwing could in one year winter in west Europe and in the other winter between the Black Sea and the Caspian Sea! The British ringing material, however, like that for Scandinavia (table 21), demonstrates a pronounced predominance of autumn and winter recoveries from southwest Europe and Italy (92% of the British recoveries) in comparison with recoveries from the area from Greece eastwards (8%).

Approximately the same proportions apply incidentally to the recoveries from Redwing-ringing at Rybachi in the Baltic States, where three winter recoveries have been reported from the region between the Black and Caspian Seas and 26 from southwest Europe. This distortion in distribution of recoveries probably does not reflect at all the significance of the different wintering areas for the Redwings, but rather is due to the predilection of the southwest Europeans for hunting thrushes (for mixing in their pâtés etc). A second important source of error in the reporting of ringed birds is the different writing in parts of southeast

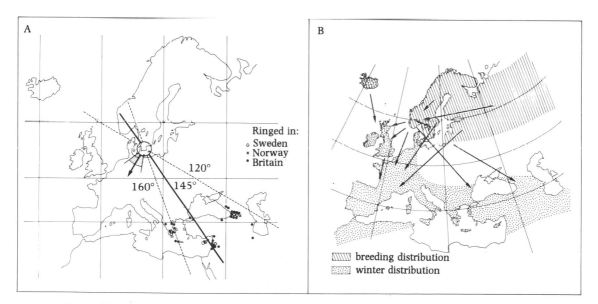

Figure 69   A. Mean direction and scatter of thrush passage southeastwards over Scania, south Sweden, according to radar studies. The normal direction of the southwestward passage is also indicated for comparison. Recoveries in southeast Europe and the Middle East include Redwings ringed at breeding sites in Sweden and Norway (three individuals, cf. table 21) while on autumn passage through these countries (three individuals) and during previous autumns and winters in Britain. A February recovery in Georgia of a Redwing ringed as a juvenile in Södermanland, southeast Sweden, in midsummer of the preceding year shows that not only adult but also first-year Redwings winter in the southeast. Based on same sources as table 21, and also Simms (1978).
B. The Redwing's migration pattern in Europe during the autumn.

Europe and the Middle East, where the local inhabitants cannot read and understand the addresses in west European characters on the bird rings.

Ringing recoveries are therefore nothing to be relied upon when it comes to determining how many Redwings winter in the various regions. Judging from the radar observations in south Sweden, the majority winter in the southeast, but this picture is misleading since the large numbers that leave Norway over the North Sea heading for west Europe completely escape the radar's cover.

As a summary of the Redwing's autumn migration pattern in Europe, the conclusion must therefore be reached that the winter quarters both in the southwest and in the southeast are of great importance (figure 69B).

During late summer and early autumn we can, on the basis of ringing recoveries, trace a tendency in certain Redwings from Sweden, Finland and even farther east still to move westwards towards Norway. The supreme record is held by the Redwing, ringed in May as a nestling at Tomsk in Siberia, which was recovered in early October of the same year not far from Oslo, more than 4000 km to the west. The reasons for this westward migration, which incidentally also occurs in the Fieldfare and others, are unknown.

When the Redwings have put on sufficient fat later in the autumn to face the crucial journey south, the winds on nights suitable for migration can determine what the winter quarters will be. Redwings which are ready to make a move on nights with north or east winds choose to migrate towards west Europe. Should northwest winds be blowing, then they instead steer a course for the wintering area at the Black and Caspian Seas. In this way chance circumstances in the state of the winds can govern that an individual Redwing in one winter lives in the vineyards in a southern French valley and in the next winter among the grapevines on hillsides in the Caucasus. The autumn gathering of Redwings in Norway accounts for the exceptionally intensive southeastward passage in the major route over southernmost Sweden.

The advantage of choosing, as the Redwings do, winter quarters depending on which way the wind is blowing is of course that the migratory journey takes place quickly and at little cost in terms of energy. In their efforts to benefit from the wind, migrants regularly await tail winds for each stage of the migration. The probability of various winds has, over the course of time, influenced the evolution of many bird species' migration routes so that the chances of tail winds along the route are furthered without the circuitous path to the migration destination becoming too long and the suitable stopping-off sites too few. But few species have gone so far as the Redwing in their dependence on the winds. The winds not only influence migration time and route but in addition determine the winter quarters.

Why does the Redwing lack fidelity to the wintering sites? (It seems to be faithful to the breeding sites to a more normal degree.) In winter Redwings probably feed to a large extent on berries, a very variable food source and one of irregular occurrence. The closest parallel case to the Redwing's migratory habits can be found in, for example, the Siskin, the Brambling, the Waxwing and the Pine Grosbeak, species whose key food in winter is seeds and berries of perennial plants. The food supply fluctuates greatly between different years and regions, and this compels the birds to lead a nomadic life in winter. This is discussed further in the next section, which deals with seed-eating birds.

We should remember that the above conclusions on the Redwing's wind-dependent choice of winter quarters are still not positively proven. One question that remains to be settled is why so comparatively few migrants, according to the Scanian radar studies, return from the southeast in spring yet such large numbers depart in that direction in the autumn. After having been at the north tip of Öland in the west Baltic in a couple of springs and witnessed massive concentrations of migrating Redwings, many of them maintaining a northwesterly course, I hazard a guess at the answer to this question: the Redwings return in the spring from winter quarters in the southeast, not to the autumn gathering area in southern Norway, but instead direct to the breeding sites. This means that the spring passage from the southeast does not pass the areas near to Scania but most Swedish and Norwegian Redwings fly northwest over the central and northern Baltic Sea and over the Gulf of Bothnia.

Among the migratory birds the Redwing represents an uncommonly dynamic adaptation through radically changing its flight destination according to the winds for the benefit of speed and energy. Listen more closely for the calls – are they moving in the direction of the summits of the Pyrenees or towards those of the Caucasus?

Winter life in the tropics

The majority of our insectivorous birds that are long-distance migrants can be found during the winter months in tropical Africa (table 20). Most move south past the Mediterranean Sea and the Sahara at some time in September. They then have before them about seven months of life in the tropics. Which environments and food sources do the migrants seek out? How do they get on in the competition with the native African species?

The first thing we think of when it comes to the tropics is perhaps luxuriant, humid and dark rainforests with greenery the year around and a stable climate. The birdlife is rich and diversified in the rainforest, but exceedingly few migrant birds resort to this environment. In the second place we perhaps think of the tropical deciduous forests which burst forth into lush greenery during the rainy season south of the Sahara, approximately at the same time as the migrants arrive from the north. But most northern migrants reject these food-rich regions, too.

No, it is not in the most fertile and resplendent tropical forests that the majority of migrants spend their winter life, but in the most *unstable* environments. The map showing the number of insectivores that winter in different parts of Africa (figure 70) reveals that the savanna areas north of the equator and in eastern Africa harbour the most species.

North of the equator the rainy season ends at roughly the same time as the northern migrants arrive. At the beginning of their winter sojourn they thus meet here a green savanna with floodlands around well-filled river courses and lakes. The rains do not return, however, until after the birds' departure in the spring. During the course of the dry season the savanna grass is scorched brown and is ravaged by fire, riverbeds dry up, and the cracks of drought unfold in the ground. Despite the drought, many migrants winter within the Sahel zone. The central area lies in a belt around 13–15° N. This is dominated by shrub savanna with scattered low trees, except locally on the river banks where the acacias grow big and sturdy. To the north this savanna type gradually changes into desert steppe with poorer and poorer grassland and ever sparser growth of bushes and trees. Despite this, there are surprisingly many northern winter visitors all the way to the limit in the north which the Sahara desert imposes on the tropical distribution of migrants.

The maximum number of species that come from afar to winter is found in the dry savanna lands that surround the Nile in Sudan and in western Ethiopia and also around Lake Victoria in East Africa. In the latter region there is a rich mosaic of tableland savannas, shrub savannas and forest savannas, dry as well as humid and luxuriant. Two annual rainy seasons occur here, one as the tropical convergence zone crosses the equator in one direction and one as it is on its way back. A great many migrant species penetrate south of the equator, mainly to savannas of an open nature where acacias dominate. On these savannas rainy season prevails for a large part of the wintering period.

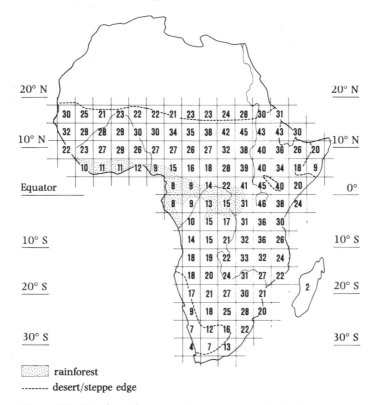

rainforest
-------- desert/steppe edge

Figure 70   Number of species of insectivorous birds from Europe and Asia which winter within different parts of tropical Africa. Basic data are from the African distribution maps for 90 different species within this category shown by Moreau (1972). The list of species is not complete since reliable distribution data are lacking for certain species. Only two species, Eleonora's Falcon and Sooty Falcon, regularly winter in Madagascar (see section 4.9).

The migrant birds therefore consistently prefer tropical environments with substantial variations between rainy season and dry season. This is an important manifestation of their general adaptation to unstable environments. Further instability is provided in many wintering habitats by the big differences in amount of rainfall between different years and regions. The shorter the rainy season and the smaller the average rainfall, the greater the relative variations between different years become. This instability is an important factor and explains the accumulation of higher numbers of migrant species in the dry savanna in such places as the Sahel zone. The severe drought that befell the Sahel between 1968 and 1974 shows how catastrophic the consequences can be when the rains fail for several years in succession. Human beings and cattle starved to death. The water level at Lake Chad dropped by several metres. And as for the birds, the course of events could be followed in Europe, where the populations of the normally very common Whitethroat, a 'Sahel winterer', were drastically reduced.

Many migrants lead a nomadic life in search of concentrations of food which occur regionally or locally where the rain has chanced to provide extra favourable conditions for growth. In certain parts there is an abundant supply of berries and fruits in the trees and bushes of the savanna, something which the migrant birds eagerly exploit. In other areas the vegetation blooms in profusion, and the flowers attract insects which in turn attract birds. One observer has reported how Lesser Whitethroats in Sudan were picking insects from the yellow flower catkins of the acacias; the little warblers were completely yellow around the bill from the flowers' pollen. Many trees and bushes bear foliage during the dry season, more vigorous on the borders of rivers or on damp ground. Migrants gather here in search of insects. It is of great advantage to the birds in the savanna areas north of the equator that the savanna vegetation buds and establishes new greenery even before the rains break. Insects attack the tender young verdure. This is just what the migrant birds require for putting on fat in the spring before the long migration north across the Sahara.

In East Africa the first migrants from the north arrive immediately before the rainy season in November and December. The odd thundershower has already fallen here and there and in places stimulated the vegetation and the insect life. At these small rain patches Rollers, bee-eaters, flycatchers, shrikes and wheatears gather. A few days later they continue on to the next verdant patch, perhaps guided by the tremendous thunder clouds and flashes of lightning which can be seen over long distances. Before the next rainy season, which begins in February or March, the same procedure is repeated but in the reverse direction; now, of course, the rain front moves northwards, and the migrant birds follow the scattered showers that precede the front.

There are lots of reports on how the migrants rapidly congregate at temporary and local food surpluses: migrating locusts, termites which fly when the rainy season begins, swarming mosquitoes, butterflies, winged ants and beetles and insects which are forced to flee grass fires.

During the dry season fires ravage large areas of savanna and are an exceptionally important factor in so far as reducing stability in the savanna environment is concerned. The fires are sometimes triggered spontaneously by strokes of lightning, but usually they are the work of the local cattle-rearing inhabitants. An isolated fire as a rule spreads along a front of between 1 km and 10 km and travels as much as 20 or 30 km in the direction of the wind before it dies out. The scale of the savanna fires is incredibly large. Over vast regions more than half the area is laid waste by fire every year. The extensive fire devastation within the whole of the 1000-km-plus width of the savanna zone between approximately 5° and 12° N, from Nigeria in the west to Ethiopia in the east, has been followed with the Landsat satellite (formerly ERTS, the Earth Resources Technology Satellite). According to satellite pictures, the first fires are already starting to spread at the end of October. In December a fifth of the land surface is ravaged by fire, and towards the end of the dry season, in February and March, fires have laid waste more than 40% of the land surface in most places. A special in-depth study within a 30 000-square-kilometre savanna region at 7° N in the Central African Republic showed that 20% of the area had been fire-ravaged in December, 80% at the beginning of January, and more than 90% as January ended. Widespread savanna fires have raged during the dry seasons for many thousands of years, at least to judge from the savanna's bushes and trees, for these have evolved special powers of resistance and are capable of surviving the fire's assault.

The migrant birds find food not only where the fire is advancing at any given time but also in its wake. Wheatears are frequently attracted to fire-ravaged terrain to pick surviving insects from the charred grass. Where the dry, brown grass is burnt to the ground new healthy grass shoots soon sprout, for the benefit of insects as well as for birds.

Not only wintering migrants take advantage of all kinds of temporary and local food sources, but many native tropical species also flock around them. Here, however, the migrants and their tropical competitors meet on equal terms, something that is not so in the competition for the stable resources, which of course the tropical residents can lay claim to within their territories before the migrants arrive.

By no means all insect-eaters lead a nomadic life during their winter sojourn in the tropics. Some establish territories and live within these throughout the winter months. This is of course possible only in fairly stable environments where the food supply is reliable. Migrants which lead a sedentary winter life of this sort differ from their tropical competitors in that they have to put up with marginal habitats, often heavily influenced by man's interference in nature. The migrants behave as opportunists at the forest edge, at windfalls, on fields, roads or in clearings, in growing sapling forest, in thinned-out vegetation in plantations, in gardens and on erosion land. The Willow Warbler is a good example of such an opportunistic species. Its winter distribution encompasses almost the whole of Africa, from 10° N to as far south as South Africa. In dry country it lives a nomadic existence, but where the environment is more stable it remains sedentary on the edges of forests or in open clumps of trees.

Similar winter habits are exhibited by the Willow Warbler's sibling species, the Greenish Warbler, which has been closely studied at wintering sites in India. The study area included young forest in felling areas and adjacent open land with scattered isolated trees. The Greenish Warblers arrive there in September and immediately set up individual territories, which they retain right up to the time for the return journey northwards at the end of April. When establishing the autumn territory, females and males sing with equal assertiveness to mark out their domains. A territory comprises on average six trees, or roughly the equivalent amount of greenery if the birds have mostly bushes within their territories. Judging from observations during the severest drought in January, the territories appear to hold exactly the amount of food that is needed for the warblers to keep their life intact. At this time they begin foraging before dawn, and are still on the go when it has become too dark to observe them in the evening. All the time is taken up with hunting for food. Despite intense watching in January, the observer saw only five occasions when the birds allowed themselves time to preen their plumage, and then never for more than one minute at a time!

In the marginal habitats the migrants escape competition from the tropical species. A major reason why many tropical insectivores which live in deciduous foliage avoid making use of open clumps of trees is that there they would be compelled to waste valuable energy in long flights to feed the young in the nest.

Palaearctic Swallows have a very wide winter distribution in Africa. They are in addition extremely numerous and outnumber their tropical relatives many times over. One observer, on a longish journey through South Africa in February and March, found that 97% of all the swallows he saw were northern Barn Swallows; the other 3% were made up of seven different native species. The House Martin's wintering habits are a bit of a mystery. House Martins are observed in Africa much more rarely than other northern species of swallow, which is presumably due to the fact that they hunt for food at extremely high altitudes and in mountainous areas. The northern hirundines do not suffer any disadvantages from competition in Africa in comparison with the native species. Quite the reverse: as they do not have any ties to breeding sites, they can freely transfer their hunting to those places where insect supply and winds provide the feeding opportunities that are best at any time.

A further distinctive survival strategy is used by northern migrants in the tropics: migration between various established winter refuges each of which has food resources that will not last throughout the winter but should well do for a few months. As has been pointed out earlier, the northern savanna along the southern edge of the Sahara is green and hospitable when the migrants arrive. Many species therefore stay on for several months in this zone. Before the wetlands have dried out and the greenery withered, they put on new fat reserves in order to fly thousands of kilometres south to another wintering centre where the rains have afforded the opportunity for a further few months' stay. Thus the Great Reed Warblers do not reach their territories (the same individuals often return to the same territory year after year) in southern Zaire until December. Before this they have spent approximately three months in the northern savanna region, where they probably have territorial habits as regular as those in Zaire. They stop in Zaire for between three and four months, until it is time for the spring migration in late March/early April.

A similar situation obtains with many Reed Warblers and Sedge Warblers, which do not arrive on the shores of Lake Victoria until December, the Sedge Warbler often not before January or February. Sedge Warbler populations that winter far south of the equator, however, make for their final destination without lengthy delay: in South Africa they are on site as early as October and November.

Extensive ringing controls at Lake Victoria in Uganda have revealed that considerable numbers of, among others, Garden Warblers, Red-backed Shrikes and Willow Warblers migrate through the region around late November and early December. The birds apparently break their migratory journey for a couple of months in Sudan or west Ethiopia before continuing to new winter quarters south of the equator. Many Garden Warblers that pass through Uganda have fairly large fat reserves, sufficient for a flight of more than 1000 km further southwards.

The most interesting revelations of all concerning the northern migrants' passage between different tropical winter quarters come from Mount Ngulia in the Tsavo National Park in southeast Kenya. The summit of Ngulia is over 1800 m high. In 1969 a tourist hotel was built on a prominent mountain ridge with a wide view over Tsavo's flat bush plateau a couple of hundred metres below. Outside the hotel, which faces north, powerful floodlights were set up directed towards nearby waterholes where big game pay nightly visits to drink. The most intense rainy period occurs in November, December and January. At this time thick mist regularly envelops Ngulia's summit, and during the late night and at dawn the mist creeps down the mountainsides and sometimes surrounds the whole hotel. During the rainy and misty nights many migrants on passage from the north are drawn towards the powerful floodlighting. This is the same phenomenon as that which has been observed in north European latitudes when, on hazy or foggy nights, nocturnal migrants fly into illuminated masts or lighthouses (cf. section 4.9).

In the very first year ornithologists noticed that amazingly large numbers of migrants appeared at the Ngulia hotel in the middle of winter, on Christmas Eve as well as on New Year's Eve if the weather was right! Ever since then visiting migrants have been caught and ringed each year. In 1981 the total ringed was almost as high as 52 000. Most are captured in mistnets during the hours of dawn as they disperse through the bushlands away from the hotel. Some are caught during the dark hours of the night in nets which have been erected in front of the hotel's open dining terrace. Sometimes it is also worth going in search with a torch on the terrace or in other open hotel buildings where the birds have ended up after losing their way.

About 30 different species occur regularly each season on southward passage through Ngulia. The seven most numerous are given in figure 71. The list is emphatically dominated by the Marsh Warbler, the Whitethroat (the eastern race *icterops*) and the Thrush Nightingale, with more than 10 000 individuals of each having been captured. River Warblers and White-throated Robins (*Irania gutturalis*) are also among Ngulia's specialities; over 1000 individuals have been ringed. A particular exclusivity is the rare Basra Reed Warbler (*Acrocephalus griseldis*), over 400 of which have been caught. The list of regular midwinter passage migrants at Ngulia also includes, among others, Nightjar, Roller, Golden Oriole, Garden Warbler, Barred Warbler and Spotted Flycatcher.

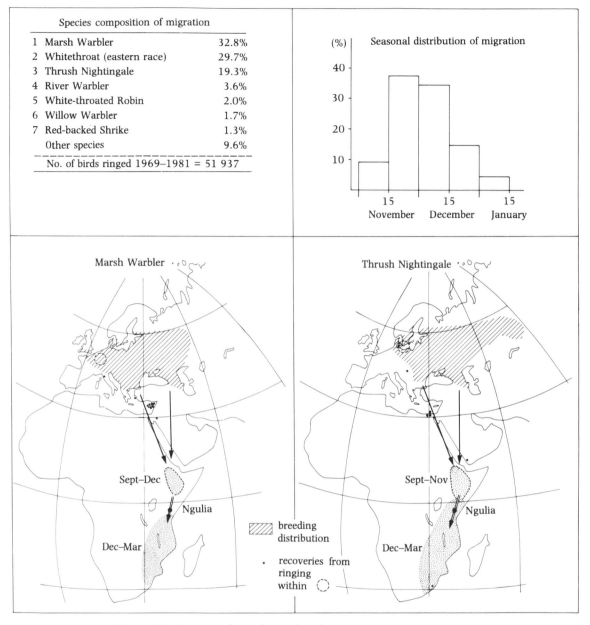

Figure 71 Passage through Ngulia of nocturnal migrants which change their winter quarters in Africa. The geographical migration pattern for the Marsh Warbler and the Thrush Nightingale is shown on the two maps. Recoveries of birds ringed in Europe show that both species migrate southeast via the easternmost Mediterranean Sea area and the Middle East. So far as is known, the species are not present in winter either in Africa west of the Nile valley or in Asia. To all appearances the entire populations of both species assemble in initial winter quarters in eastern Ethiopia; the birds then migrate further to second winter quarters south of the equator. Of the 12 months of the year, these species thus spend two to three months at the breeding sites, an equal length of time at various stopping-off sites where they renew fat reserves while on migration, around three months in east Ethiopia and four months in southeast Africa. Based on Backhurst & Pearson (1979, 1981), Pearson & Backhurst (1976) and Zink (1973).

The large numbers of Marsh Warblers, Thrush Nightingales and River Warblers are particularly surprising, since these species have not, despite reliable studies, been found to stop off at all on any large scale either around Lake Victoria or on the Indian Ocean coast of Kenya. The main migration thoroughfare for these species must therefore proceed through east Kenya in a corridor that is less than 400 km wide. All three species reach Africa on very easterly routes – there are practically no autumn recoveries west of the Nile. To all appearances Marsh Warblers, Thrush Nightingales and River Warblers gather at a first wintering site immediately after the rainy season in the regions of eastern Ethiopia; during late autumn and midwinter they leave to migrate through the Ngulia corridor and to settle down instead thousands of kilometres to the south, in the humid rainy-season climate with its verdant bushlands south of the equator.

The northern insectivores' pattern of life in the tropics – a life of roaming around in unstable environments, shunting between different fixed wintering stations, staying in marginal habitats – points to a strenuous winter existence. We ought therefore to be able to expect that the energy-demanding annual moult of the flight feathers would take place very soon after breeding, while the birds are still back in northern latitudes with food surpluses. This is indeed what happens with some species, but by no means all. The hirundines, the *Acrocephalus* warblers, the *Locustella* warblers and the *Hippolais* warblers, the Garden Warbler, the Wood Warbler, the Greenish Warbler, the Red-backed Shrike, the Spotted Flycatcher and the Golden Oriole postpone the moult until they have reached winter quarters in the tropics. The Willow Warbler seems to be unique in that it changes its flight feathers twice a year, once in the summer before the autumn migration and once more in the tropics before the spring migration.

Habits vary greatly between those species which moult in the tropics. Great Reed Warblers, Reed Warblers and Sedge Warblers moult during the late autumn while they are still living within the northern savanna region and before they continue south to second winter quarters at or immediately south of the equator. Those Sedge Warblers which winter in South Africa on the other hand postpone the moult until after the New Year so that they have completely new flight feathers at just the right time for the beginning of the spring migration in April. The Garden Warbler also times its moult for the early spring, as do, among others, the Marsh Warblers, Whitethroats and Red-backed Shrikes that migrate south through Ngulia on as yet unmoulted wings. Note that western European Whitethroats, unlike the eastern race which passes through Ngulia, moult in northern latitudes before they begin the autumn migration. The Thrush Nightingales which move through Ngulia have changed their flight feathers during the late summer in Europe.

One reason why some species or populations undergo their moult in the tropics instead of in northern latitudes may be lack of time during the comparatively short northern summer. This applies in particular to those species which, like the hirundines, have two broods. Even for insectivores with only one brood the time margins are sometimes so tight that it can be more beneficial to moult at a more leisurely pace in the tropics, despite the fact that the food base there is not so good. As a rule, in the tropics it takes the warblers between 65 and 83 days to renew all their flight feathers; on average there are no more than two feathers per wing growing out at the same time. Those species which moult in the north during the late summer do so at twice the speed: the whole process is normally completed within 30–45 days, and four feathers are changed simultaneously on each wing. A full explanation of just why certain species moult in northern regions, others during the autumn in the tropics and yet others not until the spring in the tropics requires a much deeper understanding of the different insectivores' living conditions in the various localities than exists today.

3.8   Seed-eaters

The seeds of plants have an extremely high food value and are a desirable food for many kinds of animals. The seed specialists among the birds that occur regularly in north Europe are listed in table 22.

In order to understand properly the migration patterns of the different birds it is important that we distinguish between species that feed on seeds of short-lived, often annual, plants and species that gather seeds from long-lived, perennial trees or bushes. By way of introduction let us try to see the situation from the plants' point of view. We must now stop looking at plants as inanimate set-pieces in nature and instead see them as the living creatures they in fact are, creatures with varied maturing periods, age of first blossoming, length of life and so on. The life of a plant does not actually differ that much from that of an animal; this we can see once we have made allowance for the fact that the 'food' is different – light and nutrient salts for the plants, organic matter for the animals.

Why do some plant species live for only one year? One reason may be that there is a great probability of their habitats not being suitable in the following year. The wind, the water erosion during autumn and spring high tides, breaking up by the frosts of winter etc radically alter the local growing conditions. Or else the plant has such poor availability of nutrient salts that these will not suffice for more than one year's strong growth; it can then take many years before a sufficient nutrient base has been concentrated again at the same locality. In order to be able effectively to exploit exposed environments where wind and rain ransack the area, where the local nutrient stockpile is rapidly used up or where competition from other plants soon becomes too much, annual plants must relinquish their habitat after the summer flowering and put all their strength into producing and disseminating seeds for their survival. Annual plants exploit habitats where resources for growth exist during the vegetation period but where almost total lack of resources for survival prevails during the intervening period up to the next vegetation period.

Annual plants occur commonly on shores and beaches, in gravelly soils, in windfall areas, heaths and steppes. As 'weeds' they have colonised agricultural country on a large scale, and here not only the forces of nature but also the activities of farmers contribute to a great extent to constant changes in the local growing conditions. The annual plants are 'obliged' to produce an abundance of seeds during their single year of life. This has resulted in their being, for the purposes of seeding, far less susceptible to annual climatic variations than perennial plants are. Annual plants invest in total more of their assimilated energy resources in seed production than do plants with a long lifespan. In weed seeds, the ability to germinate is often maintained for an extremely long time. This guarantees that all opportunities for flowering, in the short term as well as the long term, are taken. Seeds of the fat-hen nearly 2000 years old, found in the top-soil at excavations of Iron Age settlements in Denmark, were found on planting to grow into robust plants. In Canada, plants of an arctic species of lupin were successfully raised from seeds which had been suspended in frozen earth for at least 10 000 years.

The paradoxical consequence of the short-lived plants' adaptations is that, despite the plants' swift mortality in a fickle environment, they offer seed-eating birds a fairly stable food base year after year. In areas where the supply of weed seeds is abundant in one year, there is a strong probability that it will be so also during the following year. Birds which feed on weed seeds in winter therefore exhibit stable and regular migratory habits and site fidelity in their winter quarters.

For birds which specialise on seeds or fruits of long-lived plants such as trees and bushes, the situation is totally different. The long-lived plants are adapted to a stable and safeguarded environment in which it is possible to achieve a dynamic equilibrium between the plants' outtake of nutrient salts from the earth and a continuous supplying of them via surface and subsoil water. Individual trees may be hundreds, in some cases even thousands of years old. The individual specimens of certain perennial plants grow through vegetative propagation of clones. These can become very bulky and old, as for example a clone of reed whose rhizomes it has been possible to trace back 6000 years in time. The long-lived plants use a very large proportion of the energy resources for their survival, so for a long time to come they can stick to their favourable habitat and with this as a base produce and spread their seeds. For the mature plant it is more important to safeguard its survival and competitive strength than to ensure a regular yearly seed production. The trees and bushes in consequence exhibit a great sensitivity to climate in their annual flowering and seeding. Years with a good measure of seeds and berries alternate with years in which many trees and bushes are poor bearers of fruit. Some plant species amplify these yearly variations in fruiting. Through this adaptation they combat and 'shake off' the populations of insects, birds and mammals that consume their seeds.

For the seed-eating birds the result is that the trees and bushes, despite a long lifespan and stable growing environments, provide a very variable and uncertain food base. The fact that various species of trees and bushes in certain regions yield an abundance of winter food for the birds in one particular year does not mean that these regions are especially suitable as winter quarters in the following year, too. On the contrary, they are then often wholly unsuitable since many long-lived plants, after a peak year in fruiting, take a 'year of rest' with minimal seed production.

Bird species which live on seeds from trees and bushes do not therefore show fidelity to fixed winter quarters. They often lead a nomadic winter life and sometimes allow their varying migration routes to be determined by the direction in which the wind makes it most economical and quickest to fly. Since the winds vary, the same bird can seek its winter food in one year in west Europe and in the other in east Europe. The birds penetrate different distances south and concentrate in differing regions in different years, depending on where the food supply is most plentiful. Most of the conclusions and speculations concerning the association between the seed production of trees and the irruptive migration of birds which Gunnar Svärdson put forward in a classic paper in 1957 still hold good and are equally inspiring even today.

The difference between the 'ordered' migratory habits of birds which feed on seeds from short-lived plants and the dispersed migratory habits of birds whose main winter food is seeds from trees and bushes is illustrated for two pairs of closely related species in figure 72. Linnets from Sweden migrate southwest in a fairly narrow corridor to winter in the agricultural country of southern France and eastern Spain; there they live, as during the migration period, on weed seeds of many different sorts. The overlap in winter distribution between Linnets from separate breeding areas is limited. Thus, English Linnets winter rather more to the west than Swedish ones, the central point lying in southwest France and central Spain; Baltic Linnets on the other hand spend the winter mostly in southeasternmost France and in Italy.

Alder seeds are the main winter food of Siskins; birch and spruce seeds also play an important part, especially during autumn and spring. Recoveries of Swedish-ringed Siskins seeking alder stands with plenty of cones have been reported from Portugal in the west to the Caucasus in the east. The Siskin's regular winter distribution extends northwards to central Scandinavia, where the birds congregate in flocks varying greatly in size from year to year in the lines of alders along lakes and rivers.

Table 22. *Seed-eating birds which occur regularly in north Europe. The data on wintering areas apply primarily to the species' populations in northwest Europe, particularly Scandinavia*

Basic food	Species	Largely resident	Occurs in N Europe in winter	Important wintering areas				
				W & N Europe	Med. region	Africa S of Sahara	Southern Asia	Invasion migration
Seeds of short-lived annual plants	Quail					+		
	Collared Dove	+	+	+				
	Turtle Dove					+		
	Stock Dove	+	+	+	+			
	Crested Lark	+		+				
	Woodlark			+	+			
	Skylark			+	+			
	Shore Lark*		+	+				
	House Sparrow	+	+	+				
	Tree Sparrow	+	+	+				
	Chaffinch		(+)	+	(+)			
	Serin				+			
	Greenfinch	+	+	+				
	Goldfinch		+	+	+			
	Linnet			+	+			
	Twite		+	+				
	Scarlet Rosefinch						+	
	Lapland Bunting*		(+)	(+)				
	Snow Bunting*	(+)	+	+				
	Yellowhammer	(+)	+	+				
	Ortolan Bunting					+		
	Rustic Bunting						+	
	Little Bunting						+	
	Yellow-breasted Bunting						+	
	Reed Bunting		(+)	(+)	+			
	Corn Bunting	+	+	+				
	No. of species	6	11	16	7	3	4	—

Seeds/fruits of bushes and trees (long-lived plants that flower repeatedly)						
Jay	+	+	+				+
Nutcracker	+	+	+				+
Brambling		+	+				+
Siskin		+	+	+			+
Redpoll	(+)	+	+				+
Two-barred Crossbill	(+)	+	+				+
Common Crossbill	(+)	+	+				+
Parrot Crossbill	(+)	+	+				+
Pine Grosbeak	(+)	+	+				+
Bullfinch		+	+				+
Hawfinch	(+)	+	+	(+)			?
No. of species	2	11	11	1	3	—	10
Total no. of species	8	22	27	8	3	4	10

All categories

*Shore Lark, Lapland Bunting and Snow Bunting have extensive winter quarters in the steppe zone in eastern Europe and central Asia.

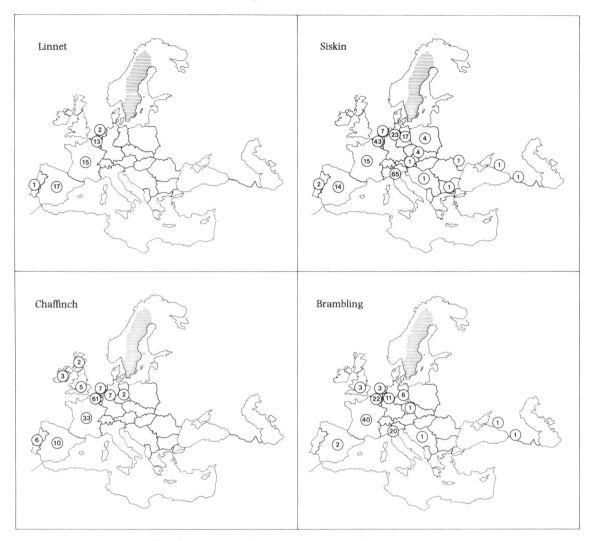

Figure 72   Recoveries in September–April in different countries outside Scandinavia of four different finch species ringed in Sweden in 1960–1971. Note that the numbers of recoveries do not accurately reflect the species' relative occurrence in the various countries. The high figures for Belgium and Italy and, to a lesser extent, also France and Spain are due to the generally widespread hunting of small birds in these countries. Based on annual reports of the Central Ringing Office, Sten Österlöf, Stockholm Museum.

A similar overall picture is presented by the migration pattern of the Brambling. Its main winter food is beechmast. An illuminating example of this species' lack of site fidelity during the winter is provided by the adult male which was ringed in one winter in the county of Halland in south Sweden and which was found again two winters later immediately south of the great mountain ridges of the Caucasus. The Brambling's winter nomadism has given rise to the most striking concentrations of birds that have ever been witnessed in Europe. In the beech woods of south Sweden gatherings of several million Bramblings have been found; but the record counts come from Switzerland, where concentrations of Bramblings at communal roosts have been estimated at 11 million, 50 million, and in one case even 70 million birds! The last-mentioned estimate was made in winter 1951/52 near the town of Hunibach, where the floor in the beech woods was covered with a carpet of nutritious mast; in the same winter there were at least a further 30–40 million Bramblings in other parts of Switzerland. The 100 million and more Bramblings that were concentrated in Switzerland that winter would almost have been enough to fill the species' breeding area in upland birch forests and northern coniferous forests from Norway in the west to the Urals in the east to its normal density.

The Chaffinch is the Brambling's sibling species in appearance and in generic relationship, but decidedly not in migration ecology. True it readily eats beechmast, but essentially it relies on the agricultural landscape being able to provide it with weed seeds and spilled grain for winter food. Its migration routes and winter quarters are firmly fixed year by year. The winter distribution of Scandinavian Chaffinches encompasses Britain and the farming districts along the Atlantic coast of France, Spain and Portugal.

A few years ago I studied bird migration in autumn in southeast Scania and found that, in northwesterly winds, considerable numbers of Siskins and Bramblings moved off southeastwards across the Baltic Sea in the morning and at dusk; in other winds the passage was often in a south or southwest direction. The migration of the Siskin and the Brambling is thus supremely reminiscent of that of the Redwing in that it involves extensive wintering area, lack of winter site fidelity, dramatic variations in migration routes during different winds, and nomadic lifestyle during one and the same winter season. (The ecological parallels between the two species pairs Chaffinch–Brambling and Song Thrush–Redwing are, incidentally, astounding in number.) Now, if the Siskin's and the Brambling's migratory habits are explained by these species' dependence on fruits from long-lived trees and bushes, cannot the same be considered to apply to the Redwing?

An account from England provides important clues to the answer to this question. In a largish area of woodland at Oxford, the total available supply of different types of berries was calculated; likewise the birds' consumption from August to April inclusive was worked out. The most avid berry-eater among the birds was the Redwing; it consumed twice as many berries within the area as the number two species on the list, the Fieldfare. (The Song Thrush incidentally proved not to be a particularly great eater of berries: the Redwings' berry consumption was 70 times higher than that of the Song Thrushes.) The Redwings ate mainly berries of hawthorn, supplemented by a small amount of sloes. We may therefore presume that the Redwing's striking migratory habits are conditioned by the search mainly for those hawthorn areas which have the year's most plentiful supply of berries.

If this conclusion is correct, the Redwing must be placed in a group of species whose wintering in Europe is founded largely on a berry diet; the rest of the group consists of the Fieldfare, the Waxwing and the Pine Grosbeak. The regionally varying fructification of the berry bushes in different years is in this case the common denominator for these species' split migration pattern, for their nomadism and invasion-type appearances in certain years.

Most seed-eating birds, particularly those which feed on seeds from annual plants, change over from a seed diet to an insect diet in summer. Insects are used for instance as food for the nestlings. The summer surplus of easily obtainable insects is opportune for the seed-eaters, coming at just the time when most seeds are germinating. Not until late summer have the year's sprouting plants had time to set new seeds, ready to be harvested by the birds.

All but two of the seed-eating species moult their wing feathers immediately after completing breeding. The two exceptions are the Scarlet Rosefinch and the Yellow-breasted Bunting. These arrive late in the spring on their northern breeding grounds and leave early, as early as late summer. Not until they have arrived back in their winter quarters in southern Asia does the moult of the flight feathers take place. The reason for these two species' rush to return to their wintering areas is not known.

A great majority of the seed-eaters remain within the confines of Europe during the winter. Among those species which feed on seeds from annual plants, however, there are examples of real long-distance migrants. Four species fly to southern Asia and three winter in the dry savanna lands in Africa immediately south of the Sahara. The most striking ringing recovery is of a Yellow-breasted Bunting ringed at its breeding site at Oulu on the Gulf of Bothnia in Finland which was recovered in winter in central Thailand. In Thailand the Yellow-breasted Bunting is a very common winter visitor living in large flocks on the arable fields and in the sugar-cane plantations. The wintering buntings in Thailand come not only from northeast Europe but also from Siberia. This is shown for example by a ringing recovery from the districts around the middle reaches of the River Lena.

The Shore Lark, the Snow Bunting and the Lapland Bunting have a curious migration pattern. The shores and coastal regions around the North Sea play host to a minor part of the winter populations, but the central point in the winter distribution is in the steppe zone in the Soviet Union, today largely a cultivated landscape. North American populations winter within a corresponding zone around the Great Lakes and along the border between the United States and Canada. The winter climate in these inland regions is characterised by severe cold and much snow. The Shore Larks and the Lapland and Snow Buntings are forced to be on constant stand-by to move rapidly to places where the snow does not prevent them from getting at weed and grass seeds. During the harshest winters, large flocks move far south of the normal wintering area.

The Shore Lark's breeding distribution, unlike that of the two buntings, is not restricted solely to arctic mountain and tundra country. In the Old World, Shore Larks breed also in alpine habitats in southeast Europe (at altitudes above 1500 m), on desert steppes in North Africa, Asia Minor and the southern Soviet Union, on stony upland plains in Tibet, and in the mountain massif of the Himalayas right up to 5000 m altitude. The larks' ability to exploit the most exposed and poor environments, places where only a sparse vegetation cover of low plants exists, has proved to be as successful in cold tundra as on hot desert edge. This tells us that the seed plants' adaptations and dispersion dynamics are very similar in these environments, despite all the differences in temperature and altitude. In North America the Shore Lark is a very common bird; the breeding range encompasses virtually the whole continent, including Central America and local parts of the grass savanna on the high plateau of Colombia. In North America it is found in most of the exposed environments, on tundra and prairie, in agricultural country, on desert steppes, and so on. The reason for this universal occurrence is probably that in North America the Shore Lark fills the larks' niche all by itself; in Europe and Asia, many competing species of lark combine to restrict the Shore Lark's distribution.

The Snow Bunting (*Plectrophenax nivalis*)

The Snow Bunting lives in the boulder lands of the Arctic. In this magnificent desolate environment its soft jingling song phrases can be heard ringing out. The males fly up into the air, flashing white as they tumble around in song flight. The breeding range extends in all directions around the North Pole, on the plains of the tundra, in the lichen regions of the mountains, on the edge of glaciers, along the edge of the Greenland inland ice, even on nunataks where vegetation can be found interspersed among the rocks. The Snow Bunting is incidentally Greenland's most abundant bird by far. As a result of intensive ringing of more

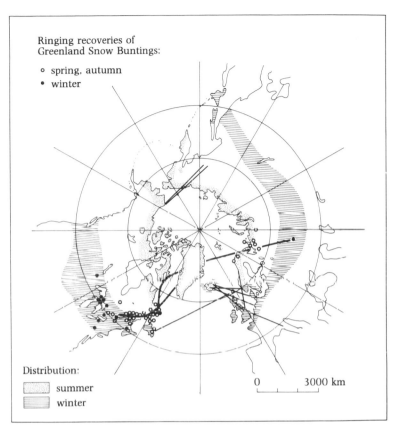

Figure 73 Summer and winter ranges of the Snow Bunting. The symbols show ringing recoveries of Greenland Snow Buntings. Recoveries in North America are of Snow Buntings from west Greenland and recoveries in Europe are of those from northeast Greenland. Migration routes are shown by thick lines. The thin lines show long distance recoveries of buntings ringed in Alaska or ringed while on passage or wintering in Europe. These recoveries are discussed in more detail in the text. Based on Nethersole-Thompson (1966) and Salomonsen (1967, 1971, 1979).

than 10 000 Snow Buntings, some extremely interesting migratory habits have been revealed (figure 73).

West Greenland's Snow Buntings cross the Davis Strait and migrate via Labrador and the coasts around the Gulf of St Lawrence to winter quarters on the lowlands around the Great Lakes in North America. In the French-speaking parts of Quebec in Canada, considerable numbers of Snow Buntings are caught in snares even today. This explains the high percentage of recoveries of Greenland-ringed Snow Buntings from these parts. The buntings are caught mainly in early spring, when the males migrate through in large flocks while the snow is still deep on the ground in many places.

Snow Buntings from northeast Greenland have a totally different migration. They cross the Greenland Sea – flocks have occasionally been observed resting on the pack-ice – and pass northernmost Norway and the White Sea area, with the Russian steppes as their wintering destination. (One individual which had flown a more southerly course landed up in the

Shetland Islands in September.) The same winter quarters are incidentally used also by Spitsbergen Snow Buntings judging from three recoveries of birds from Spitsbergen on passage through the White Sea. In the spring the buntings return by the same route. Periodically this results in gigantic flocks being concentrated in north Norway. Perhaps Greenland Snow Buntings are also represented among the many flocks which stop off in spring around the Gulf of Bothnia and on airfields in Norrland in northern Sweden. Resting Snow Buntings thrive particularly well on airfields, cleared of snow by the wind and with large quantities of suitable grass seeds. In earlier days many were trapped during the spring in north Norway. Thus, in April 1889, between 4000 and 5000 Snow Buntings were supplied to a merchant in Oslo; 95% of them were males. The reason for this large predominance of males is that in spring the males migrate north almost one month ahead of the females; this passage often results in conspicuous concentrations of huge resting flocks. The females migrate when the thaw is quite far advanced and stop-off sites have become available in abundance; their migration therefore passes off less noticeably.

Hunting of spring flocks of Snow Bunting males is known from widely separated regions. One hundred years ago the birds were hunted in the far north of Scotland, and even today, besides in Canada, the species is hunted in eastern Siberia. On a hunt in Siberia, two Snow Buntings were found which had been ringed at breeding sites at the north tip of Alaska; these recoveries show that some North American Snow Buntings winter in eastern Asia.

Where do the Snow Buntings that breed in southeast Greenland migrate to? There are three ringing recoveries from the Newfoundland area showing that some of them join the west Greenland birds and winter in North America. There are, however, also indications that others fly southeast, to the North Sea countries. Reports from weather ships of considerable spring passage over the Denmark Strait between Iceland and Greenland strengthen these suspicions. There have been recoveries in Iceland in April and May of Snow Buntings ringed in winter in England, the Netherlands, Denmark and Norway; these birds may very well be passage migrants on their way to southeast Greenland via stop-off sites in Iceland. We cannot, however, exclude the possibility that Icelandic breeders may also be involved, even though a large proportion of Iceland's Snow Buntings are residents.

Additional interest is aroused by the Snow Bunting which was ringed while on migration in April in the Shetland Islands and which was recovered in the following spring-migration period in Newfoundland. We may, I think, venture to assume that the bird had bred during the intervening summer in Greenland, probably in the southeastern parts. In that case the recovery suggests that the breeding area in southeast Greenland is occupied not only by Snow Buntings that have wintered in North America, but also by birds that come from northwest Europe. The ringed Snow Bunting had evidently joined up with migrating flocks on their way to widely separated winter destinations in the two years.

Where the migration of Scandinavian Snow Buntings is concerned, there is only one ringing recovery: a bird ringed in winter in England was found in April in Norway. This indicates wintering in the North Sea countries. It remains to be ascertained to what extent birds from Scandinavian mountains also migrate southeast, and if there are some north Russian Snow Buntings that head for wintering grounds in northwest Europe.

Snow Buntings have many different sides to their nature. They have been seen by polar travellers on the Arctic Ocean pack-ice, on one occasion very close to the North Pole itself. They breed in the most barren arctic wastes. They are the 'house sparrows' of Eskimo communities. They moult the flight feathers so rapidly during the brief summer in the extreme north that they temporarily lose the power of flight. They migrate thousands of kilometres over ice and sea, often at night (many lighthouse 'falls' on dark autumn nights in

Norway). They brave the snow and the cold for as long as possible in constant readiness to migrate in order to reach the seed plants.

In the wind over the sandy shore or the grass meadow, flocks of Snow Buntings glisten like drift snow in the sun. This must be what lies behind the bird's English colloquial name of Snowflake. Reflected in the beauty of the Snow Bunting is the life that is lived in the Arctic – in spite of, but at the same time because of, the barrenness, the ice, the snow and the wind.

Co-operation and conflict between plants and birds

'The region wherein she has fixed her place to live she seldom forsakes; nevertheless, this happens during unspecified, more or less discrete periods when, like the mountain lemming, she undertakes emigration, in larger or smaller flocks, at times in immense droves, which appear almost simultaneously over a large part of the earth. The last major emigration took place in 1844, when the Nutcracker appeared in large numbers almost simultaneously, not only throughout the Svea and Göta land, Norrland and Lapland, Norway, Finland and neighbouring parts of Russia, but also in Denmark, Germany and as far away as the land of Badia, in a large part of France, Belgium, and signs of it were even noted in England. So far as is known, all these flocks moved very slowly generally in a north-to-south direction. But from where, almost in *one* go, did this numberless multitude which spread across the land come? We cannot just assume that it had come from the Lapp countries, as none of our itinerant ornithologists had found the bird anywhere there in summer. In Finland and around St Petersburg it was supposed that the flocks had come from Siberia, but no doubt without grounds.' This quotation is translated from Sven Nilsson's *Skandinavisk Fauna: Foglarna*, volume 1 (1858).

Sven Nilsson's scepticism regarding the Finnish and Russian assertions was unwarranted: the Nutcrackers in fact really did come from Siberia, from where they emigrated when the cone production of the cembra pine failed. Similar invasions are a recurring phenomenon in the Nutcracker. It was the Siberian race, known as the Slender-billed Nutcracker (*Nucifraga caryocatactes macrorhynchos*), that was responsible for the 1844 invasion. Russian ornithologists have recorded extensive eruptions of this bird from the normal breeding range (often to south Russia and west Europe, but hardly over a large part of the earth as Sven Nilsson so dramatically put it) on more than 30 different occasions during the period 1753–1933, in other words a little more often than every sixth year on average. The Nutcracker's migratory habits are the result of an intricate interplay and counterplay with the plants that are at the centre of its living niche.

The Nutcracker is associated mainly with those species of pine which have extra large seeds. The seeds are characterised by their size and by the fact that they are wingless. They stand in sharp contrast to the small, winged seeds of the Scots pine. In all there are more than 100 different species of pine (genus *Pinus*) on the earth, all in the northern hemisphere. Only about 15 of the species have seeds that are completely wingless or have rudimentary wings. These species have given up using the wind for seed dispersal and rely instead on various animals, among them the Nutcracker, to disperse their seeds. An adaptation which attracts the Nutcrackers is the pines' large and nutritious seeds. These are easily obtainable in the cones during a remarkably long period of the autumn.

The distribution of pines with wingless seeds (the most important species are named) is compared in figure 74 with the occurrence of the two nutcracker species. I have used 'cembra pine' as a common name for all those species of pine emanating from the species, *Pinus*

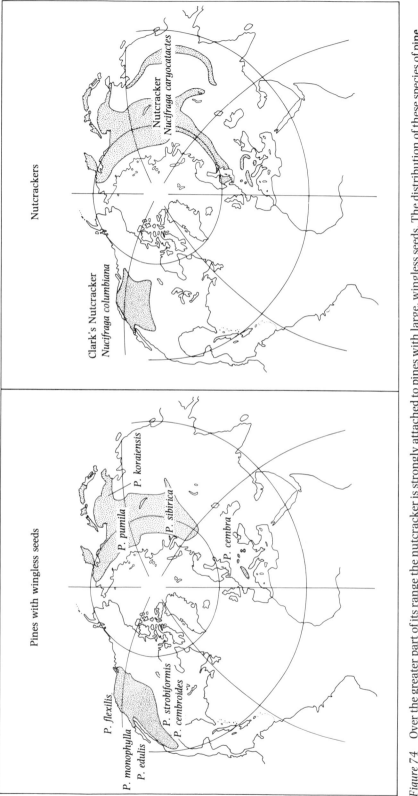

Figure 74 Over the greater part of its range the nutcracker is strongly attached to pines with large, wingless seeds. The distribution of these species of pine (genus *Pinus*) is taken from Mirov (1967); the names of the commoner species are given in the figure. The nutcracker's distribution is taken from, among others, Dementev *et al.* (1954) and Voous (1960).

cembra, which is found nearest to north Europe, in local stands on mountain slopes in the Alps and the Carpathians. Over large parts of north and east Asia, in western North America and locally in central Europe the nutcrackers are wholly tied to the cembra pines. The American nutcracker is regarded as a distinct species, Clark's Nutcracker, very closely related to the Nutcracker of the Old World, but it has quite different plumage of grey, white and black. The Old World Nutcracker has white-spotted brown plumage.

What do the cembra pines gain by providing the nutcrackers with nutritious seeds? The birds of course destroy the seeds by eating them! Well, the nutcrackers sooner or later consume most of the seeds that they pick, but not all. During the autumn they gather seeds frenziedly and make constant shuttle flights of several kilometres, sometimes even of 10–20 km, in order to hide them within their territories or in other suitable localities. The nutcracker has a special throat sac, with an opening beneath the tongue. In this the bird can transport up to 100 cembra pine seeds in a single flight. The nutcrackers conceal the seeds for future needs in the ground under moss or in rotted stumps. In an autumn with a rich seed supply, a single nutcracker can lay up well over 1000 of these hoarding caches with an average of a dozen seeds in each. It has been calculated that the total amount of stored seeds per bird per season may be almost as much as 30 000 (about 7 kg of seeds).

The individual nutcracker has an incredible ability for finding its own caches (but not those of others), even when the snow is lying thick on the ground. It is certainly not the case, as Linnaeus put forward, that 'since she takes the clouds as a landmark, as the countryman says, she does not in fact find her hoard again, as the landmark has disappeared'. In spite of the nutcrackers' unusually good memory, it is of course inevitable that some of their many seed caches get forgotten altogether. Furthermore, in good autumns many nutcrackers hide seeds in such large quantities that there is not only sufficient for survival during the winter months and for breeding and the feeding of young in the spring, but even a surplus remaining. Other seed caches are left untouched, for example because the owner happens to die.

From these forgotten and leftover caches the cembra pines' young saplings sprout. Thus the pines have adapted themselves to dispersal through caches. At the same time the trees provide the nutcrackers with their livelihood. A good example of co-operation or interplay in nature between plants and animals. But elements of conflict also exist.

In some years the cembra pines' cone crop fails. This often occurs simultaneously over wide areas. The synchronisation, which is largely accounted for by large-scale annual differences in climate, is amplified by an intense exploitation of those trees which are 'out of step' with the prevalent seeding rhythm. In years with a dearth of seeds, seed-eaters – birds and mammals – concentrate on the few seed-bearing trees. Practically all the seeds are consumed, and there is not much left over to cache. The seed-bearing trees rarely get the chance to spread their inherited traits farther. The result is that evolution fosters a synchronisation in the trees' annual seeding.

Large numbers of nutcrackers are forced to leave their home regions in autumns of poor seeding as they cannot find enough to collect for winter stores. Adult birds with well-established territories and some surplus caches of stored food from the previous season are often able to stop behind. It is mainly the young of the year, newly independent, that are forced to depart on long-distance migration of the invasion type. The invasions are most sweeping when the food shortage has been preceded by a peak year in seed production and with this also a peak in the nutcrackers' breeding success. The chances of the birds surviving the winter months by setting off on long-distance migration in order to find regions with passable stocks of cembra pine cones or to find alternative food in areas with a mild winter climate are doubtless small, but nevertheless greater than the chances of surviving in the home districts. The return flight to the normal breeding range in the late spring following an invasion year shows that the emigration has a survival value, even though the decimation in numbers of the invading hordes of birds from autumn to spring is very substantial.

The conformity in distribution between the cembra pines and the nutcrackers is not 100% (figure 74). Thus, the Nutcracker is found in parts of Europe where the cembra pine is entirely absent. Here its staple food is hazel nuts. It lives where there is access to good numbers of hazels within hoarding distance of spruce forest. In the spruce forest it holds a territory and conceals its nest. The Nutcracker's early breeding, long before the trees are in leaf, implies a habitat that provides adequate cover for the open nest. If we look at the distribution of the hazel and the spruce and encircle those regions where both occur, then we are also encircling the area of distribution of the European Thick-billed Nutcracker (*Nucifraga c. caryocatactes*)! Its lifestyle and dependence on hazel nuts have been mapped out through in-depth studies in southwest Sweden by P.-O. Swanberg. A Nutcracker's throat sac can hold around 20 hazel nuts, and these the bird conceals beneath the moss in its territory. On a single day in autumn a pair of Nutcrackers can transport and cache about 700 nuts. The autumn's hoard of stored nuts constitutes the main food base not only for the birds' winter survival but also for the rearing of the young during the spring.

In autumns when the hazel-nut crop fails, the young Nutcrackers are forced to emigrate. In a stand of ten hazel bushes, which in a good year bears over 2000 nuts, there may be only a single hazel nut in a bad year. Such a bad year occurred in 1975. In that year Thick-billed Nutcrackers moved south over southernmost Sweden in large numbers; 1500 individuals were counted on passage at Falsterbo. A great many Nutcrackers, however, stopped as soon as they reached Scania and other parts of south Sweden where they came across local nut-bearing stands of hazel or gardens with other nuts, berries or food on birdtables. At some of these sites the birds managed not only to overwinter successfully but even to store so much food that they were able to start breeding in the spring.

Further races of the Nutcracker occur outside the range of distribution of the cembra pines, for example in the Tien Shan mountain chain in the southernmost Soviet Union, in the Himalayas and in China. In the former area the Nutcracker is considered to be highly specialised in exploiting and storing spruce seeds. Which species of spruce are involved (it is not the familiar Norway spruce), how the co-operation between the spruces and the Nutcracker works, and what migratory habits the birds exhibit in these regions are for the most part unknown.

In the Himalayas and China the Nutcrackers are said to gather nuts from 'hazel-like bushes'. At a guess these bushes are very close relatives of the European hazel (*Corylus avellana*), namely the Chinese (*C. chinensis*) and the Tibetan (*C. tibetica*) hazels. Apart from its distribution in the Himalayas and in easternmost Asia, the hazel genus is absent from the greater part of the Asiatic continent.

A parallel case to the Nutcracker and the cembra pine, or the hazel, is the Jay and the oak. The oak serves up large, nutritious acorns. These the Jay gathers in the autumn and hides in the ground, often in woodland glades where the planting conditions are ideal for oak. The oak saplings quickly fix themselves in the ground with deep roots, and these they need, for as late as the early summer the Jays continue to search through their stored caches in order to feed the fledged young. When the Jay retrieves an acorn that has germinated, it pulls up the delicate sapling but only far enough to enable it to nip off the two seed-leaves. The oak sapling often survives this treatment, thanks to its well-developed roots. The oak and the Jay therefore have a common interest in the planting of the acorn turning out a success.

Table 23. *Method of seed dispersal of different trees and bushes, with examples of various bird species which feed on the plants' seeds, fruits or berries*

Method of dispersal	Plant	Bird species which feed on the plants' seeds, fruits or berries
Seeds dispersed by the wind	Spruce	Common Crossbill Siskin Coniferous-forest tits
	Pine	Parrot Crossbill Great Spotted Woodpecker Coniferous-forest tits
	Larch	Two-barred Crossbill
	Birch	Redpoll
	Ash	Bullfinch
	Hornbeam	Hawfinch
Seeds dispersed on water	Alder	Siskin
Seeds dispersed through hoarding	Cembra pine	Slender-billed Nutcracker
	Hazel	Thick-billed Nutcracker Rook
	Oak	Jay
	Beech	Brambling (does not hoard) Great Tit (does not hoard) Blue Tit (does not hoard) (Jay, Nutcracker, Marsh Tit, Nuthatch)
	Walnut	Rook
Seeds dispersed directly by birds	Rowan, hawthorn and other berry- and fruit-bearing trees and bushes	Pine Grosbeak Waxwing Fieldfare Redwing (also various thrushes, warblers, starlings etc.)

In autumns when acorns are scarce, the young Jays are forced to emigrate. According to observations at Falsterbo bird observatory, major invasions occur on average at intervals of four to six years. So far, the highest total of emigrating Jays at Falsterbo, in autumn 1977, is of 16 000 individuals. According to ringing results, the Jays generally do not get farther than Denmark; many of them refuse to fly over the sea, but turn at Falsterbo and are later found at different places in south Sweden. The emigration of Scandinavian Jays is surpassed by that of the Baltic and north Russian Jays: these move westwards across Continental Europe and regularly reach as far as West Germany, the Netherlands and Belgium (note the similarity to the Great Tit's migration pattern). In October 1983 a record movement of Jays was recorded in Britain, when huge numbers, including flocks of thousands, passed in a westerly direction through southern England; both Continental and British Jays were involved, and again this huge movement was apparently the result of a failure of the acorn crop in 1983 in Britain and on the Continent.

Apart from the cembra pine, the hazel and the oak, well-known trees that are served by seed dispersal through hoarding are the beech and the walnut (table 23). Common to them all is that they have comparatively large and nutritious 'seeds', well suited to enticing birds and mammals to hoard them (figure 75).

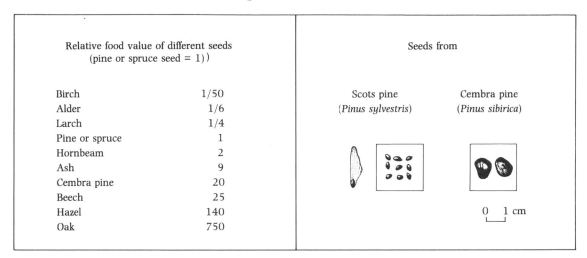

Relative food value of different seeds (pine or spruce seed = 1))	
Birch	1/50
Alder	1/6
Larch	1/4
Pine or spruce	1
Hornbeam	2
Ash	9
Cembra pine	20
Beech	25
Hazel	140
Oak	750

Seeds from

Scots pine (*Pinus sylvestris*)   Cembra pine (*Pinus sibirica*)

0   1 cm

Figure 75   The food value of seeds differs considerably among different tree species. The biggest seeds are largely adapted for dispersal through hoarding; smaller seeds have wings or 'floats' and are dispersed by wind or water. The table shows the food value in relation to seeds of Scots pine or spruce. A pine or spruce seed contains at least 30 calories of high-value bird food. On the right, the size difference between seeds of Scots pine, which bear wings (shown for one seed), and wingless seeds of cembra pine is illustrated. Based on Grodzinski & Sawicka-Kapusta (1970) and Mirov (1967).

In the adaptations to dispersal through caches the plants are pushed into a balance between co-operation and conflict with the animals. Co-operation provides the beneficial storers among birds and mammals with desirable fruits. Conflict – wide annual variations in fruit-bearing – makes things difficult for the ones that cause damage. The losses of seeds caused by the animals' consumption are of course immense. For the population of for example the oak to be maintained, however, all that is required is for each individual tree to produce an average of one successor during the whole of its long lifetime. Thus it is enough if a single one of its acorns germinates in some abandoned cache and grows into a new mature oak!

Some trees use the wind to disperse their seeds; others, such as the alder, use water. The wind-dispersed seeds have hairy plumes or wings. The alder seeds have floats. These seeds are relatively small; this furthers their 'powers of flight' and at the same time makes them less attractive as food for the birds. These trees pursue a more out-and-out conflict with birds and other seed-eaters. Sallow, aspen, poplar and willow have such small seeds that they provide poor fare as bird food.

In trees whose seeds are dispersed by the wind, seed production fluctuates widely. Lowest years occur, usually at intervals of one to three years, simultaneously over extensive regions. Large-scale climatic changes provide the signals which govern the seeding rhythm. The 'bottom years' in the trees' seed production, and the bird invasions associated with them, tend to occur simultaneously in regions as far apart as North America and Europe. The drastic fluctuations in seeding are a major weapon in the trees' efforts to limit the populations of seed-eaters. Complete synchronisation of the seeding fluctuations in separate areas is of course impossible to achieve, and this the birds exploit in their counterweapon, which is a preparedness for rapid and long-distance migration.

Migratory preparedness and the nomadic life connected with it are restricted to the winter months in the Bullfinch and the Hawfinch (cf. table 23). During the breeding season these species feed not only on seeds but also on buds and insects. They are rarely, therefore, forced to desert their traditional breeding localities as a result of acute food shortage. The Siskin's split migration pattern and roving winter existence are explained mainly by the species' dependence on alder seeds during the winter months. In spring and summer, too, the Siskin is variable in occurrence, however, for during this period the birds eat spruce seeds to a considerable extent and shift nest site from year to year depending on the availability of ripe spruce cones. In Finland the number of breeding Siskins within a defined area has been monitored over a lengthy succession of years, and it has been found that the density of Siskins is more than five times as high during very good years for spruce cones as during poor ones.

In winter as well as during the breeding season, the Redpoll seeks out regions with a surplus of birch seeds. In the mountain-birch forests of Sweden it breeds in fairly sparse populations, between two and 30 pairs per square kilometre, in normal years. During invasion years the picture changes drastically. In the autumn of 1967 the mountain birch had a particularly good seed production in southern Lapland. Redpolls accumulated and overwintered in large numbers. In the spring there were still masses of birch seeds everywhere on the ground, and seeds lay in thick drifts on thawing snow that was still left. In that year the Redpolls bred everywhere in the birch forest, almost 100 pairs per square kilometre. In the very next year, however, the number of Redpolls was back down to the usual low level; there was now a scarcity of birch seeds, and the great majority of the Redpolls left to search for regions where seeds were more plentiful. One of the Redpolls that bred during the invasion in Lapland and which was ringed then in June was reported again five months later from the Soviet Union, immediately east of the Ural Mountains, 3000 km away. In 1971 mass breeding of Redpolls again took place in association with abundant supply of mountain-birch seeds in Lapland.

In certain regions the Redpoll has the ingenious habit of breeding a first time, early in the spring, in northern forests where spruce seeds are plentiful, and then immediately moving to the mountain-birch forest to lay a second clutch.

The crossbills are specialists in pecking out the seeds from the cones of conifer trees. Their remarkable bills are effective tools; with them the birds swiftly pull out the seeds behind the hard cone scales. Even though the different crossbill species are able to open cones of all the common coniferous trees, each species demonstrates a particular preference for a certain kind of cone and has its bill-tool adapted for it: Common Crossbill for spruce cones, Parrot Crossbill for pine cones, and Two-barred Crossbill for larch cones (the Scottish Crossbill, restricted to the Caledonian pine forests of that country, is intermediate between Common and Parrot Crossbills). The seeds of conifers are the foundation for the crossbills' reproduction and survival virtually throughout the year. Within the coniferous forest zone there is the constant interaction going on between the conifers' varying fructification in different regions and the crossbills' migrations to those places where the seed supply is good.

The spruce's cones are produced during the summer. Even though the new cones are compacted and hard, as early as July the crossbills are able to open them and consume the seeds. The cones thereafter offer the crossbills suitable food throughout the autumn and winter. Towards late winter and early spring the cones ripen and begin to release their winged seeds. They are then exceptionally easy to deal with, and the crossbills have the chance to achieve maximum efficiency in their foraging. If it is a peak year in spruce-cone production, then seeds are still left to be harvested as late as May and June.

The Common Crossbill has been found breeding in every month of the year. In spruce forests where there are plenty of new cones, breeding can begin in earnest as early as August and then continue right up to May of the following year. How many broods a single pair of crossbills produces during this long breeding season is not known, but there are many reliable reports of pairs rearing at least two broods of young in rapid succession. The culmination in the breeding season occurs in February, March and April, the time when the spruce seeds are most easily accessible.

In the intervening period between the old and the new generation of spruce cones, i.e. during midsummer, the Common Crossbills begin their wanderings in search of areas with good new production of cones. Greatest of all is the exodus from areas where both cones and crossbills have been at maximum numbers and where new cone production fails almost completely because, during the following year, the spruces take a rest and store up strength. In summer and autumn there is an intensive migration traffic. Crossbills fly between different regions within the northern coniferous forest in Europe and Siberia. Unfortunately, we have no knowledge of how the geographical migration pattern within this zone varies between years with different central points in the distribution of crossbills and cone supply. On the other hand, it is well known that in certain years crossbills, both adult and young birds, leave the northern coniferous forest zones and reach west Europe in considerable numbers; between 1800 and 1985 this occurred on over 70 occasions, sometimes in several years in succession and sometimes at intervals of as long as 11 years. The first crossbills reach west Europe in June or July, in some years as early as May. The main bulk, however, does not generally arrive until several months later, in August, September and October (according to migration counts at Falsterbo). The invasions generally take place in years of poor or moderate spruce-cone production in Scandinavia. Many times the invading crossbills doubtless originate from breeding sites in Scandinavia. In other years crossbills have probably emigrated westwards from north Russia or Siberia and failed to find fertile spruce-cone regions during their migration to Scandinavia; the migration is therefore extended to other parts of Europe, where the crossbills seek out coniferous forests in mountain regions and extensive conifer plantations. In the absence of seeds from conifers, they are also able to get by on berries or fruits.

Ringing of Common Crossbills on passage through Switzerland has shown that the birds often continue on to southernmost France and to Spain and Portugal. Many survive well in west Europe and return in stages to the northern coniferous forest, judging from recoveries from Russia several years after ringing. In the wake of the invasions there are also a lot of crossbills that stay behind for a year or two in west Europe and successfully breed in conifer stands which have a sufficient number of cones.

The Parrot Crossbill and the Two-barred Crossbill also invade west Europe in some years. The invasions are more infrequent in occurrence than those of the Common Crossbill, but when the birds do invade they generally come at the same time as the Common Crossbill. The reason why the three species' irruptions often coincide is probably that both the Parrot Crossbill and the Two-barred Crossbill, in the absence of their preferred food, manage well on spruce seeds and are therefore forced to emigrate to west Europe only at those times when, like the Common Crossbill, they have failed to find spruce cones in the coniferous forests of northwest Europe. The Parrot Crossbill requires a plentiful supply of pine cones for breeding. It takes the cones of pine nearly two years to ripen; not until the second year, during the period from December to May inclusive, are the cones soft enough and easy enough to deal with to provide adequate crossbill food. The Parrot Crossbill's breeding season therefore normally occurs in March–April.

The northern trees are far from being the only plants whose conflict with seed-eating animals gives rise to highly variable seeding with peak and low years. An even more drastic development has taken place in various tropical species of bamboo. A bamboo clone grows and stores up energy reserves for many years, and completes its life by flowering and producing great masses of seeds. The seeding years occur simultaneously over fairly wide regions. Most bamboo species live for roughly 15, 30, 45 or 60 years before they finally seed; there are even species which come into flower at intervals of 120 years. In areas where the bamboo seeds, hosts of mammals and birds of all kinds gather. Owing to the synchronisation in seeding, the bamboo is, despite the many seed-eaters, guaranteed a considerable surplus of seeds which can germinate and grow into a new generation of bamboo. Note that synchronisation in flowering within a determined area need not be impaired through there being various species with differing seeding rhythms if, as we can see, the different flowering intervals have evolved into multiples of each other, such as the bamboo's!

Let us turn from conflict to co-operation. Co-operation predominates between plants which produce fleshy berries or fruits and birds which eat the berries and which thereby contribute to dispersing the hard seeds, which are voided undigested. It takes on average half an hour for, say, bilberry pips to pass through the alimentary canal of a thrush (sometimes the time delay can be a great deal longer). By that point the undamaged seeds, if they are 'lucky', can be dispersed a fair distance and to a good habitat.

There are trees, mostly in the tropics, that produce fruits which are dispersed with the help primarily of fishes, reptiles and bats or other mammals. The fruits are generally dark in colour but with a strong smell. Brightly coloured berries on the other hand usually attract birds.

Since the plants compete with one another, the berries are to the greatest possible extent placed at the disposal of those animals which help with seed dispersal. Plants which fruit at times when there is a shortage of berries in the surrounding area are favoured, since there is a great probability that they will attract many of the berry-eating animals. Competition between the plants thus leads to a spread in the time of the various species' berry seasons. In the tropics this development has resulted in a fairly regular supply of berries and fruit throughout the year. This makes it possible for outright fruit-eaters, e.g. various birds, to support themselves the year round within one limited area.

In northern European regions with their considerable seasonal changes, there are no bird species that are wholly specialist fruit-eaters. On the other hand, tits, warblers, thrushes and others readily make use of berries as additional fare. The first berries ripen as early as midsummer and at that time are picked by small birds with newly fledged young.

Most northern plants with fleshy fruits are prepared for their berries to ripen in late summer and autumn, just in time for the migration season. The maximum number of birds are on the move at this time, eager to find food when they arrive at a new stop-off site. This is when the plants come in, with their red, blue or black berries glistening conspicuously and invitingly from twigs, bushes and trees. The plant species which propagate themselves by means of berries are considerably more numerous in the southern parts of the temperate zone, where enormous numbers of migrant birds pass through or overwinter, than farther north, where fewer birds come through on passage from the north.

The birds are not left entirely without berries in the spring. Ivy berries in fact do not ripen until March and April; they are then eaten, according to studies in England, by Blackbird, Song Thrush, Robin, Blackcap and Woodpigeon among others. It is remarkable that it is not profitable for more northern plants other than the ivy to treat the newly arrived spring birds to fresh berries. Many times, as we know, the birds arrive while the food supply is still meagre.

The berry plants protect their seeds with hard husks so that they can resist the attack in the birds' digestive system. This protection, however, is not sufficient to prevent the Hawfinch for example from cracking cherry stones and consuming the seeds. The Pine Grosbeak, too, often 'cheats' by rejecting the rowanberry's pulp to concentrate instead on the hard seeds. In the same way Hawfinch, Nuthatch and Marsh Tit behave in a way that does damage to the yew tree's fruits in that they leave the red perianth and instead crack or break open the seed shell. The thrushes on the other hand eat the fruits whole and digest only the perianth; they thus promote the spread of the yew in the way that the yew has 'conceived' they should.

Of course berry-eating birds and other animals benefit from assimilating not only the flesh of the fruit but also the energy-rich seeds in the berries' pips or stones. This benefit implies a motive for the evolving of particularly heavy treatment of the pips so that the seeds become accessible and can be digested during the passage through the animals' digestive system. On the other hand, it is a vital necessity for the plants that the seed shells resist the attack so that the seeds are left intact when the berry-eaters have finished with them. This situation can result in a race in evolution: ever heavier digestive treatment in the animals meets ever more strengthened seed shells. Such a race has in some instances led to the plant's seed case acquiring such robust dimensions that the seed is unable to force its way out and germinate unless it has first passed through the animal's digestive system so that its shell has thereby become sufficiently weakened.

The plant's defence of its seed against the berry-consumers results, paradoxically enough, in the latter eventually becoming totally indispensable to the plant's propagation and continued existence. At the same time the berries are a vital food for the animal species in question. The development, carried to its extreme, can therefore lead to the most amazing co-operation between plant and animal and to mutual dependence. Such a relationship obtains between a tomato plant and the giant tortoise on the Galapagos Islands. Yet there is one example that stirs the imagination even more.

On the island of Mauritius in the Indian Ocean, a large tree, *Calvaria major*, once grew in great numbers. The local inhabitants often made use of it for timber. Today there is only a sad remnant of 30 very old calvaria trees, close to dying, left in the forests. All are more than 300 years old. They still produce fruits, 5 cm in size, fleshy and succulent, with a central seed surrounded by a hard, woody shell. Evidently not one of these seeds has germinated during the last 300 years, during the period the calvaria trees have diminished so drastically in number. In order to save the tree attempts have been made to get the seeds to germinate in nurseries, with the best care being given, but even these have failed.

The Dodo died out in 1681, barely 200 years after Mauritius was discovered by the Europeans. Judging from the few eye-witness accounts, the 12–20 kg bird fed on fallen fruits and seeds of the calvaria tree; fossils of calvaria fruits have been found alongside skeletal remains of the Dodo. The Dodo had a well-developed gizzard with large stones to crush the food.

Can the Dodo and the calvaria tree have been so heavily dependent on each other that the seed was not able to germinate unless its shell was worked on and made thinner by the heavy treatment given it in the Dodo's gizzard?

That this may in fact be the case was indicated by an American scientist who force-fed Turkeys with *Calvaria* seeds. While it is true that the Turkey's gizzard is probably weaker than the Dodo's was, this is compensated for by the fact that the Turkey works the large *Calvaria* seeds for an exceptionally long time since they must be reduced considerably in size to be able to pass through the Turkey's relatively narrow intestinal canal. Of the 17 *Calvaria* nuts with which the Turkeys were force-fed, seven were cracked in the gizzard; ten passed through intact, even if heavily worn down. These ten seeds were planted and three of them sprouted, the first ones for 300 years!

The Dodo and the calvaria tree are a remarkable example of how close the co-operation between bird and plant can become.

3.9   Omnivorous birds

Some bird species have so broad a food spectrum and tolerate such varied feeding environments that with the best will in the world they cannot be placed anywhere in the list of ecological categories which have been presented in the preceding chapters. Now on no account should it be thought that these species only make up a residue that is difficult to classify and are therefore brought together under the loose heading 'omnivorous birds' in this section. This is in no way so. A distinctive and obvious omnivore niche exists which several species from two different bird families, gulls and crows, have occupied with great success (table 24).

The gulls have invaded the omnivore niche taking their original adaptations to a life in a sea and coastal environment with them; the crows' evolvement into omnivores has by contrast come about entirely on the basis of their adaptation to a life on land.

It is often said that the role of the gulls, and especially the Herring Gull, as omnivores is probably a phenomenon of our times, having evolved concurrently with the massive increase over the last few decades in the amount of refuse from modern industrial society. This is quite wrong: for a long time past the Herring Gull has been adapted for the role of 'natural' omnivore. Just look at the fantastic flexibility in food choice that the species exhibits in non-industrialised regions of East Europe, Siberia and neighbouring parts of Asia. Here the Herring Gull occurs as a breeding bird both on the tundra along the Arctic Ocean coast and beside steppe lakes in Kazakhstan and Sinkiang. On the tundra it feeds on fish, crustaceans, lemmings, and eggs and young of birds; sometimes the gulls collect around food remains and droppings from polar bears and arctic foxes. In the steppe zone of the south the Herring Gulls catch small mammals of various kinds, especially the young of sousliks and hamsters. The birds often fly 30 or 40 km from the breeding colonies to the vast expanses of steppe to catch small rodents. Many kinds of insects are also important food; when grasshoppers are present in quantity the gulls feed largely on these. Further regular items on the menu are carrion, lizards, various small birds and nestlings, small steppe tortoises, fish, crabs, shrimps etc. We can add in parentheses that many of the steppe Herring Gulls winter on the Black Sea and the Caspian Sea, while part of the northern population extends the migratory journey to coasts in India, the Middle East and northeast Africa, around the Red Sea south to Somalia.

With these impressive qualifications as an omnivore it is hardly strange that the Herring Gull has quickly and effectively been able to build up its populations in west Europe and North America along with the increasing amount of easily available food waste from the throw-away society. Offal from the rationalised and rapidly expanding sea-fishing industry and the increasing protection from hunting and egg-gathering by man which the gulls' breeding colonies enjoy have also improved their position.

A study at the largest Swedish Herring Gull colony (Hallands Väderö in the Kattegat), where about 8000 pairs breed, revealed that refuse, which was regularly taken from rubbish tips up to 60 km from the nest site, accounted for roughly half of the gulls' diet. This was supplemented by fish offal from fishing boats, by earthworms from fields inland and by spilled grain from spring sowing. The gulls continually changed between the various food sources, depending on when the farmer started ploughing, when the fishing boats passed within reach

Table 24. *Omnivorous birds which occur regularly in north Europe. The data on wintering areas apply primarily to the species' Scandinavian populations. The Ivory Gull, the Iceland Gull and the Glaucous Gull occur only uncommonly as winter visitors in most areas of Scandinavia, though important winter populations of these three species are found along the north Norway coast*

Species (NB = does not breed in Scandinavia)	Largely resident	Present in N Europe in winter	Important wintering areas		
			Arctic regions	NW & W Europe	SW Europe, NW Africa
Black-headed Gull		+		+	+
Common Gull		+		+	
Lesser Black-backed Gull (North Sea, Skagerrak, Kattegat)				(+)	+
Herring Gull	(+)	+		+	
Great Black-backed Gull	(+)	+		+	
Ivory Gull (NB)		+	+		
Iceland Gull (NB)		+	+		
Glaucous Gull (NB)		+	+		
Magpie	+	+		+	
Jackdaw		+		+	
Rook		+		+	
Carrion/Hooded Crow	+	+		+	
Raven	+	+		+	
Total no. of species	3	12	3	9	2

of the colony, when rain and dew made the earthworms accessible, and when refuse was emptied out at the rubbish tips. The greatest population growth in the Herring Gull happened during the 1950s and 1960s; in northwest Europe the populations reached a peak during the early 1970s.

What has been said above about the Herring Gull illustrates clearly what is characteristic of all the species of omnivore. Their food is extremely varied, generally everything from live prey, carrion, refuse, insects, earthworms and all kinds of other small animals to seeds and berries. The omnivorous birds sometimes behave as pirates and steal food from other birds. Black-headed Gulls, for example, rob the Lapwings on the fields of their worms, and the large gulls attack wintering Whooper Swans to 'sponge' their recently caught bivalves from them (though note that swans are mostly herbivorous). The rubbish tip is the best place for anybody who wishes to observe all the various species of 'black birds' and 'white birds' at the same time.

With their opportunist eye the omnivores can easily find the means to survive, but access to suitable breeding environments is often limited. It will of course seldom do to breed on unprotected and exposed coasts, in open farmland or on rubbish tips. The birds must be able to conceal and protect their nests from various predators or breed on islands and other sites which are difficult of access and which foxes and similar nest-raiders cannot get to. In the omnivores' world there is therefore a surplus of survival resources compared with breeding resources. This leads to high survival and long lifespan but also to keen competition for breeding sites. The youngest and most inexperienced individuals are shut out. The gulls and crows therefore do not begin to breed until two years of age at the earliest (the smaller species); the Raven and the largest gulls wait until four years old or even longer.

Let us look more closely at the different species of omnivore (figure 76). Among the gulls, four common omnivorous species occur in northwest Europe. They are of different sizes: the smallest is the Black-headed Gull, the largest the Great Black-backed Gull. I shall disregard for the present the Lesser Black-backed Gull, which is not such an out-and-out omnivore as the other species.

The difference in size between the gull species tallies fairly well with the rule of thumb regarding doubling of weight in competing species. This rule was discussed in the section on birds of prey (3.6). As with the birds of prey, the pattern is complicated by the gulls' rather large differences in size between the sexes: in the gulls the male is the larger and may, as the Great Black-back male, weigh nearly 30% more than the female.

The length of the migration and the size of the different gull species are inversely proportionate. The Black-headed Gull, which is the smallest, on average migrates the farthest. Whether there is also a difference in migratory habits between the different-sized

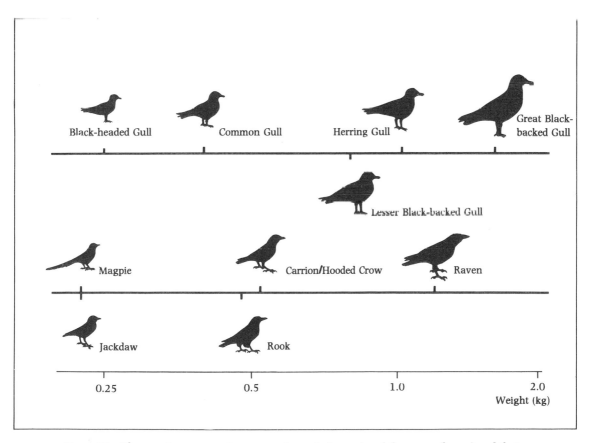

Figure 76 The omnivorous species among the gulls (upper) and the crows (lower) and their size relationships. Even though there are no species weighing less than 200 g that can unreservedly be described as out-and-out omnivores, there are candidates which have a richly varied diet and are very close to omnivores. The Starling in open terrain is a very good example of this, as are the Siberian Jay and certain tits in wooded environments.

sexes within a given species of gull has not been investigated. The wintering area of Scandinavian Black-headed Gulls includes coasts, rivers and lakes in the area from Scandinavia in the north (where the comparatively few Black-headed Gulls in winter keep to ice-free waters) to Algeria, Tunisia, Morocco and Senegal in the south; the central point is in France and Britain.

The Black-headed Gulls leave the over-exploited grounds nearest to the breeding colonies as soon as the young have reached the flying stage. The migration is in full swing as early as mid July (figure 77), and within the course of the following month some of the gulls reach as far as England and France. Many, however, make a stop after a very much shorter migration when they find good survival conditions, not least in the bountiful agricultural districts of late summer and autumn. Here the adult gulls moult their flight feathers. After a moulting period of about two months they are ready to move on to environments that are more suitable for winter survival. This is reflected in a late-autumn peak in the migration pattern in southern Sweden (figure 77).

A similar migration process, with early passage in summer and late passage in autumn combined with an intervening stop-off and moult period, is characteristic not only of the Black-headed Gull but also of the other gull species.

The Common Gull is in the main tied to coasts and tidal shores during migration and in winter. Many Scandinavian Common Gulls move to Britain and north France; only a very few wander farther south. Some migrate no farther than to south and west Scandinavia.

The Herring Gull and the Great Black-backed Gull do not as a rule extend their migration beyond the Danish waters, the most important wintering grounds not only for Danish gulls but also for Norwegian, Swedish, Finnish and Baltic populations. Quite a number of the Nordic gulls, however, stop off for the winter even before they reach Denmark. Only the odd recoveries of ringed Great Black-backed and Herring Gulls reveal that some individuals roam as far to the southwest as England, Belgium and France; in exceptional cases they even reach Spain and Portugal (ringing recoveries of a Norwegian Great Black-back and a Norwegian Herring Gull).

The gulls lead a fairly mobile life in winter and change their place of residence if the climate or the food supply deteriorates. These efficient flyers, robust and with a supreme eye for exploiting all kinds of different food sources, are easily able to roam around in the winter in a way that most other birds would find extremely trying. Gulls of non-breeding, immature, age classes often stop behind and oversummer within the basic survival areas in the south.

In the arctic regions of the North Atlantic the above-mentioned four gulls are replaced by three 'substitute' species: the Ivory Gull (about the size of a Common Gull), the Iceland Gull (almost Herring Gull size) and the Glaucous Gull (the size of a Great Black-backed Gull). The smallest and most climate-sensitive size category, represented in temperate Europe by the Black-headed Gull, has thus disappeared in the Arctic. The survival resources on the coasts of Greenland, Iceland and north Norway are good enough to provide the Arctic's omnivorous gulls, whose populations are probably to a large extent limited by shortage of suitable breeding resources, with a fairly problem-free winter life in these northerly latitudes with their winter darkness. Only occasional individuals extend their winter roaming to more southerly parts, such as to south Scandinavia, Scotland or northern parts of England.

The Lesser Black-backed Gull is very close to the Herring Gull in size and appearance. Both species exist side by side in northwest Europe. How can two such similar species co-exist without one of them being driven out by competition? The explanation is that to a large extent they have different habits. The Lesser Black-back is first and foremost a fish-eater and not an omnivore like the Herring Gull. This I have already indicated by placing the Lesser

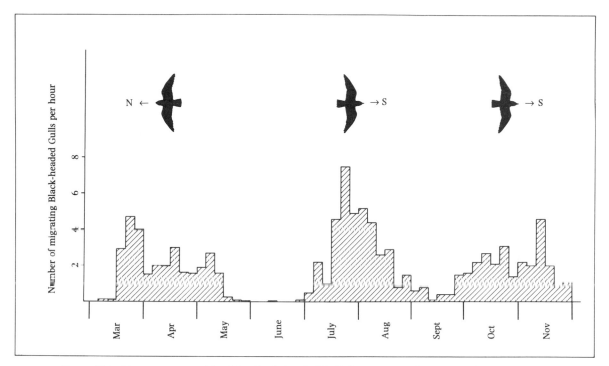

Figure 77   Migration timetable of the Black-headed Gull, from passage figures at Kalmarsund (southeast Sweden). The gulls move northwards early in the spring, with the thaw in March and April. Those passing as late as May probably belong to the northernmost breeding colonies in Scandinavia; presumably they also include immatures on their way to make preliminary visits to prospective breeding sites. The southward migration occurs in two waves, in late summer and in late autumn: this reflects the way the birds, immediately after breeding, migrate to suitable survival environments, from where they migrate again after moulting.

Black-backed Gull, primarily those populations which breed in the Baltic region, among the birds that feed on fish (section 3.4).

The western Lesser Black-backed Gulls (those that breed on west European coasts, and in the North Sea, Skagerrak and Kattegat areas) are long-distance migrants like the Baltic Lesser Black-backs – at least they were until very recently. The classic wintering area for western Lesser Black-backs is the waters off Portugal and Gibraltar and the Canaries Current along the coast of northwest Africa. Almost all juvenile Lesser Black-backs still migrate to this area and live there for a great part of the years before they reach sexual maturity. The adults on the other hand have changed their wintering habits. Since the end of the 1960s many have stayed as far north as England; here, Lesser Black-backs from England and west Scandinavia winter and live to a large extent as omnivores, especially at refuse tips. According to ringing results, the mean latitude of winter distribution of adult English Lesser Black-backed Gulls has during the ten-year period 1965–1975 shifted northwards by an average of 1.6 degrees per year, the equivalent of 150–200 km per year. Maybe in the future significant numbers of adult Lesser Black-backs will begin to winter in the harbours and around tips in Scandinavia?

The change in migratory habits is probably the result of an extremely large winter surplus of available food for omnivores during the last few decades in west Europe. The Lesser Black-backs can help themselves without coming up against serious competition from the Herring

Gulls. This is useful for the sexually mature birds since they can, if they winter fairly close to the breeding colonies, move in extra early in the spring and hold their own in the competition for a nesting site.

The crow species do not show such regular size differences as the gulls (figure 76). The three territorial species, the Magpie, Carrion/Hooded Crow and Raven, certainly follow the rule regarding doubling of weight. In addition to these, however, we have the Jackdaw and the Rook, which because of their colonial breeding habits are adapted to open landscape, to flatlands and farming country. These two species, together with the Hooded Crow, show the most pronounced migratory tendencies. By contrast, the Magpie and the Raven are residents: the adults generally live all year round within their territories and the young birds make only the sporadic excursion that takes them any more than 100 or 200 km from their birthplace.

The migration pattern of northwest European Jackdaws is illustrated diagrammatically in figure 78. The pattern is very similar for Rook and Hooded Crow. The northernmost populations of Jackdaws, i.e. the Finnish Jackdaws in figure 78, winter within an extensive area, from northern regions to as far as England and France; only a minority stop behind to winter in Finland. Of the Danish Jackdaws, on the other hand, a clear majority remain within

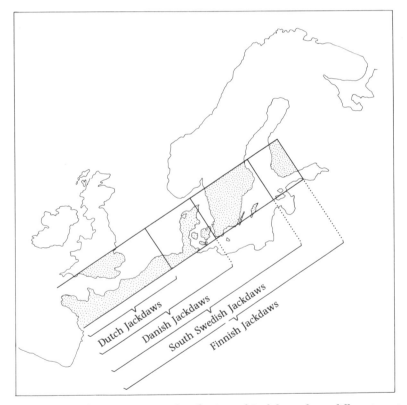

Figure 78 Schematic winter distribution of Jackdaws from different parts of northwest Europe. The birds' direction of migration within the area indicated is, like that of Hooded Crows and Rooks, strictly southwesterly. A much-used migration corridor runs from southwesternmost Finland over the Åland Sea, across Sweden to Västergötland and Halland and on over the Kattegat to Jutland; a field observer on the island of Åland or on the nearby Uppland coast of Sweden can normally count between 50 000 and 100 000 migrating crows during an autumn, on average ten times as many as pass Falsterbo.

the country during the winter; a minority, however, move southwest, to those parts where the farthest-migrating Finnish Jackdaws spend the winter. Among Dutch and Belgian Jackdaws, too, there is a minority which fly a few hundred kilometres to the southwest, to southern England or to northern France. Practically all of the English Jackdaws winter fairly close to the breeding sites.

A winter Jackdaw in north France or southern England may therefore have the most variable origin: it may belong to the local population, but may also come from Belgium, the Netherlands, Germany, Denmark, Norway, Sweden or Finland. When it comes to a wintering Jackdaw in Finland, we can, however, be pretty sure that it is Finnish in its origin (with certain reservations, as a few north Russian Jackdaws occasionally reach Finland).

The Jackdaws, the Rooks and the Hooded Crows perform their longest migratory journeys when they are juveniles. The older ones try to limit their movements as much as possible. In south Scandinavia, where there are resources enough for the Hooded Crows to live within their territories during the winter too, the adults are residents; only younger individuals without a territory migrate.

The members of the crow family are among the landbirds that are the very last to migrate in the autumn. Not until the middle or the end of October does the passage reach a peak in Scandinavia, and it continues into November. In the period between breeding and migration the adult birds have had plenty of time to carry out the annual moult of the flight feathers.

Crows, Rooks and Jackdaws from central, east and northeast Europe have basically similar migratory habits to those of the northwest European populations. Their wintering area does not, however, include Scandinavia and the North Sea region. Instead they spread out over Continental Europe, from France in the west to White Russia and the regions around the Black and the Caspian Seas in the east. In large parts of interior Russia the crows are pure migrants; adult as well as young birds move out in order to escape the severe winter climate.

What has been said above shows that the migratory achievements of the omnivores are not particularly amazing. We could hardly expect otherwise. Because of their great adaptive capacity the omnivores can find ample scope for survival in northern regions. Competition in the winter quarters is thus not particularly hard, and extreme long-distance migration is thereby both unnecessary and disadvantageous.

3.10   The evolution of bird migration

If I were commissioned to write a few sentences for a concise ecological encyclopaedia, outlining what characterises birds, then I would suggest something along the following lines:

> '*Birds* are animals with a stable body temperature. Their lower size limit is determined by heat losses to their surroundings (lowest possible weight around 2 g, as in hummingbirds) and their upper size limit is determined by factors of flight mechanics (greatest possible weight with retained capacity for satisfactory active flight about 15–20 kg). This upper size limit accordingly does not apply to birds that lack the power of flight; a small number of species, such as large penguins, Ostrich, rheas, Emu and cassowaries, have evolved in such a way that they have become heavier. The principal characteristic of birds, the evolutionary weapon that has made possible their wealth of numbers and multiplicity of species on the earth, is their ability to fly or, rather, their ability to migrate. As a result of this birds can move quickly and efficiently between different environments or niches for breeding and survival. By joining together short-term and seasonal living niches, close together or widely separated, the bird species create their different foundations for existence.'

I can imagine that many would argue against migration being regarded in this way as the central ecological factor, the heart, in the life of birds. One might easily suspect that I have become a little biased or 'maladjusted' during all this preoccupation with bird migration. Nevertheless, I certainly intend to persist in my approach. Think of the multiplicity of lifestyles that have been described in the preceding sections. The picture of opportunistic and adventurous striving for living space here on earth is kaleidoscopic. How paltry, trite and lack-lustre all that would surely become if the birds were robbed of their ability to migrate.

The question is often asked 'Why do birds migrate?'. In the light of my character sketch of birds we should hardly be surprised by this. In the ecological sense birds and migration belong inseparably together. Rather, we should be surprised that there are some bird species that are very sedentary: why do not all birds migrate?

In preceding sections I have dwelt solely on northern birds. Bird migration is a common phenomenon in the tropics, too, even though our knowledge of it there is still very incomplete. I do not intend here to go into this subject in any great depth, but will do no more than describe a few main features of migration in tropical Africa and give a few examples to illustrate it. Knowledge of the circumstances in the tropics is, of course, important if we are to be able to give a total perspective on bird migration.

The tropical environment provides ample opportunities for migration. Local downpours, floods, drying-out and fires dictate where various birds can find the best or the poorest living conditions at the time. Masses of tropical bird species perform nomadic migrations in accordance with such environmental changes. Large parts of the tropics also have regular seasons, not periods of winter cold and summer warmth like the northern temperate lands have but rainy seasons and dry seasons. Large bodies of birds migrate in time with these seasons, almost as regularly and purposefully as the migratory birds of the north. The migration distances are comparatively compressed within the tropics, which is hardly surprising bearing in mind that in some parts there is not much more than 1000 km from rainforest to desert. The modest distances travelled, as well as the birds' transient presence in or absence from different regions, help to make it more difficult to produce an efficient mapping of the migratory habits of tropical birds.

A further aggravating factor is that the migration of many tropical species results in a seasonal shift in the central point of abundance within the total range rather than in all birds simultaneously deserting certain regions.

The basic bird-migration pattern runs fairly parallel with the movements of the rains and of the sun, as shown in figure 79 (cf. the Red-billed Quelea's aberrant migration pattern which is described in section 4.6). Most species are confined to the northern or the southern tropics, where, depending on the season, they sometimes move nearer to the Tropics of Cancer or Capricorn, respectively, and sometimes retreat to equatorial regions. Some species extend their migration across the equator and thereby gain the advantage of living constantly where the rains prevail. Within equatorial parts of East Africa migration patterns are often more complex owing to the fact that two annual rainy seasons occur there.

A particularly detailed synthesis has been made of bird migration within Nigeria (see figure 80). This is especially illuminating since within the country's boundaries there are such discrete environments as humid rainforest in the south (today fragmented by man into scattered blocks which alternate with secondary forest and farming plots) and dry Sahel savanna in the north, where the rainy season lasts no longer than two months. Between these extremely wet and extremely arid environments there is the whole range of forest and shrub savannas of different degrees of luxuriance depending on the duration of the rainy season. In Nigeria almost 800 bird species occur; of these, 117 are winter visitors from Europe

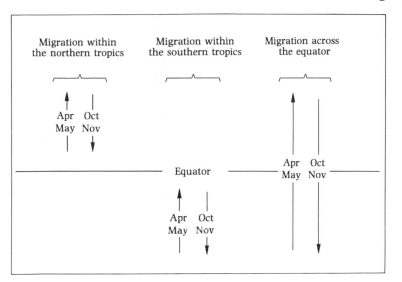

Figure 79 Basic pattern of bird migration within the tropical zone in Africa. Northward migration takes place mainly in April and May, in association with the displacement northwards of the Tropical Convergence Zone and of the rains. The migration in the reverse direction occurs chiefly in October and November.

and Asia. Of the rest, we can today distinguish 126 which are undoubtedly migratory and which perform regular seasonal movements; the number will no doubt increase in the future as our knowledge increases, especially when ringing studies have been carried out.

When the rain front advances northwards in the spring in Nigeria, the savanna grass sprouts and deciduous trees and bushes burst into leaf; the marshes, the rivers and the lakes are filled. All kinds of savanna and freshwater birds move north to breed during this period – ducks, rails, herons, kingfishers, bee-eaters, cuckoos, swallows, sunbirds, larks, weavers and others. The number of migrating warblers is surprisingly small: so far, only one tropical species of a total of 44 in Nigeria is known to be a migrant. All of 22 northern species of warbler, however, appear as winter visitors in the country. After breeding, the tropical birds retreat to wetter regions in the south in order to moult. The migratory movements of certain waterbirds run not only in a north–south direction, from and to the mangrove swamps of the Atlantic coast, but also along east–west paths, for the Senegal and Niger Rivers together with Lake Chad provide refuge sites within the Sahel zone during the dry season.

The White-throated Bee-eater is a good example of a species which breeds during the rainy season in the Sahel savanna and which then returns southwards to the areas where the savanna's luxuriance is preserved during the short dry season. In the spring the bee-eaters take on flight fuel in the form of fat in preparation for their impending migration, at least 800 km northwards.

Some species breed during the end of the dry season or at the very beginning of the rainy season in the southern savanna belt and then move north in order to moult during the short rainy season in the dry savannas. In this category of dry-country birds belong certain species of raptors, storks, one bustard, a nightjar, a couple of swallows and the Grey-headed Kingfisher (figure 80). This last species is in fact not a fish-eater at all but feeds on larger insects and smallish lizards or the like which the bird detects on the ground from look-out

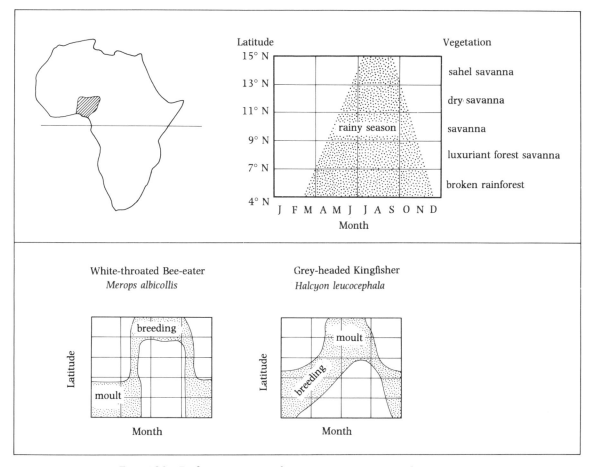

Figure 80 Bird migration in relation to rainy season and vegetation zones in Nigeria. The upper diagram shows the times in the year when the rainy season occurs at different latitudes and in different vegetation belts. One degree of latitude corresponds to a difference in distance of roughly 100 km. The lower diagram shows examples of the occurrence of two different tropical migratory birds in relation to latitude and season. Based on Elgood *et al.* (1973).

posts in the bushes or acacias in the savanna lands. In Nigeria the Grey-headed Kingfisher breeds in the spring. Some are breeding in the southernmost savannas at the same time as others are starting to migrate. As soon as the migrating ones find a suitable breeding territory on the way north they stop. After breeding, young and parents continue on to the Sahel savanna, where their moult period coincides with the rainy season.

It is worth noting in passing that many African bird species retreat south of the driest savanna zones in Nigeria for just that period, October to April, when long-distance migrants from more northerly continents invade these regions.

The pattern of migration in Nigeria has its counterpart in the southern tropics, where a good number of examples can be found both of species which breed during the southern latitudes' rainy season and moult nearer the equator and of others which do the opposite.

The seasonal variations are much subdued in the tropical rainforest; in this stable environment there are few migratory birds. Of 190 rainforest species in Nigeria, only a single one is in fact thought to migrate any distance, in an east–west direction within the West African forest belt. This is the African Pitta (*Pitta angolensis*). This species also occurs as a migrant in southern and eastern Africa, where it has for a long time been regarded by birdwatchers as being shrouded in an atmosphere of mystery and mystique. The reasons for this are several. The bird is extremely difficult to catch sight of; it lives under cover of dense vegetation on the ground, where in the gloomy darkness it picks worms and other small animals among the withered leaves of the ground litter. Birdwatchers who have been lucky enough to see it have become quite lyrical when confronted with its revelation of blazing colours. Dazzling blue-green, dark green, olive-green and scarlet, the pitta is one of the world's most gaudy birds. During migration the species turns up in amazing circumstances.

The English ornithologist R. E. Moreau lived for many years on the slopes of the Usambara mountains in the far northeast of Tanzania. Low clouds and ground mist often enveloped his house at the edge of the forest. On several occasions during the course of the year night-migrating pittas would be drawn to the light from the isolated house and fly into the windows. Moreau recounts how, on a May evening in the early 1930s, he suddenly heard a bird fluttering at the window in the living-room a couple of hours after darkness had set in:

> It was another of the thrills of my ornithological life to go outside and put my hands on this gorgeously coloured bird, unhurt as it was.

The bird was evidently on passage northwards, but where from and where to?

Today we know that the species, besides breeding in West Africa, also breeds in the tropical deciduous forests south of the equator between 10° S and 25° S. It is present here during the rainy season, from November to April inclusive, and at other times of the year it resorts to damp forests around the equator such as in the easternmost parts of the Congo Basin and in Uganda. Another area where it lives is in the forests along the Kenyan coast in the neighbourhood of Mombasa. Here the species has recently been studied at the medieval Arab town of Gedi, the ruins of which lie embedded in dazzling greenery. In this strange region, right next to the ancient stone walls, it has been possible to keep individually marked pittas under close observation from May to the end of October/early November. No doubt the pittas at Usambara were on route to this region.

Judging from discoveries in Ethiopia, some pittas migrate across the equator. There is still much to be uncovered before we know all enthralling sides of this secretive and retiring species' migration. The thought that the pittas migrate in the velvety tropical night is doubtless sufficient guarantee that interest in their migration will continue to be kept alive in the future.

Better-documented examples of tropical species which migrate across the equator include, among others, the Pennant-winged Nightjar (*Semeiophorus vexillarius*) and the Rufous-cheeked Nightjar (*Caprimulgus rufigena*). Both species breed during the rainy season in the southern tropics and then migrate many thousands of kilometres north, just in time for the rainy period in the savanna belt along the southern edge of the equator. Abdim's Stork (*Ciconia abdimii*), however, is responsible for the most spectacular migration across the equator. Unlike the nightjars, it breeds in the north and moults in the south. Using gliding flight, the birds travel in large flocks 3000–6000 km between the rainy seasons of the northern and southern savannas. In East Africa a tremendous spectacle takes place every spring when large numbers of Abdim's Storks pass over at the same time as White Storks: one species is on its way to its nests on the roofs of houses and huts in villages in the savanna zone, while the other is on route to villages and farmyards of Europe.

We might think that it should be possible for tropical birds which migrate across the equator to breed twice a year, during the rainy season in the northern as well as in the southern tropics. So far as we know today, however, there is no species that manages to achieve this.

After the above diversion into the tropics, aimed at showing that bird migration takes place commonly within all the earth's climatic zones, it is now time to come back to the concept that migration is the 'cement' with which the birds put together several different temporary detached niches to form a complete and adequate living niche. The birds migrate between environments where they can find suitable food and shelter so that breeding is successful and environments where it is easy for them to survive. Certain species have at their disposal a large surplus of survival resources in proportion to breeding resources. This applies, for example, to ocean birds, which operate over vast seas, marine birds which dive to great depths, waders which live on exposed tidal shores, and larger landbirds, such as omnivores, some birds of prey and geese which feed in open country. Such birds live in environments where breeding is impracticable. Either it is just physically impossible to breed (the birds must for example first make for land) or else predators wreak havoc among eggs and young which are all too easily accessible. Populations of these birds are limited by the shortage of breeding resources. To be able to breed at all, some species are forced to concentrate in large colonies at the relatively few protected breeding sites that exist. Other species have succeeded in opening up a breeding niche during the brief summer in high-arctic regions. Some change niche even more drastically and change from a marine existence to breeding on marshes or small lakes or, as for example the Long-tailed Skua, to becoming lemming-hunters on land during the breeding season. For species which live under these general conditions – high survival of adults in combination with keen competition for limited breeding space – the result is a relatively small clutch size and poor breeding success. Furthermore, first breeding takes place only after several years since it is most favourable for young and inexperienced birds to 'oversummer' within the survival environments and not to risk exposing themselves too soon to the intense breeding competition. Breeding birds migrate to survival environments, among other things to moult, as soon as breeding is over, if possible as early as midsummer.

Totally different living conditions are encountered by those birds which can conceal both themselves and their nest and at the same time find plenty of food in the resplendent vegetation of the northern summer. Think of the dabbling ducks and the rails and crakes in the marshlands, of the gamebirds in the dense undergrowth and of the hosts of small birds in woods and thickets! They live in luxury during the breeding season but in return are confronted with overcrowding, competition, and shortage of food during the winter months' survival period. In the north they meet the cold, the frost and the competition from the northern birds, in the south the drought and the tropical competitors. For those individuals which survive the winter months, the way ahead is open to exploit the profusion of summer for laying large clutches and producing many young. A good many species manage to raise several broods during one season. One-year-olds breed with virtually the same success as the older members of their species. The birds often have time to moult while they are still back in the fertile breeding environment and before they are forced to move away in autumn to their limited survival environments.

The above summary of the most important adaptations in birds which live with a surplus of survival resources compared with breeding resources and birds which live with a surplus of breeding resources in proportion to survival resources is based on the accounts of different ecological categories which have been given in the preceding sections. What remains to be emphasised in this connection is that migration provides the key to this view of the birds' changes between breeding niches and survival niches which we must proceed from in order to understand the different species' life histories, their reproduction rate and sexual maturity and their moult and migration timetables. My conclusion is consequently that the birds' ecology must be analysed with the significance of migration constantly borne in mind.

It is not enough to analyse only the consequences of the birds' migration and niche changes. If we wish to understand fully the evolution of bird migration, then we must look at the problem more deeply. What ecological rules determine how bird species can combine different breeding niches and survival niches? Should we succeed in clarifying these rules, then we can also understand why a certain bird species migrates but another does not. This is a big task, as yet hardly started on in the field of research into migratory birds. We can easily find ourselves at a loss as to how this far-reaching issue should be attacked. A tentative start may be to break down into categories, as in the preceding sections, different bird species and their migratory habits, in the hope that common ecological patterns and causal connections will appear.

A different angle of approach is to compare the advantages and disadvantages of migration for birds in different conceivable ecological situations. Let us, for example, start off with a species that lives in southern latitudes, where the individuals are faced with the choice of migrating a long way north during the summer in order to breed or of staying behind and breeding in the south. The advantages for those individuals which migrate are mainly that they can exploit the northern summer's extreme abundance of food and that the number of nest-raiding predators is reduced; by this means they can achieve greater breeding success than the resident members of the species that breed in the south. The disadvantages are the risks that can be involved in a long and arduous migratory flight and, not least, the lost benefit of uninterrupted site fidelity in the southern survival areas. Individuals which stay in the south naturally gain priority for the best survival territories; owing to their experience and their superior position in competition they cope better with surviving during the winter months than those birds which return after migration.

Will the species evolve into a migratory bird or into a resident bird or into both? This depends on the balance between the advantages of migration and those of site fidelity. If the migrants' gains in breeding success are much greater than their relative losses in winter survival, then the migrants will in time outcompete the resident members of their species and the species will evolve into a purely migratory bird. If on the other hand the breeding gains are not enough to compensate for the migrants' disadvantageous position in the competition with the resident individuals in the common survival area, then all migratory tendencies will soon die out and the species will become a resident. A third possibility is, of course, that the advantages of migration in the form of fortified reproductive capacity are roughly as great as the advantages to resident individuals of their site fidelity, with the result that a dynamic equilibrium between migrating and resident species populations is established. Such a balance is a common phenomenon, especially in Europe. Many thrushes and finches are for example resident in west and south Europe; during the winter they come face to face with the migrating members of their species, which return from northern and eastern breeding grounds to compete for the survival resources.

The best hot-bed for development of migration is found in areas where the advantage of uninterrupted site fidelity is greatly reduced or non-existent as a consequence of seasonal fluctuations and unstable environment with irregular variations in food supply. This is no doubt a major explanation why landbirds which migrate long distances originate in large part from savanna and steppe regions with pronounced seasonal changes between rainy season and dry season and with great instability in the environment resulting from occasional drought, fires or severe floods. In the rainforest the environment is by contrast stable and the advantages of site fidelity therefore so great that the development of migration is checked.

Table 25. *Number of species of landbirds (passerines, birds of prey etc.) which migrate from northern arctic and temperate regions to winter within various tropical regions, in comparison with the total number of species in the tropics. Based on Karr (1980)*

Tropical region	No. of landbird species		Area (millions of km²)	
	Total	Northern winter visitors	Total	With the majority of northern winter visitors
Africa (S of Sahara)	1481	118	20	17
India	1200	115	4	4
SE Asia	1198	142	2	2
Central & S America	3300	147	20	4

In Africa and India northern migrants winter in dry environments, but in southeast Asia, Central America and South America they winter to a large extent in damp and verdant terrain. The clear majority of the many American long-distance migrants have winter distributions in Central America or in northernmost South America, north of the equator (table 25). Here they live in such places as secondary forest, areas of young sapling growth, bushlands, gallery forest beside rivers and in the cloud forests on mountainsides. How do the migrant visitors cope with the competition from the many resident species in these environments? A large proportion of the migrants feed on sporadically occurring food surpluses, on fruit, nectar or swarming insects, or are forced to put up with the poorest feeding sites, with those that lie on the periphery of the resident birds' favourite environments. The reason why so many long-distance migrants come from damp environments in southeast Asia and in Central and South America is of course that extensive dry country is lacking there. This absence has not prevented numerous species of migratory birds from evolving and occupying the vast breeding grounds of North America and northern Asia that are vacant in summer. Finding out more about the differences in living conditions of the migratory birds in the different continents will be a most interesting research topic for the future.

Another important factor favouring migration but not sedentariness is that unoccupied survival niches exist, areas which are not suitable for breeding and which cannot therefore be utilised by resident birds. Migratory swallows avoid competition with resident tropical relatives by hunting at high altitudes and over rough terrain, migratory terns fish far offshore, and migratory waders live on exposed shores where breeding is impossible. Examples of migrants which have adapted to particular survival environments and thereby freed themselves from competitive disadvantage in relation to resident relatives are, as evidenced in previous sections, extraordinarily common.

Have the migratory birds and the migration patterns which we observe today existed for many millions of years past? No, not at all. Today's migrants are to a large extent products of the ice ages of the Quaternary epoch.

During the last million years, ice ages, four or five of them, have occurred in quick succession, interrupted by warmer interglacial periods. The last ice age, the Wurm or Weichselian Glaciation, began about 70 000 years ago and did not end until quite recently, 10 000 years ago. It was then succeeded by the interglacial period in which we now live. The climate during the last glacial period was not always of equal severity; periods of several thousand years when the ice melted and the ground in ice-covered regions was laid bare alternated with extremely harsh periods of glaciation. The most rigorous period of the Wurm

Glaciation began 25 000 years ago and lasted for 10 000 years. The mean temperature in central Europe at that time was about 15 °C lower than today; in south Europe it was on average 9 degrees colder than nowadays, and in tropical Africa it was 5 degrees colder. Large parts of the northern continents were covered by vast areas of inland ice (figure 81). The mountains of ice arched 3000 m above the sea. The water level in the seas was about 100 m lower than today. The shallow sea areas in the North Sea and the Bering Strait therefore constituted firm ground at that time. In the Caspian Sea, today the world's largest inland sea, the water stood considerably higher during the ice age than it does now: 20 000 years ago it reached approximately 30 m above today's level. During earlier stages of the ice age the Caspian Sea was even bigger, enclosing the Aral Sea and being almost amalgamated with the Sea of Azov and the Black Sea.

During the height of the ice age large parts of Europe and Asia were covered by arctic tundra. On the ground simple mosses, lichens and grasses grew and, in the damper hollows, purple saxifrages, mountain avens and bushes of dwarf willow. At the very edge of the inland ice the tundra was, generally speaking, a completely barren stone desert. The tundra extended a long way down into south Europe; in Spain and Italy it was overgrown with willow and dwarf birch in the lushest places. In the Europe of that time there was also room for man, who lived by hunting and fishing in the open landscape. Of the big ice-age mammals he hunted reindeer, bison and mammoth. He left to posterity his impressions of hunting, nature and the fauna in cave paintings, as for example at Lascaux in France and Altamira in Spain. During the earliest phases of the retreat of the ice, approximately 12 000 years ago, when the edge of the ice was still in Denmark and northwest Germany, reindeer-hunting parties operated as far north as the Hamburg region. Hunting-and-fishing peoples were found even in the 'North Sea country'.

Europe of those days certainly provided the birds with entirely different living conditions from those of our day. Birds of the forest in particular were cramped for space. Coniferous and deciduous forest existed generally speaking only on coasts and islands of the Mediterranean region. In North America, which had a continental climate, the extent of the forest was not so greatly reduced; the forest limit there was in some places only about 100 km from the edge of the inland ice.

In the tropics, too, the climate and the environment during the ice ages were radically different from what they are now. As a result of the low mean temperature, upland vegetation in Africa (today restricted to altitudes above 1500 m) spread out over wide regions above the 500-m contour. In addition, the tropical climate fluctuated greatly between periods of dampness and periods of drought. These revolutionary variations in climate have continued in Africa and South America for the greater part of the ice ages of the Quaternary and are still going on.

At the very time that the last glaciation reached its greatest proportions, about 18 000 years ago, fairly humid climatic conditions prevailed in tropical Africa. The belts of rainforest and savanna had a wide distribution around the equator. The Sahara desert was not much more than half its present width from north to south (figure 81). At Lake Chad the water was 50 m higher than today's level, and the lake was roughly the same size as the Caspian Sea is today. This immense prehistoric inland sea in the savanna zone immediately south of the Sahara is frequently known as Mega-Chad.

Fifteen thousand years ago a period of drought occurred in the tropics. This lasted for 5000 years. Mega-Chad shrank markedly and the Sahara desert advanced southwards. Far to the south in Nigeria, 500 km south of the present limit of the Saharan sand dunes, fossil desert dunes from this period have been discovered. At the same time the drought prevailed south of the equator, the Kalahari desert expanded and most of what is today tropical deciduous forest

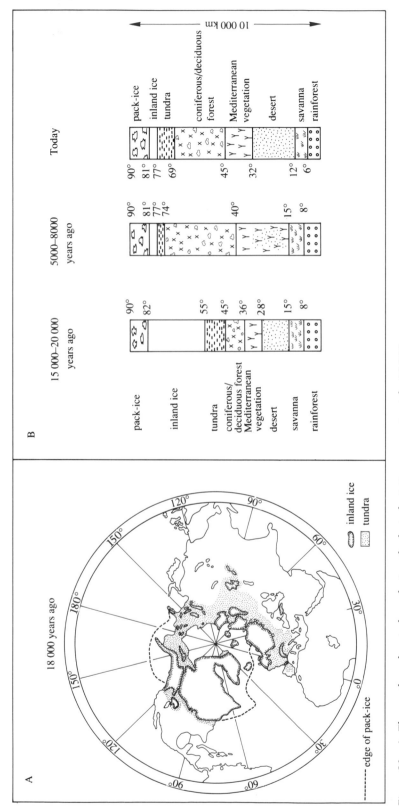

Figure 81 A. The northern hemisphere during the last glaciation, approximately 18 000 years ago.
B. Climate and vegetation zones at different latitudes within the European and African sector of the northern hemisphere during the height of the last glaciation, during subsequent warm and humid periods (warm period in Europe, humid period in Africa), and today. Based on Lamb (1977).

was covered by steppe or dry savanna. During this period the lowland rainforest was substantially compressed into isolated pockets in the Congo Basin and on the West African coast.

This dry period was superseded, 10 000–8000 years ago, by another humid period, even more supportive of life than the previous one. The rivers swelled, new lakes were formed and Mega-Chad reassumed its vast bounds. The tropical savanna vegetation advanced northwards and came into contact with Mediterranean plants which had colonised large parts of the Sahara. Suddenly, then, there was no longer in northern Africa an extensive desert barrier to birds, mammals or human beings! Pollen analysis at the great mountain massifs in the Sahara has shown that Mediterranean shrubs, oak and lime have grown in areas which are now pure desert. Hunting-and-fishing peoples moved in. The rock paintings they left behind tell us of the fauna in the Sahara at that period: elephants, giraffes, rhinoceroses, aurochs, deer, antelopes, ostriches, hippopotamuses and crocodiles. Relics of fishing harpoons in the middle of the Sahara intimate that fishing success was hardly any poorer than hunting success for the Sahara's inhabitants of earlier times. It is hardly likely that the Sahara was at that stage completely covered with vegetation; rather we have to imagine that regions with a rich plant life, especially around lakes and watercourses, alternated with areas of a desert character. Clear evidence for colonisation of the Sahara by animals and man can, however, be found from the Nile in the east to Mauretania in the west.

About 7500 years ago a shorter dry period occurred. Man was at that time forced to leave parts of central and southern Sahara. Even before a thousand years had passed, however, he was able to move in again during another wet period which began 7000 years ago and lasted for a couple of thousand years. The new immigrants were not hunters and fishermen to quite such an extent as the earlier ones. They brought with them domestic cattle, sheep and goats, for supporting themselves. At the same time the first farming civilisations were established in Egypt, where people began to cultivate oats and wheat. To Egypt came the new practices from the Middle East, from the regions around the Euphrates and Tigris Rivers. This was the beginning of the agricultural revolution, one of the most radical stages in man's cultural history.

The plant life in the Sahara, the well-filled watercourses and Mega-Chad still existed, then, only 5000 years ago. We associate the Sahel zone along the southern edge of today's Sahara desert with extreme drought and severe privation for the animals and the human beings that live there. It may seem a paradox that the word Sahel, of Arabic origin, means shore. This is, however, fully explained if we look at the environmental conditions that prevailed there 5000 years ago. At that time the equatorial rain zone was considerably broader than it is now and produced copious quantities of water which collected in the Mega-Chad depression, which lacks outflows. At the same time the subtropical high-pressure zone was displaced northwards. This resulted in the climate becoming unusually dry in Mediterranean latitudes. The Caspian Sea contracted and became much smaller than Mega-Chad. The water level was a good 20 m lower than today.

During the last 4500 years the drought has spread in the northern tropics. The Sahara has again become desert, and man and animals have been driven into retreat. The migratory birds which travel between tropical Africa and Europe or Asia have been forced to develop special adaptations in order to tackle the long passage across the desert barrier. However striking these adaptations may be, they are consequently no more than a few thousand years old.

At the same time as wet periods and dry periods alternated in the tropics, the ice was continuing to melt in the north. Roughly 8000 years ago the inland ice in Europe had melted away. It took a further couple of thousand years before the total thaw had been completed in North America. The fact that an extensive body of inland ice persisted for a long time in North America resulted in the cyclone paths being diverted northwards over the North Atlantic. This produced anticyclonic and warm weather in Europe. During this warm period the forest stretched considerably farther northwards than today. The pine, which together with the birch was the quickest to colonise Europe when the ice had melted, reached all the way north to the Arctic Ocean 6000 years ago, and the distribution of the oak extended as far as the White Sea. The hazel, which is nowadays found in Scandinavia only in the southern third, grew right up in the northern parts; it has been possible to verify this from caches containing hazel nuts left by mice, squirrels and perhaps also Nutcrackers. For a good 3000 years the climate has again been getting colder and the forest limit shifted southwards.

Nature has certainly not been characterised by stability over the past thousands of years. Climate and vegetation zones have been displaced and broken up on a revolutionary scale. The migration patterns which the birds follow today have had an evolutionary period of less than 10 000 years, in most cases not more than 5000 years.

Adapting to new conditions evidently happens so quickly that it is hardly to be hoped that we shall find in today's migration patterns many traces of the former passage and immigration routes that the birds followed in northern regions during the period when the inland ice melted.

The ice age's reshaping of nature has not only made the birds change migratory habits but also led to new bird species being established. The creation of species takes place when different bird populations are isolated and evolve independently of each other for a lengthy period of time. The conditions of the surrounding world are as a rule different in the separated isolates. The populations therefore evolve along different paths. With time the differences become great. If the populations meet again, young produced by parents which come from different stocks are not so vigorous and well adapted as young whose parents belong to the same population. By this means natural selection militates against hybridisation, and evolution leads to separate species coming into existence.

For how long isolation of different bird populations must be maintained for it to lead to the formation of species is a vexed question. An earlier school of research says hundreds of thousands of years, but new findings point to much shorter time periods, of the order of 20 000 years or, under favourable circumstances, an even shorter period. Studies of the distribution of South American and African rainforest birds, especially the passerines, indicate that many of the species have been formed during the last ice ages when the rainforest has during periods of drought been fragmented into isolated refuges, separated by extensive savanna lands.

This brings us face to face with dramatic perspectives. If species formation has taken place in this way in rainforest refuges, then the same process should also have worked in many other environments which have been splintered and redistributed during the climatic changes of the ice ages, in tropical deciduous forest, in savanna, in temperate deciduous forest, in coniferous forest etc. Many of the bird species we have around us were presumably formed during the last ice age, some perhaps as recently as during its melting phase. Of course not all bird species are of equally young origin – some forms have probably maintained their competitiveness, despite all natural changes, for hundreds of thousands or millions of years.

Bird migration has no doubt existed for as long as birds have been present on earth, for more than 100 million years. But the world does not stand still. Look at all that has happened just during the last 20 000 years. New migratory bird species have come into being, new migration patterns have evolved, all in a tremendously exciting and tumultuous upheaval.

4 *The migratory journey*

Imagine that you were faced with the task, as pilot of a small aeroplane, of making a journey of between 50 and 200 flying hours within a period of one to two months. Picture, too, that the wind was blowing about five times harder than normal, so that the wind strength often approached and at times even exceeded the speed of the plane. Further provisos would include that the flight be made as cheaply as possible, that no risks must be taken that are too great and that no delay be permitted.

There are certainly many problems to ponder over and unravel before a plausible flight plan could be drawn up. What speed should the plane maintain for the fuel consumption to be as low as possible? How should the flight speed be adapted to different winds and flight altitudes? Are the plane's gliding properties perhaps so good that it would pay to shut off the engine and glide in good thermals? Maybe it might even pay to make detours over land where conditions are good for gliding? Would it be best to take along a lot of fuel in large extra tanks in order to fly long distances non-stop or should we fly in short stages and refuel at each landing? To decide this question we have to know by how much fuel consumption increases with the additional fuel load and what the current fuel prices and landing fees are at different airports.

Which flight route would be best – would it be wise to choose the shortest route to the destination or is there a circuitous route, with more favourable winds and better staging sites, which ought to be taken in preference? What altitude should we choose? At what time of day should we start? One important thing would of course be to choose good flying weather, but what weather would actually be the most expedient and how particular could we afford to be in our choice of flying weather? Would we get the most from the wind by compensating for it or by exposing ourselves to wind-drift, and under what weather and wind conditions might it be advisable to fly at low level following the coast or other directional lines?

There is certainly a plethora of questions, and it would indeed be a tough job with investigations and calculations for one to be able to work out an acceptable strategy for the flight.

The birds encounter the same situation in principle when faced with their migratory journeys. Bird species with differing migration goals and with differing requirements at stop-off sites are adapted in different ways to solve the innumerable problems surrounding economy and safety of the migration in the best way. Natural selection has produced the birds' migration strategies and continues to refine the degree of adaptation in the birds' behaviour. The migration strategies do not consist solely of stereotype behaviours but on the contrary flexibility in behaviour is great, something which is essential if the birds are to be

227

able successfully to meet the changing and unexpected situations that constantly crop up during migration flights. Migratory birds are much more than mere flying machines.

Before describing how migrant birds accomplish their journeys, I shall provide a summary of the most important methods used to study and to analyse bird migration.

4.1   Methods of studying bird migration

The simplest method of studying bird migration is to arm oneself with a pair of binoculars and from a suitable observation point quite simply to count passing birds and study their behaviour. If observations of this kind are carried out systematically, for example daily throughout the migration seasons and for several years, large amounts of important basic information can be obtained on various migrants' daily and seasonal timetables, on their susceptibility to weather and reactions when they meet directional lines etc. Interest in such field observations awoke during the 1930s and 1940s particularly in Britain, the Netherlands, Germany, Finland and Sweden, and many bird observatories were set up with a programme of long-term field studies of migratory birds. In the north a pioneer in the detailed analysing of the behaviour of migrants on the basis of systematic binocular studies is Gustaf Rudebeck, who made daily observations of the migrants at Falsterbo at the southwest tip of Sweden during three autumn seasons at the beginning of the 1940s. These studies inspired the setting-up a few years later of the Falsterbo bird observatory. Under its direction, counts of migrating birds were organised every autumn throughout the 1950s. After a break of just over a decade they were taken up again in 1973. This renewed interest was due to two factors. On the one hand the problem of collisions between birds and aircraft had become so aggravated that the aviation authorities were keen to give further support to studies of when and how the birds migrate. On the other hand the nature conservancy department realised that continued bird counts at Falsterbo would provide unique opportunities to monitor changes in the populations of different species since the 1940s; they could thus become a major component in an environmental monitoring programme. The Falsterbo studies have been pursued since 1973 by Gunnar Roos, who not only assists aviation authorities and the nature conservancy with the information they require but who also to a great extent takes further the research, which was started by Gustaf Rudebeck, to describe and interpret the behaviour of the various species on passage.

The early studies at Falsterbo inspired annual migration counts at Ottenby, the southern tip of Öland in the Baltic. These went on for ten summer and autumn seasons between 1947 and 1956. The results of the Falsterbo and Ottenby counts have been published in detail. The complete Ottenby material presented in a book by Carl Edelstam (1972) is particularly precise and lucid.

In the matter of scope and completeness, as well as of presentation and analysis, the migration studies at Falsterbo and Ottenby are unique in the world. In table 26 I have shown a 'top 25 list' of the most numerous passage migrants at these two localities. The annual totals at the two sites are not wholly comparable owing to differences in daily and seasonal observation routines, yet it is obvious that the passage in general is much heavier at Falsterbo than at Ottenby. The overall average of migrants counted annually is nearly 2 million at Falsterbo; the figure at Ottenby is 'only' 400 000. The reason for this is of course that most of the migrants that pass south Scandinavia in the autumn have a southwesterly direction to their migration and that many are therefore guided along the coasts via the Falsterbo peninsula which projects out to the southwest, but not via the southeasterly Ottenby headland. This applies particularly to birds of prey which soar and glide over land.

Table 26. *Year totals for the 25 most numerous bird species during autumn passage at Falsterbo and Ottenby, Sweden. The figures for Falsterbo are mean values for the years 1973–1978 with daily counting 11 August to 20 November from dawn to 14.00 hours (from reports from Falsterbo bird observatory by Gunnar Roos); those for Ottenby are mean values for 1947–1956 with migration-counting throughout daylight hours 1 June to 31 October (from Edelstam 1972)*

Falsterbo		Ottenby	
Chaffinch/Brambling	1 000 000	Swift	110 000
Woodpigeon	180 000	Starling	60 000
Starling	180 000	White Wagtail	23 000
Eider	75 000	Chaffinch/Brambling	18 000
Linnet	46 000	Dunlin	17 000
Yellow Wagtail	36 000	Linnet	16 000
Swallow	32 000	Wigeon	11 000
Greenfinch	27 000	Eider	10 000
Jackdaw	26 000	House Martin	9 800
Siskin	23 000	Oystercatcher	6 600
Tree Pipit	22 000	Black-headed Gull	6 400
Fieldfare	13 000	Sand Martin	6 300
Buzzard	11 000	Swallow	5 600
Redwing	10 000	Shelduck	4 000
Meadow Pipit	9 900	Yellowhammer	3 600
Rook	9 200	Siskin	3 400
Carrion Crow	9 200	Common/Arctic Tern	3 300
Blue Tit	8 800	Greenfinch	3 200
Stock Dove	7 800	Skylark	3 000
House Martin	7 300	Pintail	3 000
Black-headed Gull	6 700	Tree Pipit	3 000
Honey Buzzard	6 500	Stock Dove	2 700
Sparrowhawk	6 100	Common Scoter	2 300
Sand Martin	5 200	Curlew	2 200
Yellowhammer	5 100	Lapwing	2 200
Total of all species	1 800 000		390 000

Scandinavian White Wagtails, however, unlike the majority of species, migrate southeastwards. Consequently, many more White Wagtails pass via Ottenby than via Falsterbo. The high numbers of Swifts at Ottenby for the most part reflect what are termed adverse-weather movements and not the actual migration south.

Migration observations over several years, even though less extensive than those at Falsterbo and Ottenby, have been made at many different places in the world. Other Scandinavian sites for such studies include Signilskär off Åland, Väddö on the Uppland coast of Sweden, the Kalmarsund channel between Sweden and Öland (seabird migration), the Onsala peninsula south of Gothenburg, and Blåvandshuk and Skagen in Jutland. Britain has a number of good observatories, among them Fair Isle, the Isle of May, Spurn, the north Norfolk coast, Sandwich Bay, Dungeness, Portland, Lundy, Bardsey and Walney, not forgetting the famed seawatching site of St Ives in Cornwall.

The biggest limitations of binocular and telescope observations are that birds which fly at high altitude often escape detection and that passage movements following directional lines are over-emphasised since they are particularly easy to observe. The fact that masses of small birds pass at high altitude and on a broad front out of sight of a field observer was discovered very early on in the Netherlands by gazing up at the sky through high-powered telescopes. No possibility of effectively monitoring the high-altitude migrations was afforded until it was possible to begin to use radar regularly as a facility for migration-watchers.

A large proportion of migrants travel at night. Getting a general view of nocturnal migration by simple means is, of course, considerably more difficult than following diurnal migration. Listening for calls is not a particularly good method since some species do not call at all and since birds call most intensely in mist and poor visibility or when they are flying at low altitude and are distracted by powerful light sources.

As a result of the catching of night-migrating species that takes place for ringing purposes at dawn and during the early morning at many bird observatories, we are able to obtain information on the species composition and the scale of nocturnal passage. In this case, however, it is not actively migrating birds that are caught but birds which have landed after their night migration, and this can lead to complications when it comes to interpreting the results. Thus, for example, a record morning catch of nocturnal migrants may very easily occur in poor weather and be due to the fact that the birds have broken off their migration and landed when they have come up against this poor weather. Locally unfavourable conditions for migration are probably the main cause of large morning catches being made, especially at bird observatories which are situated on exposed headlands and small islands (and most of them are). This is because in good weather conditions most nocturnal migrants continue to migrate until they can land in the most suitable stop-off habitats inland. That morning catches at coastal bird observatories do not accurately reflect the night-time passage is evident from, among other things, the fact that a disproportionately large percentage of the captured birds are juveniles. Adults probably have the experience and competitiveness to select the best stop-off environments inland and to avoid migration in uncertain weather situations when they risk being forced to land on exposed headlands and islands.

A unique opportunity to catch migrating birds in mid flight, by day and by night, has been taken advantage of at the Col de Bretolet pass in the Alps on the border between Switzerland and France. The pass is situated at almost 2000 m above sea level and is flanked by mountain peaks of between 2500 m and 3000 m. It lies exactly in a southwest-orientated connecting corridor between the Swiss lowland/low Alps and the Rhône valley in France. Catching nets 6–8 m high are set up in the pass and checked at frequent intervals around the clock. It is therefore only birds which 'slip' over the brow of the pass at minimum flight altitude that are caught, and consequently birds which keep a low flight altitude in a head wind or reduce altitude before landing are heavily over-represented. Examples of the results from the catches at Col de Bretolet are shown in figure 82 for Robin and Dunnock, two ecologically closely similar species of which one is a nocturnal migrant and the other a diurnal migrant. Most Robins are caught at dawn immediately before sunrise, when they are about to land after the night's migration. The paucity of Robin catches during the first three hours after sunset is due not to the migration having not got underway but to the fact that the birds are at that time flying fairly high up and that there are not yet any which are ready to land. The Dunnocks begin to migrate before sunrise, often while it is still dark (the calls can be heard), but in the dark and the dim light of dawn they fly high up and not until sunrise and during the four hours that follow it do the catches become good.

For the best visual picture of nocturnal migration we must look to the moon! I mean this literally. 'Moonwatching' was introduced in 1951 by the American George H. Lowery. The method consists of observing night-migrating birds through a telescope as their silhouettes pass across the moon's disc. With the moonwatching method one can see the birds within the

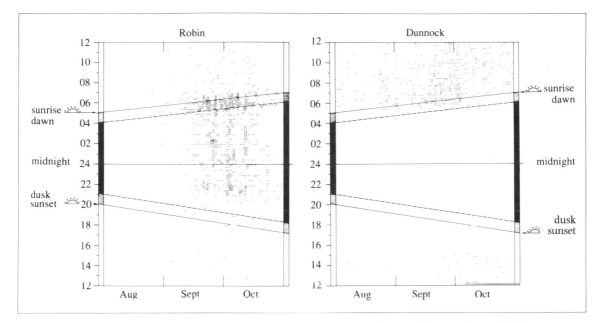

Figure 82 Temporal distribution of catches of migrating Robins and Dunnocks at Col de Bretolet, Switzerland. Each dot represents the catching of one bird, in relation to time of the autumn season and time of day. The times of sunrise and sunset together with hour of dawn and hour of dusk are also shown in the figure. From Dorka (1966).

narrow cone of sky between oneself and the moon. The moon's diameter occupies an angle at the earth's surface of half a degree and at a distance of 1000 m from the observer the width of the cone of sky is only 9 m. It takes less than one second for the birds to pass across the face of the moon, even for those which are very high up. Hardly ever is more than one bird at a time seen passing, for most birds do not fly in dense flocks at night. With the moonwatching method the number of birds passing across the face of the moon is counted and how each bird moves across the moon's disc is noted, often by imagining the moon as a clock face (with 12 o'clock at the top) and noting the nearest 'hour' of a bird's entry and exit on the moon's disc. If we take into account the height of the moon in the sky, and presuppose a certain altitudinal scatter of the night migration and that the birds are flying in a horizontal plane, then we can calculate both the intensity of the migration and the individual birds' flight directions. The disadvantages of this method are that the calculations are quite complicated, that the uncertainty in determining directions becomes great when the moon is low in the sky and that the birds' altitudinal scatter is more often than not unknown. Lowery originally assumed in his calculations that the birds were evenly distributed up to an altitude of 1500 m. More recent radar studies have shown that the altitudinal scatter varies considerably depending on the weather. The measurements of migration intensity which are obtained with the moonwatching method are therefore only approximate.

One of the strong points of the moonwatching method is its long reach. The birds' silhouettes are clearly visible against the bright face of the moon. With a 20 × telescope the smallest songbirds can be detected at a distance of 2 km, and thrushes can be seen at twice that distance. In normal nocturnal migration over land, few birds fly so high that they are missed with the moonwatching method.

This method has been used quite regularly in the United States but to only a minor extent elsewhere. Some observations have been made in southernmost Spain and in Scotland but no other systematic studies from Europe or Africa exist. The method naturally requires cloud-free weather. This partly explains why it has been so little tested in western and northern Europe, but what is remarkable is that it has so far not been utilised for studying the exciting migration conditions in the tropics.

The most grandiose effort to chart bird migration using the moonwatching method was made during the course of four consecutive nights in October 1952, when more than 1000 birdwatchers and astronomers simultaneously recorded the nocturnal migration at 265 observation points spread across North America. The results after one of these nights are illustrated in figure 83; the heavy southward stream consists of night migrants travelling in a northerly tail wind behind a cold front which has pushed down from Canada.

For the first time the moonwatching method gave reliable measurements on the extent of nocturnal migration. The results exceeded the wildest notions. The nightly intensity of traffic of migrants, expressed as the number of birds passing on a front 1 km wide, often proved to be between 2000 and 4000 in good migration weather. On the very best passage nights the figures can shoot up to between 10 000 and 50 000 birds per kilometre of front per hour!

Let us reflect for a moment how fantastic these figures actually are. On a night when migration intensity is at a peak (let us say 25 000 birds per kilometre of front per hour), the number of nocturnal migrants leaving Scandinavia in autumn over southern Sweden and Denmark on a front of at least 200 km will be 5 million per hour; when, after ten hours, the night has come to an end, 50 million birds have left south Scandinavia! If the birds' flight speed averages 50 km per hour, the density of migrants on such a night will be 500 per square kilometre, or one migrant per 2000 square metres. Over every area the size of the plot of a largish detached house, therefore, there is a bird moving over on passage throughout the entire darkness of the night. If the birds are evenly distributed across the night sky, the horizontal distance between individuals is less than 50 m. It is no wonder that, when at dusk on the eve of a night such as this the migration begins, on the radar it looks as if the country with all its coastlines has suddenly blossomed out and started to glide out over the sea – it is millions of migratory birds all taking off at the same time from fields, woods and shores.

Night-flying birds can also be observed in searchlights directed straight up at the sky. The most suitable one to use is a cloud-altimeter (ceilometer), the light beams of which are powerful enough to carry quite high up. By lying down and gazing through binoculars straight up into the beam of light one can count the birds as they twinkle past in the light, and their flight direction can also be determined. The range is limited and as a rule the observer is able to detect only the birds below 500 m, and the very smallest songbirds only below 300 m. The range can, however, be improved considerably, to roughly the standard of the moonwatching method, if the birds passing through the light beam are recorded on an image-intensifier; this increases the intensity of the light that is reflected back towards the ground by 50 000 times. The technique has recently been tried out in the United States, and on this occasion the registrations from the image-intensifier were filmed with a video camera. The 'birdwatcher' was then able to run the video tape later in comfort on his television set and count and gauge direction of the spots of light from the birds which passed in the glare of the searchlight.

The facility which has revolutionised research into migratory birds during the last two or three decades is radar. Radar makes it possible for us to track the birds over wide areas when they are moving at night and above the clouds and at the highest altitudes. The use of radar is of such central importance in what we know today of the process of bird migration that a

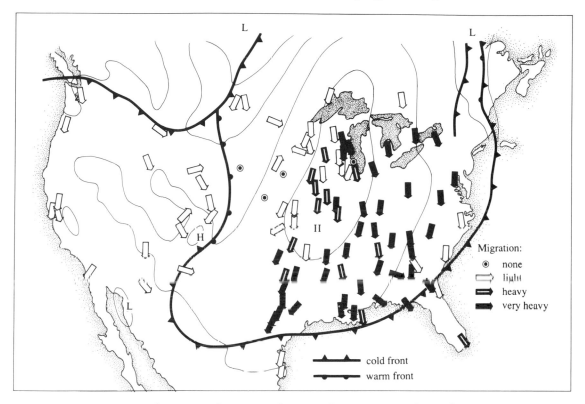

Figure 83 Mean direction and intensity of nocturnal migration on 2/3 October 1952 over different sites in North America according to observations using the moonwatching method. Observations could not be carried out in the northeastern parts owing to extensive areas of cloud and rain in a low-pressure system. The wind is blowing roughly parallel with the isobars (the fine lines in the figure connecting places with equal air pressure), clockwise around high pressure and anti-clockwise around low pressure. Migration is most intensive within and east of the central high-pressure area, where north or northeasterly following winds are blowing; passage is light and variable in direction west/northwest of the high pressure where the winds are southerly. From Lowery & Newman (1966).

more detailed presentation of the principles and development of radar ornithology is called for.

The way radar (= RAdio Detection And Ranging) works is that radio waves are sent out from an antenna in very short pulses, usually of between 0.1 and 5 microseconds; the number of pulses per second is many hundreds. After each pulse transmission the radar is reversed for receiving the echo in the form of radio waves which are reflected back from any object, e.g. from aircraft, ships or birds (figure 84). The radio waves travel at the speed of light, approximately 1000 million km per hour, and the distance to the object is determined from the time lapse between the transmitting of the radar pulse and the receiving of the echo. If the object is 50 km away, it takes the radio waves 0.3 milliseconds to reach it and to return to the radar antenna. The radar's antenna concentrates the radio beam within a restricted fan. The direction of the antenna shows the direction to the object. Thanks to the detailed information on distance and on lateral and altitudinal angle (azimuth and elevation), the object's position and height are known in full.

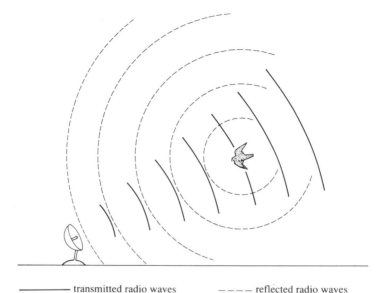

——————— transmitted radio waves   — — — — reflected radio waves

Figure 84   The principle of radar. A radar beam is sent out from an antenna and received back after reflecting against an object such as a bird. The time difference between the radar beam's transmission and its being received back gives the distance to the bird; the antenna's lateral angle and angle of elevation show the direction to the bird.

Each pulse corresponds to a 'parcel' of radio waves which is as long as the distance the radio waves have time to cover during the pulse duration; 0.1 microseconds corresponds to 30 m and 5 microseconds 1500 m. The 'pulse-volume' is determined in addition by the angle of opening of the fan of the beam and increases with increasing distance from the radar station. Several objects within the same pulse-volume, e.g. the individual birds in a flock, all contribute to a common echo and cannot be separated by the radar. A radar with good resolution therefore has a short pulse duration and a narrow fan.

The radar's range depends among other things on the receiver's sensitivity, on the antenna's amplification of the concentration of the beam, and on the transmitter's power (the pulse power can vary between 25 kilowatts for ship's radar, which is used over short distance, and several megawatts for surveillance radar for air traffic). In addition, the size of the target is of course important. How is it possible to detect such small targets as birds with radar equipment that is designed to pick up aircraft and ships?

Watery tissues and blood in the birds' bodies reflect the radio waves; feathers and bones are almost totally insignificant in this respect. We get a good average measure of a bird's radar target area if we look at the bird as a sphere of water of equivalent mass. Water has poorer reflecting properties than metal. A sphere of water reflects 56% of the radio beam that a similar-sized metal sphere would. Birds therefore produce approximately half the echo that they would have if their surface had been completely covered by metal instead of feathers. Calculations of the relative radar target area of birds are complicated by the fact that birds are often in the same size scale as the radio beam's wavelength (the commonest radar wavelengths are 3, 10 and 23 cm, respectively). Then the relative target area varies fairly widely depending on interference, for not only the radio beam that hits the target area directly but also the radio waves that have 'crept' around the edge of the target are reflected. If

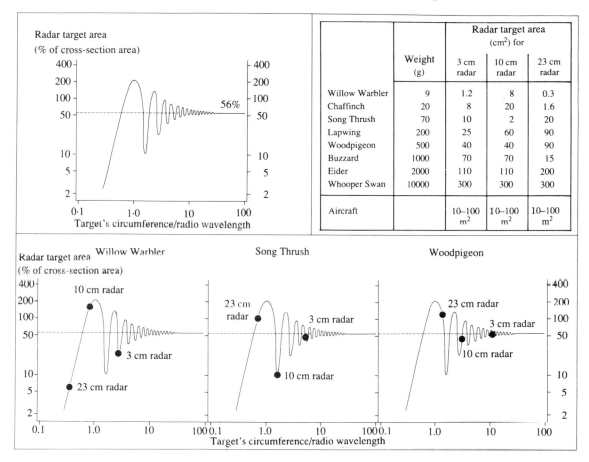

Radar target area (cm²) for:

	Weight (g)	3 cm radar	10 cm radar	23 cm radar
Willow Warbler	9	1.2	8	0.3
Chaffinch	20	8	20	1.6
Song Thrush	70	10	2	20
Lapwing	200	25	60	90
Woodpigeon	500	40	40	90
Buzzard	1000	70	70	15
Eider	2000	110	110	200
Whooper Swan	10000	300	300	300
Aircraft		10–100 m²	10–100 m²	10–100 m²

Figure 85   Radar target area of birds. The radar target area or echoing area of a water sphere varies in relation to the ratio of the sphere's circumference to the radar wavelength, as per the upper diagram. The average radar target area for birds is calculated in such a way that a bird is looked at as a sphere of water of equivalent mass. Radar target areas, calculated in this way, are shown in the table. The lower figures give examples of how the relative radar target area alters with different radar wavelengths for birds the mass of Willow Warbler, Song Thrush and Woodpigeon. The figures in the table are only approximate mean values; experiments have shown that the radar target area varies according to whether the bird is seen head-on, from behind (produces weak radar echo) or straight from the side (gives powerful echo). Based on Eastwood (1967).

the bird is very small in proportion to the radar wavelength, a disproportionately small part of the radar beam is reflected (known as Rayleigh scattering). The radar target areas of birds are shown in figure 85.

The figure shows that the radar target areas represented by the birds in most cases vary between 10 and 100 square centimetres and that corresponding areas for medium-sized aircraft are between 10 and 100 square metres, in other words ten thousand times greater. Amazingly enough, the radar's range, despite this big difference in reflecting surface, is only about ten times as far for aircraft as for birds. If the radar registers an aircraft at a distance of 200 km, it therefore picks up a bird 10 000 times smaller at 20 km! If 16 such birds are flying together, then the flock can be tracked at a distance of up to about 40 km. The explanation for this astonishing fact is that the intensity of the radar beam decreases by the square of the distance both from the radar antenna out to the target and from the target back to the antenna. The result of this is that the range raised to the power of four is proportional to the radar target area. For a doubling of range, therefore, a target area $2 \times 2 \times 2 \times 2 = 16$ times bigger is required, and for a tenfold extension of range a target area $10 \times 10 \times 10 \times 10 = 10\,000$ times bigger is needed. This is the reason why radar is such an excellent means of recording even such small objects as birds. Indeed, even when it comes to following the migration of insects – locusts in Africa, coniferous-forest nun moths in Canada and aphids in Europe – the technology of radar has proven its strength. With suitable equipment individual moths can be registered at a distance of 1–2 km. Heavy concentrations of flying insects can be tracked for several tens of kilometres. Radar ornithology has in recent years had the good company of radar entomology.

Two different types of radar have been used for bird studies. *Surveillance radar* allows mapping of the birds' movements over wide areas, sometimes at distances of up to 100 km or even more from the radar station. The beam's fan is usually broad vertically. This makes determination of altitude impossible, but on the other hand objects at widely different flight altitudes can be tracked simultaneously on the radar screen. In this way a picture is obtained of the collective bird migration, with the exception of bird movements at the very lowest altitudes, which find themselves in the shadow of the radar beam. *Tracking radar* has a very narrow beam; this allows automatic radar tracking of individual birds or bird flocks and continuous and exact information on position, altitude, flight direction and flight speed in both vertical and horizontal planes. By analysis of modulations in the reflected radar signal even the bird's wingbeat pattern can be determined, and this provides important clues to identifying which bird species the radar is tracking. Different species have different characteristic 'echo signatures'.

Radar technology was devised in Britain during the Second World War, but not until the end of the 1950s was it made use of for large-scale studies of migratory birds. First was Ernst Sutter in Switzerland, who used the surveillance radar station at Zürich airport to map nocturnal migration. Thereafter radar quickly began to be used in ornithology. David Lack in England, who mapped the migration traffic over the southern North Sea, assembled many research scientists around him. These workers later extended the studies to the English Channel, northern England, Scotland, the Hebrides and the waters around the Shetland Isles.

In North America, the radar work of William Drury and Ian Nisbet in the coastal districts of Massachusetts, at Cape Cod, broke new ground. North European researchers kept up with things: a minor study was made as early as 1960 from the Arlanda radar station off Stockholm. A few years later came Göran Bergman and Kai Otto Donner's detailed radar investigation of the spring migration of the Common Scoter and the Long-tailed Duck across the Gulf of Finland and southern Finland. The number of radar studies has increased greatly since then. In figure 86 I have indicated most places where radar stations have been utilised up to now in studies of migratory birds.

In some areas bird migration has been mapped in an exceptionally comprehensive way by means of radar. In Switzerland, Sutter's pioneer studies have been followed by a good many investigations of both diurnal and nocturnal migration. Tracking radar has also been put to use, and in recent years the studies have moved from the lowlands up to the Alps. I visited the Swiss migration researcher Bruno Bruderer myself when he was studying birds with tracking radar at an alpine pass 2000 m up. It was a great experience to track small birds in the middle of the night as they passed thousands of metres above the alpine landscape and at the same

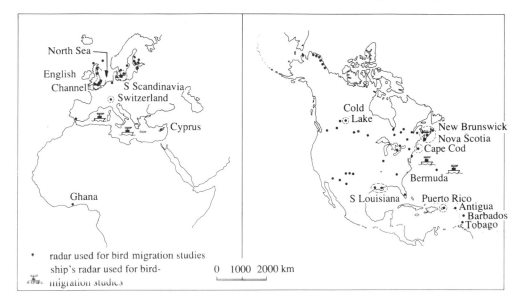

Figure 86   Places where radar has been used for mapping bird migration. Areas with the most extensive and comprehensive mapping operations have been circled. Ship-based radar has been used to track bird migration across the Mediterranean Sea and the western North Atlantic.

time to 'hear' their wingbeat pattern. This was possible because the reflected radar signal was coupled to a tape recorder so that through this we could directly hear how the signal varied in time with the birds' wingbeats.

Thanks to many years' studies, bird migration over the North Sea has been accurately documented. In southern Sweden and Denmark, during the last decade and more, several radar studies have been carried out with field observations made simultaneously. I have myself taken part in studying autumn passage of Woodpigeon, Redwing and small diurnal and nocturnal migrants, and spring passage of Eider, Crane and night-migrating thrushes. The winter movements of various seabirds have also been mapped.

In North America radar studies have been made over many years, in the classic Cape Cod area and in central Canada, where John Richardson has analysed the continental inland migration. The same scientist has in addition scrutinised with radar the exciting bird migration over the New Brunswick/Nova Scotia area, from where many waders set out in autumn over the Atlantic and fly non-stop to South America. John Richardson has also used radar to study birds in Puerto Rico, since large numbers of small birds pass through there in the autumn on their way across the open ocean from the east coast of North America to South America. The picture of this interesting bird-migration traffic over the western North Atlantic is complemented by other radar studies from the east coast of North America, Bermuda and the West Indies, and by studies using marine radar. In Louisiana, Sidney Gauthreaux has used radar to track bird migration across the Gulf of Mexico.

Among the most exotic locations for bird studies with radar are the north coast of Alaska, where ducks and geese have been recorded on passage at the edge of the Arctic Ocean, and Ghana in Africa, where investigations on a smaller scale have provided information on the migration of waders and terns.

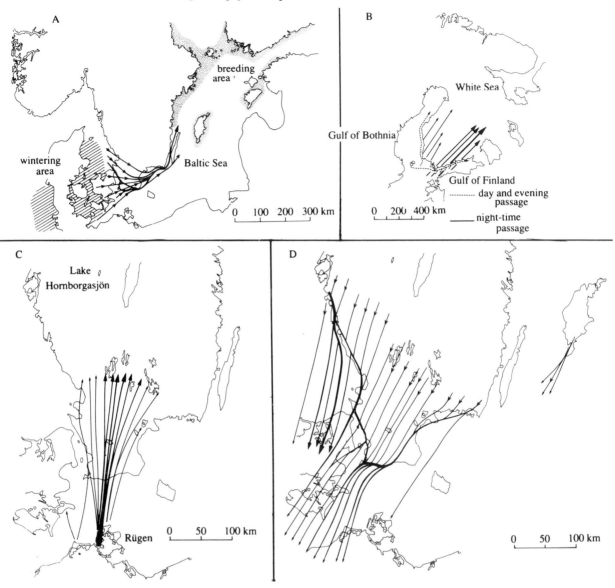

Figure 87 Four schematic 'radar portraits' of the migration patterns of different Scandinavian migrants. A. The spring migration of the Baltic Eider population (based on Alerstam *et al.* 1974). B. Spring migration of Long-tailed Ducks and Common Scoters over southern Finland (based on Bergman 1974 and Bergman & Donner 1964). C. Spring migration of Cranes over the southern Baltic and southernmost Sweden (based on Alerstam 1975). D. Autumn migration of Woodpigeons over southernmost Scandinavia (based on Alerstam & Ulfstrand 1974).

Four 'radar portraits', examples of mapping of the bird-migration pattern with surveillance radar, are shown in figure 87. To a certain extent it is possible to distinguish the different categories of birds that we see on the radar screen; this is because different types of echo differ in strength and speed. If we are to be able to analyse the migration of individual species, as in figure 87, however, it is necessary to combine the radar studies with simultaneous field observations at several different places. By counting the number of radar echoes of migrating flocks and combining them with field observers' reports on average flock size, we can estimate the size of the migrating species' populations.

In this way it is possible to calculate that in spring roughly three-quarters of a million Baltic Eiders leave the wintering waters in Denmark and West Germany; see the migration pattern in figure 87A. Of especial interest is the fact that the Eiders from the Kattegat migrate southeastwards over south Sweden (Scania) or southwards through the Öresund; there one can thus witness the remarkable fact of the birds migrating south in spring. East of Scania the Eiders turn north again; the heaviest concentrations of migrating Eiders in spring are formed in Kalmarsund, the narrow strait between Öland and the mainland of southeast Sweden.

In May, between 1 million and 2 million Long-tailed Ducks and Common Scoters pass over the Gulf of Finland and Finland itself towards the White Sea, as in figure 87B. Roughly 90% of the flocks fly via the Gulf of Finland, while the remainder move northwards via the Gulf of Bothnia. During the daylight hours the birds fly at between 100 m and 250 m above the sea and swing eastwards along the south coast of Finland; many Long-tailed Ducks fly eastwards all the way from staging posts on the Swedish coast of the Åland Sea into the Gulf of Finland. At night the ducks migrate overland on a large scale, some from bays and gulfs where they have got 'caught up' during the coastal passage of the day and some on a broad front straight from the Gulf of Finland and the Baltic Sea. During the night the flight altitude varies between 200 m and 4200 m, the mean being 1000 m.

Cranes breeding in Norway and Sweden stop off in spring on the southern Baltic coast, especially at the island of Rügen. At the end of March and in April they set off straight for the north (figure 87C). Many fly direct to their breeding sites, while others gather to make a stop-over northeast of Gothenburg at Lake Hornborgasjön, renowned for its dancing Cranes. The Crane flocks' course across the Baltic Sea south of Scania is influenced by the wind: the birds move in over the Swedish countryside more to the east than usual in west winds and more to the west in east winds. Overland, however, their routes are largely unaffected by the winds. The Scandinavian population of Cranes is between 10 000 and 30 000 birds.

Considerably more numerous, at roughly $1\frac{1}{4}$ million, are the Woodpigeons which leave south Sweden in autumn following the migration pattern shown in figure 87D. They are on their way to wintering areas in France and Spain. Approximately half of the pigeons leave Sweden via the southwesternmost Baltic Sea, a fifth via the Öresund and approximately the same number over the Kattegat. Woodpigeons are ideal for tracking with radar. They fly in large, dense flocks – the mean flock size is about 100 birds – at good speed and often at high altitude, 1000–2000 m. A clear autumn day with intensive pigeon migration is one of the most delightful experiences to be had in the field and at the same time one of the most magnificent sights to be seen on a radar screen.

If we wish to illustrate flight-mechanical and aerodynamic theories on the way migrating birds fly, such as their speed and altitude in relation to winds and flock structure, tracking radar is the most appropriate. In order to give an idea of what such radar trackings of individual bird flocks look like, I have selected a few examples from my own studies in figure 88. Here the birds' position and altitude minute by minute are shown. Thrush flocks (Redwing and Song Thrush) beginning to migrate in spring at dusk from northern Öland

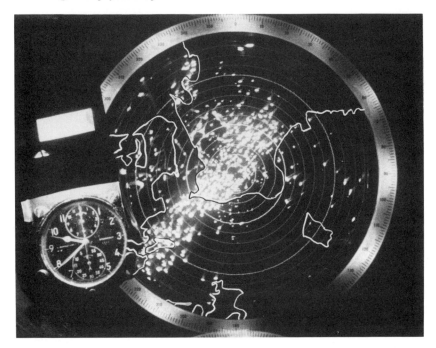

The radar screen on an autumn morning at 09.40 hours with intensive Woodpigeon passage. The radar station is situated on Romeleåsen in south Scania, Sweden. The circles on the screen indicate 10-km distances. The large and fast-moving pigeon flocks produce powerful radar echoes which can be followed to the edge of the radar screen, more than 100 km from the radar station. The southwest-orientated migration is massive over Scania and the southwesternmost Baltic Sea. A number of echoes from pigeon flocks can also be seen along the Swedish west coast at the upper edge of the radar screen.

climb at an average 75 m per minute. Of the three flocks in figure 88A, flock no. 1 and flock no. 3 stop climbing and adopt migration altitudes at 1700 m and 1000 m above sea level; flock no. 2 is still climbing at 1800 m when the radar loses contact. Flock no. 1, which is flying in almost totally calm weather, maintains an average speed of 33 km per hour, except during the minutes 12–16 when it swings to the northwest, temporarily breaks off its ascent and reduces speed to 23 km per hour. Flocks nos. 2 and 3 are flying in moderate southeast and southwest winds and increase speed to about 58 km per hour as they approach flight altitudes where the tail winds increase. Shown on the same map for comparison is the way a flock of Bean Geese passes in V formation: the altitude was all the time constant, at 220 m, and the course almost straight as an arrow (heading for southwest Finland); the speed in the fresh following wind was 118 km per hour!

In figure 88B I have shown how complicated the behaviour of seabirds can appear when they are caught up in a bay and have to move in over land. Red-throated Divers arrive in autumn over Hanöbukten bay in the south Baltic and fly west across Scania. The figure shows how two flocks were tracked on the same day, a day of fresh southeast winds. Both flocks initially fly southwards immediately off the coastline. Here flock no. 1 travels at a height of 300 m, continues in over land in the inner part of the bay, turns a little upwind and climbs 250 m in three minutes; then it swings southwest and crosses the country, slightly increasing height to around 600 m, at a speed of 56 km per hour. Flock no. 2 behaves in a totally

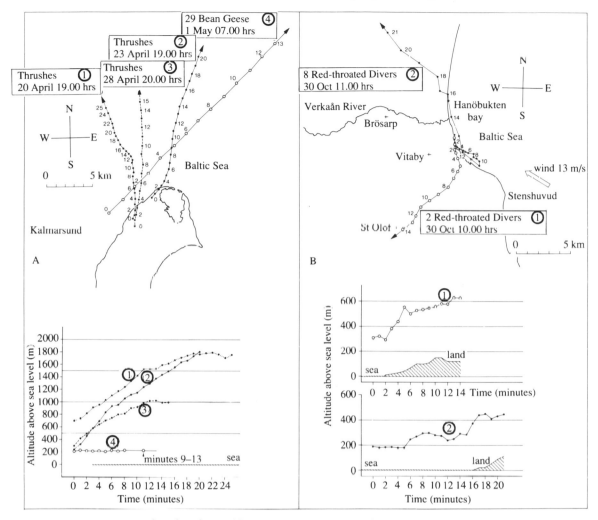

Figure 88   Examples of tracking of flocks of migrants with tracking radar. The dots show the flocks' position minute by minute and their flight altitude (lower diagrams); every second minute is numbered. A. Trackings in spring of thrushes on outward passage from northern Öland (tracking nos. 1–3) and passage of Bean Goose flock in V formation (no. 4); the radar station was positioned near the north tip of Öland. B. Trackings of Red-throated Divers migrating in over Scania from Hanöbukten bay in autumn; the radar station was positioned at the inner part of the bay east of Vitaby.

different way. It stays off the coast and passes south into a direct head wind; while flying in this head wind the mean speed is only 21 km per hour. At the same time the divers increase their flight altitude by 100 m. After eleven minutes the flock turns right around, and now off it speeds in the tail wind. The divers steer in against the coast heading northwest. When they pass the shoreline they stop, circle and increase altitude by 160 m in two minutes; then they head off northwest overland at around 450 m altitude at 121 km per hour in glorious tail winds! The first flock will, if it continues on its southwesterly course, reach the sea off the south coast of Scania. The second flock will reach the southeasternmost Kattegat after approximately the same length of time.

Tracking radar has also been used to track migrants that have been released for experimental purposes. The experimental birds are placed in a lightweight box the seal of which can be broken by a fuse. The box with bird is fastened to a balloon and the fuse, which is of just the right length for the balloon to rise to a suitable height, is lit. The whole contraption is tracked by radar when it is released upwards. When the fuse has burnt to the end, the bottom of the box falls out and the bird suddenly finds itself out in mid air, where it is followed by the radar. The flight behaviour of small birds has been studied by releasing them at dusk, in the dark, below cloud and so on. The White-throated Sparrow (*Zonotrichia albicollis*), a common migrant in North America, has been used in many such instances. Other species of small birds have been released individually in a similar way so that their wingbeat pattern could be recorded on the radar. A 'reference library' of characteristic wingbeat patterns is of great value in that it increases our chances of identifying different night-migrating birds with radar.

Tracking of migrants which have been fitted with radio transmitters has so far been tried out only in North America. Here radio receivers in light aircraft have been used to track Tundra (Whistling) Swans. The birds left their winter quarters in Maryland on the east coast, moved northwest via staging posts on the Great Lakes and in Minnesota and North Dakota, and then continued north across the Canadian forest to the breeding sites on the tundra. Owing to variable winds, the swans' flight speed varied between 45 and 135 km per hour and their flight altitude between 100 m and 2400 m above the ground.

It is not always necessary to use aircraft to track the birds: many successful trackings have been made by car with the receiver aerial mounted on the roof. The range reached, from the radio transmitter on the bird to the car's receiver, is roughly 50 km when the bird is flying at more than 1000 m, less than 20 km when it is flying low and no more than 1 or 2 km when the bird has landed on the ground. The object therefore is all the time to keep as close to the bird as possible with the tracking car, generally a little in front. It is important that the correct choices of road are made, and one must be prepared to put one's foot down really hard to keep up with birds in good following winds!

A nocturnal flight of a Canada Goose with radio transmitter during the autumn migration is shown in figure 89A. The Canada Goose was in a flock which had stopped for quite a time at the same staging site. In order easily to get prompt warning when the geese set off on migration, the radio receiver had been connected to the tracking car's horn so that the car automatically began to toot when the power of the radio signal increased, i.e. when the goose with the transmitter rose from the ground and began to fly. It was quite late on a November evening, the map-reader had already gone to bed and the driver was in the shower when the car began to toot at 22.00 hours. Now a dramatic chase was enacted, not only because the tracking car was inevitably delayed but also because the geese were flying at 95 km per hour in a following wind. It took a long time before the car had caught up with the geese. The observers never succeeded in finding suitable roads close enough to the birds to enable their flight altitude to be estimated. Choice of road was restricted by, among other things, the fact that the bridges across the Missouri River and the Platte River had to be found quickly. Nevertheless, it was possible to maintain radio contact with the geese the whole time until, after $5\frac{1}{2}$ hours, roughly 500 km, of flying, the flock landed on the River Platte at 03.34 hours. The tracking car was parked beside the river, the horn alarm was switched on and the occupants went to sleep. After only just over two hours the car tooted again, and the chase continued. After a further 80 km flying the radio signal suddenly grew weaker at 07.05 hours. At dawn the goose flock was relocated on a stubble field. So far as the Canada Geese were concerned this brought one nocturnal staging flight to an end.

Radio technology now makes it possible to construct transmitters which weigh no more than 2 g. Such a lightweight transmitter was attached to the Veery that was the source of the tracking shown in figure 89B. The Veery weighs no more than 30 g, about as much as a Skylark. The tracking car's night-time journey lasted for 11 hours and extended over 740 km before radio contact with the bird was lost over Lake Michigan. The bird quite simply flew away from the tracking car, which dropped behind when the morning traffic on the roads increased and the crew grew more and more tired after the long night. Presumably the bird turned off towards the eastern shore of the lake, since attempts to relocate it later in the terrain along the northwest shore failed, despite careful searching with radio receivers in both car and aircraft. The winds were favourable during the flight. The bird's mean speed was 64 km per hour. Despite the fairly high speed, good roads for the tracking car were successfully found, and several reliable altitude measurements could be made. The tracking gives a fine insight into what a flight stage can look like for a small night-migrating bird.

Various methods of monitoring bird migration complement each other in an admirable way. Studies with field binoculars are most suitable if one is to follow birds flying at relatively low altitude within a couple of kilometres' distance. For detailed measurements over distances of up to 20 or 30 km, tracking radar is ideal. Radio-telemetry techniques provide

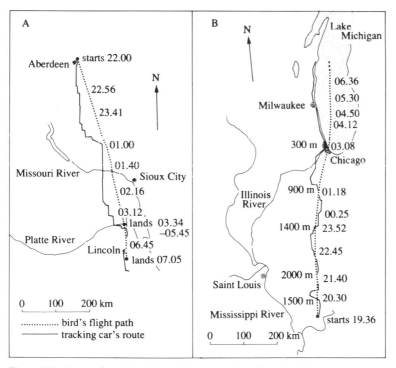

Figure 89   Examples of tracking of migrants fitted with radio transmitters (radio telemetry). The trackings were made from a car. The birds' flight path and the car's route are shown in the figure. Exact times when the bird's position was determined are shown to the right of the bird's flight path, the flight altitude (in B) to the left. A. Overnight migration of Canada Goose in November from the Aberdeen district in South Dakota to Lincoln in Nebraska; the flock of Canada Geese made a temporary halt on the River Platte. B. Overnight migration of Veery on 15 May from southernmost Illinois to Lake Michigan. Based on Cochran (1972).

accurate data on individual birds' flights over hundreds of kilometres; surveillance radar allows a broad view of the migration pattern over areas almost as wide. To follow the birds even farther, to reveal flight paths, stop-off sites and migration destinations over distances of thousands of kilometres, we have to rely on ringing studies. Radio-tracking of individual migrants over really long stretches should also be possible. Surely it will not be long before such phenomena as the large seabirds' migratory wanderings over the oceans can be tracked by satellite.

4.2   Flight speed

The supreme flying capability of birds has fascinated man since time immemorial. It is a very rare event for birds to collide while in flight or to crash-land, despite the fact that they often perform giddy aerobatic manoeuvres and fly in large and densely packed flocks.

Birds have served as models for man in his efforts to learn the art of flying. From the Renaissance until 1903, when the Wright brothers carried out the first successful flight with a motorised aeroplane, many studies were made of bird flight with a view to understanding how flying machines should be built. Attempts at flying with 'machines' the wings of which the pilot operated using the power of his arms and legs failed miserably since people knew nothing of the forces which influence a flying bird or an aeroplane. Knowledge of the elementary principles of flight was not achieved until the middle and latter part of the 1800s. The German Otto Lilienthal's book in 1889, *Der Vogelflug als Grundlage der Fliegekunst* (Bird flight as a basis for the art of flying), is a milestone in this respect. Lilienthal studied the flight of birds in detail. He removed or bound together various wing feathers on the birds in order to see how their flying capacity was affected. He made more than 2000 successful gliding flights from mountain tops, whereby he tested carrying capacity and stability of different wing shapes. Up to this point the bird studies went hand in hand with technical development.

Things changed in the 1900s, when great advances were made in flight technology. The birds had seen their best days as models. Sheer curiosity on the part of ornithological researchers now became the main driving force for further analyses of bird flight. An important tool in this branch of research today is high-speed photography, by which means we can follow the movements of various parts of a bird's wings millisecond by millisecond. The results can then be used to measure the forces that influence model profiles and sections of bird wings in experiments in wind tunnels. In this way attempts are made to piece together forces and air currents around a bird in different phases of its flight. Studies of this kind have so far shown at least one thing very clearly: bird flight is an extraordinarily complicated phenomenon.

A body that is met by a current of air is naturally influenced by an air resistance. If the body has a special design and is situated in a suitable position, it can be influenced simultaneously by a lifting force. This applies both to the bird wing and to the aircraft wing. The air current around a wing section (figure 90) happens in such a way that the speed of the current increases along the upperside of the section and decreases along the underside; at the same time the airstream in its entirety is deflected downwards a little. A law of physics, Bernoulli's Law, states that the sum of the kinetic pressure and the static pressure is constant. When the velocity of the current, and therewith the kinetic pressure, increases the static pressure drops, and vice versa. Thus, on the upperside of the wing a lower air pressure and on the underside of the wing a higher air pressure than in the area around is established. This results in the wing being influenced by a lifting force. A bird in the air is therefore 'resting on' small high-pressure points beneath the wings and at the same time is 'suspended from' low-pressure

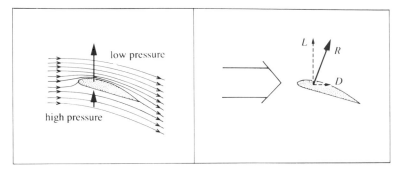

Figure 90   The airflow around a wing section leads to low pressure above the wing and high pressure beneath the wing. This gives rise to a lifting force (L) at right angles to the air current which, together with the air-resistance force (D, from drag), forms the resultant force (R) which acts on the wing.

bubbles above the uppersides of the wings. In actual fact it is more correct to say that a bird is suspended in the air than resting on it, for the lower pressure above the wings contributes approximately two-thirds of the lift and the higher pressure at the underwings only one-third.

Does it sound remarkable that a bird can be suspended in the air simply as a result of a rapid current of air passing along the upperside of the wings? Just do a quick experiment of your own: hold one end of a small slip of paper and let the other end just hang; blow along the upperside of the paper and see how the hanging end lifts up in the rapid stream of air along the paper's upper surface. The same principle explains the supporting capacity of birds' wings.

A good wing section produces a good amount of lift; at the same time the air resistance becomes as little as possible. The phenomenon known as induced drag is due to vortices that are formed at the wingtip, and these even out some of the pressure difference between the upperside and the underside of the wing. The wingtip vortices are of particular interest when it comes to working out the advantages to birds of flying in flocks and in formations. Long, pointed wings result in small induced resistance. The induced drag is the price that must be paid for lift.

The forward drive of the wing through the air gives rise to further air resistance owing to friction and to 'back-eddies' on the upperside of the wing. The air current in the layer that is closest of all to the wing, known as the boundary layer, is a key factor that determines these types of air resistance. The most effective way of limiting the air's resistance to aircraft wings is to reduce friction by making the wing surfaces as even as possible. For birds' small wings, which are to function at low speeds, the back-eddies are the biggest problem. Back-eddies are best avoided with small-scale turbulence in the boundary layer. Such turbulent flow is brought about by, among other things, fairly pointed leading edges to the wings and slight irregularities on the upperside of the wing. The special sections and feather structures of bird wings help to minimise the back-eddies. Measurements in wind tunnels suggest that the airflow over certain parts of the birds' wings is supercritical, i.e. gives good lift and little resistance, and that the flow over other parts of the wings and over the body is subcritical and has decidedly poorer lift effect and greater resistance effect as a consequence. The airflow around a bird wing is therefore much more difficult to analyse than the airflow around an aircraft wing.

Lift (L) and drag (D) when air flows over a wing together produce a resultant force (R) (figure 90). This resultant force the flying bird makes use of to balance on the one hand its

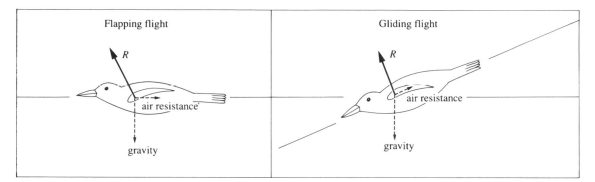

Figure 91   During flapping flight as well as during gliding flight, the resultant force (R) that acts on a bird's wings serves to counter the vertical gravitational force and the air resistance to the bird's body.

weight and on the other the air resistance to its body. This is illustrated for the two main types of flight, powered or flapping flight (i.e. using wingbeats) and gliding flight, in figure 91. In this section I shall discuss only powered flight; in section 4.4, I shall return to how migrating birds use soaring-and-gliding flight.

A flying bird generates motive force mainly over the outer parts of the wings, which cut the air obliquely downwards on the downstroke. In doing this the bird twists the outer-wing sections diagonally downwards in order to maintain the most favourable angle of attack against the air current along the whole length of the wing (figure 92). The inner wings generate lift almost solely as they move at low speed vertically. Lift and propulsive force are produced not only through the wings' downstrokes. More recent detailed studies have shown that the net effect of the wings' upward movement gives some assistance – even though slight – in lift and pull. This happens because the inner wings generate lift and the outer wings push off with an upward-and-backward movement.

Studies of the aerodynamic forces that act on various parts of a bird's wings and body during different wingbeat phases and during different types of flight are still too fragmentary for us to be able to use them as a basis for predicting which methods of flight and which speeds are most favourable to the birds. For this purpose we must instead adopt an overall perspective and estimate a bird's total power consumption at different flight speeds. Such a theory has been worked out by the English ornithologist Colin J. Pennycuick (figure 93A).

A flying bird uses up energy to accelerate air downwards by beating its wings and by that means to compensate for its weight. This induced power (power = energy consumption per unit of time) increases of course with the bird's weight but decreases with flight speed, with wingspan and with air density. The more air the bird gets beneath its wings, the less the power required for sufficient lift becomes.

The power that is used up in order to overcome the drag of the body, the 'parasite power', by contrast increases considerably with the flight speed.

The basal metabolism, including the extra power that is required for the bird to ventilate its lungs and circulate its blood with the greatest efficiency, and the power that is needed for the air's resistance to the beating wings to be countered, are considered to be independent of the bird's flight speed.

If we add up these different power components, we get the total power curve for a flying bird (figure 93A). The curve is U-shaped and shows that it costs a lot of energy to hover and fly

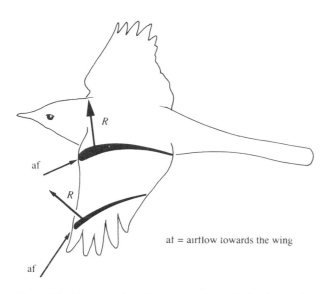

Figure 92 Downstroke of wing on a flying bird. The combination of the bird's forward flight and the wings' downstroke results in the air flowing diagonally from below towards the wings. This is particularly marked for the outer parts of the wings, which have a relatively high speed of downstroke. On the downstroke the outer wings are turned obliquely downwards so that the angle of attack becomes equally favourable for the airflow at the wings' outer and inner parts. The resultant force that acts on the outer wings will therefore be directed forwards and contribute effectively towards propulsion; the force on the inner wings produces mostly lift.

slowly and also to fly very fast, and that the power consumption is less at intermediate speeds. That the flight power curve really does have this shape has been borne out by wind-tunnel experiments. Figure 94 shows an experiment with Budgerigars that were trained to fly in wind tunnels with a mask over their head so that the oxygen consumption could be measured. Different airflow speeds in the wind tunnel simulated different flight speeds of the birds, which all the time were 'flying on the spot' in the tunnel. Both the breathing rate and the oxygen consumption of the Budgerigars clearly illustrate the U shape of the power consumption curve. Similar wind-tunnel experiments have been carried out more recently with Starlings but have given results that are difficult to interpret.

Colin Pennycuick's theory, complemented by measurements of the drag coefficients for avian bodies and of the birds' basal metabolism, makes it possible to calculate the flight power curve using the bird's weight and wingspan as a starting point. In figure 93B I have calculated this curve for a Woodpigeon with a mass of 0.5 kg and a wingspan of 75 cm. Note that the power axis shows the mechanical power (in watts). The efficiency of a bird's muscles and other body functions is around 23%; for a mechanical power of 10 watts to be achieved it therefore requires that the bird converts food the equivalent of around $10/0.23 = 43$ watts. A similar efficiency applies, incidentally, to the petrol engines of cars: for a mechanical power of 50 horsepower ($= 37$ kilowatts) to be developed requires burning of petrol equivalent to a total power consumption of nearly 400 hp. Both the car and the flying bird lose the difference in the form of excess heat.

What flight speed is most favourable for a migrating Woodpigeon? Surprisingly enough, it is *not* that speed which gives it the minimum power consumption. This is because a migrating

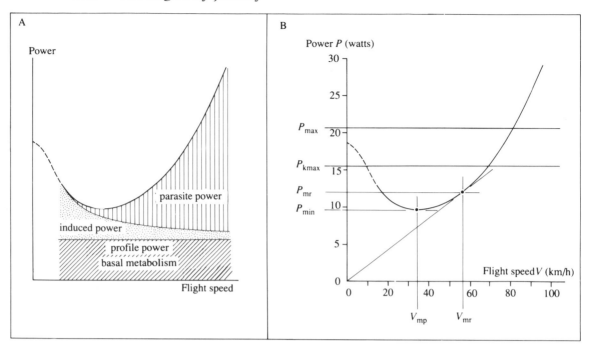

Figure 93   Power curve for a flying bird.

A. The sum of the parasite power, the induced power, the profile power to counter the air resistance to the beating wings and the basal metabolism gives the total power consumption for a flying bird at different flight speeds. The various power components can be estimated and added together on the basis of Colin Pennycuick's theory on the mechanics of bird flight (Pennycuick 1969, 1975, 1989). This theory does not allow accurate calculation of the flight power for hovering and very slow flying, which have been estimated using other methods of calculation (the broken line on the curve in the figure).

B. Flight power curve for a Woodpigeon (mass 0.5 kg, wingspan 75 cm) calculated according to Pennycuick's (1975) method. The usual symbols for power (P) and speed (V, from velocity) are used. Flying at speed V_{mp} (mp = minimum power) leads to minimum energy consumption per unit of time = P_{min}. The energy consumption per distance flown (= the quotient P/V) is minimum at the speed of V_{mr} (mr = maximum range). The working capacity of the flight muscles sets an upper limit on how high a power consumption a flying bird is able to maintain. The available power is higher for temporary all-out exertions, P_{max}, than for continuous muscle work, P_{kmax}.

bird is not in the first place favoured by the energy consumption per unit of time being kept to a minimum, which happens at speed V_{mp} (mp = minimum power, cf. figure 93B), but instead by the energy consumption per distance covered being minimised. The quotient of power/flight speed (P/V in figure 93B) corresponds to energy per distance of flight, and this is as small as possible at the flight speed where a straight line is tangential to the flight power curve as shown in the figure. This speed is termed V_{mr} (mr = maximum range) since the bird reaches maximum distance with a given energy reserve. A Woodpigeon ought therefore to fly at approximately 57 km per hour (= V_{mr}) if it is to get as far as possible on its limited energy reserves. If it were to fly solely in order to gain time, it ought to slow down to about 34 km per hour (= V_{mp}).

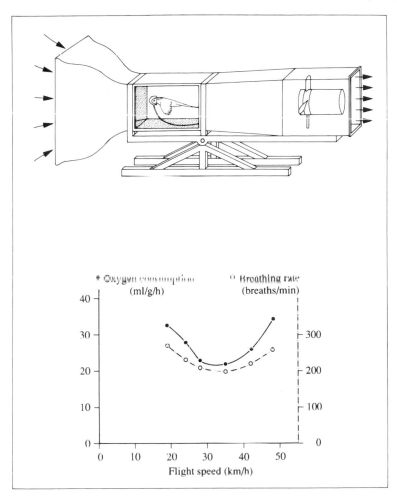

Figure 94   The energy consumption for flying Budgerigars in a wind tunnel. The speed of the airflow in the wind tunnel corresponds to the bird's flight speed. The bird's energy consumption is reflected in the oxygen consumption or in the breathing rate, which can be measured since the flying bird is breathing inside a transparent plastic mask. The rate of oxygen consumption as well as that of breathing show the U shape of the flight power curve. Based on Tucker (1968, 1969).

How, then, do speed measurements of flying birds tally with the predictions according to this theory? In table 27 I have compared theoretical calculations of V_{mp} and V_{mr} with migration speeds of various birds according to radar studies. The effect of the wind on the measured flight speeds has been compensated for; in the table the birds' speed is shown in relation to the surrounding air (it therefore corresponds to observations in dead calm). As we can see in the table, the birds' migration speeds agree well with expected V_{mr} but generally are distinctly higher than V_{mp}. The migrating birds seem therefore to adapt their flight speed in such a way that they minimise the energy consumption per distance flown. The only exception is the Whooper Swan, whose migration speed is closer to V_{mp} than to V_{mr}. This may possibly be due to the fact that the Whooper Swan is so big and heavy that it is not capable of flying with the power required for V_{mr}.

Table 27. *Comparison between theoretically calculated flight speeds, V_{mp} and V_{mr}, and observed migration speed for various species according to radar studies. The theoretical flight speeds have been calculated from the flight power curves, which in turn have been determined using the different species' masses and wingspans as a starting point as per Pennycuick (1975). Observed migration speeds are for active flight with wingbeats. The speed for Crane refers to migration over open sea (the Baltic). The data on migration speed of Chaffinch and Swift come from Bruderer (1971) and Bruderer & Weitnauer (1972); remaining data are from personal studies*

	Mass (kg)	Wingspan (m)	V_{mp} (km/h)	V_{mr} (km/h)	Observed mean flight speed in relation to air (km/h)
Chaffinch	0.020	0.30	18	34	39
Swift	0.043	0.45	19	34	40
Redwing	0.065	0.40	24	41	46
Woodpigeon	0.50	0.75	34	57	60
Oystercatcher	0.55	0.80	34	57	50
Eider	2.0	1.1	45	73	74
Bean Goose	3.5	1.6	45	73	72
Crane	5.5	2.4	42	70	67
Whooper Swan	10	2.4	52	85	60

Reliable visual speed measurements using stopwatches agree quite well with the results from radar studies. The migration speeds of Eiders through the Kalmarsund have been measured visually at 70 km per hour and those of Oystercatchers at 62 km per hour. In Germany, migrating Chaffinches have been clocked at a speed of about 44 km per hour.

Swifts have been tracked with radar not only during migration but also during the breeding season, when they 'sleep in flight' – they spend the nights flying back and forth, often at an altitude of several thousand metres. The mean speed during the night-time period is distinctly lower, only 23 km per hour, than that during the migration flight (table 27). The Swifts evidently adjust their flight speed so that they come close to V_{mp} during the nocturnal flight in order to minimise energy consumption overnight. In this situation the flying distance does not of course matter like it does during migration.

White-throated Sparrows, which were sent up in a balloon in North America on clear nights and tracked by radar, immediately set off in their normal migration direction at a mean speed of 33 km per hour, which agrees closely with V_{mr}. Under a night sky with total cloud cover, however, the birds had obvious difficulties in orientation; they flew in circles and in different directions and the flight speed was no more than 24 km per hour. We may guess that they adjusted their flight speed downwards towards V_{mp} in order to gain time and save energy until they were sure that they were on the correct course. Birds appear to be well fitted to adapt their flight speed to different situations in order to minimise energy consumption.

This was recently questioned in connection with a radar study in the United States of flight speeds of gulls and terns flying to and from their breeding colonies. The mean speed of the various species varied between 30 and 40 km per hour and without exception was below V_{mr}, below the speed we should expect if the birds were to minimise their flight transport costs. Gulls that are breeding do not, however, find themselves in the same situation as birds on migration flight. We have very good reason to suppose that migrating birds single-mindedly put a premium on the chance of getting as far as possible at minimum energy cost. When flying to and from a breeding colony, it may on the other hand very well be expedient to compromise between a flight speed that allows the birds to remain a long time on the wing (and perhaps at the same time keep an eye out for feeding sites while flying slowly) at a low price in energy and a speed which results in low transport costs.

More detailed study of the choice of flight speed made by birds under various circumstances is a very interesting area for future research. Even now we can make theoretical predictions on how birds ought to adapt their speed in relation to wind, to flight altitude and to flying in flock formations.

V_{mr} increases in a head wind and decreases in a tail wind. A Woodpigeon in a fresh tail wind, say $10 \, m/s$ ($= 36 \, km$ per hour), should thus not fly at $57 \, km$ per hour in relation to the surrounding air as in dead calm but ought to reduce its flight speed to $47 \, km$ per hour ($= V_{mr}$ in this tail wind). The total ground speed then becomes $47 + 36 = 83 \, km$ per hour. On the other hand, a Woodpigeon in a similarly strong head wind ought to increase its air speed to $78 \, km$ per hour so that the ground speed becomes $78 - 36 = 42 \, km$ per hour. This is due to the fact that V_{mr} represents the speed that gives the lowest energy consumption per distance flown *over the ground*. When determining V_{mr}, the effect of the wind must therefore be weighed in in such a way that the straight line which touches the flight power curve in figure 93B at speed V_{mr} is drawn from a point to the left or right of zero, depending on whether a tail or a head wind is blowing. Of course it is always more favourable from an energy point of view for a bird to fly in a tail wind than to fly in a head wind. The adjusting of the bird's air speed helps to make the flight transport as energy-sparing as possible in a given wind situation. The transport cost for a Woodpigeon flying in dead calm at $57 \, km$ per hour ($= V_{mr}$) is approximately 800 calories (total chemical energy) per kilometre. If the Woodpigeon maintains the same air speed in a head wind of $10 \, m/s$ the transport cost becomes $2200 \, cals/km$, but by increasing the air speed to V_{mr} in this wind, i.e. to $78 \, km$ per hour, the bird reduces the transport cost to $1700 \, cals/km$. In a tail wind the bird's air speed is not so important to energy consumption. By reducing the air speed from 57 to $47 \, km$ per hour in a tail wind of $10 \, m/s$ the Woodpigeon decreases the transport cost only by 3%, from 485 to $470 \, cals/km$. We can therefore predict that migrating birds are exceptionally eager to keep their air speed close to V_{mr} in head winds and that they are not quite so particular in tail winds. Perhaps in tail winds it may be worth a few percent extra in energy costs to get ahead more quickly by flying a little faster than V_{mr}?

The fact that birds really do migrate at a higher air speed in a head wind than in a tail wind has been shown in several radar studies. The air speeds of White-throated Sparrows released and tracked with radar in head wind were generally between 30 and $40 \, km$ per hour; in tail wind the birds flew for the most part at 25–$30 \, km$ per hour in relation to the surrounding air. According to Swiss radar studies, Chaffinches migrate in tail wind generally at a speed slower than $40 \, km$ per hour in relation to the air, but are faster in head wind. However, we still lack more detailed analyses comparing observations with theoretical calculations which would enable us to see clearly how precisely or how broadly birds alter their flight speeds and how much energy they save by so doing.

V_{mr} is also dependent on the air density; the birds ought to raise their migration speed by approximately 5% per $1000 \, m$ of increasing altitude in order to save as much energy as possible. A few radar studies suggest that migration speed increases at higher altitudes. In spring, Long-tailed Ducks and Common Scoters migrate more quickly at high altitude over Finland than at low level over the Baltic Sea. Chaffinches in Switzerland fly at a distinctly higher air speed at several thousand metres' altitude than low over the ground.

Furthermore, the birds ought to increase their migration speed when they have large fat reserves and are heavy, and they ought to decrease the speed when they are flying in dense flocks and formations. The latter prediction has been frustrated when timings were taken of Oystercatchers, Knots and Dunlins moving past the Danish bird observatory at Blåvandshuk, for the measurements did indeed reveal a positive correlation between flock size and migration speed: an Oystercatcher flock of 20 individuals flies a good 1 km per hour faster than a single Oystercatcher, and a flock of 20 Knots or Dunlins flies approximately 5 km per hour faster than respective solitary individuals. This may be interpreted as follows: that the waders 'extract' the aerodynamic gain of flying in flocks in the form of increased speed instead of in the form of reduced transport costs. The saving of time seems in this case to be more important to the waders than the saving of energy. Continued field studies are required in order to clarify these remarkable and most interesting results.

The birds' flight speed in relation to the air varies in general between approximately 30 km per hour, for the smallest birds, and 80 km per hour, for larger birds. The rule of thumb is that the speed roughly doubles when the mass of the bird increases 100 times. If a 10-g Willow Warbler flies at 30 km per hour, then a Raven of 1 kg flies, in round figures, at 60 km per hour. When the bird's mass increases 100 times, then 200 times as much flight power is required. The muscle power cannot, however, increase much more than the weight. Provided that the proportions are the same, the wing area is only 20 times as great in a bird that weighs 100 times more than another. The limited muscle power and wing area of heavy birds, in combination with the very high flight power that is required, sets a size limit above which flying is no longer possible. This limit is estimated to be around 15 kg. This corresponds well with the weight of the largest animals in the world that can actively fly – swans, bustards, albatrosses and condors.

The capacity of the flight muscles sets a 'ceiling' to how much flight power a bird can cope with – a lower ceiling for continuous power outtake and a somewhat higher ceiling for temporary all-out bursts. After this sort of brief 'muscle spurt' the muscles have to wind down while the lactic acid which is formed in the muscle tissue when energy is produced without sufficient oxygen supply is carried away. In figure 93B I have drawn in these two ceilings such as they can be estimated for the Woodpigeon (P_{kmax} and P_{max}). The pigeon has the capacity to start its flight from zero speed and can stop briefly in the air and hover. It cannot, however, manage for any length of time either to hover or to fly faster than 70 km per hour. If a migrating Woodpigeon wishes to increase its altitude, it has, when flying at V_{mr} in dead calm, a power scope between continuously available and spent power which it can use to climb at a good 40 m per minute. (Through radar studies I have established that the climbing speed of migrating Woodpigeon flocks is normally 30–40 m per minute.)

For birds that are smaller than the Woodpigeon, the distance between the flight power curve and potential maximum power increases. Sunbirds and hummingbirds hover for an unlimited period of time even in still weather; thrushes on nocturnal migration are able to climb with ease at more than 75 m per minute (cf. figure 88). For larger birds the margins shrink; they become incapable of taking off without wind assistance or without running to gain speed. The climbing speed for large birds is extraordinarily low. Eiders fly around for a long time over the bays before they have gained sufficient height to migrate in over land. The biggest birds of all are generally speaking reduced to using thermals or other rising currents of air to gain a good height in a reasonable period of time.

Colin Pennycuick's theory is of indispensable value for an understanding of how migrants adapt their flight and as a stimulus for further research, but it does not of course say it all. The birds' structural proportions influence the style of flight to a great degree. Birds with large wings and relatively low weight, i.e. with low wing-loading, find it easy to take off and can manoeuvre efficiently at low speeds. At higher flight speed, however, big wings result in considerable air resistance. For many passerines and woodpeckers with proportionately big and broad wings, the energy cost for straight rapid flight (at speeds around V_{mr}) becomes so great that it pays better to use undulating flight: the bird alternates between a phase of accelerating obliquely upwards while beating its wings and another of reducing the air resistance by 'body gliding' for a period with folded wings. Larger birds, such as hawks, with proportionately narrower wings and thereby better gliding abilities, instead alternate between periods of flapping and periods when they glide on extended wings. Birds with high wing-loading are disadvantaged when flying slowly, but can on the other hand maintain good speed with continuous wingbeats.

The American scientist Crawford H. Greenewalt distinguishes three main groups of birds on the basis of wing-loading. The group whose members have relatively low wing-loading include passerines, as well as herons, storks and raptors. Waders, pigeons and doves, parrots, geese and swans form an intermediate group. Ducks, divers, grebes and gallinules and coots come in the group in which the wing-loading is highest. The auks, however, constitute a small group of their own; their wing-loading is even higher than that of the 'duck group'. For birds in the 0.5-kg class, the 'passerine model' has wing-loadings of between 25 g/dm² (harriers) and 40 g/dm² (crows), the 'wader model' around 65 g/dm² and the 'duck model' 100–130 g/dm² (higher for diving ducks than for dabbling ducks). An auk of this weight has a wing-loading of around 200 g/dm², almost ten times greater than a harrier. Harriers are masters at manoeuvring gracefully at low speed; auks are reduced to flying as straight and fast as an arrow!

What clues to their migratory habits can be found in the birds' dimensions and flight capability? The clues are surprisingly modest. Migratory birds can be found among the smallest hummingbirds and among the largest swans and albatrosses. And migratory birds can certainly vary amazingly in shape: migration is performed by small, short-winged wrens and slender swallows and terns, and by broad-winged storks, cranes and eagles. One gets the impression that birds in general are such good flyers that migration does not involve any really critical flight problem. The birds cope well with their migratory exploits simply with the flight capacity that has been shaped by other requirements, by the need to be able effectively to exploit different feeding niches, to avoid predators, to build nests and rear young.

If we look very closely, however, we find small differences in the birds' appearance which are adaptations to different migratory habits. In particular, the shape of the wings is different in long-distance and short-distance migrants. Long-distance migrants as a rule have longer and more pointed wings than those of their close relatives that travel shorter distances or are residents. A useful measure for showing the degree of pointing of the wing is the extent to which the primaries of the wingtip project beyond the secondaries/tertials, expressed as a percentage of the total wing length. This 'wing-pointing index' (normally termed primary projection) is only 15% for the Wren but 56% for the Swallow and all of 72% for the Swift. (An even better measure of the wing shape is what is termed the aspect ratio, i.e. the quotient between the wing's length and its average breadth. This measurement is directly practicable in aerodynamic theory. The aspect ratio for a Wren wing is roughly 3 : 1 and for a Swift wing 10 : 1. Unfortunately, detailed measurements of the aspect ratio are lacking for most birds' wings.)

A few examples of the adaptations of the wing shape to the birds' migratory habits are worth mentioning here. The Wheatear winters mainly in the steppe areas south of the Sahara. Wheatears from Greenland and adjacent parts of Canada have an exceptionally long migratory journey; their primary projection is 35%. Those that breed in Europe have on average a primary projection of 32%; north African Wheatears, which have only the Sahara to cross on migration, by contrast have a primary projection of 27%. The Somalian race of the Wheatear *O. o. phillipsi* is resident south of the Sahara; its primary projection is only 22%. Other illuminating examples are provided by the pipits: the Meadow Pipit, a short-distance migrant, has a primary projection of 27%; the corresponding value for the Tree Pipit and the Red-throated Pipit, both tropical migrants, is about 34%. The Chaffinches on the Canary Islands are residents and have a primary projection of around 22%, while central European Chaffinches, which are short-distance migrants, have more pointed wings, 29%; the closely related Brambling, which migrates even farther, has a value of around 33%. Compare also the primary projection of the Purple Sandpiper, a short-distance migrant, about 51%, with that of the long-distance Curlew Sandpiper, 54%. Or the Herring Gull's, about 54%, with the Lesser Black-backed Gull's 60%.

There is no definitive lower limit to primary projection for long-distance migrants. Reed, Marsh and Willow Warblers, and Whitethroat are successful tropical migrants despite the fact that these species have a primary projection that is no more than 26–29%. Nevertheless, the Willow Warbler has less blunted wings than the Chiffchaff, a short-distance migrant whose primary projection is only 21%.

A pointed wing produces a low air resistance and at the same time a good supporting power (owing to a small induced drag). The energy cost of rapid flight, such as migration flight at around V_{mr}, consequently becomes relatively low. Bird species which spend a great part of their life in the air are always served by low transport cost and by the opportunity to fly quickly. This is the reason why the swallows have pointed wings. Big, heavy birds in general have proportionately longer and more pointed wings than light birds (this does not apply to swallows and similar small birds with extremely pointed wings). A low transport cost becomes all the more important the heavier the bird is; this is due to the reduced scope between available muscle power and necessary flight power. The primary projection of cranes and storks is, despite their broad wings, approximately 45%.

Could we expect that most migrants would have wings like the swallows, wings that took them quickly to the migration destination at little cost in energy? No; when it comes to the shaping of birds' wings, other demands compete with migration efficiency. With pointed wings the birds lose some of their manoeuvrability and flight capability at low speed. The wings of migrants are as pointed as they can be without the disadvantages in foraging and competition with blunter-winged resident birds becoming greater than the energy gains during migration.

A neat method of analysing the flight of birds is to study the vortices that are formed in the wake of a flying bird. By flash-photographing of finches which were enticed to fly in cages through air which was filled with floating dust of wood-flour or paper-flour, these vortices have been charted. With advanced flow mechanics and theories on vortex-formation, the kinetic energy in the vortices can be determined. This energy is transferred to the air from the flying bird and is therefore exactly equal to the energy that the bird uses up for its flight. The power consumption for a bird can be calculated 'backwards' in this way, starting from the energy content of the air vortices. Such calculations have recently been presented in the literature. Further studies will provide interesting opportunities for improved understanding of the flight of birds. This theory is indispensable not least if we are to be able to determine what advantages and disadvantages in terms of energy migratory birds get from flying in flocks and formations.

4.3   Migration in flocks

For many people bird migration evokes thoughts of a flock of birds passing across the canopy of the sky towards a far-off destination, perhaps of a V formation of geese, cranes, or swans, of a billowing bow-shaped flock of Brent Geese or Eiders or of a dense 'clumped' flock of pigeons or starlings.

Why do some migrants fly in such characteristic flocks? An important explanation is probably that they save flight energy in this way. As I have pointed out in the preceding section, the air stream that passes the wings of a flying bird is deflected downwards. The downward-directed movement of air is compensated for by an upward-directed airflow off the end of the bird's wing (figure 95), strongest in the wingtip vortices just off the wingtips. The birds in a flock can therefore obtain a certain amount of lift for nothing by flying close to each other in the upward-directed airflow around each other's wings. It is most beneficial to fly as near to each other's wingtips as possible. On the other hand, it is unfavourable to fly in the downward-flowing air directly behind, above or below other birds.

The total energy gain that birds can make by flying in a flock depends mainly on how close to each other the birds fly, on the distance from wingtip to wingtip, measured at right angles to the flock's direction of travel (i.e. the distance between the birds in a flock, seen directly from in front or directly from behind). Positional shifts in the direction of travel determine how the total energy gain is shared out among the members of the flock.

The total energy saving for birds which fly close to each other in the same horizontal plane with a given distance between their wingtips is therefore independent of whether the birds are flying on a straight front, in an oblique line, or in V or bow-shaped formation. Even three birds flying 'wingtip to wingtip' can save almost 40% of the lift that they would need if they were flying on their own. As aerodynamic terminology puts it, the induced drag for such a

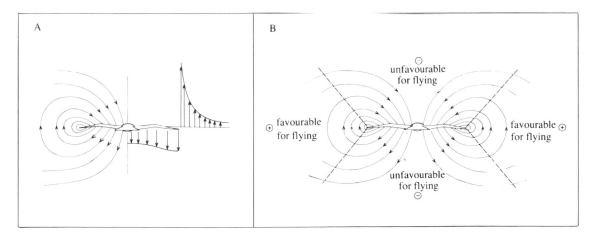

Figure 95   The airflow around a flying bird, seen directly from behind. A. The deflection downwards of air flowing over the bird's wings is compensated for by an upward-directed net flow off the end of the wings (right-hand side of figure). This flow comes about in the form of air vortices with a central point at or just inside the wingtips (left-hand part of figure). B. Lift power can be gained or lost depending on whether birds flying close are flying to the side of, above, beneath, or directly behind each other. Based on Lissaman and Shollenberger (1970).

flock is reduced by 40% of the total induced drag for three birds flying on their own. A flock of nine birds can save more than half of the lift power and a large flock can save two-thirds. If the space between the wingtips of the birds in the flock increases, then the energy gain diminishes greatly. For a large flock in which the distance between wingtips of birds adjacent to each other is the equivalent of one whole wingspan, the saving becomes only about 10% of the lift power.

Induced power is only a part of the flying bird's total power budget (see figure 93). For birds the size of a pigeon or larger, this part takes up at least one-sixth of the total flight power at normal migration speeds (V_{mr}). The proportion is less, down to one-tenth, for smaller birds. The total energy saving when flying in a flock is therefore significantly less for small birds than for large ones. As an example of energy saving for large birds let us take a flock of 25 Bean Geese flying wingtip to wingtip (e.g. in V formation). By flying in a flock of this sort the geese save 67% of the induced power, which is equivalent to a reduction of at least 12% of the *total* transport costs of birds flying on their own. A reduction such as this in the price per kilometre of the migration journey is, after all, no poor return for the bother of flying in an orderly flock!

Since the induced power decreases when flying in a flock, the optimal migration speed (V_{mr}) will become a little slower for bird flocks than for solitary birds. Although the most favourable migration speed for a solitary Bean Goose is 73 km per hour in still weather, it is only about 66 km per hour for a flock of 25 Bean Geese. Apart from the speed measurements of wader flocks in Denmark which were mentioned in the preceding section, there are no in-depth studies of differences in speed between bird flocks of differing size and lone individuals. Detailed speed measurements using tracking radar should be able to provide important information on the energy gains when migrating in flocks.

The above figures on energy saving when flying in flocks are based entirely on theoretical calculations. Some scientists doubt that bird flocks save any energy at all in practice. This doubt is founded mainly on the knowledge of how insignificant the energy gains made are when aircraft are flown close together; the wingtip vortices slightly disturb the air-current situation around nearby aircraft and lead to instability in the distribution of lift. The aircraft are therefore flown in formations that give the pilots control over the air space and over the nearest aeroplane and which guarantee the best flight safety. It has been suggested that the same thing would apply to the flock formations of birds.

Birds, however, have much greater opportunities than aircraft to take advantage of energy gains from uneven air flows. They can change the shape and angles of attack of the wings so that free lift can be exploited in the best way in different flock positions, something which is of course impossible for aircraft with fixed wings. The birds are doing the right thing if they adapt the wing shape and wing angle in such a way that they have the same 'feeling' in the wings' carpal and scapular joints (the feeling of stable distribution of lift) as in normal solitary flight.

Other grounds have been given for casting doubt on the fact that birds save energy when flying in formations. Some scientists have considered that a prerequisite for energy saving would be that the wingbeats of the birds in a flock be synchronised, with regular changes of phase. This condition is not met according to most studies, which indicate instead that the birds within a flock formation fly with totally independent wingbeats. More recent assessments, however, argue that it is of no major importance to the chances of energy gains whether the wingbeats of the birds in a flock are synchronised or not.

The objections to the idea that migrants save energy by migrating in flock formations are thus weak, and there is good reason to give credence to the theoretical predictions.

What theoretical predictions can we make regarding the distribution of the energy gain among the various members of the bird flock? In flocks in which the birds fly side by side on a straight front, the ones in the middle profit more than those at the ends; the latter of course obtain extra lift assistance only from birds on their one side. When flying in a diagonal line the middle birds are again favoured, while the bird which is flying at the front obtains the least energy gain. Even in straight V formation the leading bird is at a disadvantage (figure 96). In the bow formation which is shown in figure 96, the energy gain is, however, shared equally among the flock members. A V with an obtuse angle and a slightly rounded point also gives a power consumption about equal in amount for all members of the flock.

How 'democratic' are the formations of flying birds that are observed in the wild? Certainly the bow-shaped formation in figure 96 is reminiscent of the flock structure made use of by migrating diving ducks, Brent Geese and Barnacle Geese. Flocks of Canada Geese flying between feeding sites and roosting sites in North America have been photographed so that their formations could be determined more exactly. More than 100 flocks (mean flock size 27 geese) have been analysed. In as many as 40% of the flocks the geese formed an oblique line. Most of the rest flew in V-type formations: these were designated either as V formations or as J formations, depending on whether the 'legs' of the formation were roughly the same length or not. The V-type formations had acute angles, between 27 and 44 degrees. In another study in the United States, short-range radar was used to take direct measurements of the angles in V-type formations of Canada Geese: in this study, the angles for more than 50 flocks varied between 38 and 124 degrees, the mean value being 72 degrees. The fairly acute V angles and the many oblique-line formations suggest that the geese do not spread the energy savings fairly within the flocks but that the leading goose is disadvantaged. Nevertheless, we must not take it for granted that the flocks have a similar shape when the birds are migrating long distances and flight energy saving is at its most important and when they are making short local morning and evening movements at stop-over sites.

An interesting observation from the photographs of Canada Goose flocks was that the

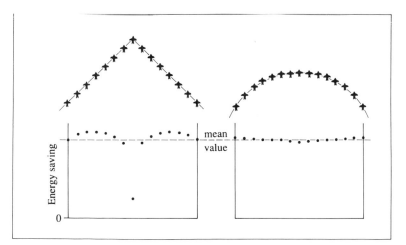

Figure 96   Energy saving for the individuals in an 'undemocratic' formation, the V to the left in which the leading bird is disadvantaged, and in a bow-shaped formation with equal distribution of energy. The average energy gain is the same for both formations, since the number of birds (15 individuals) as well as the distance between wingtips (one-fifth of a wingspan) are identical. Based on Hummel (1973).

lateral displacement among the birds amounted to only 1.2 m on average. Since the Canada Goose's wingspan is about 1.5 m, the birds' wingtips overlapped each other by a few tenths of a metre. According to theory and experiment, the aerodynamic wingtip (the central point of the wingtip vortex) lies a little inside the actual wingtip proper. From an aerodynamic point of view, therefore, the Canada Geese (with the exception of the leading goose) were flying in very favourable positions for energy saving. An advantage of pointed V formations and oblique lines over blunted Vs and bow-shaped flocks is that the liberal longitudinal displacement between the birds allows the wingtips of birds flying next to each other to overlap laterally to an optimal degree.

If the leading bird in a flock is at a disadvantage from the point of view of flight energy, there is always of course the chance that equity is arrived at through its being relieved. When and how such changes of leader take place, however, has not been studied. The closest we have got to this problem so far is a study of small flocks of homing pigeons (three to five pigeons in each flock). The pigeons were different in colour and therefore easy to distinguish individually. The birds were followed by helicopter and the flying arrangement within the flock was logged at regular intervals. Homing pigeons do not fly in stable formations but the flock's shape and the birds' positions change frequently. Despite this, one or two pigeons showed leadership tendencies and usually, though by no means always, flew at the head of the flock. Among the remaining pigeons, the sequence altered randomly. Surprisingly enough, the most competent homing pigeons, those which when flying alone find their way home the quickest, did not demonstrate greater leadership tendencies than the other pigeons.

Even though the energy gains are greatest when the birds are flying side by side in fixed formations, birds that fly in dense 'clumped' flocks can also, as new theoretical studies have shown, save flight energy. Two basic rules determine the flight power for the birds in such flocks.

1 A prerequisite for energy saving is that the cross-section shape of the flock, i.e. the flock seen directly from in front or directly from behind, show a greater width than height. The more drawn-out horizontally the flock is, the greater the energy saving becomes. A flock which is taller than it is broad is associated with corresponding energy losses. A circular flock shape does not result in any change at all in the birds' flight energy if we compare with flying alone (figure 97). The explanation for this can be traced back to figure 95, where it is shown that it is favourable to the birds to fly in the sectors to the side of each other but not in positions above and below. If the side positions predominate over above and below positions in a flock, then the net result is an energy gain; inversely, the birds suffer energy loss if several individuals in the flock are stacked up vertically rather than displaced horizontally. The energy saving is in actual fact proportional to the ratio $(B-H)/(B+H)$, where B is the flock's breadth and H its height. Thus, for a flock with a breadth/height ratio of 5:1 the energy gain amounts to approximately two-thirds of the gain for a flock all birds of which are flying side by side in a single plane. At a breadth/height ratio of 2:1 the gain becomes only one-third, and for flocks with a greater height than breadth the 'energy gain' is negative, in other words it becomes an energy loss.

2 The energy gain increases with increasing bird density in the flock. In this context the density is as viewed in cross-section area: the flock seen directly from in front or straight from behind. If the flock is drawn-out in the direction of travel, this density can be greater than one imagines. If the mean distance between the birds in the flock (measured on cross-section area) decreases by half, then the energy gain is quadrupled.

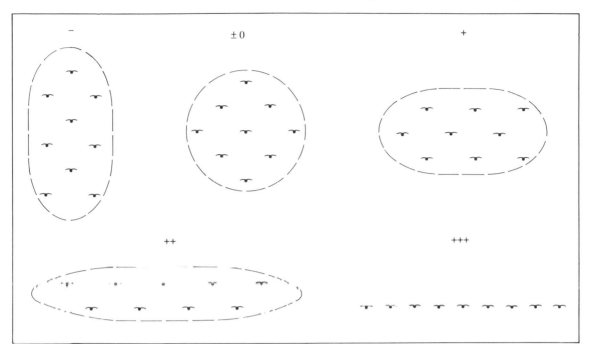

Figure 97   Examples of different shapes of bird flocks, seen directly from behind or directly from in front, and of how the flight energy changes in these flocks in comparison with flying singly. Energy gains and energy losses are indicated by plus and minus signs. Broad flocks result in energy gains; tall flocks give losses. Based on Higdon & Corrsin (1978).

Calculations suggest that the savings made by birds in dense clumped flocks amount to only a few odd percent of the total flight power. Can such modest savings really be a contributory reason why certain birds migrate in such flocks? Yes, I firmly believe so. Many times I have watched through binoculars flocks of migrant birds, especially Woodpigeon flocks, flying away, directly from behind. It is striking how often the flocks are very much drawn-out horizontally while their height is minimal – just what we could expect from an energy-saving point of view. And surely it is pretty rarely that we see birds flying in a way that is unprofitable energy-wise – directly behind, above or beneath each other or in flocks that are drawn-out vertically – for it to be sheer chance?

One detail that is worth noting when it comes to flight energy gains when flying in flocks is that occasional smaller birds that 'wedge themselves into' a flock of larger birds can make disproportionately big energy gains. They can exploit the considerable free lift next to the large birds without 'paying their way' with equally strong upward-directed airflow for the benefit of the other flock members. A big bird in a flock of many smaller birds instead makes a poorer energy gain than the majority of the flock. This is probably the reason why we so often see mixed flocks with odd birds of smaller species interspersed among the many larger birds, but rarely mixed flocks with many small and few big birds. Think how regularly we find the odd little Teal or Wigeon in migrating Eider flocks or a small number of Stock Doves in a flock of Woodpigeons! A disadvantage for the few small birds in these mixed flocks is that they are forced to abandon their most economic migration speed (V_{mr}) and adjust to the predominant species' flight speed. The little Teal flying in a flock of Eiders about seven times as heavy as itself is forced to increase its flight speed from the species' V_{mr}, around 55–60 km per hour, to normal Eider speed, a good 70 km per hour. The energy costs for this increase in speed are probably compensated for by the gains provided by flying in a flock among the large Eiders.

For migrants which travel in loose flocks or in open formation the energy savings are negligible; other reasons must be behind their keeping together. Perhaps flocking sometimes offers protection against birds of prey. Many species benefit by living in flocks at staging posts and wintering sites, and for that reason the birds keep together during the flight stages. A contributory reason for flocking may also be that the orientation of flocks is better and more reliable than that of birds flying alone. The flock's flight direction is perhaps a mean value of the various directional tendencies of all the individual members of the flock? The orientation of homing pigeons flying individually and in flocks has been compared in two different studies. In one study no difference whatsoever was found between the orientation success of the inviduals and that of the flocks; in the other it was revealed that the flocks were less inclined to show a scatter of flight directions than the separate individuals. The question of whether flying in flocks leads to more reliable orientation than flying alone is therefore as yet still unresolved.

A much-discussed problem is the extent to which night-migrating passerines travel in flocks at all. Larger birds, wader and duck type, keep together in their characteristic dense flocks both on nocturnal and on diurnal migration. But thrushes, buntings and other passerines never migrate at night in such dense flocks as one sees by day. Radar echoes from small birds migrating at night are weak, diffuse and 'flickering'; they suggest that the birds are flying at a fair distance from each other, spread out in the air space. At dawn many of these diffuse radar echoes are transformed into a fewer number of more distinct echoes at the same time as similar echoes from day-migrating flocks appear. This shows that many small birds moving at night which continue to migrate after dawn (e.g. when they find themselves over the sea) join together in groups when it gets light. Conversely, we can sometimes observe how radar echoes from flocks of small birds migrating late in the evening spread out and become diffuse roughly half an hour after sunset, at the same moment as the radar screens often 'blossom out' with diffuse echoes from the general departure of nocturnal migrants.

Studies using the moonwatching method and with tracking radar show that at night there are at the very least 20–50 m between nearest individuals among small birds in flight. According to radar studies in Switzerland, the commonest distance between night-migrating passerines is 100–300 m. The question then is whether small birds keep together in loose units at all or whether they are scattered at random in the air space at night. Here the results and the interpretations from different studies are at odds. Ian Nisbet draws the conclusion, on the grounds of simultaneous studies using radar and moonwatching in the United States, that a large proportion of the nocturnal migrants fly in loose units, most often of between two and 12 birds. A similar conclusion is reported by Bruno Bruderer from his tracking radar studies in Switzerland. A statistical analysis in the United States of small night-migrating birds which passed through the cone of light from an upward-directed searchlight indicates, however, that the majority of the birds were observed in the light beam at random time intervals. During the spring only 7% and during the autumn only 4% of the birds were recorded in such close temporal association with each other that the conclusion could be ventured that they were keeping together in a unit; most often there were fewer than six birds in each unit. Still other researchers interpret their results as indicating that generally speaking all small birds migrate independently of each other at night.

What are we to believe with any confidence when different studies have led to such differing conclusions? I shall try a guess. I think that the species of small birds that utter calls at night migrate in loose groups and that those which are silent fly as individuals. This is because I consider that the calls serve to keep contact between the birds in a flock and that they perhaps also help the flocks to find the altitudes where the tail winds are best. By listening to calls from birds at lower and higher flight altitudes a bird can work out if it is being 'overtaken' and, if such is the case, change its flight altitude to levels with better tail winds so that it obtains increased migration speed in relation to the ground. Thrushes which depart on night migration at dusk set off in distinct flocks; and the thrushes are among those species which most frequently utter their calls in the dark of night.

Not all researchers agree that calls at night serve as a contact between birds that migrate in flocks or units. According to one imaginative speculation, the birds use the calls for echo-sounding. Guided by the echo's time lapse and character the birds would, according to this theory, be able to determine their altitude and the nature of the ground below. In support of this suggestion it has been pointed out that balloonists hear surprisingly well at high altitude. A train whistle can be heard at a height of almost 7000 m, a barking dog at over 2000 m; croaking frogs are clearly audible at 1000 m. If a person himself shouts downwards at the ground, then the echo can be heard several hundred metres up, and if he blows a hunting-horn then the echo is audible at an altitude of over 1000 m.

4.4   Soaring flight

Lie on your back on a sunny hillside and watch how the cumulus clouds constantly change shape, puff out and break up into thinner clouds against the clear blue background of the sky. If we allow ourselves to follow individual clouds closely, then we shall find that the mean 'lifespan' of typical cumulus is little more than 20 minutes. No sooner do they thin out and evaporate than new clouds immediately bubble up and fill the empty spaces. Many different birds soar over land, birds of prey, crows, gulls, swifts, storks and cranes. They circle on outstretched wings in rising warm air. The same warm air brings with it the water vapour which is condensed in the cool air at higher altitudes to form the white bubbling clouds. Perhaps that is a migrating bird that we see circling up and up? For birds with the right adaptations soaring flight can be an amazingly efficient method of migrating. Storks and Buzzards soar and glide back and forth between Europe and South Africa every year. Let us examine the conditions for the soaring flight of migratory birds; we shall then perhaps have more to reflect upon the next time we lie on our back on a sunny hillside.

By way of introduction I shall confine myself to soaring flight over land. Seabirds use special methods to soar over the ocean waves, something which I shall touch on in concluding this section. By soaring the birds make the most of the free energy in upwinds. These they can find at hills and mountain slopes where the wind is deflected upwards, as well as in 'lee waves' which are formed beyond the lee sides of mountains. Birds which soar and glide over longer distances, however, as a rule rely not simply on upwinds at hillsides and in lee waves but benefit primarily from the rising air of thermals.

A warming morning sun following a clear night heralds good thermal conditions. The sun's rays warm up the ground, which in turn warms the air near the ground. For the forming of thermals really to accelerate the stratification of the air must be unstable. This is because warm air which rises through unstable air, despite expanding and cooling with increasing altitude, remains all the time warmer and thereby lighter than the surrounding air; the warm air therefore continues to rise. On the other hand, an inversion layer, in which the temperature increases with increasing altitude, effectively prevents further thermals: rising air expands and cools, and it then becomes colder and heavier than surrounding air and starts to sink.

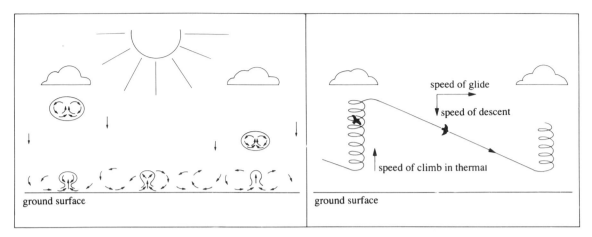

Figure 98   Above the sun-warmed ground turbulence is created, and from certain places warm air rises in the form of thermal bubbles, which grow in size with increasing altitude. The upwinds are exploited by soaring migrants, which circle in the thermal bubbles and then glide in their travel direction, gain height again in thermal air, and so on.

After several hours of morning sun, when the air is warmed up at the ground, an unstable layer with the right conditions for thermals is developed in the middle of the morning. This thermal layer extends only a few hundred metres up to begin with, but owing to the continued solar radiation grows and in the afternoon can reach several thousand metres' altitude. An inversion layer as a rule creates an effective 'roof' to the thermal layer. The sun's rays warm the ground and through that the air to different degrees of effectiveness in different places. Where the air nearest the ground becomes warmest and lightest it tends to rise in cooler air layers while cold air in other places 'streams down' towards the ground. Only the strongest pockets of warm air avoid being wiped out in the surrounding turbulence near the ground. These grow like mushrooms and 'bud off' as bubbles of warm air and rise upwards through the thermal layer (figure 98). The air circulates within the rising thermal bubble. The strongest upwinds are formed in the centre – which is where birds, and glider planes, should circle to avoid the risk of gliding out and losing contact with the thermal bubble. The thermal bubbles expand at increasing altitude and grow to several hundred metres in diameter; in good thermals at high altitude they can even reach 1–2 km across.

If the upper limit of the thermal layer is below the condensation level for water, then no clouds are formed; we then use the term 'dry thermal'. Often, however, the thermal bubbles reach the condensation level and give rise to the characteristic cumulus clouds. A single thermal bubble results in little more than a short-lived puff of cloud in the sky. A series of thermal bubbles is required for a substantial cumulus cloud to develop. With condensation of water, heat is released, and this leads to the thermal air rising with additional vigour in the lower and central parts of the cloud while the upwinds are reduced at the cloud's upperside where water vaporisation has a cooling effect. Downwinds are formed in association with this cooling effect along the sides of the cloud.

The very biggest cumulus clouds continue to develop without being reached by thermal bubbles from the ground. They become self-supporting with air, which is sucked in from below; the water vapour condenses, the air is warmed up and rises farther up through the cloud, after which new air is sucked in from below, and so on. Such clouds may endure for several hours, sometimes long into the evening when the thermals from the ground have ceased altogether.

In a damp climate, cloud formation associated with thermals can contribute towards raising the humidity of the surrounding air above saturation level. The cloudiness then spreads out from the sides of the cumulus clouds so that there are continuous layers of stratocumulus formed which prevent the sun's rays from reaching the ground. This phenomenon is called overconvection. In unfortunate cases it can take several hours before gaps in the clouds appear that are big enough for renewed formation of thermals. This can again lead to overconvection, and so on.

Thermal air does not always rise from the ground in well-defined bubbles. In dry tropical regions especially, what happens is that warm air rises while rotating powerfully in an unbroken column. The central core of the column is often narrow, sometimes only 10 m. The birds must soar in tight circles in order to keep themselves in upwinds all the time. The vortex formation that arises when air is sucked into the thermal columns at the ground's surface is intense: sand and dust are carried up in the upwinds. The thermal columns are therefore often clearly visible. They are commonly termed 'dust devils', this of course being due to the discomfort the violent whirlwinds with their whipping sand cause to the people on the ground. These thermal columns usually extend no more than 300–500 m up, but in extreme cases, e.g. when they are triggered by a grass fire, they can reach several thousand metres in height with stupendously strong upwinds.

The principle of bird migration using soaring flight is shown in figure 98. The birds gain height by circling in thermals, then glide off in the direction of their journey, circle upwards again, and so on. This can be done without energy needing to be wasted on a single wingbeat!

In the same way as the flight power curve is central to an understanding of the birds' active flight using wingbeats so the *gliding curve* (or glide polar) is important for an understanding of soaring-and-gliding flight (figure 99). A bird that glides sacrifices potential energy in order to compensate for its weight and the air resistance to its body and wings. The consumption of potential energy is directly proportional to the bird's speed of descent. The derivation of the gliding curve is directly comparable with that of the flight power curve – the induced power decreases while the propulsive power increases with increasing gliding speed. The sum of the two powers at different speeds gives a U-shaped curve. The gliding curve is usually shown as an upside-down 'U' curve in that the speed of descent is put down as a negative speed as per figure 99. When we calculate the gliding curve, we do not need to take into account as we do when calculating the flight power curve air resistance to beating wings (in gliding flight the bird keeps its wings still) or the basal metabolism (which does not affect the speed of descent).

The gliding curve for a certain bird, or for a certain type of sailplane, determines the soaring qualities. There is a limit to how slow the gliding speed can be; a sufficient amount of air must flow over the wings for the force of gravity to be balanced out. If it falls below the minimum possible gliding speed (V_{min}), then so-called stalling occurs: the bird or sailplane crashes towards the ground. The lower the wing-loading is, the slower the bird or plane can glide. A sailplane with a wing-loading above 20 kg/m² cannot glide at speeds below 60 km per hour. V_{min} for a Crane with a wing-loading of 8 kg/m² is a good 30 km per hour and for a harrier with a wing-loading just above 2 kg/m² only 15–20 km per hour.

Of especial interest is the gliding speed, V_{bg}, that results in the best glide ratio. The glide ratio is the ratio between gliding speed and speed of descent and is the equivalent of the horizontal gliding distance that the bird attains for every 1 m loss of altitude. Modern sailplanes glide for up to 50 m with a 1 m loss of altitude. Albatrosses attain the highest glide ratio within the animal world, around 24:1; buzzards, eagles, vultures, storks and cranes

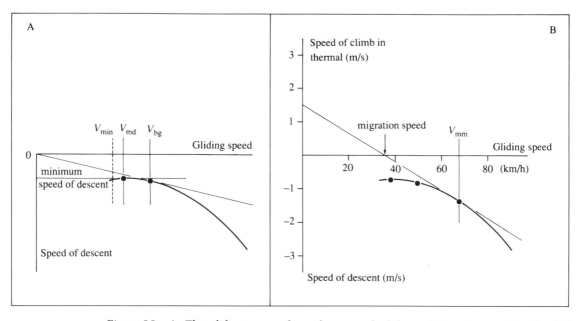

Figure 99   A. The gliding curve shows how speed of descent (negative as it is directed downwards) is dependent on the horizontal gliding speed in straight gliding flight. V_{min} is the minimum possible gliding speed (stalling speed). V_{md} is the gliding speed which gives the minimum speed of descent, and V_{bg} the gliding speed that gives the best glide ratio.
B. The gliding curve for the Crane. The resulting travel speed in soaring-and-gliding flight can be read off directly as the point where the horizontal axis is intercepted by a straight line which connects the climbing speed in thermals, set on the vertical axis, with actual gliding speed on the gliding curve. The maximum migration speed is obtained by drawing the tangent to the gliding curve, as shown in the figure; the gliding speed associated with this is denoted by V_{mm}. The maximum possible migration speed for soaring-and-gliding Cranes at a climbing speed in thermals of 1.5 m/s is thus just under 40 km per hour; the Cranes glide between the thermal bubbles at not quite 70 km per hour $= V_{mm}$ (the speed of descent during the glide is then 1.4 m/s). Based on Pennycuick (1975).

have a glide ratio of about 15 : 1. The glide ratio is determined by the aspect ratio of the plane's or the bird's wings. The longer and narrower the wings, the better the gliding properties. With constant aspect ratio, the gliding speed V_{bg} increases with increasing wing-loading. Migrants wishing to glide long distances and to make rapid progress as well are therefore favoured by long narrow wings and high wing-loading. In strong thermals glider pilots generally increase the weight of their plane by filling water tanks in the wings. In this way they can glide extra fast without adversely affecting the glide ratio; at the same time they can of course always dump the water if the thermals should weaken.

A serious disadvantage of high wing-loading occurs when sailplanes or birds circle in weak thermals. The higher the wing-loading is, the wider are the circles the plane or the bird is forced to soar in. A sailplane requires approximately ten times as large a circle diameter to circle with the same angle of bank and rate of descent as a harrier. An increasing banking angle permits tighter turns, but the rate of descent increases at the same time. Glider pilots who circle in company with birds find that in weak and narrow thermals the birds gain height much more quickly than the sailplane. The birds circle all the time in the central core of the thermal, where the upwinds are most concentrated, while the sailplane must make wider circles in poorer upwinds or pass to and fro through the core of the thermal once every revolution.

In summary we can therefore conclude that a low wing-loading is favourable to climbing capacity, especially in weak and narrow thermals, and that high wing-loading promotes a high gliding speed between the thermal bubbles. The resulting speed of travel with soaring-and-gliding flight is of course dependent both on the rate of climb in thermals and on the speed of glide: it can be calculated graphically, as shown in figure 99. Bird migration using soaring flight generally takes place when thermals are fairly well developed, when the speed of glide is of decisive importance to the resulting migration speed. In these circumstances birds with high wing-loading have the advantage. The aspect ratio is roughly the same for a crane as for a harrier, but the crane's higher wing-loading allows it to be able to reach, with good thermals and with a climbing rate of around 3 m/s, a migration speed close to 50 km per hour; the harrier manages only just over 30 km per hour. The light weight of harriers makes it possible for them to hunt elegantly in slow gliding flight, to make tight turns and to make use of weak and temporary upwinds, but their lightness is a disadvantage when it comes to soaring and gliding on migration. If a Willow Warbler with its broad wings and its low wing-loading were to take to soaring and gliding (which it does not do), then it would travel barely any faster than 20 km per hour.

A soaring-and-gliding migrant can choose various tactics, depending on whether it wants first of all as high a migration speed as possible or whether it finds it more important to glide as far as possible between the thermal bubbles and thereby to avoid the risk of needing to use flapping flight in order to reach new thermals. The gliding speed that gives the maximum migration speed, V_{mm}, is clearly higher than V_{bg} and varies with the rate of climb in the thermal bubbles. Following a thermal bubble with good lift the bird should increase its glide speed, but after a slow climb it should glide more slowly. Should the bird glide instead at the speed V_{bg}, then the gliding distance will be at a maximum at the price of lowered speed. Glider pilots usually choose a compromise between these two possibilities. Their tendency to take down the gliding speed so that it approaches V_{bg} increases the more thinly scattered the thermal bubbles are. How, then, do soaring migrants behave?

Before I set about this question I must just mention that what has been said up to now about speeds and about the behaviour of soaring birds is based on theoretical predictions and estimates. The theory of soaring flight in combination with wind-tunnel measurements of birds' gliding properties enable us to calculate gliding curves for birds of different mass, aspect ratio and wing-loading. The leading researcher in the field of soaring-flight theory for birds is Colin Pennycuick. Using the theory I have calculated the gliding curve for the Crane; this is shown in figure 99B. Now, however, it is high time to leave the theoretical arguments to examine instead whether the predictions really do agree with the birds' behaviour.

Let us start with the Crane. Colin Pennycuick, the meteorologist Bertil Larsson and myself have made a combined study using aircraft and radar in order to ascertain the soaring behaviour of the Crane. The predictions on the Crane's alternative soaring-and-gliding tactics are illustrated in figure 100.

Swedish and Norwegian Cranes winter mainly in southern Spain, southern Portugal and Morocco. During the spring migration they gather in large numbers to stop off in northern East Germany, especially on the island of Rügen. In suitable weather in April they set off northwards, across the Baltic Sea, in over south Sweden (many migrate via Hornborgasjön Lake, cf. figure 87) and on to the breeding bogs. Over the open sea there are no thermals. The Cranes are therefore forced to fly over the Baltic Sea using continuous flapping flight; radar

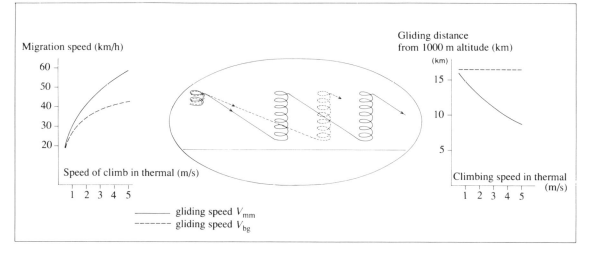

Figure 100 Different soaring-and-gliding tactics of migrating birds. The figures in the diagram are calculated using the gliding curve for Crane as a basis. At the gliding speed V_{mm} (increases with increasing speed of climb in thermal) the birds always reach maximum possible migration speed, but at the same time the gliding flights are on a fairly steep downward path. The gliding distance on the other hand is at a maximum if the birds glide at the speed V_{bg} (independent of speed of climb in thermal), at the price of reduced migration speed. The best glide ratio for the Crane is a good 16:1 and the maximum possible gliding distance from a height of 1000 m thus just over 16 km.

observations show that their flight speed there is around 67 km per hour (table 27). When they have crossed the Scanian south coast, however, they immediately change over to soaring flight. It was here that we awaited them for our study. The radar operator was in radio contact with the pilot and navigator in a motorised light aircraft and he directed the plane towards an echo from a flock of Cranes. When the plane had made contact with the Crane flock, it accompanied the latter at close range. Altitude, time and the places where the Cranes started or finished their thermal soaring were logged and marked in accurately on a map. During this time tracking radar was used for similar measurements of passing Crane flocks.

The results from tracking a Crane flock by plane over 150 km are shown in figure 101. The tracking demonstrates superbly how soaring-and-gliding migrants constantly move up and down in the air during their journey and how dependent they are on favourable thermal weather. I was myself navigator during the trackings in Colin Pennycuick's aeroplane and was able to enjoy the view from the window. The Cranes took hardly any notice of the inquisitive company. For kilometre after kilometre we kept close to the Cranes, over forest and lake, field and meadow, villages, lone farmsteads and churches. Up towards the sun and clouds, then long glides down to the next thermal bubble. That is just how the birds migrate, that is the way it looks in bird perspective! We surely never altogether understand that the migration is so splendid until we actually fly with the birds.

All the same, did the engine noise from the aircraft perhaps detract from the experience somewhat? No, there was something soothing about the plane, a 'Piper' several decades old which had been with Colin Pennycuick for a long time, for example during mapping of big game in East Africa and surveying of Roman archaeological remains in England. A large

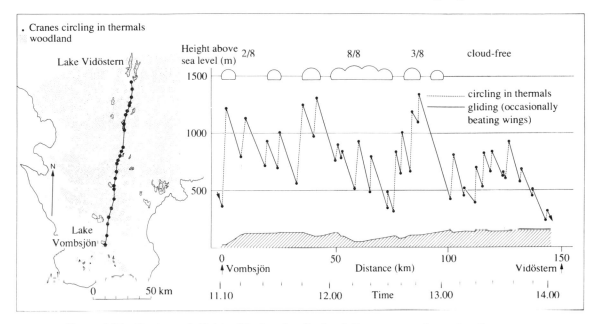

Figure 101 Route and flight altitude of a flock of Cranes on spring migration over southernmost Sweden, recorded by following in an aircraft. The flock was picked up south of Lake Vombsjön in Scania and was left approximately 150 km and three hours later just south of Lake Vidöstern in Småland. The dots on the map show the places where the Cranes circled in thermals. The graph shows a lengthways section along the migration path, with altitudes of entry into and exit from thermals. The soaring conditions varied considerably during the journey. To begin with the circumstances were favourable, with sun and scattered cumulus clouds (base around 1500 m), and the Cranes maintained good speed with pure soaring-and-gliding flight in the altitude range 500–1500 m. Thereafter the Cranes passed through an area of overconvection, and in the weaker thermals they successively lost height, right down to 300 m, before coming out of the overcast area and finding good upwinds which brought them back to a good altitude. Later the cumulus clouds petered out altogether and the air became hazy – a sign of stable air stratification. The Cranes attempted to make use of the weak and temporary upwinds but gradually lost height and used flapping flight to an increased extent. When we left them, they were only some hundred metres above the ground and were forced to continue with flapping flight.

opening had been made in the floor by my seat so that we could take photographs. There was a fair breeze in the plane, which travelled like clockwork, slowly and confidently. For somebody wanting to track migrating birds it was ideally suited.

Anyway, here are some data on the Cranes' soaring-and-gliding migration. In total we tracked different Crane flocks using plane or radar for 15 hours and 38 minutes over a combined distance of 741 km. The average ground speed was thus 47 km per hour. More often than not there was a light tail wind blowing. The mean speed in relation to the air was therefore slightly lower, 40 km per hour. Since the speed of climb in thermals averaged 1.5 m/s, the speed of travel agreed with the theoretically expected maximum migration speed (cf. figure 99). The Cranes' average gliding speed was almost 70 km per hour, close to the expected V_{mm} but clearly higher than the gliding speed for the best glide ratio ($V_{bg} = 49$ km per hour). To all appearances the Cranes' primary preoccupation is to maintain a good migration

V-type formation of Cranes gliding between thermal bubbles: a view from the migrating birds' perspective. The Cranes' upperparts are surprisingly pale, almost white with a fine grey-blue cast in the sun; jet-black trailing edges and outer wings provide a contrasting frame. The forehead is a glowing red. Photographed from an aircraft by C. J. Pennycuick while following Cranes on spring passage over south Sweden.

speed, something which is also borne out by a tendency towards increased gliding speed following the thermal bubbles that produced the highest rates of climb, up to 3 m/s.

So far everything indicates that the Cranes employ the tactics for highest migration speed as per figure 100. But that is not the whole truth. The Cranes often 'cheated' by making wingbeats while gliding between the thermal bubbles. In this way they prolonged the gliding phase and at the same time kept their speed well up. When thermals are weak, with climbing rates below or around 1 m/s, we can expect the Cranes' speed of travel in soaring-and-gliding flight to be 20–30 km per hour (figure 100). But the Cranes did not accept so low a speed. During the weak thermals they made use of their wings to a great extent and instead maintained migration speeds of around 45 km per hour, roughly the same as with pure soaring and gliding when thermals are good.

The Cranes' migration technique therefore consists in actual fact of a combination of soaring/gliding and flapping flight. Even with poor thermals they stop to circle and gain height in the light upwinds, but they use wingbeats when flying between the thermal bubbles, to an increasing extent the weaker the thermals are. This procedure guarantees a migration speed that is always quite good. The slowest Crane flock we followed was travelling at 33 km per hour in relation to the air and the fastest at 50 km per hour. Flapping flight is very energy-demanding for a bird as big as the Crane, which thus pays dearly for keeping up its travel speed when thermals are weak. Why good speed is more important to the Cranes than low transport costs is a completely open question.

Cranes travel in flocks, usually of between five and 50 birds, though occasionally flocks of several hundred individuals occur. During straight flight and gliding flight a Crane flock is arranged in V or J formation. The Cranes frequently fly in formation when circling, too, and all the birds in the flock then circle in the same direction. The aerodynamic advantages of formation flying, which were discussed in the previous section, do not of course apply only to flapping flight but also to gliding and circling flight.

How do the Crane flocks find the thermal bubbles? One possibility is that they keep an eye on the landscape so as to find places which look promising as sources of thermals. From the plane I tried myself to pick out ground with a special character, ground over which the Cranes ought to start circling, something which, however, I soon found to be completely hopeless. Sometimes the Cranes soared over enclosed forest, sometimes over open fields, at times directly above villages, at other times immediately above the outskirts of the villages. The difficulty of using the appearance of the landscape as a guide to predicting where thermal bubbles develop is confirmed by experienced glider pilots. A complicated interplay exists between the various factors that determine where the air near the ground will rise up in thermal bubbles. Of importance are, among other things, the angle at which the sun's rays hit the ground, how much of the sun's radiation is reflected by the substrate, the substrate's heat-absorbing capacity, the dampness of the ground, wind cover and wind turbulence near the ground, cloud shadows etc.

The chance of successfully locating thermals is generally considerably greater for those glider pilots who study the clouds instead of the ground. With experience one can distinguish those cumulus clouds that are on the point of expanding with the supply of thermal bubbles and those clouds that are beginning to break up in the absence of thermals. It is probable that Cranes, and also many other soaring birds, have a good eye for such cloud studies.

Another good method of finding thermals is to watch out for circling birds. During our tracking flights we saw several instances showing that soaring Cranes attract Crane flocks that are in the neighbourhood. Likewise it was common for us to find soaring birds of other species where the Cranes started circling. The Cranes were probably attracted to these indicators of the presence of upwinds. At on average every fifth one of the thermal bubbles that the Crane flocks utilised, we detected from the plane more soaring birds, such as Common Gull, Black-headed Gull, Buzzard, Rough-legged Buzzard, Sparrowhawk, Osprey or Red Kite. Many of these species were, like the Cranes, on spring passage and were making their way by soaring and gliding.

Colin Pennycuick has studied the soaring-and-gliding flight of various species in East Africa by following the birds in a motorglider. A motorglider is quite simply a glider which is fitted with an auxiliary engine for taking off and also of course for using when the thermals give out. His studies provide a basis for interesting comparisons with the Cranes' soaring-and-gliding technique.

In East Africa the thermals are in general well developed, usually giving climbing rates for soaring birds of about 3 m/s, at times 5 m/s. Soaring-and-gliding birds therefore have little reason to have recourse to extra wingbeats, and all of the species which Pennycuick studied in East Africa travelled by means of pure gliding flight. Those which are most reminiscent of Cranes in behaviour are the White Pelicans. They migrate between different lakes in the Great Rift Valley and at these times glide in well-ordered flock formations.

Vultures which follow the migrating herds of antelopes and zebras on the look-out for corpses by contrast soar and glide individually. Several vultures often use the same thermal source, it is true, but they circle in different directions and after gaining height glide out from the 'thermal screw' at different points of the compass and independently of each other. The normal travel speed of the big vultures in soaring-and-gliding flight is 45 km per hour. Rüppell's Vulture breeds in mountain regions and not uncommonly has 100 km to fly to the nearest large herds of antelope. This means a journey time there and back between the hunting area and the nest of over four hours. The time period during the day that permits efficient soaring, at the longest from 10.00 hours to 18.00 hours, is more often than not insufficient for the vultures to have time to get to and return from the hunting area in the same day. This is because they must have not only time for the soaring and gliding in transit but also time to make reconnaissance flights in search of food and then time to fill their crop. Vultures that are fetching food for their young therefore stay away for the night and do not return to the nest until the next day when the thermals permit; then nest relief takes place and the second parent bird soars and glides away on another search for food. When the big vultures have filled their crop, they are reduced entirely to soaring-and-gliding flight; they are by then so heavy that they quite simply are unable to fly using wingbeats.

White Storks on passage through East Africa use an interesting method of finding thermal bubbles. The storks usually travel in large flocks of between 200 and 500 individuals. When gliding between thermals they spread out in loose formation on a front of several hundred metres. When birds in some part of the flock come across upwinds they start to circle; all the birds then quickly come together at the thermal. The storks evidently rely 100% on a few birds in the widely spread flock finding upwinds somewhere during the glide, for they follow straight migration paths and do not seem to take any notice at all of whether the appearance of cloud formations points to better soaring conditions just a little to the side of the prescribed flight course. In a 'stork spiral' there are different individuals circling simultaneously both clockwise and anticlockwise; as the storks each reach the top of the thermal layer, or the cloud base, they glide off in the migration direction. Members of Crane and pelican flocks on the other hand all break off circling at the same time. In East Africa the storks' migration speed varies between 40 and 47 km per hour. The normal flight altitudes are between 1000 m and 2500 m above land (equivalent to 2500–4000 m above sea level).

Do soaring birds make use of the upwinds inside the clouds? Colin Pennycuick made interesting observations by circling in his motorglider immediately below the cloud base with the air brakes fully open so that he would not be dragged up into the cloud. The storks often soared up to a point directly above the aircraft but usually broke off their climb just when they reached the first veils of cloud. Sometimes they disappeared temporarily in the lowermost cloud veils, but as Pennycuick glimpsed them through the haze they were gliding at high speed with legs dangling, a clear sign that they were trying to avoid being pulled further up into the cloud. What happened on a few occasions was that the storks disappeared into the cloud and were not seen any more from the aircraft. This need not necessarily be interpreted as meaning that the storks really did soar far up into the cloud: without having been seen they may have broken off circling and glided rapidly away while they still had visual contact with the ground through gaps in the veils of cloud.

The studies in south Sweden of soaring/gliding Cranes afforded similar experiences. In general the Cranes break off soaring well below the cloud base, though occasionally they disappear, usually briefly, behind light curtains of precipitation and cloud. We saw Cranes circle up to an altitude of 2000 m. According to our calculations, the cloud base was also at this level.

Elsewhere in Europe, according to reports, Cranes have been sighted considerably higher up from aircraft or helicopters, at up to 4000 m. The most remarkable report of all comes from an airline pilot. At 18.30 hours on 16 September 1950, he saw a V-type formation of 80 Cranes which were flying with constant wingbeats (take note!) due south over the Strait of Dover in the English Channel; the altitude was 4300 m. Over England lay a broken cover of stratus cloud, its upper surface about 2000 m up. The Channel area itself was free of cloud, but over France there were scattered cumulus clouds the tops of which reached between 2500 m and 3000 m in altitude. Whether the Cranes had reached their high flight altitude by soaring up through clouds or not it is, unfortunately, impossible to determine.

Thermal bubbles are formed only over land. Soaring migrants therefore avoid travelling longish distances over open sea if possible, for they are then compelled to use flapping flight. Instead they make detours over land and are therefore concentrated at the shortest sea passage-ways. In Europe the biggest numbers of soaring migrants collect at Falsterbo, Gibraltar and the Bosporus (figure 102). Large crowds of birdwatchers, who are captivated by the spectacle of the migrants soaring up into the sky, are also drawn to these places in the autumn. Who would not be tempted by the fantastic migration figures: thousands of falcons, hawks, buzzards, eagles and storks! At Falsterbo, clearly fewer species and individuals pass than at Gibraltar and the Bosporus. This is of course due to the fact that the recruitment area for the Falsterbo passage is to the north and is comparatively small. Birds from the greater part of Europe pass through Gibraltar and the Bosporus. Thus the two roundabout routes for soaring-and-gliding flight to Africa run either side of the Mediterranean Basin. Large numbers of soaring migrants from easternmost Europe and Siberia also pass between the Black Sea and the Caspian Sea on their way to Africa. This is evident from ringing recoveries (cf. the migration of the Buzzard) and is also confirmed by counts that have been organised in recent years in easternmost Turkey, at the southeast end of the Black Sea. Autumn figures reported from here include 140 000 Honey Buzzards and 200 000 Buzzards.

Raptor-watching is a popular pastime in North America. The most famous observation points are Hawk Cliff, where every autumn tens of thousands of buzzards (known in America as buteos) and hawks concentrate along the north shore of Lake Erie, and Hawk Mountain in the north Appalachians, where the raptors follow the mountain slopes that have the best upwinds. As the North American continent gradually narrows southwards the concentration of soaring migrants increases. In the narrow region of Central America the aggregation of birds of prey that migrate between North and South America is incredible. In the autumn the stream of soaring birds moves on across the Panama Canal in a corridor that is not much wider than 5 km. A single spiral of buzzards can contain more than 1000 individuals. In 1972 and 1973 counts were carried out on a large scale, mostly in such a way that the enormous contingents of raptors gliding past were photographed (powerful telephoto lenses often had to be used to enable birds at very high altitudes to be recorded). In both years the final total came to almost 1 million individuals: on average 370 000 Broad-winged Hawks (*Buteo platypterus*), 320 000 Swainson's Hawks (*Buteo swainsoni*) and 230 000 Turkey Vultures (*Cathartes aura*) were counted per autumn. In the early morning the raptors soar no higher than 400–700 m above the Panama Canal, but in the afternoon they regularly climb with the help of thermals to 3600–4000 m above sea level. Reports from aircraft over Panama tell of buzzards and vultures soaring in the convection above rain clouds at over 6400 m altitude.

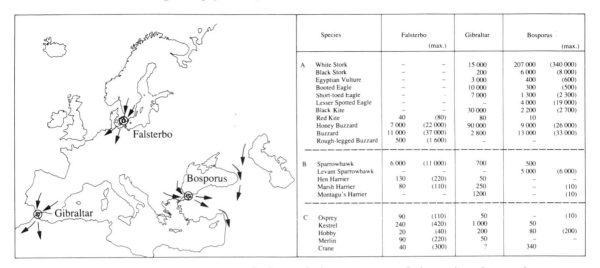

	Species	Falsterbo	(max.)	Gibraltar	Bosporus	(max.)
A	White Stork	–	–	15 000	207 000	(340 000)
	Black Stork	–	–	200	6 000	(8 000)
	Egyptian Vulture	–	–	3 000	400	(600)
	Booted Eagle	–	–	10 000	300	(500)
	Short-toed Eagle	–	–	7 000	1 300	(2 300)
	Lesser Spotted Eagle	–	–	–	4 000	(19 000)
	Black Kite	–	–	30 000	2 200	(2 700)
	Red Kite	40	(80)	80	10	
	Honey Buzzard	7 000	(22 000)	90 000	9 000	(26 000)
	Buzzard	11 000	(37 000)	2 800	13 000	(33 000)
	Rough-legged Buzzard	500	(1 600)	–	–	
B	Sparrowhawk	6 000	(11 000)	700	500	
	Levant Sparrowhawk	–		–	5 000	(6 000)
	Hen Harrier	130	(220)	50	–	
	Marsh Harrier	80	(110)	250	–	(10)
	Montagu's Harrier	–		1200	–	(10)
C	Osprey	90	(110)	50	–	(10)
	Kestrel	240	(420)	1 000	50	
	Hobby	20	(40)	200	80	(200)
	Merlin	90	(220)	50	–	–
	Crane	40	(300)	?	340	

Figure 102 Migrants which regularly use soaring flight gather during the autumn migration at the narrow sea passage-ways at Falsterbo, Gibraltar and the Bosporus. The concentration of individuals at these thoroughfares is large, moderate or poor for the species in the categories A, B and C, respectively. The migration figures from the different localities are not exactly comparable owing to differences in observation (there are for example several observation points at Gibraltar, but only one at Falsterbo). The Falsterbo figures are rounded mean values from five years in the 1970s (Roos 1978) and maximum values over 20 different autumn seasons 1942–1978. The Gibraltar figures are mean values of counts in two autumns, 1972 and 1974 (Bernis 1975). For the Bosporus the figures shown are from autumn 1966 with complete migration monitoring (Porter & Willis 1968) and maximum counts over eight different autumns 1966–1973 (taken from bird reports of the Ornithological Society of Turkey). Migration figures of less than 10 individuals have been omitted altogether. The Bosporus figures for Sparrowhawk and Kestrel are rough estimates only, owing to difficulties in distinguishing these species from Levant Sparrowhawk and Lesser Kestrel.

Different species of soaring migrants concentrate to different extents at the shortest sea passage-ways. In figure 102 I have divided the species into three main categories: large, moderate and rather poor concentration tendency. A good measure of this tendency can be had, as Gustaf Rudebeck (1950) has shown, by comparing the migration figures from Falsterbo with the population estimates from the recruitment area in Sweden and Finland. In round figures, 30% of all the buzzards and kites that migrate through southern Sweden in autumn pass Falsterbo (the figure is somewhat lower for Rough-legged Buzzard, whose migration direction is more southeasterly). For Sparrowhawk and harriers the corresponding figure is 5%; and for falcons, Osprey and Crane only the odd per cent, let us say 1%. The high passage figures for storks and eagles at Gibraltar and the Bosporus clearly show that the tendency of these species to form concentrations is at least as strong as that of the buzzards and the kites.

For buzzards, eagles and storks, the energy costs per migration distance covered are at least three to four times as great during active flight as during soaring-and-gliding flight. The biggest soaring birds of all, e.g. the storks, 'squander' up to ten times as much energy if they use flapping flight instead of soaring and gliding. For falcons, sparrowhawks and harriers, however, the transport costs with active flight are not much more than twice as high as with soaring and gliding. This means that it is advantageous for a buzzard to make a detour over land that is three or four times as long as the straight migration route across the sea. For a stork a soaring-and-gliding detour that is ten times as long can be profitable. Falcons, sparrowhawks and harriers on the other hand waste energy if they make detours that are longer than twice the direct route across the sea.

That falcons and hawks are not so keen on soaring and gliding is clearly apparent from their migratory behaviour. They migrate to a large extent using active flight, often in the morning long before thermals have yet developed, as well as on days when weather and wind are totally unsuitable for soaring and gliding. When a day with good thermals occurs, however, they readily make the most of the occasion: they are then seen circling as they join the larger soaring migrants in the upwinds.

Much of the difference in the various species' propensity to use soaring-and-gliding flight can be explained by energy calculations. But some questions remain unanswered. Why do falcons follow directional lines over land to a lesser extent than do sparrowhawks and harriers? How can the Osprey and the Crane, for both of which it costs five to ten times as much in transport energy to fly actively as to soar and glide, withstand long detours of soaring and gliding? The Osprey migrates on a broad front across both the Baltic Sea and the Mediterranean Sea (see section 3.4), and Finnish and Russian Cranes fly, sometimes even at night, across the eastern Mediterranean Sea and the Black Sea on their way to and from the wintering grounds along the Nile.

Birds about the size of a Swift are the smallest birds for which soaring and gliding is profitable compared with active flight. For still smaller birds, the power saving made by soaring and gliding is swallowed up by the loss in travel speed. This does not necessarily mean that small birds never exploit upwinds on migration. Those which fly across the Sahara desert during the daytime might for example save some energy by reducing speed (without soaring or altering course) when they fly through upwinds and increasing speed where the air is sinking. Future studies can show whether migrants ever profit by this possibility.

Soaring birds which end up in windy weather are faced with particular problems. (Here I am referring to moderate wind strengths, for if the wind is too strong the thermal air is 'holed' and the chances of soaring flight disappear altogether.) The thermal bubbles, and thereby also the soaring birds, drift away with the wind. In order to keep on course on days with variable winds, the birds must compensate for the drift by gliding up into the wind between the thermal bubbles to a greater or lesser extent depending on the wind direction and strength. Whether this actually happens or whether the birds are subjected to wind-drift is a contentious question. The radar studies of Cranes indicate that this species at least compensates fully. Radar has also been used in southern Canada to track buzzard migration. In windy weather the upwinds are often encountered in corridors running more or less in the direction of the wind, so-called thermal and cloud lanes. Soaring birds of course often follow these 'lanes', and radar echoes of migrating buzzards are therefore often in distinct lines. The 'echo lines' are in different places on those days when the winds are different; sometimes they shift sideways gently in one direction or the other. Whether the variations in the buzzards' migration pattern in different winds are a consequence of inadequate compensation or only a result of local adjustments of the flight path to those places where the upwinds are most favourable it has not hitherto been possible to determine. Recent radar studies of migrating raptors in the United States have added much new knowledge on the birds' flight strategies, but the effect of wind-drift in soaring migration has proved to be a difficult and complex question (Kerlinger 1989). A penetrating analysis of how the appearance of the landscape and cloud lanes and winds influence migration patterns of migrants still remains to be made.

When it comes to the passage of soaring birds along coastlines and at headlands in different winds we know considerably more thanks to Gustaf Rudebeck's observations at Falsterbo. With offshore winds the thermal air drifts out over the sea. Soaring migrants which follow the coast are seen to circle and glide a bit out over the sea or right above the shore. Onshore winds on the other hand push the thermals in over land. The birds then follow the coast a good distance inside the shoreline. Over the Falsterbo peninsula the passage of soaring birds in autumn is poorest in east and southeast winds. This is probably due to the fact that birds which follow the west coast of Scania southwards circle in thermal air which drifts out over the Öresund sound; the birds then choose to glide the fairly short way to Denmark and thus cross the sound before they reach the Falsterbo peninsula.

Soaring-and-gliding flight is practised not only by landbirds but also, supremely, by gulls, fulmars, gannets, shearwaters, albatrosses and other seabirds. Since thermals are not as a rule formed over sea, the seabirds must use a completely different soaring technique from that of the landbirds. This technique consists of a combination of (1) gliding in the upwinds above the windward side of the waves and (2) dynamic soaring.

1 It has been calculated that in a wind strength of 10 m/s the upwinds above the waves are sufficient for an albatross, which glides low along a wave, to attain a speed that is enough for a free altitude gain of 20 m. Up to and from this height the albatross can glide in the desired direction of travel before it must again accelerate along a wave to gather momentum for another gain of altitude. Using the same flight technique and in equivalent conditions, the Fulmar should be able to climb to approximately 8 m after each low-level acceleration in the wave troughs.

2 The possibilities of dynamic soaring increase with increasing wind strength. The real experts in this special flight technique are the albatrosses. What, then, is dynamic soaring? It is a technique which is the total preserve of the birds, for no human pilot has yet succeeded effectively in exploiting it. The principle consists of the bird gaining flight energy from the variation in the horizontal wind strength – no upwinds are therefore involved. Owing to the friction of the air against the waves, the wind blowing at low level over the sea is not so strong as that blowing higher up. This wind gradient is often obvious for up to about 20 m altitude. A bird gliding along through the air at a certain speed can of course convert its kinetic energy to potential energy by gliding upwards. On climbing, the bird rapidly loses gliding speed and soon has to make a downward dive to avoid stalling. If, however, it glides *upwards into a head wind* which increases with increasing altitude, then the increasing wind strength provides a free addition to the gliding speed in relation to the air. The bird can thus conserve its kinetic energy and continue to glide upwards up to the point when the increase in the wind becomes too slight to counterbalance a reduction in gliding speed with further climbing. Then the bird turns and glides *downwards in tail wind.* During this phase, too, it can assimilate free energy from the wind gradient. When the bird glides downwards the tail wind decreases, and this leads to the bird's gliding speed in relation to the air increasing. When it approaches the water's surface it often glides for a distance in the upwinds in the wave troughs before again swinging into head winds and gliding upwards so that the whole procedure can be repeated (figure 103).

The net result is that the bird, without beating its wings, makes its way efficiently across the ocean in the direction of the tail wind. By gliding up and down in zigzags, diagonally against and with the wind, it can also move along in a cross wind. Theoretical calculations suggest that in realistic wind strengths albatrosses can even travel in straight head winds

Figure 103 Dynamic souring in which the seabirds extract flight energy from the wind gradient above the sea's surface. By gliding upwards in a head wind the birds can gain height at no cost (i.e. without their air speed being reduced). By then gliding downwards in a tail wind they also gain additional kinetic energy from the wind gradient. Based on Rüppell (1977).

using simple dynamic soaring. In that case, the birds, after having gained height while gliding in head winds, must dive very steeply in tail winds and speed up for a long glide against the wind low over the waves (where the wind strength is weakest). The resulting speed of travel with this process is, however, very modest; without doubt it is much easier and more energy-sparing for seabirds to move in the directions in which tail winds are blowing. This is evident not least from the fact that there is a clear association between the migration routes and the geographical pattern of the winds. Using dynamic soaring, the large albatrosses travel around the Antarctic in strong tail winds within the zone of westerlies.

A different opportunity for dynamic soaring exists in the bird gaining free energy from the wind increase in gusts of wind. The Bateleur Eagle has developed this technique as its speciality. The bird looks like no other. It has big long wings with white undersides bordered with black, a black underbody, vivid red legs, and a short rust-red tail: an unforgettable silhouette in Africa. It glides on raised wings over the savannas, constantly banks to one side or the other, and maintains its height for as far as one can follow it to the horizon, all without beating its wings and without soaring. It moves at good speed. It has been estimated that every day of its life a Bateleur searches for prey by gliding for up to 300 km, every day for perhaps 15 years. By banking and turning its upperwing towards the wind gusts, the bird assimilates energy from the horizontal wind; no doubt it also makes use of quite a few upwinds.

The Bateleur is a good symbol of the amazing ability of birds to exploit the winds for soaring/gliding flight. At the same time it also shows how much we have left to learn of all the fine ingredients that go to make up the soaring and gliding capability of birds.

Let us try to look at the familiar sky, the clouds and the soaring birds with a new eye.

4.5   Flight altitude

In autumn 1975, a huge irruption of Coal Tits occurred in south Sweden. Gunnar Roos, the observer of migration at Falsterbo, counted more than 15 000 Coal Tits heading southwest out over the sea. On some days the emigration took place in fresh head winds; the tits' extremely low flight altitude, only a few tenths of a metre, caused astonishment among the birdwatchers on the outermost headland of the Falsterbo peninsula. Gunnar Roos (1975) reported:

> 'The Coal Tits often came "creeping out" over the golf course really low down. Right out on the Point they were seen literally slalom-flying between birdwatchers and telescopes, only to launch out over the waves after landing briefly in the very last, exceedingly small rose bush. Here we got a fascinating impression of tremendously inflamed migration fever the like of which I have seldom, if ever, experienced.'

One afternoon in July 1962, an airliner collided with a bird over north Nevada in the United States. The collision was felt as a faint thud in the plane. After landing, it was found that the bird had struck one of the stabilisers on the tail section. Here there was a dent the size of a football, and a hole had been torn in the metal. Inside the hole a well-preserved wing feather from a Mallard was found. The flight altitude when the collision happened was 6400 m.

Around 15.00 hours on 9 December 1967, a radar operator in Northern Ireland detected a powerful echo moving southwards over the sea near the Hebrides: its altitude was estimated roughly at 8000 m. The radar operator checked whether there had been any radiosonde or something similar released in the area which might account for the mysterious echo. However, he received negative responses and therefore made radio contact with the pilot of an aircraft in the vicinity and requested that he make a detour via the position of the unidentified radar echo. When the aircraft had reached the area the pilot reported that he could see a flock of about 30 swans at a flight altitude just above 8200 m! The radar operator was then able to track the echo from the swan flock for more than 100 km farther south, until it disappeared just off the Northern Ireland coast. The swans had by then gradually reduced their flight altitude, and finally disappeared at low level beneath the radar's cover.

The swans were in all probability Whooper Swans on their way to Ireland from Iceland. Whooper Swans regularly overwinter in Iceland, but periods of severe cold have for a long time been known to cause many to move south in the middle of winter. On the day when the high-flying swans were discovered, a tremendous burst of cold air arrived in the north, and fresh north-northwest winds swept down across Britain. At higher altitudes the wind blew at jetstream force – above the Hebrides, north winds at 8000 m were measured at 50 m/s (=180 km per hour). Meteorological measurements showed that the swans in this area were flying just above the tropopause, at a level where they could pass above heavy snow-squalls over the sea. The temperature at this altitude was −40 °C. The strong tail winds probably made it possible for the swans to reach a ground speed of up to 200 km per hour. With this they should have been able to cover the 1300-km route across the Atlantic from Iceland to Ireland in only seven hours.

On 29 November 1973, an aircraft collided with a bird at a good 11 000 m above the Ivory Coast in Africa. The bird wrecked one of the aircraft's engines, though the plane managed to land without further mishap. Feather remains in the wrecked engine showed that the bird was a Rüppell's Vulture.

The above examples illustrate the extremes of altitude for bird migration: a few tenths of a metre above the ground and altitudes at the tropopause. It may perhaps be thought that the records made at the extremely high altitudes are unique chance occurrences, that for some unknown reason the birds had lost control of their situation? Yet this is not the case. The examples above of high flight altitudes are not the only ones of their kind; there are other scattered reports of collisions between birds and aircraft at similar altitudes and also direct observations from aircraft, though in these cases the bird species could not be identified. Furthermore, there are observations which show that many migratory birds regularly travel over the Himalayas at fantastic heights.

How is it possible for birds to stay alive, and moreover to work flat out by flying, at such great altitudes? The lack of oxygen makes it hard for a human being who has not become acclimatised over a long period of time to move about even above 3000 m, and at 5000 m most people who are not used to altitude lose consciousness. The highest-lying permanent settlements, in the Andes and in Tibet, are situated at just above 5000 m. Not even people belonging to these mountain communities would be able to survive more than a few hours in the oxygen-deficient air at 8000 m. The oxygen content of the air is about 21%, independent of altitude, in the troposphere; the oxygen pressure consequently decreases in parallel with the decreasing air pressure at increasing altitude. At 6000 m the oxygen pressure is only half what it is at sea-surface level; at 8000 m it is a third of that and at 10 000 m only a quarter.

The ability of birds to stay alive at high altitudes is explained by the fact that they have a more efficient respiratory system than mammals. This has been demonstrated, for example, by experiments in pressure chambers. House Sparrows (which are not migrants and which do not normally fly at great heights) and mice were subjected to pressure the equivalent of that at fully 6000 m altitude. The mice lay on their bellies and hardly moved; their body temperature was 10° below normal. The sparrows, by contrast, hopped about briskly, and they were even able to fly without difficulty. A bird's lungs function according to the through-flow principle: the inspired air collects in the bird's posterior air-sacs and flows through the lungs to the anterior air-sacs before it passes back out. In the lungs the blood is oxygenated by fine air capillaries, where the air and the blood flow in opposite directions. Owing to this counterflow, the oxygenated blood that leaves the bird lung acquires a higher oxygen concentration than that corresponding to the oxygen pressure in the expired air. An equally efficient absorption of oxygen is impossible for mammals: their blood is oxygenated in the lungs' microscopic alveolae, and these are 'blind alleys' for the air. The relatively large heart of birds may also be a contributory reason for their ability to fly at high altitudes. The bird heart's proportion of the total body weight is generally between 0.8% and 1.5%. The mammal heart weighs only about 0.6% of the weight of the body. The birds' large heart makes possible a rapid transport of blood and an intensive oxygen renewal by the body's various tissues. We should not let ourselves be deceived into thinking that the special respiratory system of birds is a necessary adaptation for the power of flight as such – bats as we know fly excellently with the 'mammal system' – but it certainly gives the birds a unique opportunity to fly at the oxygen-deficient high altitudes.

The intense muscle work during flying obviously provides surplus heat sufficient for the birds to endure the low temperatures at high altitudes (the temperature in the standard atmosphere drops 6 °C for every 1000 m of increasing altitude). The cold may even amount to an advantage in that the water balance can be preserved. At high temperatures flying birds lose more water through evaporation, via exhaled air and through the skin, than is produced during metabolic processes. According to wind-tunnel experiments, flying Starlings lose water at temperatures above $+7\,°C$; when flying in $+25\,°C$, water losses mean that the Starlings' total body weight decreases by almost 2% per hour of flying. Birds obviously run the risk of drying up during long flights at such high temperatures.

The above-mentioned altitude records are particularly noteworthy since they involve such big and heavy birds as swans and vultures. Flying at high altitude demands extra large power consumption on account of the low air density. The flight power is approximately 50% greater at 8000 m than at sea-surface level. At a guess the vulture was soaring/gliding at

11 000 m and therefore avoided the intensive work of flying. Perhaps it had reached its high altitude by circling in the strong upwinds in cumulonimbus clouds? These can extend all the way up to the tropopause. That the swans' flight work at 8000 m altitude was facilitated by upwinds appears less likely. On the other hand, it is surprising that swans, which are so heavy that in theory they should have narrow margins to their flying capacity even at lowest altitude, are able to muster the energy that is required for high-level flying.

Well, the examples above, then, illustrate the lower and upper altitude limits of flight by migratory birds, but what are the most usual flight altitudes under normal conditions? The American ornithologist Frank C. Bellrose (1971) studied night-migrating birds by observing them directly from a light aircraft. The plane was fitted with extra searchlights so that the birds could be seen when they glimmered past in the light between the plane's nose and the leading edges of its wings. Hardly any but small birds appeared in the light. This was probably due to the fact that flocks of waders, ducks and geese managed in good time to move out of the way of the slow-moving aircraft (the cruising speed was 190 km per hour). In the light, small birds were seen to dive at the last second to avoid the aeroplane. Despite thousands of small birds swishing past at only a few metres' distance from the propeller and the aircraft's wings, only three bird strikes were recorded.

In order that the birds' altitudinal distribution could be mapped, the plane was flown at stepped altitudes with 150 m between the different levels, forward and back, night after night, over the same route in central North America. The bird density was shown to be greatest at the 300-m level; the 150-m level and the 450-m level came equal second. The density of birds then decreased with increasing altitude. Few individuals were counted at the highest flight altitude, 1800 m, while 90% of the nocturnal migrants flew below 900 m. The greatest bird density registered at any one time occurred at the beginning of October during a flight at the 300-m level, when an average of 94 birds per minute was counted. An important observation was that at night the birds did not occur at the very lowest altitudes above ground, where bird migration often goes on during the daylight hours; even though take-offs and landings were made in the dark on innumerable occasions, nocturnal migrants were never seen from the plane below 30 m altitude.

The results of this study have been confirmed in rough outline by various radar studies of the flight altitudes of birds over inland and coastal regions in North America, in Britain and in the Swiss lowlands. Generally these radar studies show that on average 50% of nocturnal migrants, of which the clear majority are small birds, travel below 400–700 m and 90% below 1500–2000 m above the ground. The radar studies show that birds also migrate regularly, albeit to a minimal extent, at much higher altitudes: the records are several cases involving birds noted at around 6000 m altitude.

The mean altitude of bird migration varies regularly during the course of the night. Immediately after the start of migration at dusk the birds quickly increase height and reach a maximum altitude during the early night. The mean altitude then drops off gradually after midnight. In extreme cases the birds reduce flight altitude so substantially in the latter part of the night that they travel beneath the radar's cover and so disappear completely from the radar screens. This has been observed at radar stations in the Shetlands and on the east coast of England. The observations relate mainly to passage of thrushes which arrive in autumn from Scandinavia across the Norwegian Sea and the North Sea. The thrushes are visible on the radar when they travel across the sea on a southwesterly course in the hours around and immediately after midnight, but during the late night the radar echoes become fewer and on some nights disappear altogether. At dawn, however, something surprising happens: large numbers of radar echoes suddenly appear over the sea. Nearest the coast the echoes generally move straight in towards land; the others move mainly to the south and southeast.

After a number of years of radar and field observations, British ornithologists have been able to piece together the most likely explanation for the phenomenon. The thrush migration continues during the late night in a southwesterly direction, but then often at such a low height above the sea that it does not register on the radar. When it starts to get light the thrushes rapidly increase altitude (and appear on the radar screens). If they see land within close reach, they head in that direction in order to land and rest. If, on the other hand, they are over the open sea, they alter flight direction and swing to the southeast. This behaviour is presumably a good insurance policy against the risks to which the thrushes expose themselves when they head out across the sea on autumn nights, when changes in the weather can occur quickly and unexpectedly.

One risk for those birds which fly across the northern waters off the Shetland Islands and which have still not reached land by the late night is that they miss north Scotland and its islands and carry on westwards over the open Atlantic. By dropping down to low flight altitude over the sea the birds reduce the risk of being carried too far west by strong tail winds. If no land is visible from high up when daylight comes, then southeast is an appropriate flight direction that should bring the birds back in towards land. The fact that the thrushes do run the risk of 'flying past' north Scotland is confirmed by radar studies over the sea off the Hebrides, on the northwest side of Scotland, where both birds climbing at dawn and birds reorienting towards the southeast have been observed on repeated occasions; these birds have originally set out from Norway.

Another risk for the thrushes is that over the sea they will run into a cyclone with strong head winds, so strong that fat reserves and flying strength are not sufficient for them to be able to reach Britain. If they have not reached land before the late night, then it is certainly a good idea to fly at low altitude, where the head wind is as light as possible. Should there not be land within close reach during the dawn ascent, then it is just too risky to butt further against the wind. By cancelling the migration to Britain altogether and instead swinging to the southeast, the thrushes can be fairly sure of reaching land somewhere in the North Sea region quite soon and of preserving their lives in so doing.

Radar observations show unanimously that migrants usually avoid moving up to altitudes of several thousand metres. Under special circumstances, for example on long flights and crossings of mountain chains, the flight altitudes deviate from this normal pattern. Several such cases have been mapped as a result of radar studies. Birds that reach the south coast of the United States in spring across the Gulf of Mexico more often than not fly at between 1000 m and 3000 m above sea level. They have started their journey from Central America immediately after sunset. Since they have over 1000 km of sea to cross, they do not reach the Louisiana coast until the next day, and with unfavourable winds their arrival can be delayed until the following night. If the birds arrive early in the day they continue northwards overland in Louisiana at gradually increasing altitude, but, if the arrival is delayed until the dusk, they generally land as soon as they reach North American terrain.

In autumn, both waders and passerines fly from northeast North America directly over the Atlantic to the West Indies and northernmost South America, a flight distance of 3000–4000 km non-stop (cf. next section). The wader flocks that travel out over the Atlantic at Nova Scotia in Canada fly at a mean height of 2000 m. One-tenth of the flocks fly at over 3900 m, and the highest-flying flock was recorded at 6650 m above sea level.

Over the Atlantic the migration usually takes place at altitudes between 1100 m and 5000 m. Over Puerto Rico, where passerine migration is most intensive, as well as over Antigua and Barbados, where waders predominate, the *mean altitude* can on some days reach almost 5000 m. Over Puerto Rico migrants have been recorded at up to 6800 m above sea level. When the birds approach South America they prepare for their landing by drastically reducing flight altitude: over Tobago, immediately off South America's north coast, all the migrants generally speaking pass at less than 500 m.

The tallest mountain peaks of the Alps reach about 4000 m above sea level. The dominant mountain ridges are 3000 m high and the major passes lie at between 2000 m and 2500 m. Large numbers of migrants cross the Alps. Radar studies in a Swiss alpine pass at 2000 m have shown that by no means all of them fly through the passes. In tail winds it is, on the contrary, usual for the migration to move on a broad front and without the birds allowing the mountain topography to determine their flight paths; the birds' flight altitudes are then between 3500 m and 5500 m above sea level.

Birds also migrate over the world's highest mountains, the Himalayas. No radar studies have been undertaken there, so we must base our opinions on reports from ground-based observers. This of course implies that considerable bird passage can very well be supposed to take place out of reach of human eyes and ears. Many migrants fly down the long valleys of the Indus and the Bramaputra, west and east of the mountain massif. Others concentrate in river gorges where they can fly between the Indian lowland area and the Tibetan high plateau, which lies at an altitude of 4000 m and more, without needing to climb above the highest mountain ridges. This applies, for example, to Demoiselle Cranes, which migrate between breeding sites on the steppes of Kazakhstan, Siberia and Mongolia and wintering grounds in India. One observer counted 30 000 Demoiselle Cranes over ten days in October passing south through a river valley in Nepal; the distance to the peaks over 8000 m high on either side of the valley was no more than 15 km. On some days the cranes encountered such strong head winds in the valley that they could not overcome them; they were then seen circling to a high altitude, to at least 5000 m, and then gliding further southwards high above the valley and the head winds. Another observer saw 63 000 Demoiselle Cranes on migration through the same Nepalese river valley during 12 days at the end of September and early October 1978. In good thermal conditions the cranes often soared to 5000 m and 8000 m. Crane flocks at lower altitudes were regularly attacked by Golden Eagles. In the valley, along a distance of 20 km, there were at least six pairs of Golden Eagles, and in some cases these were seen to attack cranes at more than 1000 m above the ground. Of 67 observed eagle attacks, only four ended with the eagles succeeding in striking a migrating crane in the air. High flight altitudes may therefore sometimes be a good protection against attacks from stooping birds of prey.

Many birds even migrate over the Himalayas' highest massifs. They appear here especially in mountain passes, approximately 6000 m above sea level. Birds observed in such passes include Black Kite, Sparrowhawk, Temminck's Stint, Common Sandpiper, Green Sandpiper, Hoopoe (seen at up to 6400 m above sea level), White Wagtail (common), Black Redstart (common) and Greenish Warbler. Several Asiatic species of lark and pipit can be seen in small flocks. The eastern Steppe Eagle also migrates through the Himalayas; one has been found dead at 7900 m above sea level.

The most astonishing observations relate to geese, flocks of Greylags and, above all, Bar-headed Geese (*Anser indicus*), which fly over the highest mountain peaks. The Bar-headed Goose breeds at mountain lakes in the Tibetan highlands, where in times past it was protected by the local people, who considered it to be sacred. It winters on the south side of the Himalayas at the low-lying wetlands around the Indus and the Ganges. Mountaineers have seen goose flocks flying over the top of Mount Everest, 8849 m above sea level. One observer managed to see a goose flock through his telescope at night passing across the face of the full moon above the mountain landscape around Mount Everest and estimated the birds' altitude at 9000 m. Another informant tells of an April night which he spent beside a glacier at 5000 m in the same area. The dense silence was suddenly broken by the distant cackling of a flock of Bar-headed Geese. The sound revealed that the flock, invisible against the dark firmament, was moving north. Judging by the 'aurally perceived flight path', the flock flew straight over a mountain peak of 8500 m altitude, only a few kilometres from Mount Everest. These high flight altitudes appear even more fantastic when we bear in mind that in spring the geese have only 150 km to fly from the Indian lowlands, just under 200 m above sea level, before they pass over the world's tallest mountain. The time that the birds have to acclimatise themselves to high altitudes is really minimal.

It is worth pointing out in passing that mountaineering expeditions to Everest have provided interesting information not only on migratory birds but also on the highest altitudes for resident mountain animals. Mammals such as mountain sheep and yaks can be found at up to 6500 m. Lammergeiers have been observed soaring at 7600 m, and Alpine Choughs have visited camp bases 8200 m above sea level.

What factors determine the altitudes of migrating birds over sea and lowland? Here of course they can freely choose to fly low or high and are not obliged to travel at the very highest levels in order to pass mountain chains. The air density decreases with increasing altitude, which affects the birds' flight. When the air density diminishes, the air resistance that the flying bird meets also becomes less; the same amount of propulsive power is therefore no longer required. On the other hand, the power needed for the bird to get sufficient lift increases. These changes affect the flight power curve (figure 93). A consequence is, as already mentioned, that the most favourable flight speed for a migrant, V_{mr}, increases with increasing altitude. If the best migration speed for a Woodpigeon is 57 km per hour at low altitude (table 27), then it is 70 km per hour at 5000 m and 88 km per hour at 10000 m. From a speed point of view, therefore, it is an advantage for the birds to fly high.

A prerequisite if the birds are to be able to exploit this gain in speed is that they have the stamina to increase the flight power to a sufficient degree. This is because P_{mr}, i.e. the power required to fly at the best migration speed, increases with increasing altitude. P_{mr}, however, does not increase to quite as great an extent as V_{mr}. This leads to the resulting transport costs, i.e. the energy expenditure per flight distance covered ($= P_{mr}/V_{mr}$), decreasing with increasing flight altitude. Here, then, we find a further advantage with flying high: the energy consumption for the flight becomes less. The reduction in transport costs is, however, quite modest, only 1% or at most 2% per 1000 m of increasing altitude.

Judging from the flight theory, therefore, it is advantageous from a speed point of view as well as an energy point of view for migrants to fly as high as possible, i.e. where the oxygen supply is still sufficient for the birds to be capable of reaching P_{mr}. For birds which fly long distances non-stop, the argument can be developed further. At the start of the flight the birds are extra heavy since they have large fat reserves, and they are then not able to fly particularly high; as the fat is used up, so their weight decreases and therewith also their flight power at constant altitude. On the other hand, the capacity of the flight muscles does not diminish but allows the birds to increase their altitude accordingly as their flight fuel is burnt. The farther the birds get while flying, the higher then their flight altitude should be. We can theoretically calculate that a small bird flying in still weather should, according to this principle, gradually increase its flight altitude by about 1300 m over a distance of 1000 km.

This has been seen as a possible explanation for the high flight altitudes that have been observed on just those occasions when birds do fly long distances, for example across the Gulf of Mexico and the west Atlantic.

Another possible reason, already mentioned in passing, is that, during a long flight lasting several tens of hours without stopping, birds seek the cold at high altitudes in order to reduce water evaporation and avoid desiccation.

In my opinion, we may seriously doubt that the effect of the air density on the birds' speed and energy consumption would be of any major significance to the altitude of bird migration, for there is another factor that influences the birds' speed and economy of energy to a much greater degree than the air density, namely the wind. Even a head-wind strength amounting to half of the birds' air speed results in halved ground speed and doubled transport costs. The same tail-wind strength gives a 50% speed increase and a saving of one-third of the transport costs. I am of course making comparisons with the situation that exists in still weather. The wind often varies considerably at different altitudes. This leads us to expect that this factor above all will be of decisive importance in the migrants' choice of flight altitude.

Radar studies have shown unanimously that the wind really is a key factor when it comes to explaining the altitudinal dispersion of bird migration. The birds fly mainly at heights where the head winds are the minimum possible or the tail winds the best possible. Since the wind strength usually increases with increasing altitude, we get a general rule of thumb: bird migration takes place at low altitude in head winds and at high altitude in tail winds. In fresh head winds many species migrate by day at the very lowest altitude so that the birds can fly sheltered from the wind in valleys, behind hills and clumps of woodland etc. In figure 104A and B, examples are given of how altitudinal distribution of bird migration at one and the same place can change from one night to another, one with head winds and one with tail winds. Birds have a striking capacity for finding those altitudes where the most favourable winds are blowing. When both wind direction and wind strength vary greatly at different altitudes, birds on route to migration goals in separate directions concentrate in different altitudinal strata so that the different categories of birds find just their most advantageous winds.

Even extremely high altitudes for migrants can in many cases be explained by the wind conditions. One example is given in figure 104C, which shows how passerines on long-distance flight from North America to South America are distributed at different altitudes over Puerto Rico. The West Indies lie in the trade-winds zone, where east winds, troublesome for migrants on their way south, generally prevail. The easterly trade-winds diminish in strength at high altitudes, where the wind direction is more variable. The birds' flight levels in figure 104C accord well with the altitudes at which the cross winds were lighter and therefore less of a hindrance to the birds.

The birds' flight altitudes are also affected by cloud. Birds avoid flying in cloud. In total overcast weather with dense clouds the migration is therefore concentrated at fairly low altitude, even if very favourable winds exist at high levels. The birds generally travel above low banks of mist or fog. Radar studies have shown that in addition bird migration is sometimes concentrated both below and above cloud layers of moderate or minimal thickness. The fact that birds avoid flying in cloud does not mean that they never do so. On the contrary, several radar studies have shown that migrants may fly and maintain both course and height, at least during the short periods the radar has tracked them, in and between compact layers of cloud. A human pilot is not able to fly in cloud without technical assistance; how the birds' senses help them with orienting in this situation is an unsolved mystery. That even birds at times come up against orientation problems, however, is demonstrated by a number of radar observations of 'disoriented' bird migration. At such times echoes from birds move irregularly this way and that in all different directions in weather with thick cloud, fog, haze and drizzle. The reason for the birds' incoherent behaviour may perhaps be that great care is required if they are to avoid dropping into the sea or colliding with the ground when they reduce flight altitude in order to attempt landing without visibility in unfamiliar areas.

The influence of cloud on flight altitude has been studied in a particularly elegant fashion by a Swedish research team (Gustafson, Lindkvist, Kristiansson, Gotborn and Gyllin 1977)

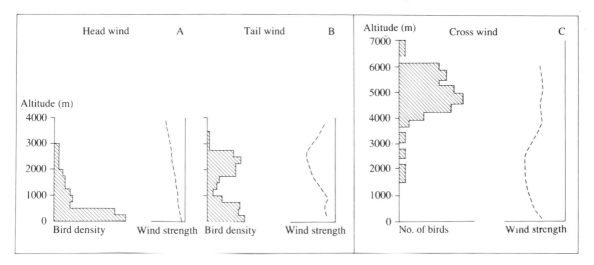

Figure 104   The altitudinal distribution of migrating birds changes markedly under different wind conditions. A and B show a comparison between the flight altitudes of birds over the Swiss lowlands on two nights, one with head winds and the other with tail winds. The birds fly mainly at those altitudes where the head winds are lightest and the tail winds strongest. C shows the altitudinal distribution of small birds on long-distance flight over Puerto Rico in the West Indies on one autumn morning; the birds' extremely high flight altitudes coincide with the level at which the easterly trade-winds weaken and become less troublesome as cross winds. Based on Bruderer (1971) and Richardson (1976).

using Swifts on return flight to the nest. The flight altitudes of the individual birds were recorded with an altimeter 2 cm long and weighing less than 0.5 g. This was pasted on to the bird's back feathers and removed when the bird was recaptured at the nest. (The paste was mixed with a water-absorbing substance so that the altimeter would fall off after a few days if recapture proved unsuccessful.)

The principle of the altimeter was based on the case that a component of radioactive radiation, known as *alpha*-rays (ionised helium atoms), has a range of only a centimetre or so in air, and on this range being dependent on the air's density and thus on altitude above sea level. The atomic nuclei of helium are slowed down by the electrons in the air's molecules. If the air density is reduced, the helium nuclei travel farther before the same number of electrons slow them down; the range of the radiation therefore increases with diminishing air density. The altimeter consists of a radioactive preparation and a photographic plastic film on which the *alpha*-rays emitted from the preparation are recorded. After the film has been developed, the range of the radiation can be measured and the bird's maximum altitude immediately determined. Through more detailed analysis one can also work out how long the bird has spent at different altitude levels.

The highest altitude for Swifts which flew home in clear weather was 2300 m on average. Ten out of a total of 50 Swifts had flown above 3000 m, and those which flew highest of all reached 3600 m. In sharp contrast to this, the average highest altitude in overcast weather was only 700 m. Some of the 19 birds which were released in cloudy weather never flew above 50 m; the one that went highest reached 1700 m.

A particularly exciting experiment was made when 14 Swifts which had been captured at their nests outside Örebro were freighted 400 km to the south, to Lund, where they were fitted with altimeters and set free. When the Swifts were released in Lund in the morning, they climbed steeply, past their local conspecifics, and soon disappeared. The next day they were recaptured in Örebro. The weather was fine and clear on the day of the journey; light head winds increased in strength at high altitude. Farther north it was almost dead calm. The altimeters gave information of a similar kind. The total flying time varied between ten and 17 hours, the mean being 13 hours. The time was divided up into flying in the main at two discrete levels, a maximum altitude of around 2000 m, where the birds flew for between four and 13 hours (the Swift flying highest flew for a good five hours at 2350 m), and a lower altitude, just above 1000 m, where the flying time varied between three and eight hours. We can guess that the Swifts, owing to the head wind, flew at the lower level to begin with and that, when the head wind abated in the north, they climbed to a high flight altitude. The Swifts' rather low average speed, 31 km per hour, is of course explained by the fact that head winds were blowing during part of the flight.

Swifts sometimes climb to surprisingly high altitudes on short flights. Several of those that were released with altimeters only 40 km from their nests flew for between just under one hour and two hours at 2000–3300 m. Something similar was observed when homing pigeons with the same type of altimeter were released. The pigeons' commonest flight altitudes were a few hundred metres, but one particular pigeon chose to fly unusually high whether it was released 300 km (flying time 3 hours and 38 minutes at 1550 m above sea level), 40 km (40 minutes at 1700 m), 35 km (38 minutes at 1350 m) or 10 km (35 minutes at 1600 m) from the pigeon loft.

How can the birds determine their flight altitude? Experiments with pigeons in pressure chambers indicate that birds are very sensitive to small differences in air pressure. Most pigeons are capable of detecting air-pressure changes of one millibar (mb), the equivalent of an altitude difference of only 10 m, within ten seconds. The high-flying pigeon persisted in flying at altitudes with an air pressure of 825–827 mb; a low-flying pigeon made the same four flights where the air pressure was 970–1005 mb (its altitude varied between 100 m and 450 m). These figures make it even more likely that the birds actually determine their flight altitude by means of an inbuilt physiological 'barometer' (this sense organ is presumably located in the birds' middle ear).

4.6   Fat as flight fuel

Migratory birds store up fat in thick layers beneath their skin. The fat is used as flight fuel during the migration. The burning of 1 g of fat releases an energy amount of a good 9 kilocalories. The energy value in carbohydrates or protein is not even half as much. The advantage of storing fat is reinforced by the fact that fat can be stored without extra water. For each gram of carbohydrates (glycogen) that is stored in the cells, at least 3 g of water are required. In order to extract a certain amount of energy, therefore, in practice about eight times the weight of fuel reserves of carbohydrates are needed as of fat. No wonder, then, that migrants, whose main objective is to take on as much energy-rich flight fuel as possible without becoming too heavy, are adapted to make use of fat reserves.

With the flight power curve as a starting point, we can calculate the total distance and the length of time a bird is capable of flying with different amounts of fat reserves. In figure 105 I have shown the results of such calculations. The quantity of fat at the beginning of the flight is given as a percentage of the bird's total body mass. Note that a doubling of fat percentage, calculated in this way, does not mean twice the amount of fat reserves: in order to attain a fat percentage of 50% the bird must put on additional fat equivalent to the whole of its normal

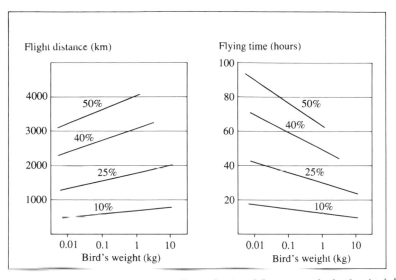

Figure 105   Flight distance (in still weather) and flying time for birds which begin the flight with various fat percentages (as percentage of total body mass) and which fly until the fat has been completely used up. The calculations are based on the theory of bird flight according to Pennycuick (1975).

fat-free body mass; a fat reserve one-third as large gives a fat percentage of 25%. The lines in the figure show the approximate values for birds of 'average appearance'. Efficient long-distance flyers with narrow, pointed wings and streamlined body, such as swallows and certain waders, achieve slightly longer distances and times than shown in the figure; the opposite applies to blunt-winged birds.

Large birds are not able to fly for such a long time as smaller birds with the same fat percentage, but owing to their higher flight speed they nevertheless reach a greater distance than small birds before the fat runs out. The flight distance that has been specified in figure 105 is valid for still weather; to calculate flight distance in different wind conditions, we must estimate the birds' average ground speed and then multiply by the flight time for the actual fat percentage.

As is clear from the figure, a fat load of 10% is sufficient for migrants to be able to fly for between 10 and 20 hours and, in still weather, to reach 500–750 km. Birds which set out with half their body weight consisting of fat can fly for three to four days and nights and reach 3000 or 4000 km before the fat is spent. The birds have the chance to prolong the flight time by a quarter to a third above that shown in figure 105 by reducing their flight speed so that the flight power becomes minimal (i.e. by reducing the flight speed from V_{mr} to V_{mp}). This, however, has a negative effect on the flight distance and should therefore be used only if the bird wants to gain extra time in a critical situation; if for example landbirds, on a long flight over open sea, have run into poor weather and are forced to fly around while awaiting better weather and better conditions for orienting.

The reduction in body weight of migrating birds due to the consumption of fat should, according to theoretical calculations, be around 0.7% per hour of flying for the smallest passerines and between 1% and 1.5% per hour for birds in the half-kilogram class and above. Several different field studies have been carried out in order to check the weight reduction of migrants. The results are shown in table 28. The studies were based on widely differing

Table 28. *Estimates (field studies) of weight reduction in birds during migration flight. Most data have been taken from Nisbet et al. (1963); for studies nos. 8–9 from Hussel (1969), for no. 10 from Hussel & Lambert (1980), for no. 4 from Fry et al. (1972), for no. 11 from Raveling & LeFebre (1967) and for no. 12 from Dolnik & Gavrilov (1973)*

Species	Weight loss (% body weight per hour)	Basic data for calculation	
1. Robin *Erithacus rubecula*	0.9	Estimated weight change when flying direct	Norway–Scotland
2. Goldcrest *Regulus regulus*	0.7	Estimated weight change when flying direct	Norway–Scotland
3. Wheatear *Oenanthe oenanthe*	0.8–1.3	Estimated weight change when flying direct	Greenland– Scotland
4. Yellow Wagtail *Motacilla flava*	0.8	Estimated weight change when flying direct	N Nigeria– Tripolitania
5. Blackpoll Warbler *Dendroica striata*	0.6	Estimated weight change when flying direct	Massachusetts– Bermuda
6. Song Sparrow *Zonotrichia melodia*	0.7	Weights before and after migration night	Massachusetts
7. American thrushes *Hylocichla* spp.	1.8	Weights at different times during the night	Television mast, Illinois
8. Veery *Hylocichla fuscescens*	1.3	Weights at different times during the night	Lighthouse, Ontario
9. Ovenbird *Seiurus aurocapillus*	1.0	Weights at different times during the night	Lighthouse, Ontario
10. Nine different American passerines, 8–33 g weight	0.9	Weights at different times during the night	Lighthouse, Ontario
11. Tennessee Warbler *Vermivora peregrina*	1.8	Weights at different times during the night	Television mast, Wisconsin
12. Chaffinch *Fringilla coelebs*	2.2	Weights at two ringing stations 50 km apart (birds' air speed 40 km/h)	Baltic States

methods: (1) calculation of mean weights for populations of migrants before and after a lengthy direct flight; (2) calculation of mean-weight changes in birds which collide with or are caught at television masts and lighthouses during different times of the night (and which by then have presumably been flying for different lengths of time since beginning to migrate at dusk) or which are caught in the evening before and in the morning after a migration night; (3) calculation of mean-weight differences between birds that are caught at different points along a migration route on which the birds do not make a stop-over between the trapping stations. There are many uncertain factors in the various studies, but despite this the results from most field studies agree fairly well with the theoretical predictions. Some investigations, however, give surprisingly high values for weight loss in migrants, especially one study of the Chaffinch's autumn passage along the Baltic Sea coast in the Baltic States. Has the Chaffinch, in its capacity as a short-distance migrant, a poorer fuel economy than species which migrate in protracted flight stages? This question must be left unanswered. A contributory cause of rapid weight reduction may be the water losses that the birds incur if they migrate at low altitude where the air temperature is high.

Which factors limit the amount of fat a bird can store? Here two things are of great importance: the capacity of the flight muscles and the flight transport costs.

Increasing fat percentage is accompanied by increased weight and greater air resistance (drag). Both these factors contribute to raising the bird's flight power. The capacity of the flight muscles sets a 'ceiling' on how much flight power the birds can stand. Large birds have narrower margins than small birds in this respect and therefore cannot store up extremely large fat reserves. Fat percentages as high as 40% and 50% do not occur in big birds, so calculations of flight distance and flying time for these situations have been omitted in figure 105. Flying with 35% fat content requires twice as much flight power as flying without fat reserves; 50% fat percentage calls for three times as much power.

In order to cope with the increased flight-power demands imposed by high fat percentage, the birds strengthen their flight muscles before the migration. In spring, large numbers of Yellow Wagtails stop off at Lake Chad south of the Sahara in Africa to put on fat reserves prior to the long flight northwards across the desert. The birds increase their body mass from an average of 14 g to 25 g. A detailed study has shown that the mass increase cannot be attributed to the fat reserves alone. A small proportion, approximately 1 g, is due to strengthening of the flight muscles. The flight muscles in Yellow Wagtails without extra fat have a mass of 2–2.5 g, but birds with large fat deposits have built up the flight muscles so that the mass reaches a good 3 g. Similar muscle accretion has been demonstrated also, though not in such detail, for other migratory species, for example Willow Warbler, Sedge Warbler, Reed Warbler, Whitethroat, Sand Martin and Redpoll. The strengthened flight muscles may also be good to have as reserve energy, when the fat has been used up and the birds no longer have need of extra powerful flight muscles. Whether the birds actually do break down the 'excess muscle tissue' as they use up their fat reserves during the migratory flight has so far not been investigated.

Too high a fat percentage means poor transport economy, for the birds must pay extra energy for the transport of the fat reserves themselves. For a bird which already has a fat deposit equivalent to 50% fat percentage, 1 g of fat lasts for less than half as long a flight distance as it would if the bird had not had any fat deposits. Owing to this 'incremental rate of taxation' on energy, a migratory journey requires the least total energy when it is split into many short flight stages so that the birds need to deposit only small fat reserves before each stage. A bird which builds up 50% fat percentage and then flies until the fuel runs out consumes approximately 40% more total energy than a bird which covers the same distance by building up 10% fat content, flying as far as this fat reserve lasts out, putting on the same amount of fat again and then flying another stage, and so on until after six or seven stages the whole distance has been completed.

This method of saving flight energy cannot of course be used when migrants are travelling across ecological barriers, over seas, deserts or glaciers where they cannot stop off and renew their fat supplies. In this case the birds are compelled to deposit as much fat as will be enough for a safe direct flight.

The method of saving flight energy by migrating with small fat reserves in short stages has several drawbacks. Great demands are put on the birds' ability continually to find food-rich stop-off sites along the migration path. Getting properly established at a new stop-over site can in addition take time and involve various risks. Many studies show that migrants generally lose weight during the first days at a new stop-over site. A detailed study at one such site in Texas for Northern Waterthrushes (*Seiurus noveboracensis*) showed that the resting birds established a territory and defended it against members of their own species. Birds with a territory put on weight and increased their fat deposits, while newly-arrived individuals which had not yet succeeded in finding an unoccupied territory went down in weight. Normally it took two or three days before a newcomer was able to set up a territory and begin to 'fill up' with fuel for the next flight stage.

The size of the fat supplies that the birds deposit when they migrate across 'ecologically hospitable' areas is governed by a compromise between the need for the total flight energy to be minimal and the need for the loss of time and the extra risks that using many fat-depositing stations brings to be avoided. Different species choose different strategies. Some 'pick their way forward' in small stages on modest fat resources, others make use of only one or a few fat-depositing stations.

When the migrants have got underway with putting on fat they eat in earnest and increase in weight by up to 10% of their normal body weight per day. As a rule, however, the daily rate of weight increase is between 3% and 6%. At this rate it takes a good week to attain a fat percentage of 25% and about three weeks or a month to double the body weight to 50% fat percentage.

The ability of migratory birds to fly non-stop across ecological barriers is one of the most impressive within the animal kingdom. Not so long ago bird migration across the North Sea, a sea-crossing of between 500 km and 800 km which forces landbirds to make an unbroken flight for 10 to 20 hours, was regarded as a minor miracle. With today's knowledge this no longer appears so impressive. Radar studies have revealed that a multitude of different birds take part in an intensive migration traffic in all directions across the North Sea, night and day during all the months of the year. The North Sea has shrunk when seen in the perspective of later studies in other parts of the world, where migrants fly without a break for three or four days, between 3000 km and 4000 km. To give an insight into the circumstances surrounding long flights and the fat-storing associated with them, I have selected a few examples.

Crossing the Mediterranean Sea and the Sahara

Migratory birds which travel to and from winter quarters in Africa have to cross vast areas of desert and sea (figure 106A). On the way south in late summer and autumn the landbirds meet with a Mediterranean area that is characterised by the heat and drought of summer. The North African coastal regions and the Middle East in particular are severely dried out and have little nourishment to offer those birds which require large quantities of food in order to put on fat before the impending flight across the Sahara desert. The landbirds therefore stop off mostly in the northern Mediterranean area. In Portugal and on the Italian mainland, for example, there are plenty of lush woodlands and marshlands.

Here the birds put on fat deposits sufficient for the journey across both the Mediterranean Sea and the Sahara. As is clear from figure 106A, it is a matter of distances of between 3000 km and 3500 km before they can renew their fat resources in the savanna regions south of the Sahara. When the birds reach there they are met with fairly favourable feeding conditions. After the summer rains north of the equator, the lands along the southern edge of the Sahara desert turn green.

The birds usually cover the whole distance from the northern Mediterranean region to the African savannas in a single long direct flight (but see below). Only if they encounter unfavourable winds are they forced to make an interim landing in North Africa or at one of the Sahara desert oases. But more often than not the winds are on the birds' side. Over the Mediterranean Sea in autumn the winds are variable, with a slight predominance of northwesterlies. Here the birds have the chance to await a favourable wind situation before they set out. Over the Sahara, which lies in the trade-winds belt, north and northeast winds

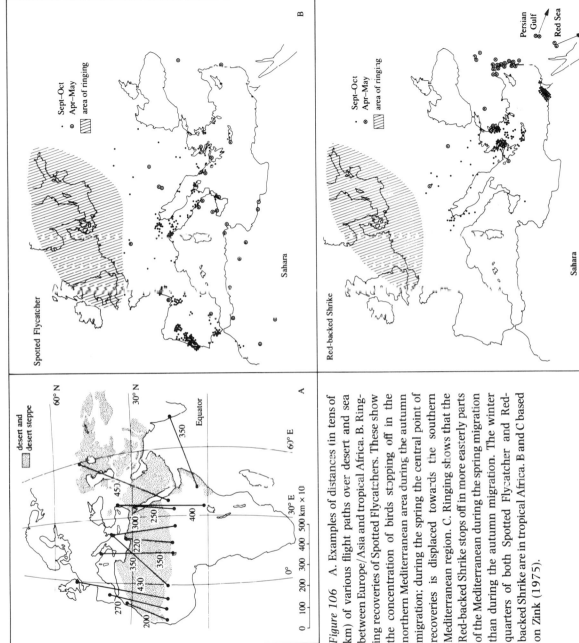

Figure 106 A. Examples of distances (in tens of km) of various flight paths over desert and sea between Europe/Asia and tropical Africa. B. Ringing recoveries of Spotted Flycatchers. These show the concentration of birds stopping off in the northern Mediterranean area during the autumn migration; during the spring the central point of recoveries is displaced towards the southern Mediterranean region. C. Ringing shows that the Red-backed Shrike stops off in more easterly parts of the Mediterranean during the spring migration than during the autumn migration. The winter quarters of both Spotted Flycatcher and Red-backed Shrike are in tropical Africa. B and C based on Zink (1975).

predominate. The birds' prospects of locating good tail winds, between 5 m/s and 10 m/s, at an altitude layer between the ground and 2000–3000 m up are good. Consequently a small bird with an air speed of 35 km per hour can generally count on getting wind assistance for an average travel speed of 50–60 km per hour. This means that the entire long-distance flight across the Mediterranean Sea and the Sahara takes 50–70 flying hours and requires that the birds set out with at least 30–40% fat percentage (see figure 105).

A fascinating discovery was reported recently that calls into question the commonly accepted view of a single non-stop flight across the Sahara: a great many migrant birds have been found to alight in the desert during the daylight hours and hide quietly and motionless in the shade behind rocks, in crevices and holes, and in the few bushes and trees that are to be found (some have even been found in wrecked cars and behind petrol-barrels along the desert roads). In an experiment, a few tamarisk bushes were brought along and temporarily 'planted' in the desolate Libyan sand desert during the autumn migration period. In these bushes, many small migrants, on some days more than 100 individuals, mostly Willow Warblers and Lesser Whitethroats, landed at daybreak to seek shade during the day, only to resume their flight at nightfall. They all had large fat deposits, more than sufficient to complete the journey across the Sahara. Hence, it was not on account of energy problems that the migrants landed, but presumably in order to maintain their water balance during the long desert crossing (Biebach *et al.* 1986).

In the spring the birds are faced with a more troublesome situation. The drought is now severe on the steppes south of the Sahara, and migrants concentrate in the verdant lands still remaining around rivers and lakes, such as in Senegal, Niger and Chad, in order to put on fat before the Sahara crossing. These limited areas, however, will not suffice for all, so many are forced to start flying from areas a good bit south of the drought-ravaged steppe zone along the Sahara's southern edge. Some species of small birds and waders put on such exceedingly large fat reserves in spring in equatorial regions of East Africa (Kenya, Uganda) that it is probable that they make the long flight northward from as far south as these areas.

The biggest problem during the spring passage across the Sahara is the winds. This is because over the desert north and northeast trade-winds predominate, as in the autumn. The fact that these head winds predominate does not, however, mean – luckily for the migrating birds – that they monopolise the wind situation. Occasions arise when the wind turns. Particularly good chances of getting tail winds occur if the birds are flying on a northeasterly course above 2000 m altitude. In Egypt and the Middle East, influx of southerly desert winds is a regular spring phenomenon.

Great demands are placed on the birds in spring. They must set off across the Sahara at the correct time, just when the winds are with them. The whole time the risk of encountering head winds is imminent. If this happens it can mean disaster, for small birds are not able to negotiate flying across the Sahara in head winds. A last resort for birds that run into unfavourable winds may be to fly at low altitude, only a couple of hundred metres or less above the desert, where more often than not the winds are still, particularly at night.

Of great value to the spring birds is the fact that they can land and rest as soon as they have passed the desert, since the winter rains in the Mediterranean region provide greenery and good stop-over sites the moment they reach North Africa and the Middle East. There are many reports of large flocks of migrants arriving completely emaciated in North Africa – telling examples of how hazardous the flight across the Sahara is. The commonest distance for this spring flight is 2000–2500 km. If the birds' speed is normally between 35 km per hour (in neutral winds) and 50 km per hour (periodic tail winds), then the flying time becomes 40–70 hours. As in the autumn, therefore, it is necessary for the birds in spring, despite the shorter flight distance, to begin their long-distance flight with a fat percentage of 30–40%. Birds which venture on a 4000-km flight from East Africa to the eastern Mediterranean, even if they start with a 50% fat content, must rely on a certain amount of assistance from tail winds.

An example of the concentration of resting landbirds in the northern Mediterranean area during the autumn and in the southern Mediterranean area during the spring is given by ringing recoveries of Spotted Flycatcher in figure 106B. The distribution pattern is also highlighted by the fact that the hunting of small birds is intensive in the autumn in many south European countries but in the spring in North Africa. The seasonal differences in hunting habits are of course due for the most part to the fact that the migrants stop off at different places in spring and autumn. The Red-backed Shrike, which is a bird of dry country and which feeds on insects and occasionally also on small birds, does not behave in the same way during the autumn crossing of the Mediterranean area as it does during the spring (figure 106C). In autumn, for the journey south to East Africa, it puts on fat mainly on the islands in the Aegean archipelago, on Crete and in the Nile delta at Alexandria. In spring, however, the shrikes pass through staging-posts in the Middle East, some even flying in the direction of the Persian Gulf. The explanation for this difference is seemingly twofold. On the one hand, the eastern regions, which in summer and autumn are extremely arid (normal rainfall during June–September in Jerusalem is 1 mm and in Baghdad 0 mm; this compares with for example rainfall in Athens, 46 mm), offer exceptional abundance of food in the spring when 'the desert blossoms' following the life-giving winter rains (rainfall in December–April in Jerusalem is 431 mm and in Baghdad 125 mm). On the other hand, the spring winds over the Sahara and the Arabian peninsula are generally more favourable for flying northeast rather than due north.

Unfortunately, analyses of weights and fat percentages of tropical migrants have not been made at as many staging sites as we might wish for. The picture we get of the different species' fat-storing habits is therefore fragmentary. In figure 107 the maximum fat percentages that some warblers regularly achieve in different staging areas are shown diagrammatically. The Garden Warbler may serve as a typical example.

In north Europe Garden Warblers store only moderately large fat reserves in late summer and autumn; the fat load reaches a maximum of 20%. In consequence most fly in stages to the north Mediterranean area; here they must 'fill up' substantially with fat, but no detailed data exist on this. The Garden Warbler's highest fat percentages have been recorded in central Nigeria and at Lake Chad in spring: the species' normal body mass without fat is about 17 g, but individuals are regularly encountered here which weigh 30 g, occasionally even a little more – the heaviest Garden Warbler at Chad weighed 35.5 g. These big fat reserves are used up entirely on the long flight northwards over the Sahara. In the southern Mediterranean region the birds' mean mass in spring is as low as 15–17 g, and emaciated individuals weighing only 13–14 g occur regularly; in this region the spring birds 'bunker' only a couple of grams of fat before, with a 10–15% fat percentage, they continue on northwards, obviously in short stages and with several stop-over periods in Europe until they reach the breeding sites.

The Garden Warblers which arrive in Chad and Nigeria in autumn have on the whole used up all fat after the long flight across the Mediterranean Sea and the Sahara. They confine themselves to 2–3 g of refuelling before continuing south. Many of the birds which pass through Lake Victoria in East Africa in November–December are on their way to distant winter quarters, perhaps as far south as Botswana and Zimbabwe judging from regularly occurring fat loads of around 25%. When the birds stop off in the same area on their way

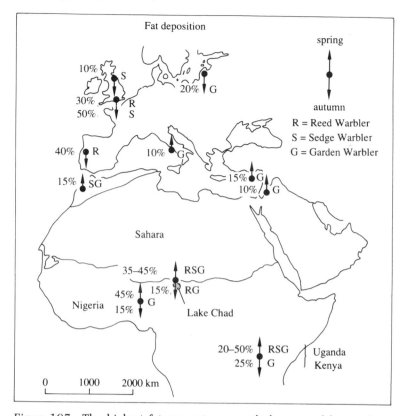

Figure 107 The highest fat percentages regularly attained by Garden, Reed and Sedge Warblers at different staging sites during spring and autumn migration. In some studies the amount of fat reserves has been determined using physiological measurements, but in most cases the fat percentage has been estimated using the birds' body weight as a guide. Many Sedge and Reed Warblers leave East Africa in the spring with modest fat reserves. At certain staging-posts with exceptionally good food supply, such as Lake Nakuru and the Ahti River in Kenya (Sedge Warbler) and Lake Edward in Uganda (Reed Warbler), birds are also found which build up their fat resources to around 50% fat percentage; with so much fat the birds have the means to fly in tail winds non-stop direct to the Mediterranean region.

north in April, some put on even more fat and reach weights of around 25 g (equivalent to over 30% fat percentage). These birds have so much fuel that they can reach a staging area along the southern edge of the Sahara in one single flight stage. Many other Garden Warblers leave East Africa at the same time with more modest fat loads to fly north in stages towards the Sahara.

The Garden Warbler illustrates clearly a characteristic feature possessed by most tropical migrants. They fly across 'ecologically hospitable' regions in stages and on modest fat reserves. The very highest fat percentages are found only when the birds are faced with unavoidable long-distance flights.

There are, however, some very interesting exceptions to this pattern, one of which has been revealed by a detailed comparison of the autumn migration habits of British Reed Warblers and Sedge Warblers. These two species are closely related and virtually the same size; their normal body mass is about 10 g. They live in similar habitats, in reedbeds and bushy areas in marshes and on lakeshores. Despite this, their migratory habits differ radically. In England many Reed Warblers reach a fat percentage of up to 30%, enough for them to be able to fly direct to the next big fat-storing region, which according to ringing results is in Portugal. Some, however, mainly juveniles, leave Britain with more limited fat resources and travel to Portugal in several stages. This is illustrated by a good number of ringing recoveries of juveniles during the autumn along the whole route from England to Portugal. At the Portuguese staging sites the Reed Warblers steadily increase their weight; the daily average weight increase is 0.25 g. When they weigh between 15 and 20 g they are ready for a long flight to West Africa south of the Sahara. Many travel along such a westerly flight route that they pass over the sea west of the Strait of Gibraltar before they meet the desert in western Morocco.

In the matter of distribution of ringing recoveries during the autumn migration British Sedge Warblers differ markedly from British Reed Warblers. Many Sedge Warbler recoveries have been reported from areas immediately south of England, in northwest France. Further south, in southwest France and throughout the Iberian Peninsula and in Morocco, however, recoveries are very few and far between. As early as the beginning of the 1960s, a ringer had got on the track of the explanation (Gladwin 1963). He found that some Sedge Warblers in southern England increased their weight during August to as much as 20 g, occasionally even 1–2 g above that, the equivalent of a fat load of around 50%. He drew the conclusion that these Sedge Warblers fly in a single hop from southern England to West Africa south of the Sahara, more than 4000 km without a stop.

This conclusion has gained support from later studies (Bibby & Green 1981), in which it has been shown that southern England and northwest France is a region where the Sedge Warblers rapidly put on weight and reach extremely high fat percentages. Interestingly enough, Sedge Warblers in northern England leave their breeding sites with negligible fat reserves, obviously to make for the central fat-depositing region approximately 500 km to the south.

One prerequisite if the Sedge Warblers are to succeed in flying direct from southern England to West Africa is that the flight take place for the most part in tail winds. The most difficult part is probably choosing the right occasion for the flight over west Europe. By the time the birds are flying across the Sahara they have generally, as already mentioned, good prospects of obtaining tail winds. As the mean speed is 50–60 km per hour and the total distance 4300 km, the flight lasts between 70 and 90 hours. For more than three days and three nights, during which the earth turns more than three revolutions on its axis, the Sedge Warblers fly non-stop over land, sea and desert.

Why does the Sedge Warbler, unlike the Reed Warbler, make such a magnificent long flight, direct from 50°N to tropical West Africa? In the studies mentioned above, the researchers have managed to come up with the interesting explanation that in late summer and autumn Sedge Warblers specialise in feeding on aphids in order to build up fat reserves. The species concerned is the plum-reed aphid, which in summer disperses from blackthorn bushes and plum trees to reeds. In the reedbeds the aphid numbers explode in suitable weather, and thousands of aphids can be found on a single reed stem. When the reeds have flowered and come to maturity in late summer, they no longer provide food for the aphids. The latter then return to their primary hosts, mainly to the blackthorn, where the autumn's last generation of female aphids lays eggs before the winter.

The reeds develop early in south Europe; in Portugal, for example, the aphid supply begins to diminish as early as the middle of July. When the time comes for the Sedge Warblers to migrate in August, there are consequently not many aphids to be harvested in south Europe but in more northerly areas, such as southern England and northern France, the aphids are at their peak of abundance. The growth progress of the reeds and the aphid presence associated with it thus explain in a nutshell the extreme long-distance flight of the Sedge Warblers.

The Sedge Warbler is a terrific consumer of aphids while it moves from stem to stem through the reedbeds. It has been timed to take an average of just over 40 aphids per minute where aphid density is high; Reed Warblers that concentrated on catching aphids in the same clump of reeds managed only half as high a catch rate. The Sedge Warblers' efficiency when it comes to catching aphids is also reflected in their rapid depositing of fat. In good aphid years the birds increase weight by a good 0.5 g per day, a rate which permits them to achieve a 50% fat percentage in less than three weeks; some individuals put on up to as much as one whole gram of fat per day. The Reed Warblers' more modest rate of fat-deposition was mentioned above. At Lake Chad, where midges in the bulrush stands are the Sedge Warblers' most important food in spring, they increase in weight by around 0.2 g per day.

The Sedge Warblers concentrate in late summer in those clumps of reeds which hold the very highest numbers of aphids. Here there are no territorial divisions; the birds sometimes perch only a few metres from one another. At a lake in southern England with a reedbed barely 40 hectares in extent, up to 1000 Sedge Warblers were present during one period in August.

The Reed Warbler's feeding ecology during the autumn migration differs markedly from that of the Sedge Warbler. The Reed Warbler has a broader food spectrum; it is certainly not averse to taking aphids at times, but it also catches a lot of flying insects which it hunts in brief pursuit flights. Since Reed Warblers feed to a large extent on prey that are sensitive to disturbance, they spread out and establish individual territories at the stop-over locations. At these times they choose not only reeds as feeding habitats but also bushy areas, at times in fairly dry terrain.

Behind the different migratory habits of the Sedge Warblers and Reed Warblers, therefore, there lie a number of interesting ecological differences. And several fascinating questions remain to be answered. How are the Sedge Warblers affected in those years when the aphid presence in the reeds is very sparse? Does this lead to many Sedge Warblers not surviving the autumn migration and to the breeding population in the following spring being smaller than normal? Which fat-storing strategy do Scandinavian Sedge Warblers put into practice? Do they perhaps move south in late summer to Germany–Austria to feed on aphids in the reeds there and then fly direct to Africa south of the Sahara? (There are very few ringing recoveries of Scandinavian Sedge Warblers made in the Mediterranean region.) Is it conceivable that some Sedge Warblers become extremely fat while still in south Scandinavia and venture on a super-long flight to the tropics?

Many bird species that breed in Asia have winter quarters in Africa. Like the birds from Europe, they must make strikingly long flights to and from the African continent (figure 106A). The flight path across the Indian Ocean between India and East Africa is used by wagtails, wheatears, bee-eaters and Eastern Red-footed Falcons among others. Presumably it is easier to fly over the sea there than we may at first imagine, for in the autumn the northeast monsoon provides good tail winds, and at an opportune time for the spring migration the wind turns so that the birds receive assistance from the southwest monsoon.

Even though positive proof is lacking, there are several factors which indicate that in autumn Siberian birds make a long flight all the way from Kazakhstan in the Soviet Union to the northeastern parts of tropical Africa, nearly 4500 km! In the intervening areas there are large inaccessible desert regions. Only refuges of forests, floodlands and marshlands escape being dried out during the summer, and these are far too limited to suffice reasonably as staging sites for the immense numbers of birds which are on their way from Siberia to Africa. There are reports of an intensive autumn passage along the desert-like shores of the eastern Caspian Sea. Marsh Harrier, Hobby, Black Kite, Tree Pipit, Redstart, Spotted Flycatcher, and several different *Acrocephalus*, *Phylloscopus* and *Sylvia* warblers have been observed for example. Mass deaths have been reported among birds which, because of poor weather, have been forced to land on the desert shores. In autumn, there are more often than not tail winds blowing all along the route from the Caspian Sea to Africa, a factor that makes it likely that the birds actually do manage to fly the whole distance without making any stops on route.

Crossing the Atlantic from North to South America

It has long been well known that large numbers of birds migrate over the Gulf of Mexico. Using the moonwatching method it was possible in early days to record the nocturnal migrants when they set off in autumn over the south coast of the United States to fly at least 1000 km over the open sea (figure 108). The birds return the same way in the spring. At that time many depart from Central America (probably mainly from the Yucatan Peninsula) at dusk. Normally they do not get to the south coast of North America until after about 20 hours' flying, i.e. in the afternoon of the following day. Here, in full daylight, warblers, thrushes, Bobolinks, tanagers and orioles can be seen arriving over the sea in flocks; sometimes they dive down steeply as soon as they have passed the shoreline and land in the woods on the coast.

The size of the fat deposits the birds have when they leave North America's south coast in autumn for a long flight across the Gulf of Mexico has been studied by collecting up and analysing birds that have collided with a television mast in northwest Florida during the night. The fat percentages often amount to between 40% and 45%. As well as the bird species mentioned above, the Ruby-throated Hummingbird also has this high fat percentage. It normally weighs no more than 2.5 g, but before it migrates over the sea it increases its weight with a good 2 g of fat. To fly more than 1000 km across open sea, freely exposed to the winds, is indeed a considerable achievement for a hummingbird.

The high percentages of fat in the autumn migrants on the north coast of the Gulf of Mexico suggest that many make a flight that extends considerably farther southwards than to the Yucatan Peninsula, 'only' just over 1000 km away. With a 40% fat load the birds should normally reach around 2500 km in neutral winds. This is the equivalent of a long-distance flight direct to Colombia in northwesternmost South America. Radar studies at Miami have demonstrated that in the autumn the majority of small birds set out from the south tip of Florida in a south-southeasterly direction, and many of them evidently fly the 2000 km direct to the South American mainland.

The long-distance flight performances over the Gulf of Mexico pale, however, in comparison with the autumn migration over the western Atlantic, from the northeastern United States and eastern Canada direct to South America. Originally it was thought that only one species, the American Golden Plover, regularly made such an Atlantic flight. Radar studies in more recent years, however, have revealed a far greater migration intensity than expected. It has been clearly established that significant numbers of passerines and waders of various kinds are represented among these far-flying migrants. Using radar studies, endeavours have been made on a large scale to map this impressive bird migration. Both high-power surveillance-radar stations and tracking radar have been made use of at the

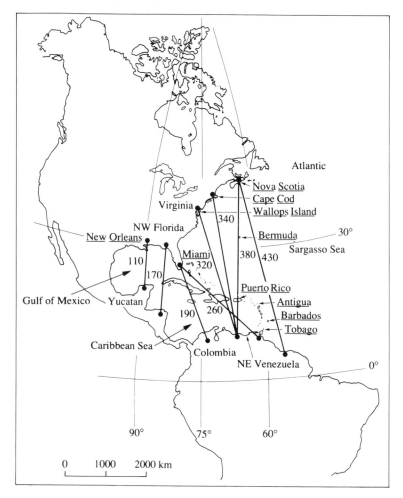

Figure 108   Examples of distances (in tens of km) of different flight paths over sea between North America and Central or South America. Radar studies of bird migration have been carried out at sites for which the name has been underlined. As regards fat deposition, birds passing NW Florida in autumn and waders at staging sites in Nova Scotia (autumn) and in NE Venezuela (spring) have been analysed.

various sites that have been indicated in figure 108. In addition, during several autumns vessels have cruised far out over the open Atlantic so that marine-radar observations could be made.

In actual fact, today's studies of bird migration over the West Atlantic have led to a rediscovery of an intensive bird passage, for it was seen long before – by Columbus! During his Atlantic voyage he arrived at exactly the right time to come across the birds over the ocean. On 4 and 7 October 1492, large flocks of landbirds were seen when the ship was approximately midway between Bermuda and Puerto Rico. Further to the southwest Columbus's crew witnessed a night during which they continually heard calls from overflying bird flocks, and sometimes they could even see the birds as silhouettes against the moon. According to bird observations in more recent times from vessels in the West Atlantic and studies of migrants stopping off in Bermuda, a good many, widely differing species take part in this migration. One major group of Atlantic travellers is waders, from Least Sandpipers right up to godwits (see below). Many species of passerine also make this long flight: more than 30 species were observed or caught during the ship's-radar studies in the waters around Bermuda. Among the passerines regularly encountered were Ovenbird, American Redstart, Barn Swallow, Bank Swallow (Sand Martin), and several species of wood-warblers of the genus *Dendroica*, including the Blackpoll Warbler, this last a well-studied species where migration is concerned. Even the Great Blue Heron, the Blue-winged Teal, the Common Nighthawk (a species of nightjar) and the Merlin have been met with far out over the Atlantic in the autumn. A Merlin which used the masts of a ship southeast of Bermuda as a resting place for a couple of days was seen on several occasions to fly low over the sea and hunt flying-fish!

The list of migratory species that cross the West Atlantic in autumn will no doubt grow when further studies are made. Several species, including gulls, rails, cuckoos and kingfishers, are among the host of those suspected of doing so.

The birds fly out over the Atlantic on a southerly or southeasterly course along the entire coastal stretch from Virginia in the south to Nova Scotia in the north. Some land and stop over in the Antilles, but most head on farther towards South America without making any stop. The total flying distance is thus generally between 3000 km and 4000 km. The flight path is not absolutely straight but runs in an arc across the sea. The small passerines usually leave North America on a southerly or slightly south-southeasterly heading, and the majority of them pass around or west of Puerto Rico, where the flight direction is south-southwesterly. The waders depart on an even more southeasterly course and make a wide sweep across the Sargasso Sea before they approach South America at the easternmost Antilles. Their mean direction is in fact due southeast when they leave Nova Scotia and Cape Cod. If we extend this direction on the map, we end up not in South America but in West Africa. When the waders are seen on the radar screens in Antigua, Barbados and Tobago, however, they have swung on to a southerly or southwesterly course.

The birds' change of course over the sea is due to the changing winds during their flight. They depart from North America immediately after a depression with its accompanying cold front has passed eastwards (cf. section 4.8). They then benefit from the northwest winds behind the cold front which give them good speed in a southeasterly direction. The mean speed for wader flocks leaving Nova Scotia is 74 km per hour. South of Bermuda the birds fly into the trade-winds zone, and here stable east to east-northeast winds prevail from which they can derive a certain benefit by veering to a southwesterly direction. We might of course imagine that the birds ought to fly even farther southeastwards out over the Atlantic, so that they could then approach South America on a west-southwesterly course and thereby obtain ideal tail winds within the belt of trade-winds. Their flight path is, however, probably the most favourable compromise from an energy point of view between good winds for flying and the least roundabout route possible to the migration goal.

As mentioned in the previous section, the Atlantic flyers regularly climb to very high flight altitudes in order to get the best possible winds for travel. Despite this, they cannot maintain throughout the flight the high speeds that they have when they set out in good tail winds from North America. On the basis of various radar studies, one researcher has estimated the mean speed for the whole flight at between 35 and 60 km per hour (higher for waders than for small birds). This means that the normal time for the migration over the sea is 86 hours, plus or minus 12 hours. For a three- to four-day flight such as this, substantial fat reserves are required.

This has been borne out by field studies of Blackpoll Warblers at autumn staging sites in

Massachusetts (see figure 109). The warblers arrive there in early or mid September on nights of northwesterly winds. At this point they are almost totally bereft of fat reserves and weigh 10–13 g (average body mass without extra fat is 11 g). During the first few days they disperse in the woodlands, in search of suitable feeding areas. Once they have found such areas they rapidly increase in weight, and after 10–20 days their body mass is 16–19 g. At this level the depositing of fat is generally suspended for a few days, but sets in with renewed vigour in rainy weather during the passage of depressions and quickly takes the birds up to the 20–23 g weight range. When they have reached these high weights they set off as soon as there is an evening of clear and cool weather. What is interesting is that the birds save putting on the last grams of fat until the weather indicates that the passage of a cold front is imminent so that suitable winds for flying can be expected. With a 50% fat load the warblers are ready to fly for about 90 hours (figure 105), just right for 'next stop South America!'

Some Blackpoll Warblers, especially juveniles, settle in less suitable stop-over environments, increase slowly in weight and leave when they weigh only 14–20 g. Very little is known of the fate of these birds. Many presumably turn back at the sea coast, where ringers periodically catch juveniles which are migrating back and landing in order to make a fresh attempt at putting on fat. Those which carry on farther out over the open sea with insufficient fuel reserves are doomed.

The autumn migration of waders over the Atlantic has been studied through ringing as well as analysis of the birds' fat resources at the big wader staging sites at the Magdalen Islands and Sable Island in Nova Scotia. The distribution of ringing recoveries (including many observations of colour-ringed birds) fits well the migration patterns from radar studies (figure 109B). Certain wader species still have much of their autumn migration left to do even after they have reached the north coast of South America; Knots and White-rumped Sandpipers regularly winter as far south as Tierra del Fuego, the southern tip of South America.

The migrants do not return over the West Atlantic during the spring; the winds are far too unfavourable for long-distance flying in a northerly direction. Waders which stop off and put on fat reserves on the Venezuelan coast in spring return instead over the Caribbean Sea and the West Indies islands. When they have reached Florida or the south coast of North America they continue north overland (some species also follow the east coast from Florida northwards). During the spring migration the waders in Venezuela are therefore faced with a long flight of at most 2500 km to Florida. Owing to favourable trade-winds, considerably smaller fat reserves are needed now than were required during the autumn's 4000-km flight from Nova Scotia.

A comparison of waders at these two staging areas shows, as expected, a significant difference in the fat percentages (table 29). During the spring in Venezuela the waders set off with a fat load of 20–25%; the corresponding figures before the autumn journey from Nova Scotia are 40–45%.

The radar screens at Nova Scotia in autumn show how the waders start to migrate out over the sea from staging sites in the coastal region. But wader flocks can also be seen arriving from sites far inland and continuing straight out southeastwards over the Atlantic without stopping. How far away have these waders begun their journey? Is it possible that they have the winds behind them and enough fat reserves to fly more than 5000 km direct from the Hudson Bay area of arctic Canada to South America?

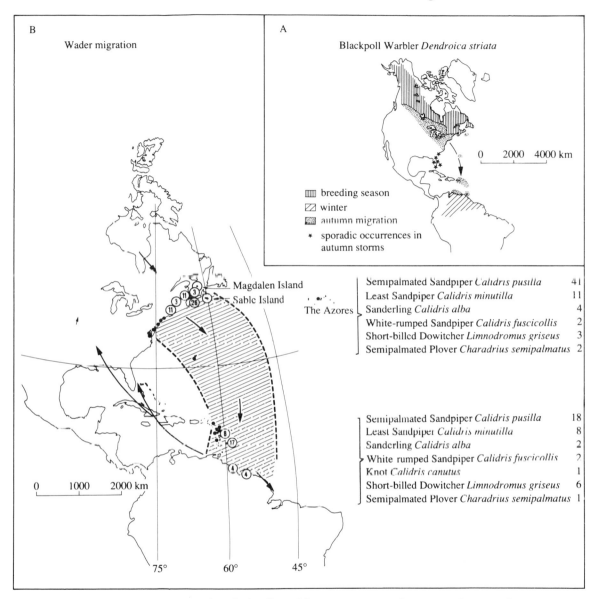

Figure 109 A. In autumn the Blackpoll Warbler migrates over the West Atlantic and is regularly found stopping off in Bermuda, in the Lesser Antilles and on the islands off the South American coast. The species is seen in Florida and the Bahamas only in association with autumn storms, when the birds are driven off course by east winds. Based on Nisbet (1970). B. Recoveries of waders ringed at autumn staging sites on Magdalen Island and Sable Island in easternmost Canada. In autumn the waders migrate over the Atlantic in the flight corridor indicated but return north in spring along a more westerly flight path via the south coast of North America. All recoveries except one are from the autumn; the exception (a White-rumped Sandpiper, found in Kansas in May) is indicated by an upward-pointing arrow. Note the recovery of a Semipalmated Plover on the Azores. Note also the occurrence of American migrants in Europe as per section 4.9. Based on McNeil & Burton (1977).

Table 29. *Highest fat percentages (% fat of total body mass) regularly occurring in waders at staging sites in Nova Scotia during the autumn migration and in Venezuela during the spring migration. In autumn Greater Yellowlegs do not migrate over the Atlantic direct to South America but follow the east coast of North America southwards. Most Lesser Yellowlegs also follow this route; a few probably cross the Atlantic during the autumn. Based on McNeil & Cadieux (1972)*

| Species | Body mass without fat (g) | Fat percentage | |
		Nova Scotia (autumn)	Venezuela (spring)
Greater Yellowlegs (*Tringa melanoleuca*)	150	20	20
Lesser Yellowlegs (*Tringa flavipes*)	70	30–35	20–25
Knot (*Calidris canutus*)	130	45	
White-rumped Sandpiper (*Calidris fuscicollis*)	30	40	20
Least Sandpiper (*Calidris minutilla*)	18	45	25
Semipalmated Sandpiper (*Calidris pusilla*)	20	40–45	25–30
Short-billed Dowitcher (*Limnodromus griseus*)	80	45	
Hudsonian Godwit (*Limosa haemastica*)	250	30–35	

The Turnstone flies a great distance

In order to underline the fact that non-stop long-distance flying by no means occurs only between Europe/Asia and Africa and between North America and South America, my idea was to give a further example of what a migratory bird is able to accomplish. Which bird should I choose? My thoughts went to Snow Buntings which fly across the Arctic Ocean, to Wheatears which travel across the Atlantic, to Brent Geese on route direct between Alaska and California, and to two species of cuckoo in the southern hemisphere which breed in New Zealand and fly more than 3000 km across the sea to winter quarters in the Solomon Islands and the Samoan Islands. Then finally I decided on another species altogether, on the Turnstone, a renowned long-distance flyer.

Can this really be right, some readers may perhaps ask themselves. When the Turnstone is brought up in conversation, their thoughts go to the tranquil rocky islets and skerries of the summer, the typical breeding habitats of the species in southern Scandinavia, or to the stony shores of Britain, where the Turnstone is a characteristic bird in the winter. But this wader also has other sides to its character which are quite different.

The Turnstone is first and foremost a high-arctic breeding bird and is found along the coasts around the whole North Pole basin, in the northern parts of the Bering Sea and along the coasts of Scandinavia and Finland. The Turnstones of north Europe migrate mainly to the coasts of West Africa for the winter. Those that are found in winter at the Cape of Good Hope (see section 3.1) probably come from the Siberian coast of the Arctic Ocean and travel to and from Africa via the Caspian Sea and the Black Sea. This long journey, about 13 000 km in all, they probably cover in only three or four stages, if we are to judge from the considerable fat deposition preceding the spring departure from the southern tip of Africa. The Turnstone's normal weight without fat is approximately 100 g. Birds arriving at the Cape of Good Hope at the end of September generally have not used up all their fat resources during the final stage of the autumn migration and they weigh on average just over 110 g. During the height of the

South African summer, in December and January, the body weight drops to a minimum; the birds' fat percentage averages only 5%. In March the storing of fat begins in readiness for the spring migration, and the body mass increases to 165 g before the departure at the end of April. With fuel reserves equivalent to a fat percentage of 40%, the Turnstones can fly at least 3000 km, and a good bit farther in tail winds, before they need to stop off for a fresh bout of fat-storing.

Many Turnstones, mainly from breeding sites in Greenland and on the islands in high-arctic Canada, winter in west Europe. Their migratory habits accord quite well with those of the Knot which were described in section 3.1. Like the Knots, the Turnstones use Iceland as a station for putting on fat during the spring migration. Before the departure from Iceland at the end of May the birds' mean weight is 158 g; the heaviest weigh over 170 g. The fat reserves are greater than those required simply for the birds to be able to fly to the breeding sites; this applies even to those which have the longest flight path, between 2500 km and 3000 km over sea, pack-ice zone and the inland ice of Greenland to Ellesmere Island in Canada. Turnstones which have just arrived at Ellesmere have a mean weight of around 115 g, equivalent to nearly 15% fat load. The surplus fat is a valuable adaptation enabling the birds to weather the initial period when food is scarce on the breeding sites; they can of course be subjected to spring frosts and snow storms. The Turnstones have flown far beyond the point of no return and have no chance of going back to areas with better spring weather. Such starvation periods occur fairly often, and in extreme cases, when the birds' additional energy reserves are not sufficient, they turn into catastrophes. In unusually inclement June weather, dead and emaciated Turnstones have been collected on Ellesmere Island with weights of between 50 g and 70 g.

East Siberian Turnstones have a particularly striking migration pattern. As a result of extensive ringing and colour-marking on the Pribilof Islands in the Bering Sea, a major fat-depositing station during late summer and autumn, an exceptionally good picture of the migration has been obtained (figure 110). (More than 16 000 Turnstones were ringed over four years.) Ringing recoveries show that Turnstones come to the Pribilofs from the coasts along the east Siberian Arctic and from St Lawrence Island and the Siberian coasts of the Bering Sea. They arrive at the Pribilofs at the end of July and gather in large flocks to feed on blowfly larvae at seal-killing sites. Thanks to this nutritious fare they increase rapidly in weight, from an average of 115 g to 155 g 14 days later; the heaviest Turnstone weighed 195 g. Some birds are said to become so heavy that they are barely capable of taking off from the ground.

The Turnstones really do need their large fat reserves. On those rare evenings at the end of August when the cloud cover over the Pribilofs breaks up with northerly winds, the birds head off due south across the Pacific Ocean. When they have passed the Aleutian Islands chain, where they do not as a rule make a stop, the Turnstones have almost 3000 km to the next contact with land – the Midway Islands. It is here that many ringing recoveries have been made at the end of August and in September. The most interesting is the recovery of a Turnstone which was ringed on 23 August on the Pribilofs; it then weighed 174 g. On 27 August it was recaptured at the same place; for some unknown reason it had decreased in weight to 155 g. (Had it perhaps made a failed attempt at migrating?) Another four days later it was shot 3650 km away on one of the Midway Islands (weight unknown). If the Turnstone left the Pribilofs in the evening of 27 August, it travelled for a maximum of 88 hours; this gives us a mean speed of 41 km per hour at the very least. If it delayed its departure for 24 hours, then the mean speed was at least 57 km per hour; if it waited for 48 hours, then the mean speed was at least 91 km per hour.

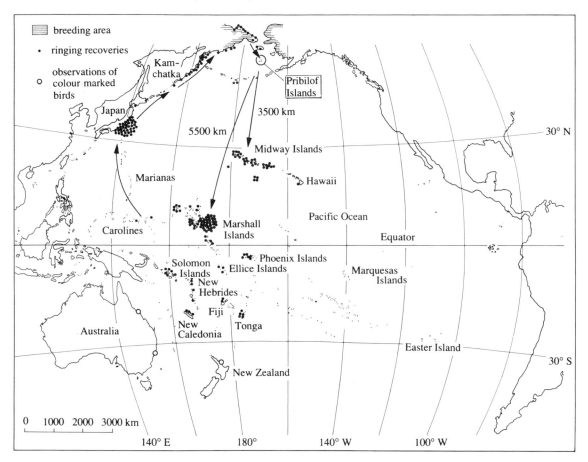

Figure 110 Recoveries of Turnstones ringed in autumn on the Pribilof Islands. The recoveries on the Midway Islands are primarily from autumn, on the Marshall Islands and in the southern half of the Pacific Ocean area from autumn and winter; the recoveries along the coast of Asia (Japan/Kamchatka) are entirely from the spring migration period. Based on Thompson (1974).

The scientists who ringed the Turnstones on the Pribilofs believe that many fly direct all the way to the Marshall Islands, about 5500 km away, from where several recoveries were reported as early as late September/early October. One bird for example was found only 16 days after the date of ringing, and quite a number of recoveries were reported only 20–25 days after ringing. Is it perhaps possible for the Turnstones to get even farther without renewing their fat resources? This is what we ask ourselves when presented with a recovery that was made on the Ellice Islands only 14 days after ringing on the Pribilofs, 7000 km to the north.

The fact that Turnstones were recovered so soon on the Marshall Islands and the Ellice Islands is worth reflecting on. It is not completely out of the question that they made a stop on the way and boosted their fat reserves for a period of at least a week on the Midway Islands. In the absence of evidence, I shall try to make a probability judgement of the Turnstones' chances of flying non-stop for 5500 or 7000 km. Suppose that only the very heaviest birds, those which on the Pribilofs weigh over 165 g and which therefore have a fat percentage of at least 40%, are capable of flying such extreme distances. The Turnstone, like other waders, has comparatively pointed wings and presumably a somewhat better flight performance than that calculated in figure 105. I estimate that with 40% fat percentage it would be able to fly for nearly 70 hours before the fat was used up. The mean speed required for the bird to get 5500 or 7000 km in this time must therefore be 79 or 100 km per hour. If the Turnstone has a flight speed in relation to the surrounding air of about 50 km per hour, then an average tail wind of 8 m/s is required for it to complete the direct flight from the Pribilofs to the Marshall Islands. This in fact lies within the bounds of possibility. The birds start to migrate when north winds sweep down across the Bering Sea. At a latitude of approximately 30° they reach the trade-winds belt and its ideal northeast winds. I would bet a small sum that some Turnstones actually do fly direct to the Marshall Islands. When it comes to flying all the way to the Ellice Islands I am more uncertain. Only an enormously fat Turnstone, with a fat load of nearly 50% and thus scope for 90 hours' flying time, could probably manage such a feat. On the other hand, it is not inconceivable that some enormously fat Turnstones such as this do in fact occur on the Pribilofs. Another reservation which should not be forgotten in this matter is that the last word no doubt has not been said on the theory of bird flight. Perhaps future modifications of the theory will result in our advancing the limits of what we today regard as possible.

The wintering area of the east Siberian Turnstones encompasses the island world of the southern Pacific Ocean, from the Marshall Islands in the north (only a few birds stay through the winter on the Midway Islands) to New Zealand and Australia in the south, where birdwatchers have observed individuals that have been colour-marked on the Pribilofs. Judging from the many spring recoveries of ringed birds in Japan and Kamchatka, the spring migration runs north along the Asiatic coast of the Pacific. In spring, too, long flight stages certainly take place, something which is indicated for example by the infrequency of ringing recoveries from the Carolines and the Marianas.

The tendency to migrate in a 'clockwise loop' in the northern hemisphere can be traced to a more or less pronounced degree in many landbirds and seabirds, and reflects an adaptation to the wind systems – to the belt of westerlies and the trade-winds zone. An adaptation which encourages migration in tail winds is of course of additional importance for species which, like the Turnstone, fly long distances over the ocean.

The rainy-season migration of the Red-billed Quelea

After all these magnificent feats of long-distance flying it may perhaps be appropriate to recover our breath. An example of bird migration over less expansive ecological barriers is the Red-billed Quelea's rainy-season migration in Africa. The beauty in the adaptations of migratory birds to a changing environment is seen no less clearly in the fine-spun effects on the small scale.

The Red-billed Quelea occurs in the African grasslands and in the open savanna forests. It never has need to fly over either sea or desert. The quelea's ecological barrier is roughly six weeks of the early rainy season. To explain this more precisely I must first recount the quelea's lifestyle.

The quelea feeds almost exclusively on grass seeds, which the savanna produces in rich measure. Only prior to the migration and when the birds are feeding their young do they also take insects: nutritious termites, larvae and grasshoppers. During the late rainy season the savanna grass sets its seeds, and the queleas pick these while they are still green and still in the ear and panicles. When the dry season approaches the seeds ripen and fall to the ground. This is the time that the birds' breeding season comes to an end, partly because the supply of insects, suitable for feeding to the young, decreases and partly because it becomes more difficult to get at the grass seeds. True there are immense amounts of seed on the ground, but they are lying underneath the withering tall grass. During this period the queleas sometimes cause great damage to harvests. Flocks of tens of thousands or hundreds of thousands of birds invade fields of rice, wheat or millet. These crops grow more slowly than the wild savanna grasses and the ripening seeds are still available. In order to remedy the damage, studies are being made in many African countries of the queleas' way of life.

As the dry season progresses and the savanna grass is broken down or is grazed by cattle, antelopes and zebras, the ground is laid bare and the seeds become available to the queleas on a large scale. Grass fires over extensive areas also contribute to this, for the seeds are left undamaged when the fire advances rapidly through the dry grass.

When the rainy season comes, the queleas' situation alters radically. The grass seeds germinate, and not until the growing grass has had time to flower and to set new seeds – this takes at least six weeks – can the queleas obtain food again. They are forced to set off on migration. During the dry season the birds gather at communal roosts in reedbeds, bushland or acacia groves, sometimes hundreds of thousands of them congregating at a single roosting site. Almost all then set off together, a few days after the first rain showers have begun to fall thus marking the start of the rainy season.

Where do the queleas migrate to? What we first think of perhaps is that they ought to make off in the direction away from the approaching rain front, so that they could continue to exploit the seed supply in the dry lands. To migrate away from the rain front, however, would be only to resort to a delaying tactic: sooner or later the birds would still be forced to meet the rainy season. Instead the queleas immediately head off straight towards the rain front! Therefore they must travel until they reach regions where the rainy season has already been going on for at least six weeks and where the grass is bearing seeds. In which quarter and for what length of time they migrate varies in different parts of Africa, depending on the direction in which the rain front is moving and its speed.

When the queleas have migrated through the rains they arrive in an environment where the rainy season has entered a later phase. They can then enjoy masses of fresh grass seeds and large insect larvae. The birds remain in this, for them favourable, zone by moving slowly with the belt of rain. As soon as they reach a suitable place they establish a breeding colony. Breeding takes place synchronously and very quickly. The incubation period is record-short, only 9–11 days, and the young leave the nest after only ten days. Not six weeks have passed since the colony was established before the parent birds abandon their fledged young in order to make their way further in the direction of movement of the rain belt. The queleas often have time for two or three breeding efforts during one rainy season, each time at different sites but always in the most favourable environment and during the latter phase of the rainy season. The exceptionally short breeding period is an adaptation that enables the birds to complete many breeding attempts during one and the same rainy season and to breed speedily at a particular site before the grass seeds drop to the ground and the profusion of insects diminishes.

The Red-billed Quelea's rainy-season migration has been closely studied at three widely separated places within the species' large range in Africa (figure 111). The rain front advances slowly across the northern savanna areas and generally reaches Lake Chad in late June/early July. The queleas that are present there need fly no more than about 300 km south, to the valley of the Benue River in Cameroon, to find green grass seeds. Over the region of the equator the rain belt moves much faster. In north Tanzania the rain front arrives from the northeast at the end of November. At this point the queleas must usually fly as far as

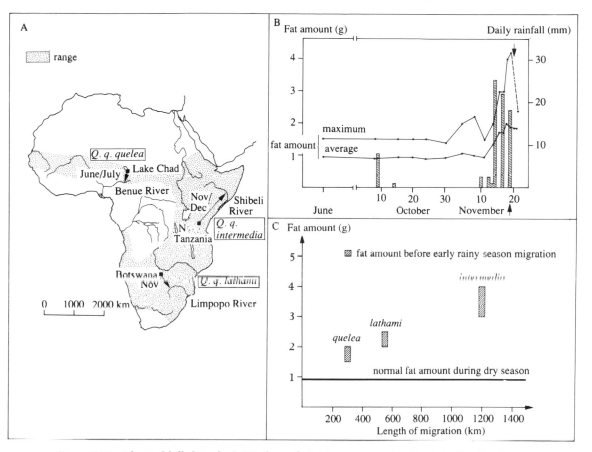

Figure 111   The Red-billed Quelea's (*Quelea quelea*) rainy-season migration is in the direction of the approaching rain front and extends to regions where it has been raining for at least six weeks and the savanna grass has had time to set new seed. A. The direction and length of the migration are shown for three different populations belonging to separate races (names shown in figure), at Lake Chad, in northern Tanzania and in Botswana. B. With the first regular rain showers the birds begin to store fat to prepare themselves for the rainy-season migration; the diagram shows the fat amounts in queleas at a roost site in Botswana and daily rainfall (vertical columns). C. Normal fat reserves for queleas setting off on rainy-season migration in relation to the length of the migration. Based on Ward (1971) and Ward & Jones (1977).

1200 km, to the Shibeli valley in Somalia, where the rains start in October and the grass has had time to set seed. The rainy-season pattern is more complex in southern Africa. The rains begin to fall in the region around the Limpopo River in September and October and then spread gradually across southern Africa during the coming months (with the exception of Angola, where the rainy season begins as early as October). When the rains reach Botswana, the queleas migrate a good 500 km to the Limpopo valley.

Since the queleas, during their rainy-season migration, travel across regions where they cannot find food, they must take with them adequate fat reserves for flight fuel. Figure 111 shows how, at a roost site of hundreds of thousands of queleas in Botswana, the birds' weight rises during the period of the first rain showers. Practically all birds left the site between 19 and 20 November (the arrows in the figure); on the following two evenings only a few stragglers assembled at the roost.

When the first of the showers which precede the rainy season fall, the termites swarm, at just the right time to provide convenient extra fare for the queleas during the fat-storing period. The queleas, however, are also capable of successfully putting on fat in other ways should the termite swarms fail. At the end of the dry season, many grass seeds are baked in and inaccessible in the cement-hard, cracked ground surface. After several rain showers, when the ground has been soaked and has dried out alternately, the topmost millimetres of the soil layer are loosened up. After having scraped away the loose earth crust with their feet, the birds can help themselves to the plentiful quantity of seeds that lie hidden there and in this way obtain extra food to convert into fat. When the queleas have to put on fat the termites are swarming and the earth crust is loosened up. What a fortunate coincidence!

Different quelea populations put on different amounts of fat depending on how far they have to migrate (figure 111C). The birds at Chad store average fat percentages of at least 5%, those in Botswana averages of at least 8%, and the birds in northern Tanzania average percentages of at least 15%. The theoretical range for these fat resources accords very well with the distances the queleas in the different parts of Africa need to fly from the rain front to the nearest areas with newly formed grass seeds behind it.

It has recently been discovered that two other species of weaver in northern Nigeria, the Red Bishop (*Euplectes orix*) and the Golden Bishop (*Euplectes afer*), also store fat in a similar manner. Both species migrate south when the rains begin and return to breed one or two months later, in the middle and at the end of the rainy season. In northern Nigeria the birds put on so much fat that it is sufficient for a southward migration of approximately 500 km. A surprising discovery was that even the Red-billed Fire Finch (*Lagonosticta senegala*), which had previously been considered a sedentary bird, regularly puts on extra fat at the beginning of the rainy season. These small seed-eating waxbills are marked dry-country birds and, unlike the Red-billed Quelea and other weavers, migrate in the direction away from the rain front and towards the desert steppes of the Sahel which lie approximately 300 km to the north.

Birds which migrate by soaring and gliding also store fat reserves before the migration. Of particular interest is that among the seabirds the parents abandon their young just before they have fledged. At that stage the young are exceedingly fat and can use the fat to travel a long way, before they have to start foraging by themselves. This applies for example to the Manx Shearwaters which breed in Britain. The young have a fat load of nearly 40% when they leave the breeding islands on their own. Many of them presumably cover the entire autumn migration across the Atlantic, from Britain to the coast of Brazil, almost 10 000 km, without renewing their fat resources. This splendid range is made possible by the shearwaters' energy-sparing method of flight. They combine gliding in the upwinds above the sea waves with dynamic soaring and obtain assistance from the winds of the northeast trades for a large part of the journey.

Honey Buzzards increase markedly in weight during the late summer, from an average 625 g to around 900 g. If the whole of this weight increase is due to storing of fat prior to the autumn migration, soaring-and-gliding Honey Buzzards should be able to make the entire journey, about 7000 km, from Europe to tropical Africa without additional food.

In contrast to the Honey Buzzards, Buzzards do not store any major fat reserves when they leave South Africa in spring to soar and glide 10 000 km northwards to east European and

Siberian breeding sites. The Buzzards increase in weight by less than 10% during the time that passes between their arrival in the South African winter quarters in October–November and their departure north in February. They obviously have good opportunities for catching prey at the staging sites along the migration route. Before crossing the Sahara and Arabian deserts, however, they must in all probability put on more fat.

When writing about the flights of migratory birds at up to 10 000 m altitude and above the earth's highest mountains, about long flights of several days without stopping, about the birds' ability to endure extremely low oxygen levels and severe cold, I have, with our human ground-level perspective, almost had to make a strong effort to realise that all this is reality. It is not a fairy-tale. I suspect that future studies will make the facts surrounding migratory birds even more fantastic.

4.7   Diurnal and nocturnal migration

The daily rhythms of different species during migration differ appreciably from one another. At least four different basic categories can be distinguished.

1 *Birds which are very flexible in their migration times and which migrate by day as well as by night.* The majority of ducks, geese, swans, waders and gulls belong to this category. The intensity of the visible migration is particularly great at dusk and during the hours immediately before dusk, as well as during the very earliest hours of the morning.

2 *Birds which are primarily nocturnal migrants.* Many species of small landbirds, such as warblers, flycatchers, chats and thrushes, are included in this group. A characteristic feature is that the migration begins suddenly, as if at a given signal over wide areas, on average 30 minutes after sunset. The whole process appears very dramatic on the radar screens, which suddenly 'blossom out' from echoes from night migrants starting out. Sometimes it looks as if the land areas, with their distinctive coastal outlines intact, are gently drifting out over the sea. In actual fact we are seeing radar echoes from thousands upon thousands of small birds all setting off at the same time on migration over the sea. I have myself watched the radar in amazement as both Gotland and Bornholm drifted southwards in autumn over the Baltic Sea at a speed of about 50 km per hour.

The night migrants' sudden departure takes place at broadly the same time in relation to sunset in different parts of Europe and North America. From radar studies, average departure times have been reported of 28 minutes after sunset in eastern Canada, 45 minutes in the northeast United States, 39 minutes in England, and 32 minutes in south Sweden. In a study in Louisiana in the United States, the night migrants' departure has been monitored by directing powerful searchlights from a tower sticking up above the tree tops horizontally over a woodland area. Birds which rose from or landed in the wood in the dark were easy to detect through binoculars as they passed through the searchlight beam. On nights with good migration activity, the first bird was seen to leave the wood and climb steeply towards the night sky at 22 minutes after sunset at the earliest and at 41 minutes after sunset at the latest. Within no more than 20 minutes after the departure of the first bird, the majority of the night's migrants had set off and left the wood.

The beginning of the nocturnal migration takes place right at the end of the twilight period. By way of definition, the twilight period is regarded as ending and the night as taking over when the sun has sunk six degrees below the horizon. At this time it is still light enough for a human eye to be able to distinguish clearly the horizon and landmarks on the earth's surface, while at the same time the brightest stars are just starting to appear. The departure time of the night migration is probably explained by the opportunities for orienting that exist in these special light conditions.

Even though the birds in this category normally restrict their migration to the night-time when flying in short stages across ecologically hospitable areas, we should still remember that many of them also fly long distances, across seas or deserts, during both night and day, sometimes for more than three whole days without stopping.

3 *Soaring-and-gliding birds (see section 4.4) which migrate in the middle of the day.* It is often four or five hours after sunrise before their migration gets properly underway. The daily rhythm is adjusted to the periods when thermals over land are most strongly developed. In good thermal conditions the migration can continue until long into the late afternoon. That soaring migrants also make use of occasions for migrating that are totally outside the normal timetable is demonstrated by a report of storks on migration in the middle of the dark of night in Algeria. The storks there were exploiting the rising warm air above burning gas flames at oil-drilling sites and along stretches of pipeline. A total of 8000 storks, split into many flocks of between 15 and 270 individuals, was seen circling above the gas flames at one such oil installation and passing northwards on one night at the end of February. Stork migration was observed on the same night under similar circumstances 200 km farther north. This suggests that the storks may carry out substantial stages of migration by night across certain parts of the Algerian Sahara desert, where the installations with gas flames are as close together as one every 5 km.

4 *Birds that are principally diurnal migrants but do not migrate by soaring and gliding.* This group includes, among others, most finches, buntings, wagtails, swallows and martins, bee-eaters, starlings, crows, pigeons and doves, and irruptive migrants (except owls). The birds normally begin to migrate immediately before sunrise, some species so early that the first birds cannot be detected by a human observer in the dim light of dawn. According to radar studies in south Sweden, the first diurnal migrants set out on average 43 minutes before sunrise. The passage is more intensive during the earliest hours of the morning and becomes weaker towards the midday period, when several species cease migrating altogether. Some species occasionally show a second migration peak in the evening, at sunset. As with the nocturnal migrants, there are among the normally day-migrating birds some species which migrate long distances and which regularly make journeys that last both day and night.

It is odd that the daily rhythm of migratory birds has been analysed and discussed to so minimal an extent in the specialist literature. The migration counts at Falsterbo and Ottenby provide a voluminous and valuable database which should make interesting comparative studies between different species possible. For rough comparisons of the timetables of different diurnal migrants the data can be used directly. When it comes to demonstrating the migration rhythm in detail, we must take into account the effect of coastal leading lines and the birds' catchment areas.

In figure 112 I have shown a calculation of the migration rhythm of the Chaffinch and the Woodpigeon, based on data from Falsterbo. From radar studies it is known that the mean direction of both these species during the autumn migration over the interior of southern

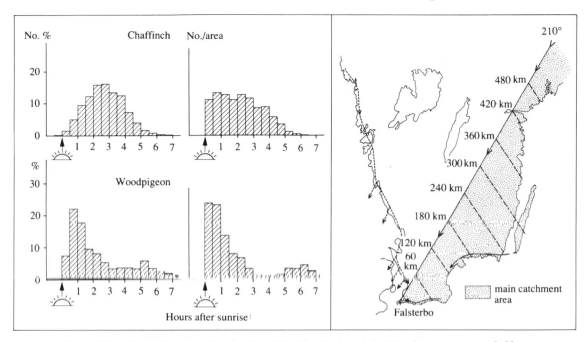

Figure 112   The daily migration rhythm of the Chaffinch and the Woodpigeon as recorded by the migration-observer at Falsterbo (left-hand diagrams) and as it appears when the number of passing birds is looked at in relation to the size of the catchment areas (right-hand diagrams). For calculation of the right-hand diagrams, it has been assumed that Chaffinches and Woodpigeons which pass Falsterbo come mainly from the eastern part of south Sweden and that the mean speed for the Chaffinches (which at Falsterbo are most numerous on days with head winds) is 30 km per hour and that for the Woodpigeons (which often migrate in tail winds) is 70 km per hour. The map shows migration distances to Falsterbo for birds which travel on a 210° heading across the interior and which thereafter follow the south coast westwards. The basic data are based on migration counts at Falsterbo from dawn until 14.00 hours (for the Chaffinch 16–30 September 1973–1975, for the Woodpigeon 26 September to 25 October 1974) and have been collated by Gunnar Roos.

Sweden is about 210°. Birds reaching Falsterbo are primarily those whose migration is diverted westwards along the Swedish south coast; they therefore originate mainly from the catchment area in the eastern part of south Sweden which is indicated in figure 112. Finches and pigeons which migrate over western Götaland and sometimes allow the Swedish west coast to act as a leading line southwards leave Swedish shores by migrating out over the Kattegat or over the narrow waters of the Öresund; only minimal numbers reach Falsterbo.

By estimating the birds' mean speed and assuming that they travel in a 210° direction across the interior before diverting to migrate along the south coast to Falsterbo, we can calculate from which parts of the catchment area those birds come which pass Falsterbo during a certain period of time after the start of the migration. The birds that arrive earliest of all of course come only from a small area in southwest Scania. Those arriving later in the morning come from considerably larger areas.

I have used these estimates and assumptions in order to convert the temporal distribution of the number of birds recorded by the migration-observer at Falsterbo in relation to the size of the catchment areas (figure 112). If the geographical dispersion of the birds is uniform before the migration gets going, then the conversion shows the actual daily rhythm of the passage. For the Chaffinch, this means that the migration intensity increases abruptly at sunrise and then is maintained at a high level over the next three hours. Thereafter it begins to fall; after a good four hours the intensity is reduced to a half, and after five hours it peters out altogether. That the observer at Falsterbo notes that the number of passing Chaffinches increases during the first hours of the morning is thus due not to heightened migration intensity but to the fact that the birds are now coming from a larger area of country.

The more irregular temporal distribution of migrating Woodpigeons is probably explained by the fact that the pigeons are unevenly distributed in the landscape. The migration intensity of the early morning begins to drop off only one hour after sunrise and sinks to a surprisingly low level within three hours. The Woodpigeons that pass Falsterbo in the early morning peak come from the deciduous woodland and agricultural country of Scania, which offers many suitable stop-off sites. The migration intensity then falls as the catchment area accordingly shifts northwards, to impoverished coniferous country where presumably comparatively few Woodpigeons stop off. Interestingly enough, we can trace a tendency towards a passage peak from five hours after the migration has got underway and onwards. This is probably caused by pigeons which have moved on from staging sites in the agricultural country of Östergötland and Södermanland to the southwest of Stockholm. Pigeons which come originally from Finland also stop off there.

Before we become absorbed in analyses of minor differences in the daily migration timetables of different species, it is of course desirable that we look into the more fundamental question: Why are certain species typical night migrants and others primarily day migrants? If we look up the specialist literature, then we find no serious suggestions whatever for explaining this problem (if we leave aside the soaring migrants, where the association with thermal formation is of course obvious). If this question has not been speculated on before, then it is, I think, high time that we started now.

I believe that the practices of diurnal and nocturnal migration are adaptations to the varying food situations of different bird species. Typical diurnal migrants are species whose food occurs patchily and in very variable amounts. Day migrants therefore appear most often in flocks when seeking food. Even if the birds have found the right type of feeding environment, it is not certain that there is in fact any food to be gathered just on that actual occasion. A Chaffinch may fail to find seeds at a seemingly suitable site: the seeds may have happened to blow away to a place where the birds cannot reach them, or a flock of finches may already have been at the site and consumed the seeds; seeding may also have failed for local reasons. For a swallow, the best feeding sites change from day to day, depending on where insects are swarming and are concentrated; this in turn depends on local temperature, humidity and wind conditions.

Some birds fly several and sometimes tens of kilometres to assemble at communal roosting sites. In ecological research this has been interpreted as an adaptation to uneven and varying occurrence of food. Birds or bird flocks which have been at favourable feeding localities the evening before return purposefully to these good food sites after roosting. Individuals which have been less successful in the search for food during the previous day 'tag on' and accompany the purposeful birds in the morning. In this way the communal roost serves as an 'information centre'.

Interestingly enough, birds which are typical diurnal migrants by and large assemble at communal roost sites but birds which are nocturnal migrants do not. Starlings, swallows and wagtails set off at nightfall on 'roost migration' to spend the night in enormous throngs in reedbeds. The Starlings also often use roost sites in city centres, where they perch in flocks in trees or on tall buildings, often in the mats of ivy on fronts of buildings. In Trafalgar Square in London, the chirping and buzzing din from the Starlings competes with the noise of the traffic. Finches gather in dense spruce stands or in solid rows of bushes. Crows seek out copses with tall trees. The Jackdaws also roost in city parks or on tall buildings; cathedrals are often used for this purpose and some, such as that at Uppsala in Sweden, are well known for their large numbers of roosting Jackdaws.

Further studies are required before it is proven that the advantage of information transfer among the birds has been an important contributory reason for the development of communal roosts. For the sake of argument, however, I shall assume that this explanation is in fact correct. Flock birds which use communal roost sites are faced with particular difficulties when, after completing a flight stage, they have to settle in a new stop-over area. It is important that very soon after arrival in the new area they fit into the social system, find the communal roost sites and thereby locate the best feeding sites.

This happens most simply and most effectively if the migration takes place in the daytime. When the flight stage begins to draw towards its end during the morning, the birds can, while still migrating, begin to look around for stopping-off sites. It is not always sufficient for a site to appear to be a suitable environment. Actively feeding flocks of their own species are the most reliable indication of an acceptable feeding site. By joining with stop-over birds already established locally, the migrating birds also gain the advantage that at the end of the day they can accompany the local birds to a communal roost in the new area. Then the newly arrived birds are fitted into the region's local information system and can stop and top up their fat reserves until the time comes for the next stage of the migration.

Birds which have evolved into nocturnal migrants generally forage individually. The food supply is fairly evenly distributed in a suitable environment. For these birds it is an advantage to migrate at night since the whole of the daylight hours can then be used for foraging. A Willow Warbler which has landed in woodland after a nocturnal flight stage can generally begin to seek food immediately after day has broken. If it has chanced to land at a site with a slightly sparse food supply, it probably does not need to move particularly far through the wood before it comes upon better feeding conditions. Similar circumstances presumably apply for example to Sedge Warblers. Even when migrating at night the Sedge Warbler can see well enough to be able to land in a suitable environment: reedbeds at marshes, lakes or sea-shores. When dawn breaks, it does not need to feel its way forward in the area for long before it finds a clump of reeds with a sufficiently good food supply. Many small nocturnal migrants feed on flying insects. Owing to the prey's sensitivity to disturbances, it is an advantage for these birds to spread out in the terrain to forage and to avoid forming flocks. Several species, such as Reed Warbler and Pied Flycatcher, even defend individual territories at autumn staging sites in the northern Mediterranean region.

A phenomenon with close reference to the above discussion on the different feeding situation of diurnal migrants and nocturnal migrants is what is known as reverse migration, which occurs particularly in regions where landbirds meet with a sea crossing while on migration. In the autumn in south Sweden, large numbers of birds are seen regularly migrating north or northeast, directly the opposite of the normal migration direction, which is out over the southern Baltic Sea. The most likely explanation is that the reverse movements consist of birds which, when their fat reserves run out and the day's migration period is nearing an end, break off the migration before crossing the sea and turn back in order to find suitable stop-over sites inland. The heavy concentration of migrants in the coastal zone itself leads to strong competition for food and to high exploitation, perhaps also to great risk of attack by predators. In addition, certain species have no suitable stop-over environments in this zone. The birds therefore have better chances of finding suitable stop-over sites if they migrate back inland. Most birds migrating back probably find suitable sites as soon as they have returned to the first major woodland areas or mixed wood- and farmland inland.

Both diurnal and nocturnal migrants are seen on reverse migration, but there are revealing differences in their behaviour. Reverse migration by diurnal migrants, such as finches, wagtails and swallows, is most intensive during the later part of the morning and therefore follows on the early morning passage in the normal migration direction. Often normal and reverse migration meet over the coastland. Irrespective of the fact that the birds are flying in diametrically opposite directions, both these migratory movements take place in a very similar manner. The birds migrating back are obviously searching out the terrain for suitable environments, in the first place to land and stop over where there are already actively feeding flocks.

Reverse migration by nocturnal migrants is not so conspicuous. In the early morning warblers, flycatchers and others travel from tree to tree, from copse to copse or along rows of bushes, consistently moving away from exposed headlands and coastal sectors. The whole time they pick their way forward in order to find stop-over sites with good food supplies and negligible competition and interference from other individuals.

In this way the birds' food situation does, I believe, influence the development of flock unity, communal roosts, different tactics when migrating back, and the practice of diurnal or nocturnal migration. I readily admit that this is speculation and that the argument should be further developed and elucidated with critical field studies.

4.8   Weather and wind

The intensity of bird passage changes considerably from day to day or from night to night, even when the migration season is at its height. The birds wait for favourable weather and then migrate in large droves, but they stay behind at their staging sites in poor weather conditions.

Despite the fact that for most bird species the migration season both in spring and in autumn lasts for several weeks, a very large proportion of the migrating birds pass during a few days or a few nights of intensive passage. The observer of migrants at Falsterbo usually, where most species are concerned, counts nearly half of the season's total of birds during the three best passage days. Species which migrate late in the autumn and which are short-distance migrants are concentrated into a few days of passage more so than are early-autumn and long-distance migrants. The most concentrated migration of all is that of irruptive species: almost two-thirds of the birds which fly past Falsterbo in autumn are counted during the three best days of passage. This suggests that short-distance migrants, especially irruptive species, are more sensitive to weather than are long-distance migrants, and/or that changes in the weather are more fundamental and of more decisive importance to the intensity of passage during late autumn than earlier in the autumn season.

The connection between weather and bird migration is of great interest from a flight-safety point of view when it comes to reducing the number of dangerous collisions between aircraft and birds ('bird strikes'). This has served as an additional stimulus for many of the studies that have been made in recent decades aimed at surveying the reactions of migrants to the weather. For most studies surveillance radar has been used, and in this way a picture has easily been obtained of the number of migrating birds over wide areas under different weather conditions. Direct field observations are a necessary complement to the radar studies. In Sweden such investigations form the basis of a forecasting system for bird migration which is used by the Swedish airforce to warn air traffic and to report when and where the likelihood of intensive movements of migrants is great. The system is based on predictions of how the weather will develop, which determines the scale of migration of the different categories of migrants.

Studies of the connection between weather and bird migration have taken place almost wholly within the temperate zone of westerlies in Europe and North America. The weather within this zone acquires its character from the itinerant low-pressure systems (the cyclones and depressions). This we in northwest Europe are all too familiar with – how unwelcome the passages of depressions with their accompanying wind, damp and rain so frequently are! Really we ought to look upon them with more gratitude. The cyclones are important in that they bring warmth from the equatorial regions towards the polar regions. Owing to the fact that the tracks of depressions run from the Atlantic northeastwards in over northwest Europe, west Europe is an unusually mild region with plentiful precipitation and fertile land.

Cyclones develop as wave disturbances on the polar front, where warm tropical air meets cold polar air (cf. chapter 2). The warm air is trapped in a sector of the cyclonic vortex and advances behind a warm front, where the raising of the moisture-saturated warm air leads to extensive cloud and rain formation. Behind the warm air sector cold air advances. The warm air is pushed away and is forced to rise, and this results in cloud and rain on the cold front, too. In a well-developed cyclone, the weather system can extend over areas several thousand kilometres across. The typical weather pattern in a cyclone is illustrated in figure 113, in which a cross-section through a warm front, warm sector and following cold front is also shown. The fronts' slope is very shallow, approximately 1 km in altitude over a distance of 100 km. (The cold front's slope at low altitude is, however, noticeably steeper than that of the warm front.) The distance between the cirrus clouds that are present on the upper parts of the warm front and the point where the front arrives at the ground surface can exceed 1000 km. Since the cyclones move at a mean speed of around 40 km per hour, it can take more than 24 hours after the first cirrus clouds of an approaching cyclone have been seen before the warm air comes along at the surface of the ground.

During the early stages of its development the cyclone has an extensive warm sector. The cold front, however, moves faster than the warm front, and the warm sector gradually becomes narrower. Eventually the cold front catches up with the warm front, starting next to the centre of low pressure, and an occluded front is formed: the warm air is lifted up by cold air from both the front and the back edge of the cyclone. With that the cyclone has reached the peak of its development – at this stage the winds are often gale or storm force. During the final phase of the cyclone's life the warm and cold fronts merge completely into a diffuse occluded front, the cyclone's travel comes to a halt and the low pressure begins to fill. The period between the cyclone's origination and the start of the occlusion process is approximately 24 hours. Cyclones which have formed on the Atlantic polar front are in their dynamic phase when they reach in over west Europe. A couple of days later they begin to die. Between the passages of depressions there may be fine high-pressure weather established, but occasionally a series of depressions develops, often four or five of them, with intervals of only a day or two between each one. Between such cyclones, which pass in rapid succession, there are only weak and short-lived ridges of high pressure.

On satellite pictures of the earth's surface we can straightaway see the characteristic cloud convolutions and cloud trails from the many cyclones which at every moment are on the move within the west-wind zones. It is hardly surprising that the birds' migratory behaviour within these zones is adapted in the first place to the weather pattern of the cyclones, to the large-scale and dramatic changes in winds, cloud cover, precipitation and temperature.

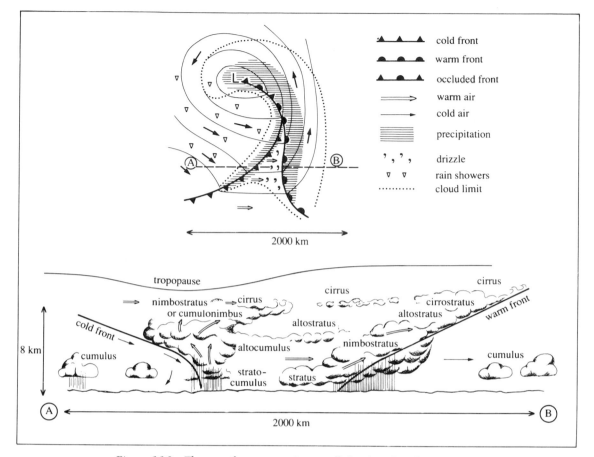

Figure 113 The weather system in a well-developed cyclone. The lower figure shows a vertical section along the line from A to B. Note that the vertical scale in the lower figure is considerably enlarged in proportion to the horizontal scale. In the northern hemisphere the wind blows anticlockwise, parallel with the isobars (shown as thin continuous lines which connect places with the same air pressure) in the low-pressure vortex. Cyclones move eastwards within the westerlies zones in both hemispheres. In the northern hemisphere the direction of movement often tends towards the northeast, in the southern hemisphere towards the southeast. Frequent paths of depressions in the North Atlantic region stretch from the northeastern United States east and northeast across the Atlantic to Britain or Iceland and farther eastwards over Scandinavia or the Norwegian Sea. Based mainly on Liljequist (1970).

Shown diagrammatically in figure 114 are the weather situations, in relation to low-pressure and high-pressure systems, in which intensive bird migration regularly takes place in spring and autumn.

Heavy spring passage occurs mainly within or west of a high-pressure area, before a cyclone arrives from the west. In the spring the birds in addition fly within the cyclone's warm sector where the clouds are not too compact and where it is not drizzling. When migrating northeast in spring they also fly behind cold fronts, provided that the cold front is followed by westerly, fairly mild winds and not by cold northwesterlies. Heavy autumn passage occurs chiefly within or to the east of high-pressure areas, generally immediately after a depression has passed. The passing of a cold front with ensuing cold northwest winds often triggers intensive autumn migration, and particularly of birds which migrate southeast. Sometimes the birds start out so soon after the cold front that they have caught up with it after a few hours. Should they then find themselves over the sea and with no opportunities for landing, the situation can become problematic, for there is a risk of wind-drift and difficulties in orienting where the cloud cover is solid.

Common to autumn and spring is the fact that passage is most intensive when the migrants have favourable *tail winds* and that migrating birds *avoid areas of precipitation.* These two factors are of decisive importance to migratory activity. Even though intensive spring passage often also coincides with rising temperatures, falling air pressure and deteriorating visibility, and heavy autumn passage generally with cold weather, rising air pressure, low humidity and good visibility, the connection is principally a secondary effect for these weather factors are in turn correlated with tail winds and dry weather. Once we have taken into account wind and precipitation conditions to explain the intensity of the migration, other factors do not play any major role.

When birds meet unfavourable weather, head winds and rain, they generally break off the migration and land as soon as they are able to. Situations such as this sometimes result in resting birds concentrating in large numbers on islands, headlands or coastlands, at places where the birds find the nearest site for landing. Such 'coastal falls' of migrating birds are fairly common in autumn, e.g. on the English east coast, when migrants from Scandinavia are caught out by poor weather over the North Sea. The phenomenon is also well known in southern Louisiana, where large flocks of birds arrive in spring from the south across the Gulf of Mexico. Should the flying weather be favourable, then the birds continue in over the coastline and past between 50 km and 150 km of marsh, open farmland and grass plains before landing in the first extensive woodlands. If they encounter head winds and rain over the Gulf of Mexico, however, they land as soon as they get to the Louisiana coast, usually several hours behind time. Large numbers are then concentrated on the outermost islands in the waterside swamplands.

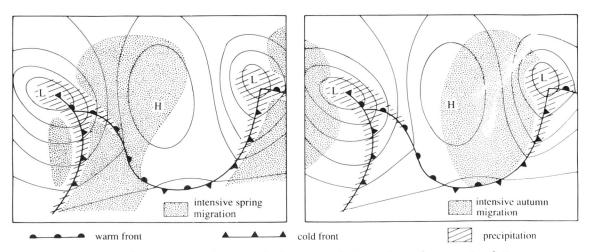

Figure 114   The occurrence of intensive bird migration during spring and autumn in relation to high-pressure and low-pressure systems. The winds blow parallel with the isobars (thin continuous lines), anticlockwise around low pressure and clockwise around high pressure. Based on Richardson (1978).

Arctic waders which migrate from the north Russian tundra to the shores of the North Sea stop off during July and August in southern Sweden in numbers that vary from year to year. In summers when the weather has the accent on high pressure and with a liberal measure of northeast winds, few arctic waders stop off in Sweden, no doubt owing to the fact that they travel to the North Sea in a single unbroken flight stage, more often than not at high altitude. In contrast, in summers when many depressions pass through, followed by areas of rain and southwesterly head winds, many waders break off their migration in Sweden and take the opportunity to stop over and feed on the shores and lakes of that country. Should the unstable weather continue, then the waders are forced to complete their journey in short stages. In such cases they fly at low level along the coastlines in order to avoid the head wind as much as possible.

Which factors govern the migrants' choice of migration weather? Three such factors can be considered to be of great importance: the living conditions in the area the birds are leaving, the living conditions in the area for which they are heading, and the flying conditions during the migration itself. If the birds' choice of migration weather were mainly an adaptation to the conditions in the area which they are leaving, then in the first place cold, frost, snow and formation of ice, weather factors which make their living conditions considerably worse, ought to trigger emigration. Similar weather in the destination area, if it could be foreseen by the birds, naturally ought to have the effect of deterring migration. Where the flying conditions during the migration are concerned, two factors are of the greatest importance to the birds: they should carry out the flight as economically as possible from an energy point of view and they should avoid weather which may lead to orientation problems. The first factor is provided for if the birds choose tail winds, and the second if they avoid flying in rain, fog and dense cloud. That precisely those points, tail winds and avoiding of areas of rain, usually are key factors for intensive migration has been confirmed. We can therefore draw the conclusion that the migrants' reactions to weather, those reactions that determine the variation in migration intensity from day to day, in general (but not always, see below) are an adaptation to good flying conditions and do not have that much to do with short-term changes in the living conditions in the area they start from or finish in.

Occasions arise, however, when birds are driven to migrate by degenerated living conditions or are held back by poor prospects in the area of their destination. They are then forced to waive the need for energy savings and safety during the flight itself.

Some birds attempt to overwinter in northern regions so long as it is mild and they can find food. When the cold, the frost and the snow step in with full force, however, they have no alternative but to leave on winter migration. This can be observed at Falsterbo, for example, during periods of severe winter weather: crows, finches, larks, Starlings, gulls, ducks and geese leave southern Sweden and head out over the sea in a southwesterly direction towards milder regions. Sometimes the passage goes on in blizzard conditions, as for example on the December morning with whirling snow, less than 50 m visibility and wind strengths of over 15 m/s when Gunnar Roos logged emigration at Falsterbo of Fieldfare, Starling, Skylark, Waxwing and Common Gull. The most numerous of the winter migrants at Falsterbo are the Bramblings: daily migration figures of tens or hundreds of thousands of individuals can occur, for during good mast-years millions of Bramblings are enticed into attempting to overwinter in south Sweden's beech woods. There they can find a good living so long as they can get at the fallen mast on the ground in the woods, but if it starts to snow heavily the finches' food is buried in the snow. They then flee the winter. This happened in January 1959, when more than 10 000 Bramblings stopped off on the outermost point of the Falsterbo headland, constantly harried and pursued by several Hen Harriers and a Peregrine, before they all headed off farther south a few days later. The largest mass occurrence of winter-migrating Bramblings happened at the turn of the year in 1976/77, when a series of heavy snowfalls forced them to migrate: for at least five days more than 100 000 emigrating Bramblings were counted every day at Falsterbo, a passage intensity easily on a par with what a peak day in autumn can offer!

Radar studies reveal that the migration traffic during the winter is greater than one might imagine. One study in southern Sweden showed that the total number of echoes from migrating bird flocks was roughly the same whether the winter movements were to the south, west, north or east. The flocks of migrants probably consisted mainly of ducks and gulls, moving away from frost and freeze-up. Winter migration over the southern North Sea has also been mapped using radar studies. Here, Starlings and Lapwings are among the participants in what we call 'hard-weather migration' (see section 3.1). If the weather fluctuates a lot, these species may commute to and fro across the North Sea several times during the course of the winter.

A further example of migration which is triggered directly by deteriorating conditions of life for the birds is the retreat migration that the earliest spring arrivals are sometimes forced to make. Especially striking is that of Skylarks, Lapwings and Chaffinches. These species accompany the first mild winds of spring in March and at the beginning of April and move northwards over Scandinavia. The spring, however, is not yet to be counted on: the risk of a setback, with wintry weather and snow, is great. When such a reversal occurs, the spring birds turn right around and make their way back southwestwards. An observer in Västergötland in Sweden tells of how the first Chaffinches came with the thaw at the end of March. After only a couple of days the spring's big Chaffinch migration was in full swing across the province, and it went on every morning for a period of a week. Then one night there was an outbreak of heavy snow. At dawn the blizzard continued until the forest was covered by snow. The Chaffinch migration was soon in full swing and went on throughout the morning, but all the Chaffinch flocks, as many as could be seen and heard in the thick falling snow, were flying in retreat towards the southwest in westerly head winds.

The songbirds in spring are in a hurry to reach the breeding area in order to stake a claim to their territories. They play a hazardous game which presupposes well-polished adaptations. Those individuals which arrive first are rewarded with the best territories and have the best prospects for breeding successfully, but at the same time they expose themselves to the risk of being hit by a reversal in the spring weather. The retreat migration is a necessary adaptation if the birds are not to perish in this situation. A similar adaptation has been revealed in radar studies of the northward migration in May of small insectivorous birds in the northeastern United States. The birds hold back and avoid migrating at times when, despite favourable tail winds, there is a risk of poor feeding conditions associated with damp and cool weather on or immediately following arrival in their destination area. The birds' reactions in the area of their departure are thus governed by how the weather will be the next day in an area that lies more than 300 km to the north! How do the birds manage to be weather-forecasters in this way? The secret is that they refrain from migrating in the moist southwest wind immediately preceding a cyclone which will soon pass the area of their destination and which is followed by precipitation and by cloudy and chilly weather. They prefer weather dominated more by high pressure. The winds then are not always so favourable for flying as they are before a cyclone, but on the other hand stable, sunny, warm and dry weather prevails. This naturally promotes the catching of insects and aids survival once the birds have arrived at their destination.

Great differences exist in various species' response to the weather. This is not very evident

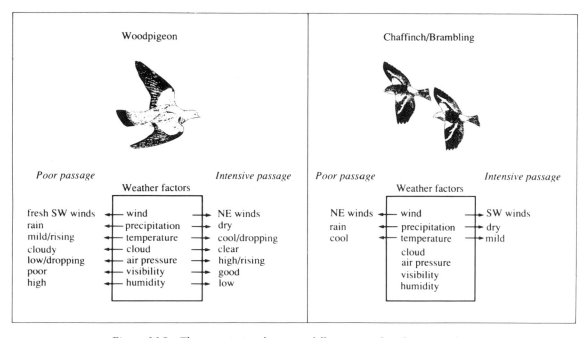

Figure 115 The association between different weather factors and migration intensity at Falsterbo of Woodpigeon and Chaffinch/Brambling. Based on Alerstam (1978) and Roos (1977, 1978).

on the radar screens, where it is difficult to distinguish migration of different species, but it is all the more clearly obvious in direct field observations. Certain species are hardy and migrate even in quite poor weather; others are economy-minded and are seen on migration only in the very best weather. Species which have a short migration journey and plenty of time to travel can naturally afford to be scrupulous in their choice of migration weather. Long-distance travellers on the other hand cannot risk being delayed too long while awaiting ideal weather. A good example of such a long-distance migrant is the Honey Buzzard: it does not migrate only when the thermal conditions are at their best but is also prepared to cover quite long stretches by using flapping flight, sometimes even in head winds. This is in stark contrast to the behaviour of Swedish Buzzards, the population of which includes only short-distance migrants; generally speaking, the Buzzards concentrate wholly on migration using thermals. When the weather is on the way to improving after a rainy and misty period, the Honey Buzzards are always the first to react and soon turn up on outward passage at Falsterbo. The Buzzards usually hang on until the approaching high pressure has really become widespread, with sun, light winds and thermal clouds.

Differences in reactions to the weather also exist between species which migrate roughly the same distances and at the same time of the season. Figure 115 shows the association between different weather factors and passage through Falsterbo of Woodpigeon and Chaffinch/Brambling (the finches migrate in mixed flocks); both pigeons and finches are short-distance migrants which fly to west Europe, and the passage of both peaks at the beginning of October. The differences in migration weather, however, are amazing. Intensive passage of Woodpigeons is clearly correlated with a long series of weather factors. All of these indicate that the pigeons migrate in the cold air mass which pushes down from the north on the back of cyclones. The number of migrating Woodpigeons at Falsterbo is on average six times as large on days with northeast winds as on days with southwest winds. The largest numbers of finches pass Falsterbo during totally different weather conditions – when mild, southwesterly head winds prevail! Eighty times more finches on average are counted on days with southwest winds than on days with northeast winds. The finch passage decreases in rain but is independent of most other weather factors, of cloud, air pressure, visibility and humidity. What interpretation are we to put on this?

In figure 116 I have indicated schematically the occasions when maximum numbers of birds, pigeons and finches, move past. I have distinguished two different basic weather situations: (1) when cyclones pass in rapid succession one after another, separated only by weak ridges of high pressure; (2) when cold air pushes down from the north and high pressure enters between two cyclones. The clear majority of days with pigeon migration occur in association with the latter type of weather situation. This largely agrees with the general results from radar studies as shown in figure 114. One discrepancy is the good pigeon-migration days that occur when high pressure is centred east or northeast of Falsterbo. In its entirety the pigeon migration over southern Sweden on days with southeast winds is markedly lighter than it is when the winds are coming from the northern quarter. The difference is obvious on the radar screens. In southeast winds, however, the geographical migration pattern changes (see figure 87 and the radar photograph in section 4.1): the pigeon flocks do not migrate out over the Baltic Sea from the south coast of Sweden but instead follow the south coast westwards in a narrow zone all the way to Falsterbo before crossing the sea on route for the nearest shores of Denmark.

Peak days of finch passage occur at Falsterbo mainly in situations with west and southwest winds between cyclones which follow rapidly on the heels of one another. This does *not* mean that the finches actually *prefer* to migrate on these days of head winds, for they also migrate extensively on days with northerly or northeasterly winds, at which times pigeon passage is at its most intensive. In tail winds the finches fly on a broad front without following coastlines, and are therefore not concentrated at the Falsterbo peninsula; in addition, their flight altitude is often so high that a ground-based field observer cannot detect

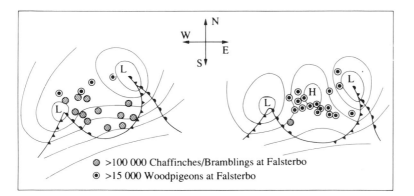

Figure 116   The weather situation at Falsterbo on autumn days with mass passage of Chaffinch/Brambling (more than 100 000 recorded) and Woodpigeon (more than 15 000). The weather situation has been determined schematically in relation to pressure centres and fronts. A few days with mass passage have been omitted as they could not be given a correct position on the greatly simplified weather maps. Data from Gunnar Roos's annual reports on migration at Falsterbo 1973–1979.

most of the small finches. How, then, can we be so certain that an intensive finch passage is actually going on when northeasterly tail winds are blowing? Radar provides clear evidence: masses of echoes, of a type that is characteristic of small-bird flocks, move southwest on a broad front, giving the appearance of a silent fall of snow across the radar screens. That finch flocks are responsible was confirmed by observations from an aircraft one autumn in the extreme southwest of Scania (i.e. not far from Falsterbo): at altitudes of between 200 m and up to 1000 m and more, flocks of finches were seen swishing past the aeroplane, one flock after another.

Both finches and pigeons, therefore, prefer to make use of days with the best tail winds for their migration. Unlike the pigeons, however, the finches are prepared to migrate on a large scale even in head winds between the passages of depressions. The migration is then conducted at low altitude and is concentrated along coastlines. This migration into head wind is probably an adaptation to those periods, frequent in northwest Europe, particularly in autumn, when series of depressions pass without any high-pressure areas developing. During the autumns when this type of weather pattern predominates, no doubt most finches leave south Scandinavia by migrating into head winds. This is presumably what happened in autumn 1975, when the Falsterbo observer counted $2\frac{1}{2}$ million finches migrating through. During the finches' migration period that autumn the winds were southwesterly on more than half of all days, and there was not a single day when winds blew from the northeast quarter. During the finches' migration period in autumn 1973, the weather was totally different: northeast winds prevailed on almost half of all days, and southwest winds on only a quarter. In that autumn probably the majority of finches migrated in tail winds, on a broad front and at high altitude; at Falsterbo, therefore, fewer than half a million were logged.

Why are finches, to a greater extent than pigeons, prepared to migrate in head winds? The finches' flight speed in still conditions is approximately 40 km per hour and that of the pigeons 60 km per hour (see section 4.2). If we start from these speed values and from wind data from Falsterbo for the finches' and pigeons' migration period, we can form a general idea of the birds' chances of maintaining various migration speeds (figure 117). During the most favourable tenth part of the period, pigeons can keep up a migration speed in a southwesterly direction of 75 km per hour or more and finches one of 50 km per hour or more. During the better half of the time, pigeons can travel at at least 50 km per hour and finches at at least 30 km per hour. On one day out of eight on average, the winds are so unfavourable that the finches are not able to steer a southwesterly course at all, but on only one day in 50 are the winds too much for the pigeons.

On the assumption that the daily migration time is the same for finches and for pigeons and that the birds choose by preference to migrate on those days when the tail winds are best, we can calculate by how much the flight distance covered increases when more days during the migration period are used for migrating. With their higher speed, pigeons of course need fewer migration days than the finches to get the same distance. In figure 117 an example is shown in which the pigeons migrate on one-third of all days and then at a speed of 60 km per hour and above. To get as far the finches must, if they travel at at least 25 km per hour, migrate on no less than 60% of all days of the migration season. The pigeons thus manage in this example if they migrate only on days when the wind improves their flight speed; the finches, on the other hand, are forced to make the most also of days with head winds for their migration.

In this way we can interpret the finches' habit of migrating not only in tail winds but also in head winds as an adaptation of a small and slow bird with a short daily migration period. The migratory journey can be carried out without delays despite unstable and often

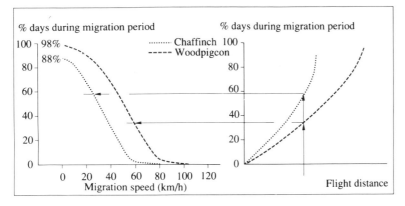

Figure 117   Calculated migration speed and flight distance in a southwesterly direction for Chaffinch and Woodpigeon in relation to number of days during the migration period which are used for migration. The birds are assumed to use in the first place those days when the winds give the highest possible migration speed. The left-hand figure shows how the lowest occurring migration speed decreases accordingly as the birds migrate on an increasing number of days with more and more unfavourable wind conditions. On 12% of the days the wind is too strong for the Chaffinches and on 2% of the days for the Woodpigeons; the birds are then unable to hold a southwesterly flight direction. The right-hand figure shows how the distance covered increases, first at a fast rate when the birds migrate on those days which have the best winds, then more and more slowly as more and more migration days with poorer winds are added. The arrows show examples of a flight distance which requires of the pigeons that they migrate on one-third of all days of the migration period; the finches must migrate on nearly 60% of the days. During these passage days the pigeons' speed is 60 km per hour or higher and the Chaffinches' at least 25 km per hour. The calculations are based on wind measurements at Falsterbo at 07.00 hours, 27 September to 17 October over 25 years (1949–1973, 525 days in total). The Woodpigeon's air speed is assumed to be 60 km per hour, the Chaffinch's 40 km per hour.

unfavourable winds in west Europe. The pigeons' greater demand for good migration winds leads in some autumns to their being forced to wait long into October in southern Scandinavia for the right weather. Should good migration weather still not come along, then the delayed pigeons finally set off in spite of cross winds and overcast weather with very poor visibility. The risk of being delayed and of being forced to migrate in poor weather anyway is a disadvantage which the pigeons must accept in exchange for the advantage that it usually means to be able to place high insistence on favourable migration weather.

Birds which fly long distances direct over seas or deserts must be even more finely adapted so that they migrate only when the likelihood of suitable weather prevailing along the whole flight path is greatest. Some very interesting aspects have been revealed in connection with the mapping of the migration between North America and South America over the West Atlantic (figure 118). The birds usually begin their journey from North America immediately after the passage of a cold front. Fresh winds are then blowing from the northwest quarter. Thanks to these winds they reach a good travel speed out over the open ocean heading southeast. The waders react more quickly to the passing of the cold front than the small birds; at coastal radar stations in easternmost Canada, echoes are occasionally seen from wader flocks in the gaps between the cold front's areas of rain and cloud. In this situation the risk that the birds will be wind-drifted is immediate (see section 4.9).

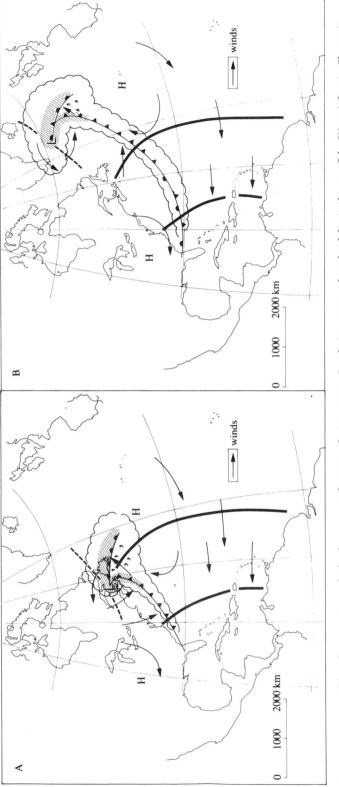

Figure 118   Typical weather in autumn for bird migration from North America to South America within the delimited zone of the West Atlantic. Shown in A is the typical weather situation at the time the birds leave North America. When the passing cyclone moves farther east over the North Atlantic, the cold front is displaced to the southeast over the sea and finally becomes stationary. The birds must fly through the frontal area far out over the open ocean (B) before they can continue through the trade-winds zone with stable and clear weather. The maps are based on data from Richardson (1979) and Williams & Williams (1978).

The cold fronts behind which the birds set out shift southeast at a slower and slower speed and finally, up to 1000 km off the coast, become stationary. The birds are then faced with a difficulty, for they must pass the area of the front. The front marks a dividing line between temperate north winds and subtropical easterly trade-winds. Within the frontal zone itself, variable winds and cloudy weather predominate. On the front, wave disturbances periodically arise with extensive areas of rain. Radar studies have shown that often the passage through this zone is critical for the migrants. Birds have been recorded flying in widely varying directions after having lost their bearings, at extremely low speeds and at unusually low altitudes. Those which do not find their way through the frontal area relatively soon are doomed to perish. Once the passage is completed the birds can count on good flying weather for the remainder of the journey; in the trade-winds zone, stable and clear weather with moderate easterly winds prevails.

A special case which cannot be passed over when we are discussing weather and bird migration is the adverse-weather movements of the Swift. Of all the northern birds the Swift is the one that is best adapted to a life in the air. Birds which are not incubating eggs or brooding the young spend the night flying around at high altitude. The Swift runs into particular difficulties when depressions pass: its food, aerial insects, disappears almost completely when extensive areas of cloud and rain appear. Strong winds produce further problems. When flying at night the Swifts' speed is about 20 km per hour, and during migration flight 35–40 km per hour. This should be compared with the normal wind strengths in a fully developed cyclone, between 15 and 20 m/s (54–72 km per hour), at times even more. The Swifts are of course capable of flying faster than the normal speeds just given, at the price of additionally intense flight work and great energy consumption. The highest air speed that has been measured with radar for migrating Swifts in Switzerland was 50 km per hour. The fact that Swifts are able in extreme cases to fly even faster is shown by observations made at the southern tip of Öland, Sweden, on a day when the southwest wind gradually increased to storm force. Swifts were still migrating out over the sea in head wind when it was blowing at 20 m/s (72 km per hour). When the wind strength increased to 23 m/s (83 km per hour), however, the passage ceased altogether. A few hours later, when the wind strength was up to 27 m/s (97 km per hour), a flock of 15 Swifts was seen to come drifting back in tail wind; they made a couple of attempts to fight against the wind so as to reach the lee behind the headland but were unsuccessful and were carried relentlessly eastwards, out over the open Baltic. Even if the Swifts thus might sometimes be able to battle against the winds in a passing cyclone, they are forced to waste huge amounts of flight energy.

Many Swifts leave their breeding sites and set off on adverse-weather movements when a cyclone is approaching. They may have travelled up to 2000 km by the time they return after the cyclone has passed. The Swifts' young may fall into a torpor-like state in the nest: their body temperature is considerably reduced, and by this means they are capable of surviving up to ten days without food. In this way the young get through periods of starvation when the weather is poor and the parents have left on an adverse-weather movement. As well as breeding adults, the majority of the first-year Swifts which are not yet capable of breeding take part in these movements. (Some parent birds stay behind in the nesting cavities in adverse weather.)

The most manifest observations of adverse-weather movements of Swifts in June and July have been made on the east coast of England and at Ottenby at the southern tip of Öland, Sweden. When low pressure crosses Scotland or northern England, southward passage of Swifts is seen along the English east coast. The Ottenby movement is clearly associated with movements of low pressure over northern Scandinavia. The big 'Swift days' at Ottenby, when tens of thousands of individuals pass, occur in fresh (wind strength between 8 and 15 m/s) westerly and southwesterly winds, at which times the birds struggle at low altitude against the wind.

The Swifts seem, therefore, to head southeast when a cyclone approaches from the west. If cross winds or head winds become troublesome, it is good tactics to travel southwards at very low altitude in the lee of the east coast. When the birds have progressed far enough south to avoid the worst of the rains and the winds, they turn up into the west winds. They then continue westwards in head wind until the cyclone has passed north of them and they can begin the return northwards on the back of the cyclone.

This picture of how the Swifts fly around the southern edge of cyclones is based entirely on direct field observations. We must be cautious in the interpretation since such observations cover mainly passage against the wind at low level; movement with the wind at high altitude is overlooked. Should it not be better for the Swifts to fly around a cyclone on its north side? They would then have tail winds for the whole journey. Yes, indeed it would, and I guess this does happen, out of sight from the ground, when the cyclones follow a more southerly track. When low-pressure centres pass as far to the north as Scotland or north Scandinavia, however, it is probably best for the Swifts to fly the 'head-wind circle'. To fly north of these cyclones would force British and Scandinavian Swifts far out over arctic seas, and here the food supply is altogether too meagre.

Interesting observations have been reported of thousands of Swifts moving north at Ottenby in association with those fairly rare occasions when summer cyclones move northwards over east Europe. Swifts then flee northwards away from rain approaching over Poland.

Bad-weather movement of Swifts is also observed regularly at Falsterbo, under almost exactly the same wind conditions as at Ottenby. The number of individuals passing through, however, is much less at Falsterbo than at Ottenby, on average only one-tenth as many. One possible explanation is that the catchment area for Falsterbo Swifts is comparatively small, encompassing perhaps only north Denmark and southernmost Norway. Nor can English passage figures come anywhere near matching the Ottenby figures, either. The cyclones usually cross Britain very rapidly. This is perhaps the reason why not so many Swifts leave on bad-weather movements; those that do so do not travel particularly far. (Though note one recovery of a Swift ringed in Oxford, probably on a bad-weather movement when it collided with a jet plane over Jutland.) In Scandinavia, by contrast, it is not that uncommon for the cyclones to slow down and for the rains, combined with cold, to become prolonged. This promotes bad-weather movements of Swifts on a large scale.

Not only the common European species of swift but also its relatives within the west-winds zone in other parts of the world make bad-weather movements. In addition, the House Martin makes flights in a similar fashion after the end of the breeding season: House Martins in their thousands have been observed at Ottenby fleeing north to escape cyclones over the Baltic States.

Swifts, then, avoid the warm and cold fronts in a passing cyclone, but they gather in large numbers at another type of front, at the sea-breeze front! This has been of great help in meteorology. Thanks to the Swifts it has been possible to detect and to track sea-breeze fronts on radar.

Onshore wind springs up on warm and sunny days when the temperature rises more quickly over the land than over adjacent sea. The cool sea air pushes in like a wedge at low level over the land and forces up the warm land air. Right where the sea air and the land air meet a sea-breeze front is formed. Here exceptionally strong upwinds exist, occasionally to over 1000 m altitude. Glider pilots in England have reported that the front is often very clearly defined. The belt of upwinds, which can vary in strength from 2 m/s to 7 m/s, is only a couple of hundred metres across. If the rising land air is moist enough, cloud is formed; the sea-breeze front is then indicated by a strip of cumulus cloud with pendent streaks of cloud. If the air is dry, then the front is invisible.

Despite the often very light or non-existent cloud at the front, at English radar stations it has been possible to detect clear linear echoes just inside the coast. Glider pilots have confirmed that sea-breeze fronts are responsible, and they have also provided the explanation for why the fronts produce such amazingly clear radar echoes: large numbers of Swifts concentrate in the belt of upwinds in order to catch insects. Thanks to the Swifts, it has been discovered, using radar, that during the course of the day the sea-breeze fronts move slowly inland, at about 10 km per hour. On the very sunniest of summer days, the fronts in England push 100 km in before they begin to break up at nightfall. The Swifts then leave them.

Do migratory birds in normal events suffer wind-drift during migration or do they compensate for the deflective effect of the wind (figure 119)? This question has been very much discussed among those doing research into migrants. It was thought that it could be answered quickly and conclusively when radar became a means for studying bird migration. If the mean direction of the migration was displaced to the right when the winds were blowing from the birds' left-hand side and to the left when the winds were coming from the right, then this should be good evidence that the migrants are wind-drifted. The earliest radar studies in fact showed such a change in the migration's mean direction with different winds. It was therefore soon concluded that the question had been settled.

Continued radar studies in different parts of the world have confirmed that there is virtually always a clear connection between the mean direction of migration and the wind direction. Notwithstanding this, it has been demonstrated that the initial conclusion was altogether premature. The fact that mean direction of migration is influenced by the direction of the wind need not necessarily be due to wind-drift but can also come about when the birds compensate totally for the wind. What we observe on the radar is then apparent or illusory drift, so-called pseudodrift. This is due to the fact that different bird species or bird populations, with migration goals in separate directions, take part in the migration in different winds. The wind favours birds to different degrees depending on the direction to the migration goal; those which in a given situation have the best winds migrate on the largest scale (figure 120).

It soon became evident that pseudodrift was a contributory reason for the correlation between wind direction and mean direction of migration which had been demonstrated in the radar studies. In a number of detailed radar investigations of both diurnal and nocturnal migration overland, it was even shown that pseudodrift provided the complete explanation. In these cases it was thus possible to draw the conclusion that the migrants compensated in full for wind-drift. Other radar studies of migration overland, however, have given conflicting results.

Despite radar having been used in studies of migrants for 30 years or so, we still have not been able to answer the question of whether the birds are wind-drifted. It looks most likely that in actual fact there is no clear-cut answer to the question. Is it conceivable that the birds compensate for wind-drift in certain regions and weather situations but allow themselves to be wind-drifted in others? One thing, however, has been shown clearly by radar studies made up to now: birds are wind-drifted when they fly over open sea. This we can be certain of, for the wind's effect on the direction of migration has always been found to be greater over sea than over land, even when the same birds have been studied in the two different situations.

Woodpigeons and Cranes which migrate over southern Sweden compensate in full for

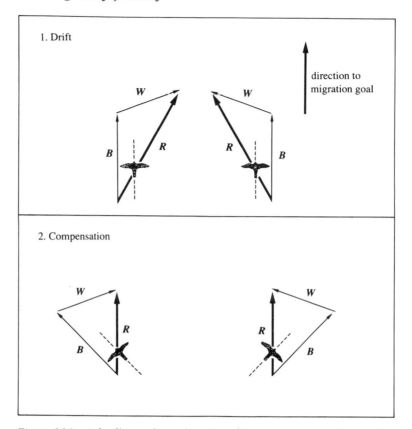

Figure 119   A bird's resultant direction of travel and ground speed (R = resultant travel vector) is determined by its heading and air speed (B = bird's flight vector) combined with the wind's direction and strength (W = wind vector). The vectors are drawn as arrows: these point in the direction of movement, and their length is proportional to the speed of movement. Birds which maintain a constant heading towards the migration goal, independent of the winds, will be subjected to wind-drift. Examples of this are given for winds from the left and winds from the right in 1. To compensate for the wind's deflecting effect, the birds must alter their flight course in towards the wind as shown in 2.

wind-drift as long as they are flying over land. In contrast, over the Kattegat or the Baltic Sea their migration direction is clearly influenced by the wind (figure 121). Amazingly enough, the direction of travel is not altered by the wind as much as it would have been if the birds had flown over the sea on a constant heading. Calculations show that the birds alter flight courses in towards the wind, but only by about half as much as would be required for complete wind compensation. The birds seem therefore to counter wind-drift over both land and sea. On the whole they succeed in compensating in full for the wind over land but only partially for the wind over the sea.

How can birds detect and counter being drifted by the wind? One possibility is that at regular intervals they mark out their desired travel direction (for this they can use their inbuilt compasses, see chapter 5) over the landscape below them; then they see to it that they fly so that the landmarks at various intervals along the direction of travel are the whole time directly behind one another. We would probably use this method ourselves if, for example, we

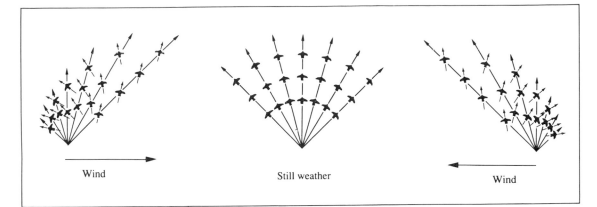

Figure 120 Pseudodrift. Different migrants which set out from a certain area have their migration goals in different directions. This results in a scatter in the migration directions. The lengths of the arrows in the figure are proportional to the birds' travel speed in different migration directions. In still conditions the birds have the same travel speed independent of the migration direction. In cross winds, however, the speed becomes less for birds which meet diagonal head winds than for those who get diagonal tail winds. This is instanced with cross winds of a mean strength which corresponds to 75% of the birds' air speed. Since the birds which can maintain the highest travel speeds set out on migration on the largest scale, the mean direction of the migration as a whole changes with different winds. This happens despite the fact that all birds compensate fully for the wind (the figure shows how birds with different migration directions 'hold up' against the wind).

were trying to row a straight course across a flowing river. At such times, of course, it will not do to row the whole time with our sights simply on the target site; if we do this, then we shall drift with the current and, when we get to our goal, we have rowed in a wide sweep over the river. If, on the other hand, we aim from the start both for the target site and for a landmark directly behind it and then row so that these are in line with each other the whole time, then we shall follow a straight path towards the goal and compensate in full for drift with the current. For migrants, this method of compensation should be especially practicable over land, where they can find many landmarks in the countryside. Over the sea, however, the situation is problematic in that the birds must rely on reference points in the landscape of the waves. Even though there are irregularities and disruptions, larger and smaller waves etc., which the birds can use, a complicating factor remains: the landscape of the waves is a moving one!

Another method which the birds could conceivably use to counter wind-drift is based on the same principle that was used in former days for determining wind when flying with an aircraft. The aircraft's navigator watched the ground over which the plane was travelling through an opening in the floor of the plane. In the opening there was a ring or disc with parallel lines which could be turned so that the ground below was moving along the lines. In this way it was possible to measure by how many degrees the resulting travel direction deviated from the aircraft's angle of course. It is by no means inconceivable that birds are in a similar way capable of determining their resultant direction of travel. In this case they would keep an eye on the ground below while flying and adjust their flight course so that they maintained the desired migration direction independent of the winds. Problems arise, however, over the sea. The crux of the problem is that the direction of travel in relation to the

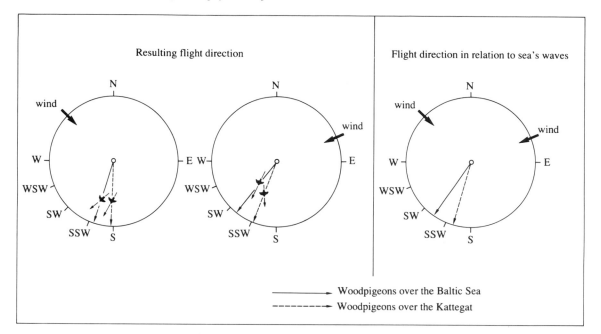

Figure 121   The migration direction of Woodpigeons is influenced by the wind direction when they fly over the Baltic Sea or the Kattegat in autumn. The pigeons alter their heading towards the direction of the wind to too little an extent for it to counter wind-drift entirely. If the migration direction is calculated in relation to the wave landscape of the sea surface, which is moving slowly in the same direction as the wind, the differences between days with different wind directions disappear. The drift suffered by the pigeons seems therefore to be due to the movement of the waves.

wave landscape is not the correct one since the wave landscape is moving. The error is not very important for aircraft, which maintain relatively high speeds, but it creates a more serious problem for helicopters. Significantly enough, it was in fact a helicopter pilot, with great experience of sea flights, who told me about this when I mentioned that I could not explain why birds find it more difficult to counter wind-drift over sea than over land. The pilot was of the opinion that compensation in relation to the landscape of the waves, despite unavoidable errors, was often the best alternative that was available. The waves move in the same direction as the wind but more slowly than the winds are blowing, and it is therefore less of a disadvantage when flying to drift with the waves than to drift with the wind!

For birds which use one or the other of the two methods described to counter wind-drift, the consequences are the same: they achieve complete wind compensation over land but incomplete compensation over the sea, where the drift is dependent on the movement of the waves. So far the argument tallies well with the observations on Woodpigeon and Crane migration in southern Scandinavia. It remains to be determined, however, whether the exact extent of the birds' deviation over the sea can be accounted for by taking into consideration the movement of the wave landscape. Fortunately, there are accurate wave and wind measurements which have been made at various lightships (the majority now defunct) in Swedish waters, including at Fladen in the Kattegat and at Falsterbo Bank in the southern Baltic. These measurements show that the waves' speed of movement on average comes to roughly half the wind strength at an altitude of a few hundred metres. If, starting with these data, we calculate the Woodpigeons' and the Cranes' flight direction in relation to the wave landscape, then we find that it is independent of the winds (figure 121). This bears out the conclusion that over the sea the birds really do compensate for wind-drift in relation to the landscape of the waves.

Now, if migrants use one of the above-mentioned methods of wind compensation, we ought to be able to expect other interesting consequences as well. When flying without being able to see the ground, e.g. above the clouds, birds should not be able to counter wind-drift at all. Trying to compensate by using a cloud landscape below them will probably help them little, for the clouds are moving at the same speed and in the same direction as the wind. Future studies will show how matters really stand in this situation.

Can it be favourable in certain situations for the birds, even if they are able to compensate, 'voluntarily' to lay themselves open to being drifted by the wind? The advantages and disadvantages of drift and compensation change depending on the wind conditions during the migration journey. Let us reflect on several different cases.

1 If the wind stays fairly constant during the flight, complete compensation is always preferable. This of course not only takes the birds straight to their destination but also gives the shortest flying time and thereby the lowest energy consumption. Birds which aim their flight course the whole time straight towards the goal expose themselves to wind-drift (compare the oarsman who steers only towards his target site on the other side of the river) and travel a roundabout way to their destination; on this roundabout route they meet winds that gradually become less and less favourable, and in the final stage they are forced to fly directly into the wind. In the example with oblique cross winds shown in figure 122A, the flying time to the goal is more than 60% longer for a bird which is subjected to wind-drift than for a bird which compensates in full. Over the sea, the incomplete compensation which is achieved using the wave landscape is of good benefit: with the same winds as above, the flying time to the target is only about 20% longer than when compensation is complete.

2 If the wind varies greatly during different flight stages of the journey to the migration goal, the birds make the fastest progress if they avoid compensating fully and instead use a tactic which I should like to call 'adapted drift'. The birds should carry through their flight stages in such a way that they minimise the distance to the goal after each stage. This they achieve if they expose themselves to wind-drift when a long way from their goal and then compensate for the wind to a greater and greater degree the closer they get to their destination; during the final stage of their flight they should compensate fully. An example with varying winds is shown in figure 122B: a bird which uses the tactic of adapted drift reaches the goal after $6\frac{1}{2}$ flight stages; eight flights of equal duration are required for a bird which compensates for all wind-drift to reach the goal in the same conditions. The disadvantage of the former tactic is that the staging sites will lie more or less on a zigzag path on the way to the goal. Consequently the birds cannot make use of the same staging sites in the following migration season since the winds will then in all probability be different. Can it be worth the gain in flight time and energy consumption to lose control of the exact flight path to the destination and of where the staging sites end up?

3 If the winds alternate along the migration path so that cross winds from one direction are succeeded by cross winds of similar proportions from the opposite direction, the birds are best helped by laying themselves open to full wind-drift. Perhaps this applies to the migrants that cross the West Atlantic from North

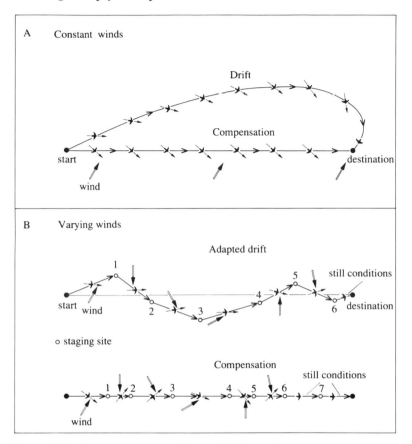

Figure 122   A. In constant winds, wind-drift results in a roundabout route and a longer flying time when the heading is aimed straight at the target destination than when compensation is complete. The figure illustrates the birds' flight paths and flight headings in diagonal tail winds (60° in relation to the direction to the destination) of a strength corresponding to 75% of the birds' air speed.

B. With varying winds during different flight stages to the migration goal, most flight time and flight energy is saved with 'adapted drift'. This means that with each flight stage the birds travel in such a way that the distance to the goal after the end of the stage is the smallest possible. In the figure travel paths, headings and staging sites are compared in the same wind conditions for birds which use adapted drift and those which compensate fully. In this example, the wind strength during the first six flight stages is 75% of the birds' air speed; thereafter the wind is still. All flight stages are of equal time duration.

America to South America? Their heading points fairly constantly southeast to south-southeast, both when they set out from North America in fresh west to northwest winds and when they later pass through the zone of easterly trade-winds. In the former area the winds lead to the resulting direction of travel shifting eastwards, and in the trade-winds zone to its shifting instead to the west (compare the arc-shaped flight path in figure 118).

4 Even if the wind direction is constant during the birds' flight to the migration goal, in certain cases they can gain by flying part of the distance at high altitude with a

certain measure of wind-drift and 'over-compensating' during the rest of the flight at low altitude (figure 123). Owing to friction, the wind strength is generally lighter near the ground than it is higher up: 10 m above the ground surface the wind is on average only about half as strong as at 1000 m. Behind hills and woods, of course, the wind is weaker still. When the birds have tail winds that are not too oblique, they should therefore as a rule fly a straight route to the destination at high altitude. If, on the other hand, head winds or cross winds are blowing, they should instead take a straight path at low level so that they avoid the unfavourable winds as much as possible. The intermediate situation, when diagonal tail winds are blowing, is of particular interest, for the birds then do best with a combined high- and low-level flight. The harder the wind is blowing at high level and the more cover from the wind the birds can get flying at low level, the more profitable this tactic becomes. Usually, however, it results in only a few percent savings in flight time and flight energy. Can such modest flight gains really be sufficient reason for the birds to employ this quite advanced tactic? One reason why we must take the possibility seriously is that over-compensation has in fact been noted for bird migration at low altitude. Autumn observations of finch passage low over inland Scania in south Sweden show that the mean direction is south-southwest to southwest on days with west winds but around south-southeast when the winds are coming from the east. Very detailed studies of Chaffinches and Starlings in the Netherlands have also shown that the mean direction of low-level migration is displaced in the direction of the wind in a similar way. These observations do not, however, provide any definite proof that the birds use a combination of high-level and low-level flight as in figure 123.

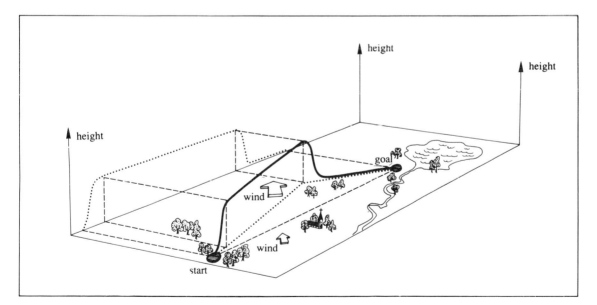

Figure 123   When diagonal tail winds are blowing and when the wind strength is greater at high altitude than near the ground, the birds can save flying time through limited wind-drift at high level for one part of the journey and over-compensation at low level for another part. From a time or an energy point of view, it makes no difference of course whether the 'high-level phase' of the flight precedes the low-level phase or vice versa. The gain in time is the same even if the flight is split into several alternating high-level and low-level periods.

The above viewpoints show that the question of advantages and disadvantages of drift and compensation is more complicated than we might at first imagine. The migrants probably use different tactics on different migration routes and during different periods of the season, depending on the extent to which drift or compensation is favourable. Exactly how they fly in different situations remains to be resolved.

We can easily see good evidence of the striking adjustment to the winds shown by migratory birds. Note how birds migrating at low level temporarily follow valleys, edges of woodlands and similar directional lines, where they manoeuvre skilfully in order to find shelter from the hampering effect of cross winds and head winds and at the same time avoid making detours that are too long. Coastlines have a particularly strong leading-line effect. This is reflected in the large numbers of birds that are guided via the south and west coast of Sweden to the Falsterbo headland in autumn. Bird migration along coastlines is a clear manifestation of adaptation to the wind. The wind strength is in general greater over the sea than over adjoining land, owing to the fact that wind friction against broken terrain is comparatively high. When the birds' migration direction is targeted out over open sea, it can be most advantageous in head or cross winds for them to choose a circuitous route along the coast where the winds do not impede them so much. The chance of flying in some degree of shelter behind hills and vegetation makes it additionally favourable to follow the coasts. A further contributory reason why birds migrate along coastlines is that by so doing they avoid the risk of being wind-drifted across the sea. When this risk arises in cross winds, in many cases it can be an advantage to follow the coast so that the sea-crossing is shortened or perhaps even avoided altogether. When the birds are flying in tail winds, on the other hand, they have little reason to follow the coasts; and this does not happen, either. Radar studies show how the migration instead heads out over the sea on a broad front and at high altitude.

4.9   Dangers during the migration

When the birds begin their migratory journey, they embark upon a hazardous venture. This applies to a particularly high degree to the juveniles, which are migrating through a completely new and unknown world. The birds are certainly small against the mighty seas and deserts and when up against the weather gods. Nevertheless, they more often than not negotiate the migratory journey without faults thanks to skilfulness and, occasionally, a bit of luck. Inevitably, however, there are times when they end up in really critical situations.

If birds set off on migration across desert regions with meagre fat reserves or encounter unexpected head winds during their flight, the consequences can be disastrous. In such situations large hordes of exhausted birds gather at oases in the desert. Others are forced to land in completely open desert with no vegetation and in desperation seek shelter behind stones and rocks in order to avoid the burning sun. In the vast sand-desert areas of the Sahara, migrants are regularly found seeking shade by the metal barrels that mark out the desert roads, by car wrecks and car tyres and in ruins of military installations. Often widely differing species are crowded together in the same shelter. A person reporting from the Sahara tells us of 24 Turtle Doves and a Scops Owl seeking shade together in a car wreck; the observer's own car was also used as a shelter by Quail, Swallow and Subalpine Warbler among others. In the shelter behind petrol drums and car wrecks there are sometimes some insects present, and resting shrikes, wheatears and Willow Warblers are not slow in taking these. Many birds seem to be in such good condition that they are able to complete the passage across the desert as soon as the cool of night reaches them and the wind becomes favourable. The weakest birds, however, have extremely low percentages of fat and water and are doomed to perish.

The winds make the passage across the Sahara particularly difficult in spring. At this time many birds arrive in the Mediterranean area with scant reserves. A striking example is provided by the Quail. In former days this bird arrived in stupendously large numbers in April and May in southern Italy after having flown across the Sahara and the Mediterranean Sea. The exhausted birds were easy to catch. From the 10th century onwards, the Bishop of Capri (the 'quail bishop') received the church's tithes for the most part in the form of Quails. Even at the end of the 1800s more than 50 000 Quails were sold every year to Rome and Marseilles!

Migration across the sea is always of course combined with great risks for landbirds. Most sailors have witnessed how landbirds make forced landings on ships in strong winds or poor visibility. Those that attempt to save themselves in this way represent only a very small fraction in comparison with all those which die over the sea without ever reaching a ship. Large 'bird falls' on ships show that even such relatively small seas as the Baltic, the Kattegat and the North Sea reap many victims among migrants which are caught out by adverse weather.

Lighthouses, masts, and tall buildings with powerful lighting represent dangers to nocturnal migrants, as is well known. In overcast, hazy and damp weather the birds lose their ability to orient in the powerful glare of the light. They are 'intercepted' by the light and fly helplessly back and forth in the beam. The result is often that they lose control of flight and fall and crash to their death on the ground. A particularly nasty menace is the gas flares that burn continuously at oil installations. Studies at oil platforms in the North Sea and the Norwegian Sea, however, have alleviated the worst fears. In good weather conditions the birds do not seem to react at all to the powerful flames. Only on cloudy and hazy nights are the migrants attracted, but there seem to be very few that fly too close to the flames and get burnt to death. The following account comes from an oil platform off the Shetland Isles on a hazy night in October. The birds began to gather during the latter part of the night. Calls from Redwings were heard continuously, interspersed with sporadic calls from Fieldfares and Song Thrushes. At dawn about 4000 Redwings, 200 Fieldfares and scattered Song Thrushes, Blackbirds and Starlings could be seen flying around the installation, lap after lap, all in the same direction, like a ring of Saturn. Most birds were flying at a safe distance, at least 100 m, from the gas flame. One or two individuals occasionally flew in towards the flame but turned away 20–30 m from it, and no bird was seen to be killed in the gas flare itself. Even though the birds do not fly into the gas flares at oil installations, these may nevertheless be a deadly hazard if they hold the birds up over the open sea so that they waste valuable flight energy by flying around for hours in the light to no purpose. The ring of thrushes around the oil platform thinned out during the course of the day. The poor weather with mist and drizzle remained unchanged, however, and in the following dawn there were still some Redwings left. They were by now so exhausted that many of them alighted on the sea, doomed to an early death. The mean weight of dead Redwings that were collected was 36 g; the degree of exhaustion is clearly obvious if we compare this with the normal weight of a Redwing *without* extra migration fat, approximately 60 g!

Similar observations have been made during misty autumn and spring nights at most oil rigs in the North Sea and the Norwegian Sea. Usually the birds leave the area around the gas flare as soon as dawn arrives and visibility improves. Not only Redwings but also Skylarks, Blackbirds, Fieldfares, Song Thrushes, Starlings and Chaffinches have been observed in their thousands circling around one and the same oil platform.

Table 30. *Numbers of species of vagrants from North America and Siberia found in Britain up to and including 1977 (official checklist of British Ornithologists' Union) and in the Nordic countries up to and including 1975 (Iso-Iivari 1976). Fennoscandia comprises Finland, Sweden, Norway and Denmark. Waterbirds consist mainly of waders, ducks and gulls. Landbirds are dominated by passerines, but also include cuckoos, doves, woodpeckers, swifts, kingfishers and birds of prey. Some species breed on both sides of the Bering Sea; individuals which have reached Europe may belong to North American as well as Siberian populations*

| | From North America | | | From Siberia | | | From N America and/or Siberia |
	Waterbirds	Landbirds	Total	Waterbirds	Landbirds	Total	Waterbirds
Britain	31	34	65	3	20	23	5
Iceland	14	14	28	2	3	5	3
Fennoscandia	19	5	24	5	19	24	5

What I have in mind in this chapter is to discuss in detail two other dangers that migratory birds face: the risk of ending up on the wrong course and the risk of becoming victims of birds of prey.

The chance of getting to see a rare bird adds an extra spice to the birdwatcher's life. The stamp-collector has the same sort of dream when he hopes to come across some really rare stamp in a long-forgotten bundle of letters in an old attic. The archaeologist perhaps indulges in images of a well-preserved hoard of gold hidden away in the days of the Romans or the Vikings. Anybody who collects, searches or observes must have hopes of this kind. Birdwatchers are in a favoured situation: it is not so uncommon for migrants, being as mobile as they are, to take the wrong route.

Of the bird visitors that reach north and west Europe, the ones that have come from farthest away belong to two different categories: migrants that breed in North America and migrants that breed in Siberia. Discoveries of these in Europe are made mainly during the birds' autumn migration. Table 30 shows the numbers of visiting species in these two categories which have been found in different parts of west Europe. The fact that there are so many birdwatchers in Britain partly explains why the number of rarities observed is greater there than in the Nordic countries. Other factors, however, also play an important part. In Iceland almost six times as many North American species as Siberian ones have been found, and in Britain three times as many. In Fennoscandia these two categories are equal in numbers. If we look only at Finland, however, then we can find that the number of Siberian vagrants there is twice as great as the number of American ones. That American species are more numerous in Britain than in Fennoscandia is due mainly to the fact that American vagrants, particularly landbirds, get farther east than Britain only to a minimal degree. Many of the Siberian rarities, however, pass Scandinavia and go farther west to England and Scotland. The reasons for the American and the Siberian birds' erroneous flights to northwest Europe are different: American birds get off course as a result of wind-drift over the Atlantic; Siberian birds which reach Europe have made a consistent mistake in orientation.

The normal distribution and migration route of two typical representatives of American and Siberian species which are regularly encountered as vagrants in Europe are shown in figure 124. The great majority of the American species found in Europe are species whose members use the flight path over the West Atlantic in autumn from northeast North America to the West Indies or South America (see figures 109 and 118). As described in the last section, the birds set out from North America in strong northwesterly winds behind a cold front. If they start too near the centre of the depression they run the risk of being carried into the warm sector and being drifted off course by strong west winds. The risk is even greater when the birds are about to pass the trailing cold front over the sea between the North American continent and Bermuda. Wave disturbances along this front often manifest themselves in secondary cyclones, intensified cloud formation and rain. The cyclones move at exceptional speed eastwards along the front towards Europe; within two or three days they have crossed the Atlantic. They often have broad warm sectors which slowly become occluded. Since the temperature difference between the air masses at the stationary cold front is considerable, the wind in these warm sectors can accelerate to strengths of around 30 m/s. These strong winds, known as 'low-level jets', are often channelled to altitudes of about 1000–1500 m above sea level, occasionally even higher. Migrants which are on route through the stationary cold front and happen to come near to one of these secondary cyclones are in a jam! In the cloudy weather they risk losing their bearings and at the same time being drifted off course by the extremely strong west winds. With their large fat reserves they can keep flying for two or three days, by which time the winds have had time to carry them almost

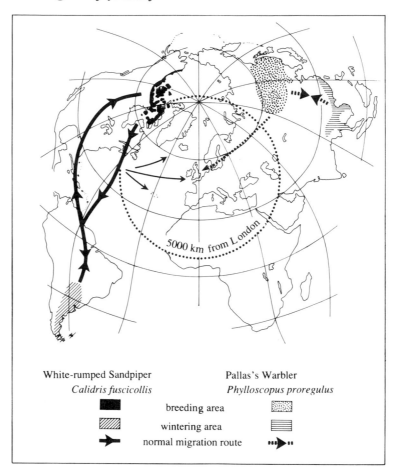

White-rumped Sandpiper
Calidris fuscicollis

Pallas's Warbler
Phylloscopus proregulus

breeding area

wintering area

normal migration route

Figure 124 Two examples of species, one American and one Siberian, which are encountered virtually every autumn as rare visitors in northwest Europe. The White-rumped Sandpiper is roughly the size of a Dunlin. It breeds on the arctic tundra in Canada and winters at both coastal and inland waters in South America, from Paraguay and Argentina south as far as the Falkland Islands and Tierra del Fuego. In Britain and Ireland it has occurred more than 300 times (up to 1988), mainly in Ireland and the south and west of Britain, during mostly September and October; in addition, there are a number of records from the Azores and scattered observations from, for example, France, the Netherlands and Iceland. Yet the species has been seen only nine times in Fennoscandia, where almost all records have been in summer. Pallas's Warbler is intermediate between a Goldcrest and a Willow Warbler both in size and in appearance. It weighs only about 7 g. It breeds in the coniferous forests in Further Siberia and is, incidentally, regarded as the taiga's finest songbird. It normally arrives in its winter quarters in southeast Asia during October and November, at approximately the same time as odd individuals are observed in northwest Europe. In Fennoscandia the number of definite observations (up to 1987) must be well over 300, and in Britain over 470 have been recorded. A couple of separate races of Pallas's Warbler also breed in Tibet; these do not migrate far and have never been identified in Europe, so their distributions have been omitted from the map.

5000 km, right across the North Atlantic. A more detailed analysis of the exact time when the American vagrants have turned up in Britain suggests that many of them were flying with the wind westwards over the Atlantic when they went off course.

An amazingly varied host of American species reaches Europe after this sort of wind-drift: plovers, sandpipers and stints, gulls, crakes, bitterns, cuckoos, nightjars, thrushes, orioles, grosbeaks and wood-warblers, indeed even sapsuckers, a small species of woodpecker. The very smallest passerines are found in very small numbers – for they have the poorest capacity for negotiating the extremely long erroneous flight with their life intact. The Blackpoll Warbler, which is such an abundant Atlantic traveller in America, has for instance been found on only 26 occasions in Britain and Ireland (up to 1988). Not only birds are wind-drifted across the North Atlantic from America to Europe during the autumn – even butterflies are hit by the wind! This happened in October 1968, for example, when not only were exceptional numbers of American thrushes and wood-warblers observed in Britain and Ireland but also over 60 monarch butterflies and even some American moths. The monarch butterfly is a well-known migrant: the summer population in southern Canada and the northern United States flies thousands of kilometres to winter in southern California, Mexico, Texas and Florida

American Atlantic migrants not only risk being drifted off course in the autumn when cyclones pass, they also run the risk of encountering tropical hurricanes during their migration over the sea. Ten such tornadoes occur on average each year during August, September and October, most of them before the autumn migration over the Atlantic has really got underway. The hurricanes form over the central Atlantic and move in over the West Indies and the Caribbean Sea. Some swing northwards off the east coast of North America and move northeast off the coast, and they are then sometimes transformed into ordinary cyclones which continue over the North Atlantic towards Europe. These storms often lead to startling 'falls' of migrants on the south and east coast of North America. Every now and then they also contribute towards the migrants being drifted towards Europe.

American birds sometimes reach Europe as free passengers on ships. This has been reported from the big ocean liners, the *Queen Elizabeth*, the *Queen Mary* and the *Mauretania*, which used to operate the England–New York route. The *Mauretania* made a particularly remarkable voyage in October 1962. The departure from New York took place in storm-force winds from the remnants of hurricane 'Daisy', which was positioned near Newfoundland and was on its way out over the North Atlantic towards Europe. In the morning after the first night, approximately 500 km out of New York, the ship was echoing with the calls of birds. Everywhere there were American species of goldcrests, wrens, wood-warblers, woodpeckers, sparrows, thrushes, waxwings and doves, in all at least 130 individuals of 34 different species! A Merlin also appeared and hunted small birds over the ship's decks. It was completely overcast and the sea was high. The west wind was very strong, but on the ship the birds could find respite since calmer conditions reigned there, for the wind was blowing not much faster than the vessel was moving. The birds gradually left the ship or died as the voyage progressed, but when the ship passed off the south coast of Ireland five days later nine birds of six different species were still on board. Some of these were seen to leave the boat and fly towards the nearest land. A Snow Bunting, a Song Sparrow and two White-throated Sparrows were still on board when the ship docked at Southampton.

An even more dramatic voyage from a weather point of view took place in September 1964, when the *Queen Elizabeth*, a day's journey out of New York, headed in towards the eye of hurricane 'Gladys', one of the most fierce for many years. In the morning there was a raging wind, high seas and freezing fog. As visibility slowly improved migrants landed on the ship, all kinds of species ranging in size from goldcrests to cuckoos.

The American migrants' long flight over the West Atlantic is thus far from risk-free but must claim millions of victims annually. Nevertheless, the birds' timetable is adjusted for as safe a flight as possible. The Atlantic migration takes place surprisingly late in the autumn, with a peak at the beginning of October. This should be compared with the fact that insectivores that are long-distance migrants leave northern and central Europe as early as August or September, and a great many of them have already flown across the Mediterranean Sea and the Sahara before October. October is the safest month for anyone who is going to fly over the West Atlantic. The risk of encountering tropical hurricanes is less then than during the months preceding it, and secondary cyclones and strong west winds at high level are not yet forming on such a large scale as later in the autumn and winter.

What happens to the American birds that are wind-drifted to Europe? No doubt most do not survive for any real length of time in the difficult climate, ever more forbidding the more winter advances. More hardy species such as waders, however, sometimes succeed in adapting to survival in the Old World, where they commute annually between northern summer habitats and more southerly winter quarters. White-rumped Sandpiper, Baird's Sandpiper and Pectoral Sandpiper are regularly reported in circumstances which suggest normal migration through west Europe. We may even speculate on whether perhaps a very small breeding population of these American species has established itself somewhere in the northern parts of the European side of the Atlantic: in north Scotland, a pair of Spotted Sandpipers laid four eggs in 1975 and Pectoral Sandpipers have been seen displaying in May.

Where autumn records of Siberian birds in west Europe are concerned, we can rule out wind-drift as an explanation. It is true that most records occur during autumns with extensive high pressure and east winds over central Russia and indeed the birds often arrive in migration 'waves' in northwest Europe in association with winds from the east, but these winds are rarely strong enough to drift the birds off course; nor is the flow of easterlies continuous all the way from Siberia to west Europe. On the journey westwards the Siberian birds have encountered several cyclones and come up against many periods of westerly head winds. When they reach western Europe they have, as a rule, been travelling for a month or two. The fact that they come to west Europe on easterly winds presumably does not indicate anything other than that they prefer, as migrants usually do, to migrate in tail winds. They appear to carry out their migratory journey in a manner that is entirely normal for birds which travel across continents without major ecological barriers. They have made only one mistake: they have migrated in completely the wrong direction! Instead of flying towards their normal winter quarters in the south or southeast in tropical Asia, they have headed westwards. The reason why larger numbers of vagrants from Siberia reach west Europe in autumns of anticyclonic weather in the Soviet Union is presumably that survival among the birds on the long flight westwards is greater on these occasions than in autumns when cyclones continually usher in moist west winds or cold north winds over the migration path. For many species the migration route to west Europe is longer than the distance to the normal winter quarters. A Pallas's Warbler which breeds on the shores of Lake Baikal has about 4000 km to its winter quarters in southeast Asia, but it is 6000 km to Scandinavia and 7000 km to Britain.

We may suppose that many of the Siberian birds are prepared to set up winter quarters when they reach Europe. They have migrated a distance that is more than the equivalent of that to the normal wintering area and they can hardly be aware that their compass faculties have gone wrong and led them astray. Many of the visitors from afar are small insectivorous warblers of the genera *Phylloscopus*, *Acrocephalus* and *Locustella* which forage in trees, bushes and tall herbaceous vegetation; they are adapted for a winter life in the tropics. The chances of their surviving for any length of time in a Europe where the leaves are already falling fast and the cold of winter is approaching are of course largely non-existent. Nevertheless, the unbelievable occasionally happens. I am thinking, for example, of the Pallas's Warbler that weathered a winter in the Netherlands, from January right up to the end of March; that it then disappeared was perhaps due to the fact that it migrated back towards the breeding area. Ground-dwelling insectivores, such as pipits, are somewhat less delicate and in a few cases manage successfully to winter in Europe. This is indicated by observations in spring in northwest Europe of Richard's Pipits (*Anthus novaeseelandiae*), probably on migration back to Siberia. For every spring observation in Britain of this species, however, there are 60 autumn observations. Practically all of the small Siberian birds which fly to Europe by mistake are juveniles. This is a clear sign that winter survival in Europe is extremely low. There are quite simply no birds left to migrate back to Siberia in the spring and, on completing breeding, to return as adults to their winter homes in west Europe, staked out the previous year. Only the most robust of all Siberian species, such as the Black-throated Thrush and the Pine Bunting, have been shown on repeated occasions to overwinter successfully in Europe. In normal events these species do not migrate as far south as to the tropics but are adapted to endure the winter climate in central Asia.

I have once discovered a very rare Siberian bird myself in my Scanian homeland in south Sweden. It was to all appearances making an attempt to overwinter and surprisingly enough had managed to survive both the late autumn and the early winter. One day between Christmas and the New Year, I was wandering across a pasture meadow at the edge of a wood. There was no snow but the ground was frozen and the meadow was icy in places. Then – from dry-twigged bushes along the fence – a bird flew a few metres in to the meadow, landed and picked something from the ground, only to return immediately to the bushes. This manoeuvre was repeated a few times while I watched the bird with growing amazement through my binoculars. It was about the same size as a Robin but its tail, which it frequently jerked conspicuously, was bright blue and its flanks a warm buff. It was a Red-flanked Bluetail, the 'Siberian taiga's bluebird'! The bird, which was a female or immature, appeared to prefer to spend its time in a dense plantation of young spruce trees nearby. It was seen three days on the trot; its fate thereafter is not known. Since the bluetail is normally found at that time of the year in southeast Asia between India and southern China, the chances of its surviving in the winter country of south Sweden are probably not particularly high. There are local breeding populations of Red-flanked Bluetails as near as the European part of the Soviet Union and in easternmost Finland. The central point of the species' distribution, however, is in Siberia, and in my view it is at least as likely that the bluetail which I observed originated from Siberia as that it came from the very sparsely occupied taiga in northeast Europe.

Apart from the Scanian record, over 20 observations of bluetails that have flown the wrong way have been reported between September and November from countries including Scotland, England, the Netherlands, Germany, Italy (from where a December record has also been reported), Cyprus and Lebanon. The Lebanon find was made in an unusual manner. A Swede on a brief visit to the country was invited one day in October by the local inhabitants to an open-air meal in an olive grove. One man in the gathering was wholly preoccupied with shooting off-passage migrants. This hunting of small birds is common in a good many Mediterranean countries, where the bird-catchers sell their spoils at market and in bazaars. The man in the olive grove was shooting Blackcaps, Nightingales and Redstarts among others. A small bird suddenly landed in an olive tree right next to the assembled diners. The bird-hunter succeeded in shooting it and passed it across to the Swedish guest, who was offered the chance to have it grilled over the fire. The Swede, a competent birdwatcher, was most surprised when he saw that the bird was a Red-flanked Bluetail!

Why do juvenile birds from Siberia misorient and migrate to west Europe instead of to south Asia? Some ornithologists, the Dane Jørgen Rabøl for example, believe that 'there is a clear method in their madness'. He argues that the juveniles have an innate navigation programme which guides them to a series of target areas along the migration path towards the ultimate winter goal. For some reason unknown this system occasionally goes wrong, so that the birds migrate directly away from their target areas instead of directly towards them. After this type of reversed migration the birds eventually end up in northwest Europe instead of southeast Asia. If this argument is correct, the birds ought not to misfly in other directions, for example towards south Europe, the Middle East or out over the Pacific Ocean. The accidental visitors from Siberia that are encountered in the Mediterranean area are, it is true, many fewer in number than those in northwest Europe, but there are not so many birdwatchers and bird observatories in the former region as in the latter. Nevertheless, a lot of birds from Siberia have in fact been observed in the Mediterranean region (the bluetails mentioned above, for example).

Recently, several years' bird observations have been put together from the treeless islands in the west Aleutians, 600 km east of Kamchatka in the North Pacific. These islands seem to be, in respect of rarities, a direct equivalent of the Shetland Isles in west Europe (see below). Siberian warblers, pipits, thrushes etc turn up on the Aleutians as regular but rare autumn visitors. In October 1978 even a Wood Warbler was found, a European species whose distribution extends not much farther east than the Moscow region (sporadic in occurrence in extreme southwest Siberia); the Wood Warbler had thus flown the wrong way and landed up at least 7000 km to the east and not, as normal, in the winter quarters in tropical Africa.

I therefore consider that there are hardly any convincing basic facts that support the theory of a particular reversed migration. The observations of rare migrants in various regions of North America do not fit this picture, either. American ornithologists argue instead that the case may be that the misdirected flights are due to the birds, when orienting, confusing right and left in relation to north or south. Such a mirror-image error in orientation, however, can hardly explain the Siberian birds' migration to northwest Europe.

I think that an adequate explanation may be as simple as this: that some birds have a defect in their sense of direction such that their course deviates at random to a greater or lesser degree from the normal migration direction. Some individuals have such a trenchant defect that they set off in immediately the opposite direction from their winter quarters, others deviate 90° to the right or left, and yet others miss the regular winter quarters only narrowly. Birds which fly in directions where they can survive and be discovered by birdwatchers will be recorded. Landbirds from Siberia which in autumn fly the wrong way eastwards over the Pacific Ocean, northwards towards the tundra and Arctic Ocean or southwest to the central Asiatic deserts will soon perish. Such obstacles to survival do not exist on the route to west Europe.

The survival factor, as well as the birds' varying degree of visibleness to birdwatchers, determines how the discoveries of rare vagrants will be distributed within northwest Europe. During early autumn, in September and at the beginning of October, many discoveries of eastern bird species are made on the treeless and windswept Shetland Islands, particularly at the bird observatory on Fair Isle. At the same time comparatively few rarities are observed in England. Later in the autumn, the ratio is reversed: the records on the Shetlands become fewer and fewer, despite the fact that Siberian vagrants are now being discovered in increasing numbers farther south in Britain. Pallas's Warbler comes all the way from the farthest parts of Siberia, and because of the long distance does not usually reach Britain before the end of October and November; in normal events, therefore, only the occasional isolated record of this species is reported from Shetland, but there are many more records from southeast England. The closely related Yellow-browed Warbler has a wider breeding distribution: this extends westwards to the Ural Mountains, considerably closer to west Europe than that of Pallas's Warbler. The first Yellow-browed Warblers therefore reach Britain as early as the beginning of September. Of the British sightings in September, roughly two-thirds have been made in the Shetlands. Yellow-browed Warblers continue to turn up in west Europe later in the autumn, at which time the birds involved are probably mainly from the farthest parts of the breeding area. From the middle of October onwards, only about one-sixth of the British records are made in Shetland or in the rest of Scotland; the remainder are found mostly along the east and south coasts of England.

No doubt one of the main reasons why the pattern looks like this is that the birds' chances of survival are greatly diminished during late autumn on the route between Siberia and the Shetlands, for this route runs across northernmost Russia and the areas around the Gulf of Bothnia and then via Norway further westwards over the Norwegian Sea. When the autumn has really gripped these northern parts, when the leaves have fallen and the cold has set in, insectivorous birds can no longer find sufficient food. They will perish before they have reached the north Scottish islands. Birds which reach west Europe on a more southerly route, over central Russia and southern Scandinavia, can on the other hand still get by a month later.

Why are there not more discoveries in England earlier in the autumn when the reports from Shetland show that many eastern vagrants come to northwest Europe? One reason is that the distance from Siberia to the Shetland Isles is shorter than that from Siberia to England via the southern North Sea countries. The birds thus arrive earlier in north Britain than in south Britain. The difference in distance seems to be a reasonable explanation also for the earlier arrival of Siberian vagrants in north Scandinavia as compared with south Scandinavia (Ullman 1989).

In North America, a lot of juvenile individuals of short-distance migrants fly from the northeastern United States east or southeast in autumn, straight over the open Atlantic. If they had behaved normally they should instead have migrated southwest over the interior. Ornithologists at bird observatories along the east coast of North America, where many of the off-course juveniles are observed and trapped, estimate that every autumn several million migrants fly out over the open sea to a certain death.

A similar example is provided by the North American landbirds that misfly to Greenland in the autumn – many of them are small insectivorous woodland species. In southwest Greenland more than 50 different species of landbird have been found which are vagrants from North America. Many have their nearest breeding sites in central Canada. Instead of migrating south or southeast in autumn towards tropical winter quarters, they have flown a couple of thousand kilometres northeast over treeless tundra and sea to Greenland. The very shortest distance from the forest limit in Labrador to southwest Greenland is a good 1000 km. It is striking that woodland birds survive a journey of 1000 km at the very least, often probably twice that or more, over such inhospitable regions. One of the reasons they do is no doubt that many of the species normally migrate in autumn over the West Atlantic towards South America and are therefore well adapted to long direct flights. This applies, for example, to the Blackpoll Warbler, which has visited Greenland on a good number of occasions.

It seems, therefore, that misoriented flights in all different directions take place. Why do migratory birds make these serious errors in orientation? Here I must admit to my own bewilderment. In previous sections I have time and again written of and expressed my admiration for the perfect adaptations shown by migrants. We might think that natural selection, which has chiselled out these fine adaptations, should long since have removed serious orientation defects. Selection combats these shortcomings in orientation in the strongest way possible: the birds so affected almost always fly to their death and do not get the chance to propagate their tendencies further. Nevertheless, errors in orientation recur in generation after generation of new juveniles. Do birds perhaps carry a latent genetic handicap which manifests itself in a percentage of the offspring in this way? Or possibly the birds' compass sense develops wrongly if they grow up at places with local anomalies in the magnetic field (see sections 5.1–5.3)?

There are many important magnetic anomaly regions, associated with deposits of magnetic minerals such as iron ore, in the Soviet Union, e.g. at the Ural Mountains, Central Siberian Plateau, southeast Siberia and, most pronounced, the Central Russian uplands around the city of Kursk. The latter region constitutes one of the earth's greatest magnetic anomalies. Birds growing up where the magnetic field is strongly deflected from normal may, through magnetic miscalibration, acquire an erroneous compass, which will lead them astray during the migration season. If these speculations are correct, one should also expect magnetic miscalibration to occur close to the magnetic north pole in high-arctic Canada, where the horizontal component of the magnetic field is erratic. Indeed, the Nearctic wader species with breeding ranges extending close to the magnetic north pole are those which most frequently show up as vagrants in Europe as well as in other parts of the world (one of these species is the White-rumped Sandpiper, cf. figure 124).

Despite the fact that the migrants' misdirected flights mean waste of life, they contribute, inevitably, to making birdwatching additionally exciting. We can admire the birds' natural abilities to become pioneers and colonisers, when opportunity arises. We can never be totally certain that we shall not suddenly, on any day at all, find ourselves standing face to face with the most unexpected bird.

One example among many: One day in the middle of December 1860, something unusual happened in the yard at Strands weaving mill right on the shores of Lake Vättern in Jönköping, south Sweden. A bird

> 'had crept in under the fence which encloses the yard on the Vättern shore. Here it was noticed by a couple of factory hands, who only after much chasing managed to get hold of it. It ran at great speed and also used its wings, but seemed not to have the sense to rise into the air in order to get away from its agonising running and escape its imprisonment. It was captured alive, but died, however, amid convulsive spasms a short while thereafter.' (Translated from Professor F. Wahlgren's report, 1867.)

The bird, which was later stuffed, proved to be a Parakeet Auklet. These birds breed almost 6000 km away on the other side of the North Pole, in the Bering Sea (figure 125). The off-course flight is even more remarkable in that birds of this species do not as a rule migrate particularly far but winter at sea in the ice-free parts of the North Pacific Ocean. Even today this Parakeet Auklet is the only record of this species in the West Palearctic, indeed in the whole of the Atlantic region also. The closest equivalent is a record of a Crested Auklet, a close relative of the Parakeet Auklet and with a similar distribution. This was shot from a fishing boat off north Iceland in the middle of August in 1912. The skipper shot the bird because of its remarkable behaviour – it was resting on the water right next to the boat – and had it stuffed

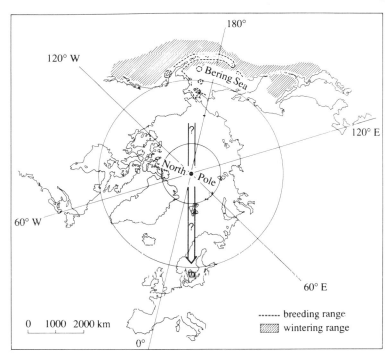

Figure 125   What is a Parakeet Auklet, with its normal range in the Bering Sea, doing on the other side of the North Pole, at Lake Vättern in south Sweden? The Parakeet Auklet breeds in colonies, sometimes together with Crested Auklets, in cliff crevices and cavities between boulders. It feeds mainly on planktonic oceanic crustaceans and often gathers food on the open sea far off the coast.

after arriving home in the Faeroes. After many complications and rumours concerning this highly remarkable bird (which was in full summer plumage, with crest, brilliant red bill plates etc), at the end of the First World War the Crested Auklet came to Copenhagen, where it was possible to establish its correct identity.

The age of sensations is far from past when it comes to rare migrants.

Migrant birds are many times sorely exposed to attacks by birds of prey. Before making long flights they sometimes put on such large fat reserves that they can gain height only with the greatest difficulty. This applies for example to the Yellow Wagtails that stop off at Lake Chad before the spring flight across the Sahara. Where the wagtails gather at communal roost sites numerous Pallid Harriers and Dark Chanting Goshawks wreak havoc. The main reason why the raptors are so successful in their hunting, however, is perhaps not always that the migrants are hampered by their weighty fat load. They also have the disadvantage of being at a new and unfamiliar stop-over place. The element of surprise in the raptors' attacks therefore secures maximum effect. In the hordes of migrating small birds there are also birds which are predators. Thus the shrikes catch quite a number of small warblers at the staging sites during the migration to the tropics.

In autumn masses of resting thrushes and Robins sometimes concentrate at the Falsterbo peninsula. Wherever one looks, the place is abounding in migrants searching for food, in every piece of shrubbery, in gardens and in parks. On such occasions birds of prey also gather in large numbers. Gustaf Rudebeck once estimated that the number of Sparrowhawks hunting at the same time at Falsterbo was at least 400, and perhaps nearer 600! The object of the Sparrowhawks' pursuit was Song Thrushes, which on that occasion had stopped off in particularly large flocks. The hawks were seen repeatedly as they chased them in furious pursuits. One thrush which had only just eluded an attack was the next second attacked by another Sparrowhawk. At times the poor thrushes were set upon from several different directions at once. Having recently arrived at a staging site, where the feeding places with the best cover were not sufficient for all of them, and exposed to this stream of attacks from Sparrowhawks, the Song Thrushes suffered heavy losses. At the same time the hawks enjoyed exceptionally good hunting success.

Migrants run the risk of being attacked by birds of prey not only at stop-over places but even on the migration flight itself. Particularly at risk are landbirds over open terrain or over sea, where, if the situation becomes critical, they have no chance of coming down and landing in the cover of vegetation. Over the open sandy areas on the Falsterbo headland Sparrowhawks and Merlins often take the chance to hunt small birds that are migrating past. The flight pursuits around the few bushes and trees into which the harried birds attempt to escape can at times be intense. The Peregrine prefers to hunt migrating Starlings, pigeons or crows at the coast, over open sandy areas as well as over the sea. From high in the air it dives towards the flocks of birds, and when a flock has split up it strikes its victim in mid air in a tremendous stooping dive. Sometimes the Peregrine also hunts over the open sea, where it has been observed to strike both migrants and seabirds, particularly storm-petrels, in the vicinity of ships.

Not even the biggest migrants are immune from attacks by raptors. As described in section 4.5, Demoiselle Cranes migrating over passes and valleys in the Himalayas are attacked by Golden Eagles. In Africa the big Martial Eagle has been seen striking migrating storks in similar fashion.

Among the most dangerous predators for small birds on migration are the gulls, especially the Great Black-backed and the Herring Gulls. The gulls often operate between 100 m and 1–2 km off the coastline, where they attack birds migrating low over the water. When a gull has spotted a victim and made an initial attack, as a rule several gulls quickly gather and join in the chase. The small bird is forced to make repeated sudden evasive manoeuvres and often drops perilously close to the water's surface. If it cannot manage to reach the shore quickly, it is forced down on to the surface of the water through exhaustion. Here the gulls tear it to pieces, or swallow it whole if it is small enough. Occasionally the small bird is knocked down in flight into the water by the wings or feet of the pursuing gulls. The gulls may also succeed in seizing their victim in the air with their bill and hurling it into the water.

This phenomenon has been observed with particular frequency in England. Many of the attacked migrants approaching the English coast after having crossed the North Sea in autumn come from Scandinavia. Thrushes and Starlings are among the commonest types of prey, but small warblers and Goldcrests also fall victim to the gulls. Occasionally giddy pursuits are enacted before the eyes of birdwatchers. Through the telescope a small bird is seen barely 1 km out over the sea and heading straight for land. It is flying about 10 m above the waves when it is attacked by gulls. Next comes a race for dear life! The small bird fends off the gulls' dives, dodges and accelerates, flies tenaciously towards the safety of land ahead of it and at the same time attempts to gain height. But the attacks from the gulls are now coming so thick and fast that it has to sacrifice height in order to manoeuvre out of the way. Suddenly the small bird receives a blow from the feet of one of its attackers and only just avoids falling into the water. There are now only a couple of hundred metres to go to land, but the small bird is getting weak and is flying less than a metre above the water's surface. The chase continues.

Will the bird make the final short stretch to land? I shall not reveal the outcome in this particular case but simply state that the migrants sometimes succeed in reaching the shore and that they sometimes come to an end only a few tens of metres from the shore. What a wretched fate to have negotiated the flight all the way across the North Sea and to have land within close reach ahead and then to die a paltry few metres from the goal!

Landbirds which migrate over the sea in daylight thus benefit by flying at high altitude (unless the head wind is too strong). Using radar, I am currently mapping the way landbirds migrate out over the sea. They climb strongly even if the wind should happen to be slightly more unfavourable at high level than low over the waves and even if only a crossing of a bay no wider than a kilometre is involved. Presumably the danger of gull and raptor attacks is the reason for this.

The gulls do not hunt migrating small birds only at the coast but also far out over the open sea; this has been observed from fishing boats and, especially, from oil platforms in the North Sea. A number of Great Black-backed Gulls and Herring Gulls spend almost all of their time at these installations; occasionally there can be several hundred of them. They have a ready-laid table on those days when the migrants are concentrated around the gas flares in misty weather. At such times the gulls are fully occupied in chasing more or less exhausted small birds flying near the sea's surface. Now and then the gulls are accompanied by migrating raptors; the odd Merlin or Kestrel then joins in the bird-chase. At night Short-eared and Long-eared Owls sometimes break off their migration to hunt small birds in the light of the gas flares – certainly a remarkable phenomenon more than 200 km from the nearest land.

The risks to migrants of meeting birds of prey are, as we can see, obviously quite big. Nevertheless, I have not yet said anything about the true specialists in hunting migrants: Eleonora's Falcon and the Sooty Falcon.

Eleonora's Falcon (*Falco eleonorae*) and Sooty Falcon (*Falco concolor*)

Eleonora's Falcon is intermediate in size between the Hobby and the Peregrine; it weighs about 0.4 kg. It has extremely long and slender wings; the wingspan of just under 1 m is approximately the same as that of the Peregrine. Eleonora's Falcon breeds in colonies on isolated islands or on steep cliff faces in an area from the Canary Islands in the west to Cyprus in the east (figure 126). The one hundred and more breeding colonies vary in size from a few pairs up to almost 200 pairs. The total population is estimated at between **4000 and 5000** pairs, almost two-thirds of which are found in the Greek archipelago in the Aegean Sea and around Crete. For breeding, Eleonora's Falcon is wholly dependent on catching small birds on migration over the Mediterranean Sea or over the Atlantic off Morocco. It therefore does not breed until the autumn. Egg-laying takes place at the end of July and hatching one month later. The busy period when the young have to be fed therefore falls at the end of August and continues until the beginning of October, during exactly that period when the long-distance flight traffic of small birds from south Europe to tropical Africa is at its most intensive.

In recent years, Eleonora's Falcon has been closely studied at two of the larger colonies, on the Mogador Islands off the west coast of Morocco and on the Greek island of Paximada, not far from Crete. Interesting differences have been shown between the two colonies. The falcons at Mogador have significantly better hunting success than those at Paximada, this being reflected in the fact that clutches are larger and the number of fledged young greater at Mogador (average clutch size 3 eggs, 2.5 fledged young per pair) than at Paximada (average clutch size 2 eggs, 1.3 fledged young per pair). The Mogador falcons sometimes catch so many small birds that they lay up a 'larder', most of which decomposes without ever coming in useful.

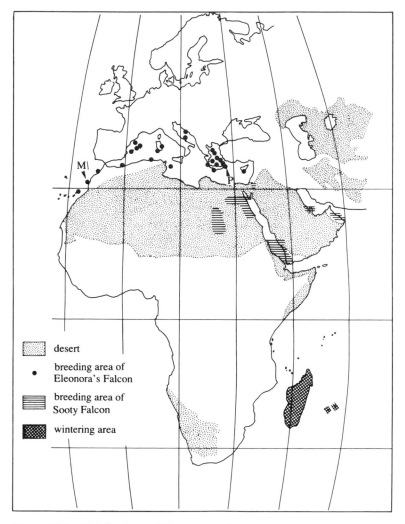

Figure 126   Distributions of Eleonora's Falcon and Sooty Falcon. M and P show where the breeding colonies of Mogador and Paximada are situated. The Sooty Falcon's breeding distribution is poorly mapped and probably includes larger parts of the Sahara and Arabia than shown on the map.

What is the reason for the difference in hunting success? The falcons hunt over the sea where the migrants cannot escape by landing. The small birds migrate primarily at night; generally, there are none that set out from south Europe over the sea after dawn. Those which have already passed the coast when morning breaks, however, do not turn back but continue their long flight in daylight. These are hunted by Eleonora's Falcons. True there are suspicions that the falcons sometimes also hunt on moonlit nights, but this is certainly very rare, if it occurs at all. Paximada is in a much inferior position to Mogador for passage of migrants in daylight. This is because roughly 160 km north of Paximada is the island group of the Cyclades, where several large islands provide suitable stop-over environments for migrants. After dawn very few small birds continue their migration past these islands. The falcons on Paximada are consequently reduced to birds which find themselves at dawn over the open sea the 160 km north to the Cyclades. If the small birds' travel speed is around 50 km per hour, the migration past Paximada in daylight will go on for only approximately three hours from dawn onwards. This fits well with the falcons' daily hunting habits. Around one hour before sunrise all the male falcons leave the nest sites to start hunting. Occasional birds return very quickly to their females with prey, while it is still almost completely dark. The falcons have by far the best hunting success, however, during the first hour after sunrise; thereafter hunting success diminishes rapidly, and after three hours hunting ceases almost totally. The falcons also have a very short hunting period at dusk, at which time they attempt to catch small birds which have stopped off during the day on Paximada Island itself and are resuming their migration.

The falcons at Mogador catch small birds migrating direct from Portugal to the very westernmost parts of tropical Africa. The distance from the southwest tip of Portugal to Mogador is roughly 600 km. The journey is over the open sea the whole way. Small birds which have just left European shores behind them at dawn therefore have about 12 hours to fly before they pass Mogador. This means that the falcons at Mogador have the whole day available to hunt migrants, right from dawn, when the time is just right for the first small birds, those that left southwest Portugal at dusk, to arrive, to late evening, when the last ones, those that migrated out over the sea immediately before dawn, pass.

Nor is the falcons' hunting technique the same at the two colonies, either. At Mogador the falcons often fly far out over the sea northwestwards to meet the migrants. They often slow down their flight, glide, soar, fly on, change altitude and turn. In this way they effectively search through a large airspace, both vertically and horizontally, on the look-out for passing small birds. Sometimes they concentrate on hunting very low down over the sea. We may assume that this happens mainly when the migrants meet head winds over the sea and are therefore flying lower than normal. The low-level chases are very reminiscent of the catching technique of the gulls described earlier. One or more falcons tire their victim through incessant attacks. Another hunting method consists in the falcons 'stationing themselves' in head-wind flight at different altitudes over the breeding colony and in its immediate vicinity and waiting for the migrants. This method is not practised so often at Mogador but is easily the predominant hunting method at Paximada.

The winds at Paximada are usually moderate northwesterlies. This favours the migrating small birds, which receive tail winds, but it also makes the falcons' hunting method of awaiting the migrants by 'hanging' in the head winds over the sea appropriate. As a rule all the 150 males at Paximada take part together in hunting. Each falcon maintains a distance of between 100 m and 200 m from its nearest neighbour. The falcons are dispersed at various altitudes up to 1000 m and more above the sea; most prefer the higher levels. Every morning around sunrise, therefore, a giant trap forms for migrating small birds at Paximada, a barrier of Eleonora's Falcons from the sea surface up to an altitude of over 1000 m, several kilometres across and several kilometres deep.

A small bird which is migrating in towards this barrier is initially attacked by a falcon which attempts to strike it from above in a steep dive. Should the small bird manage to manoeuvre out of the way, then several falcons quickly join forces in the attack. The small bird is forced to fend off, by constantly diving and zigzagging, repeated attacks from above, from the sides and from behind, at times by ten or 20 different falcons in rapid succession. As the small bird becomes exhausted, the likelihood of a falcon succeeding in striking it with its talons increases. The more attacks a bird is subjected to, the more height it loses; sometimes the hunt goes on all the way down to the sea surface. Here the falcons continue their assaults, and often the outcome is that the bird is caught after it has been forced into direct contact with the sea waves one or more times.

Small birds which are straight above the little island of Paximada when they are attacked sometimes use a special technique to endeavour to escape. They fold their wings and drop like stones as straight as an arrow towards the island, sometimes from as high as 1000 m up. The most serious element of risk comes when the bird reaches a low altitude and has to level out and slow down before landing: at this point a falcon may be at hand and strike it with its talons. Should the bird succeed in landing properly, then it must remain still and quiet in cover among bushes and cliff crevices throughout the day right up until it gets dark. Then it can continue its migration and get off with nothing more than a fright. Some small birds, however, attempt to leave the island during the day; this is usually unsuccessful since the falcons get sight of them and institute a furious chase over the sea.

When hunting over Paximada is at its height in the morning, each male falcon delivers prey to its mate, which guards the nest and feeds the young, at an impressive rate, often several items per hour. The record is held by a male which captured five birds in 35 minutes.

Before they eat or before they feed the young, Eleonora's Falcons pluck their prey at the nest. By collecting and analysing the feather remains, the choice of prey can be determined; it has been established that the species' diet includes more than 90 different bird species. The commonest prey species at Paximada and Mogador are shown in table 31. The prey composition at the different colonies provides an interesting insight into the geographical migration patterns of the small bird species. The greatest surprise perhaps is that such a large number of Grasshopper Warblers have been found at Mogador, for hardly any winter observations at all of this species, which lives a very secretive life in tall herbaceous vegetation, have been made in Africa – only a few autumn and spring observations in Mauretania and Senegal and a recovery in Ethiopia have been reported. Prey remains from Eleonora's Falcons indicate that the Grasshopper Warbler is far more common than the few field observations suggest and that we should look for its unknown winter quarters somewhere in West Africa first.

Eleonora's Falcons thus provide important contributions to our knowledge of bird migration. We must, however, be cautious in interpreting the falcons' prey composition as an entirely random selection from the passing bird traffic. Presumably the falcons have a particular eye on the shrikes, which provide two or three times as much nourishment as the small warblers. Other prey species of suitable fare which occur, in few numbers but regularly, in the falcons' diet are Turtle Dove, Cuckoo, Nightjar, Scops Owl, and various waders such as Ringed Plover, Common Sandpiper and Green Sandpiper. On Paximada occasional finds have also been made of Roller and Little Bittern, killed by the falcons. Eleonora's Falcons can hardly complain of a narrow diet when it comes to different bird types.

Each migration season between August and October, the Eleonora's Falcons in the whole Mediterranean region and on the easternmost Atlantic are estimated to catch nearly 2 million migrants. The daily average is almost 30 000. The falcons' combined consumption amounts to around 40 tonnes of bird flesh. These seemingly titanic figures should, however, be compared with the enormous total number of birds that migrate in autumn from northern regions to Africa: at least 5000 million. The Eleonora's Falcons, after all, catch only one in 2000–3000 migrants.

In the eastern Sahara and the Arabian Peninsula lives the Sooty Falcon, which, like Eleonora's Falcon, breeds between August and October and which concentrates wholly on hunting migrants. It occurs in sandy desert, where it hunts migrating birds over the dunes, in rocky mountain desert and on inaccessible rocky islands. A breeding population exists in the

Table 31. *The 15 commonest prey species of Eleonora's Falcons at two different breeding colonies, Paximada and Mogador (see map, figure 126). Feather remains of prey were collected daily between 26 August and 23 September at a selection of nest sites within each colony, to an extent corresponding to 600 'nest-days' at Paximada and 367 'nest-days' at Mogador. Species identification based on feather remains is often difficult within the genus Phylloscopus and the genus Hippolais (confusion risk with Spotted Flycatcher). The positive species identifications that it was possible to make within these categories suggest that the Phylloscopus category at Paximada is dominated by the Willow Warbler (approximately two-thirds of the specifically identified feather remains); this is followed by Wood Warbler (one-fifth) and a smaller proportion of Bonelli's Warblers and Chiffchaffs. At Mogador, only Willow Warbler (commonest) and Bonelli's Warbler were identified with certainty. The Hippolais/Spotted Flycatcher group is represented at Paximada by Icterine Warbler (approximately two-thirds) and Spotted Flycatcher (one-third), at Mogador on the other hand mainly by Melodious Warbler (two-thirds) together with Spotted Flycatcher and Olivaceous Warbler. Data from Walter (1968, 1979)*

Paximada (Greece)		Mogador (Morocco)	
Species	No.	Species	No.
Phylloscopus spp.	158	Woodchat Shrike	189
Red-backed Shrike	128	Whitethroat	176
Whinchat	81	Nightingale	158
Whitethroat	76	Redstart	152
Short-toed Lark	63	Orphean Warbler	73
Hippolais spp. + Spotted Flycatcher	55	Pied Flycatcher	57
Lesser Grey Shrike	39	Grasshopper Warbler	53
Pied Flycatcher	35	*Phylloscopus* spp.	42
Hoopoe	29	*Hippolais* spp. + Spotted Flycatcher	36
Wheatear	23	Subalpine Warbler	30
Nightingale	18	Rufous Bushchat	14
Golden Oriole	16	Quail	13
Woodchat Shrike	13	Whinchat	10
Redstart	11	Wryneck	10
Tree Pipit	9	Garden Warbler	9
Other species	74	Other species	102
Total	828	Total	1124

Red Sea basin on the many small coral and desert islands with an extremely hot and humid climate. The Sooty Falcon is not so tied to colonial breeding as Eleonora's Falcon and occurs in solitary pairs as well as in fairly loose groups of at most 100 breeding pairs. It hunts alone or in rare cases in pairs; no group hunting similar to that of Eleonora's has ever been observed. One of the few ornithologists who has visited the breeding islands in the Red Sea tells us that the Sooty Falcon often perches on the top of a rock or on a euphorbia on the lookout for passing birds. From here it makes a sudden rush attack, either towards birds flying low over the sea or towards birds passing above the falcon. In the latter case it rapidly gains height and accelerates in a long shallow dive to seize the prey from behind.

A considerable proportion of the Sooty Falcons' prey on the Red Sea islands consists of such exotic species as Little Green Bee-eater (the commonest prey of all), Bee-eater, Hoopoe and Golden Oriole. Feather remains found at the nest sites have also included Yellow Wagtail, Wheatear, Willow Warbler and Sand Martin. For Sooty Falcons breeding in the Libyan desert the prey composition is different: here *Phylloscopus* and *Sylvia* warblers predominate.

Eleonora's and Sooty Falcons exhibit great similarities, not only in lifestyle when breeding but also where migration and wintering are concerned. They both change from a bird diet to an insect diet and make their way to Madagascar and adjacent islands, where between November and May they live in flocks and catch swarming Neuroptera, grasshoppers, termites, flying ants etc. The Sooty Falcon is the commonest of the two: judging from winter observations in Madagascar, it is about ten times as numerous as Eleonora's Falcon (though some Eleonora's regularly winter in East Africa).

Why do these two species of falcon choose Madagascar of all places for their winter quarters? No other migrants from northern parts, from the Sahara northwards, winter there. The reason is presumably food competition with other insectivores and gregarious falcon species which winter in the southern part of Africa. The various species have divided up the region in such a way that the central point of the winter range is in a different place for each one. The Lesser Kestrel winters in the Republic of South Africa, the Red-footed Falcon in southwest Africa, the Eastern Red-footed Falcon in Zimbabwe and Botswana and the Hobby in Zambia. Eleonora's Falcon and Sooty Falcon thus fill up the map perfectly through their presence in Madagascar.

The huge multitudes of migrants which cross the Sahara in March and April on their way north are not left in peace by the birds of prey: the Lanner Falcon represents the greatest threat on the crossing. The Lanner breeds during the spring at scattered sites throughout the Sahara desert and North Africa (its nest has for example been found on the Egyptian pyramids). A good number of pairs are wholly dependent for their breeding success on catching migrants, for example those nesting in the Tibesti massif in the central Sahara. Around the nest sites there is not the smallest patch of vegetation as far as the eye can see. Here the Lanner Falcons hunt Turtle Doves, Quails, Yellow Wagtails and Hoopoes which because of head winds or exhaustion migrate at low level over the desert.

The earth harbours so many migratory birds that only a small proportion is caught, despite the multifarious threats from birds of prey. The migrants have created possibilities for bird-hunters such as hawks and falcons to exist. In speed, manoeuvrability and explosive flying power the latter have no counterpart within the animal world. Over this course of nature we can feel a sense of comfort and pleasure.

5 *Orientation and navigation*

How birds find the right way on their migratory journeys is one of the greatest mysteries of bird migration. It is well known that some birds return year after year to the same territory, both within the breeding range and within the wintering range, and sometimes to the same staging sites along the migration route as well. To all appearances these birds have commanding powers of navigation. These powers are not necessarily common to all birds, not even to all migrants. The degree of site fidelity (*Ortstreue*) differs appreciably from species to species, and juvenile and adult birds also have different tendencies to return. Displacement experiments have revealed differences in various species' tendencies or abilities to return home. In this chapter I use the terms orientation and navigation in partly different ways. By orientation I mean simply the capacity for determining and maintaining a certain compass heading; navigation implies that the birds will be able to determine their position in relation to a fixed goal.

Many seabirds exhibit very good navigational powers. What else can be said of the Manx Shearwater that was transported from its nest burrow on the little island of Skokholm off the Welsh coast to Boston, almost 5000 km away on the other side of the Atlantic, and which flew home in 12 days? The letter informing that the bird had been released at Boston reached Wales the day after the bird had got home! Leach's Petrels have been responsible for equivalent feats in the reverse direction: of seven individuals which were caught in their nest burrows in easternmost Canada and released on the south coast of England, four returned, the first two within 14 days. The birds that have flown the farthest in successful displacement experiments are Laysan Albatrosses that were caught on the Midway Islands in the Pacific Ocean: two birds which were released on the west coast of the United States, approximately 5200 km from the breeding site, returned after ten and 12 days; one albatross found its way home all the way from the Philippines, 6500 km away, but the return journey lasted a whole month.

Many landbirds, too, have good navigating ability, something which has been known for thousands of years. It is said that during the days of the Roman Empire Swallows were caught at their nests and taken to the chariot races; the Swallows were afterwards released and soon brought home news of the winning chariots' colours. The carrier pigeon is the bird that was first exploited for supplying messages quickly. The ancient Egyptians used carrier pigeons, as also did the Greeks and Romans, and even today they are still used. Their great use is indicated for example by reports from the siege of Paris during the Franco-Prussian War of 1870–1871: an estimated 150 000 official and 1 million private messages to and from the city were delivered via carrier pigeons. Even as recently as the Second World War, hundreds of thousands of carrier pigeons were used by the military authorities in Britain and the United

States. Near on 17 000 carrier pigeons were dropped by parachute to resistance movements in occupied Europe, and around 2000 returned to Britain with important communications. Quite a number of military aircraft crews were rescued thanks to carrier pigeons which were released with SOS messages when the planes had crash-landed.

Concurrently with the advances made in telegraphy and radio communications the carrier pigeon's importance as a message-carrier diminished and finally disappeared. Modern society, however, has another use for the carrier pigeon. When the railway networks were developed during the 1800s, the foundations were laid, with rapid transport to distant regions, for the sport of pigeon-racing. No end of pigeon-racing events are organised annually in west Europe and in North America. Events in which the pigeons are released between 500 and 1000 km from the lofts are common, and sometimes races are even arranged over distances of nearly 2000 km. In long-distance races a fairly large proportion of the pigeons fail to return home. The best homing pigeons can fly for almost 16 hours a day. The average speed to the pigeon loft in one-day flights is between 60 and 70 km per hour, and in strong tail winds over 100 km per hour.

Homing pigeons play a dominant part as experimental birds in research into orientation. The standard procedure in orientation experiments consists in following individual pigeons through binoculars when they are released and determining the direction in which the pigeons disappear towards the horizon. This 'vanishing direction' can then be compared with the correct direction to the pigeon loft. The pigeons' normal behaviour is to circle two or three times over the release site before deciding which way to travel. Many pigeons embark upon a course which is not far off the correct homeward direction within as little as one minute, by which time they have not even gone 1 km from the release site. Not only is the way in which the pigeons' vanishing directions are distributed recorded, however, but the time it takes the pigeons to reach home is also measured and the number of them that fail completely to find the pigeon loft again is counted.

Surprisingly enough, the homing pigeon does not have its genealogical roots in any species of migratory bird. Like the city pigeon, it originates from the Rock Dove, which lives as a resident on rocky coasts and steep mountain slopes mainly in southern Europe and Asia. Through man's selection, the homing pigeons have become superior to their ancestors in the art of returning home quickly. Comparative releases at distances of up to 100 km suggest, however, that it is not so much navigating ability as the driving force to return home and flight endurance that distinguishes the two types of pigeons. Where vanishing directions are concerned they are about equally accurate, but despite this it takes the Rock Doves a considerably longer time than the homing pigeons to get home (sometimes days or weeks) and many of them never return. The experiments with pigeons show, therefore, that good powers of orientation and navigation are in no way the preserve of the migratory birds.

How do pigeons and other artificially displaced birds find their way home again? How do migratory birds find their correct route? We might as well openly admit from the start that there are no conclusive answers to these questions, but there is no reason to feel disappointed and despondent at this. The fact that the mysteries are still unsolved does not mean that knowledge, research or ideas are lacking. Quite the opposite!

Research into bird orientation and navigation has experienced a revolution during the last decades. What was previously written and spoken on this subject is today regarded as hopelessly incomplete or simply as mistaken. The development of research is still going on at a breakneck rate; even the specialists within the field admit that they sometimes have difficulty in finding the time to digest and appraise all the new findings, which are often hard to interpret and conflicting.

Read what has been demonstrated on the subject of birds' powers of orientation with due consideration, but let the imagination also play a part. With a sufficient number of good ideas, perhaps somebody will succeed more quickly than we anticipate in piecing together a solution to the mystery of how migratory birds find the right direction in which to travel.

A major reason for the stimulating breakthrough in this field of research is that our perception of the sensory experiences of birds has recently undergone a radical reappraisal. Our perspective has been opened up to the fact that the world the birds know and react to is very different from our own world.

5.1   The sensory world of birds

The method that has been used most in recent years in laboratory studies of the sensory awareness of birds is based on the animals' heart reflexes. If an experimental bird is subjected to a signal and immediately after receives a quite harmless but fully perceptible electric shock, it soon learns to associate the signal with subsequent discomfort; the heart rate then increases in anticipation of the electric shock. If the bird has learnt to react to a certain type of signal, to light or sound for example, one can vary the character of the signal and by recording the bird's heart activity determine whether it has understood the signal or not. Figure 127 shows how in this way it has been possible to establish that homing pigeons are able to detect polarised light. The pigeons were subjected at irregular intervals to 20-second periods during which a projector lamp was kept switched on. During half of these periods (randomly distributed) when the light was on, a polarisation filter started to rotate after 11 seconds, and an electric shock followed after a further nine seconds; in the remaining illuminated periods the polarisation filter was still (or in additional control experiments was replaced by a rotating non-polarising glass lens) and the electric shock not given. The pigeon's heart rate in normal circumstances is approximately 140 beats per minute; before an anticipated electric shock it rapidly accelerates above 200 beats per minute. Of a total of 12 homing pigeons, four showed that they could quite positively detect when the polarisation of the light was altered.

The light from the sky is polarised in different ways in relation to the position of the sun. On the basis of the sky's polarisation pattern it is therefore possible to determine the sun's position with reasonable accuracy even when it is hidden behind clouds or has gone down behind mountains or sunk below the horizon. Bees, which orient according to the sun and which have the ability to see the light's polarisation (this ability is also possessed by many other insects and invertebrates), use this method: they manage to orient without making mistakes so long as the cloud cover is not completely solid but the polarisation pattern can be read in gaps of clear sky. Man has to use polarisation filters to be able to record the light's polarisation effectively; this kind of recording is used today for example in flight navigation in the proximity of the magnetic poles, where the magnetic compass is not reliable. According to Icelandic sagas, the Vikings already had access to polarisation filters in the form of a kind of sunstone which they used for navigating on their voyages between Scandinavia, Iceland, Greenland and North America. In one saga the story goes that Olof the Holy paid a visit to Rödulf, whose son Sigurd claimed to have the power to work out the sun's position even when it was hidden: 'When Sigurd had indicated the sun's position to the king, the latter had his sunstone brought out and watched how the rays shone. And the stone gave the same position as Sigurd had stated.' The sunstone was presumably a polarising crystal, of which there are several possible sorts that are relatively easy to find in the Nordic countries, including a limestone spar from Iceland which is used today for polarisation prisms.

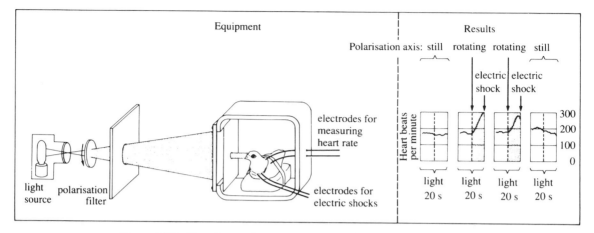

Figure 127 Experimental equipment in studies of the homing pigeon's ability to detect polarised light. On the right examples are shown of the results from a pigeon which was able to determine the light's polarisation. When the polarisation filter started to rotate, the pigeon's heart rate immediately increased in anticipation of the expected electric shock. When the light's polarisation plane was kept constant, no electric shock was given; the pigeon's heart rate was not modified either. Based on Keeton (1974) and Kreithen & Keeton (1974).

It has long been known that birds have well-developed colour vision. Their sensitivity is best in the wavelength range extending from blue to red light and which coincides generally speaking with the range that is visible to us humans. It has recently been discovered that the homing pigeon can in addition see light in a range of appreciably shorter wavelength, that of ultra-violet light. The lens system in the human eye does not allow UV light to pass through to the retina at all. By contrast, this does happen in the bird eye. The ability to see UV light, which incidentally the pigeon shares with the bee among others, is of advantage to the pigeons when they have to pinpoint the sun behind thin clouds and it allows them to see the light's polarisation pattern exceptionally clearly.

Birds' eyes exhibit many interesting characteristics, of the significance of which we still know very little. Judging from the density of visual cells in the retina for example, most birds have a visual acuity equivalent to the upper limit that man can achieve when he stares rigidly at an object (birds of prey have even sharper vision). In the human eye the density of visual cells is high in the visual pit in the retina known as the fovea; in the avian eye the density of visual cells is high over almost the whole retina. A bird's entire visual picture is therefore probably almost as sharp as the point of the human eye's fixed stare. Some species of bird have two different foveae; others have the foveal pit oddly shaped, sometimes as a ribbon-like extension. The homing pigeon has only one circular fovea, but its retina is divided into at least two different specialised areas, a central/lower part for monocular vision (the visual pictures of the two eyes do not overlap) and a rear/upper part for binocular vision (the visual pictures overlap, making judgement of distance to nearby objects possible). In these two areas of the retina there are mysterious yellow and red oil droplets, the function of which is unknown.

The avian eye contains the peculiar pecten, a pleated membrane with many blood vessels, which extends from the base of the eye into the vitreous body. Its function has long been disputed. It is thought, for example, to be important to the retina's nutrient supply. One question that has been asked repeatedly is whether the pecten might also function in facilitating orientation. The latest suggestion is that the shadow cast by the pecten in the light from the sun's image on the retina (not directly from the sun's rays) might allow a very exact position-fixing of the sun; this is because the shadow is very sharply defined and falls where the density of visual cells is greatest, in immediate proximity to the spot where the image of the horizon is projected.

Studies of the homing pigeon's hearing have recently afforded a splendid surprise. Unlike man, the pigeon can hear low-frequency and muffled sound, so-called infra-sound. Man's lower hearing limit is around the sound frequency of 20 hertz (Hz = vibrations per second). Pigeons on the other hand can hear infra-sound down to a frequency of about 0.1 Hz. And not only that – they can even distinguish amazingly small differences in frequency. When the infra-sound remains at around 1 Hz, they are able to detect a frequency difference of 7% (they thus distinguish between infra-sound of 1 Hz and 1.07 Hz), and at 20 Hz the same threshold is a 1% frequency difference. Sound travels through air at a speed of about 330 m/s; the sound wavelength at a frequency of 1 Hz is accordingly 330 m. A pigeon of course has no chance of localising such long-wave sound on the basis of the difference in phase between its two ears. By flying in a circle, however, it has a chance to work out the direction to the sound source owing to the difference in frequency (the so-called Doppler effect) that results when the pigeon flies towards and then flies away from the source of the sound. This is the same effect as that which leads to the sound from, for example, a train having a different tone when it is approaching from when it is going away. A pigeon flying in a circle at a speed of 20 m/s moves towards the sound source at + 20 m/s and away from the sound source at − 20 m/s. This is equivalent to a frequency difference of 12% (40/330), which lies well within the pigeon's discriminatory ability in the infra-sound range.

There are a great many sources of infra-sound in nature: thunderstorms, magnetic storms, earth tremors, sea waves, jetstreams, and wind currents through mountain passes and around high mountain peaks. As a result of the long wavelength the infra-sound travels a long way, sometimes thousands of kilometres. The pigeons can therefore hear a vast 'acoustic landscape' which is totally unknown to us human beings. Such a capacity can naturally be of great help to them when they have to assess approaching weather and facilitates orientation and navigation. On windy days problems may arise with distinguishing the infra-sound from the noise made by gusts of wind and turbulence. Future investigations will show whether the birds actually do orientate by listening to the world around them.

Homing pigeons have been shown to be extremely sensitive to small differences in barometric pressure. They are clearly capable of detecting a pressure change in less than 10 seconds of 1 millibar (equivalent to roughly 10 m altitude difference); some pigeons have even shown signs of being able to detect a difference of down to as little as one-tenth of a millibar. Man's corresponding sensitivity is in the region of about 4 millibars' pressure difference. This high sensitivity to barometric pressure is perhaps not so important for orientation but is probably a major explanation of why migratory birds are expert weather-forecasters and of their ability to maintain a constant flight altitude over long distances.

Are birds also sensitive to small variations in gravity? Orientation experiments with homing pigeons in the United States suggest that this possibility cannot be excluded, for the birds altered their flight directions, slightly but quite distinctly, in tact with the way the position of the moon changed in relation to the earth! The displacement, which is independent of whether the moon is visible or not, increased gradually during the 30 days of the month and quickly reverted back to the original situation at full moon (the moon and the sun are then on directly opposite sides of the earth). In some years or seasons the displacement started and returned to zero at the new moon instead (the moon and the sun are then in the same direction seen from the earth). Since the solar day (approximately 24 hours) and the lunar day (approximately 24 hours and 50 minutes) are of different lengths, the orientation experiments, which were always carried out around 12.00 hours midday on the various days of the month, fell at different times during the lunar day. The displacement in the pigeons' orientation is therefore shown to have a connection not only with the monthly cycle but also with the cycle of the lunar day. As is well known, the moon's position influences the gravitational circumstances on the earth, this being clearly evidenced by the tidal variations.

A detailed analysis of the pigeons' orientation in relation to the gravitational changes suggested that the birds may possibly have been influenced by the horizontal gravitational component in the direction of the pigeon loft. No unequivocal connection could be shown, however, and we have no explanation either of why birds should be affected in this way. The remarkable 'lunar rhythm' in the pigeons' orientation can for the time being therefore be interpreted only very speculatively as support for the idea that birds might be sensitive to small gravitational effects.

If in the future this is shown to be the case, this may be of advantage to the birds' navigational abilities. The gravitation at the earth's surface not only varies in time but is also different in different areas. These differences are partly global and partly local. The global ones are due to the fact that the earth is not perfectly spherical, the local ones to differences in the structure of the earth's crust. An invisible 'gravitational landscape' therefore exists on the surface of the earth.

Birds' sense of smell and taste are considered to be only middlingly or poorly developed. Recently, however, the idea has been launched that homing pigeons might use the sense of smell to navigate home to the pigeon loft. This possibility has also gained support in some, though by no means all, experiments (olfactory navigation is discussed further in section 5.3). It is thus important that earlier analyses of the avian sense of smell be reappraised.

The most intriguing new discovery of all in recent years is that birds have a well-developed magnetic sense! From as far back as about 1970 evidence has been collected supporting the fact that both migratory birds and homing pigeons make use of the earth's magnetic field while orientating. It has been concluded that the birds possess a magnetic compass, and some success has been achieved in ascertaining how this functions in connection with changes in and disturbances of the magnetic field (the avian magnetic compass is described in more detail in the next section). All the time, however, a nagging doubt blighted these conclusions: for not the slightest sensitivity to magnetic fields could be detected in birds which were tested using heart-reflex experiments or other conditioning experiments in laboratories. One failed laboratory test after another was reported – right up to 1977, when success was at last achieved. In that year it was shown that earlier failures had been due to the fact that the experimental birds had not been given sufficient freedom of movement. In the successful experiment homing pigeons were used. The birds were put into one of the short sides of a cage which was between 3 m and 4 m long and roughly 1 m wide and 1 m tall. At the opposite side of the cage, on the left and on the right, there were two 'nestboxes' with food dishes; food, however, was present only in one. The pigeons had to hop into the boxes to be able to see whether the food dishes were empty or not. The aim of the experiment was to find out whether the pigeons were able to learn that food was in the left-hand or in the right-hand box if a magnetic field (of approximately the same strength, 0.5 gauss, as the earth's magnetic field) existed or was missing in the cage. The pigeons failed to choose the correct food box when they walked calmly. The results were totally different, however, if they were spontaneously active, if they flew, flapped and ran: they then chose the correct food box in 80% of attempts, a good sign that the learning had been successful.

These results pointed to the possibility that the birds' magnetic sense was based on magnetic induction, i.e. on the principle that electric currents are formed in conductive matter which moves through a magnetic field. A bird could therefore detect a magnetic field only when it moved through the field and electric currents were induced somewhere in its body. Orientation experiments with homing pigeons, however, argued against this and indicated instead that the pigeons had a magnetic material in their body which, like an ordinary compass needle, sought to set itself parallel to the magnetic field. The reason for this was that the pigeons' orientation was affected by bar magnets and electric coils on their bodies; these create a stable magnetic field that does not induce electric currents but which does, of course, affect magnetic materials. The conflicting results had still not provided any consistent explanation when a sensational discovery was reported in 1979: magnetic material could be traced in the heads of homing pigeons!

This discovery meant of course that a milestone had been passed by scientists working on orientation. Within only one year of the publication of this report the findings of a second investigation were announced. These confirmed the results as regards the pigeons and in addition demonstrated that roughly similar amounts of magnetic material were present in a number of other bird species, among them the North American White-crowned Sparrow (*Zonotrichia leucophrys*), the Tree Swallow (*Tachycineta bicolor*), the Western Grebe (*Aechmophorus occidentalis*) and the Pintail (*Anas acuta*).

The birds' magnetic material consists of small crystals of magnetite (a kind of iron ore, known as black ore). The total magnetic field that is generated by the magnetite in a bird's head suggests that there are between 10 and 100 million such pieces of crystal, each one about one ten-thousandth of a millimetre in size. The combined weight of the magnetite is reckoned at less than one ten-thousandth of a milligram. With their negligible size the crystals correspond to a single magnetic domain, in other words they are just big enough to be permanently magnetic. It is important that they do not become so large that a second magnetic domain is formed, for its field would in that case often neutralise that of the first. How the birds synthesise these microscopic magnetic needles is still unknown, but in certain places in the pigeons' heads magnetite has been found in association with larger yellowish crystals. It is suspected that these may be ferritin, a protein in which iron is stored. Ferritin could be used in the formation of magnetite.

The birds' magnetic crystals are found both in the front and in the rear part of the head, between the skull bone and the membrane of the brain (in one investigation a 1-mm patch of tissue with a particularly strong concentration of magnetite crystals was found there) as well as in the musculature of the neck. This musculature contains a quantity of sensory nerves, including in the muscle spindles which record pressure and tension in the muscles. The muscle spindles' high sensitivity to small traction forces ought to make them suitable for measuring the forces that influence the magnetite crystals when these are differently orientated in relation to the earth's magnetic field. It has therefore been speculated that magnetite crystals in the muscle tissue might be coupled to the muscle spindles. This could possibly also explain why the birds have to be in rapid motion to be able to feel the magnetic field, for the muscle spindles' signals are understood, at least by us human beings, only in relation to controlled muscle movements.

Before magnetite crystals were discovered in birds, similar crystals had already been found in sediment-dwelling bacteria (various species which live on the bottom of seas and lakes and in saltmarshes) and in bees. The bacteria's magnetite crystals are arranged in a straight line and because of that form a magnetic needle which aligns the bacteria cells parallel with the earth's magnetic field. In the northern hemisphere the magnetic bacteria are north-seeking, i.e. they are directed diagonally downwards; using their flagellae they can thus efficiently find their way down to their living environment in the oxygen-deficient bottom sediment. Magnetic bacteria have also been found recently in Australia, Tasmania and New Zealand. Here they are south-seeking, which in the southern hemisphere is a prerequisite if they are to be able, using the magnetic field, to orient downwards, to the bottom mud.

The bees have their magnetite crystals in the fore part of the abdomen. Here they are arranged horizontally crossways. The crystals' north pole points either to the right or to the left, probably depending on how the honeycomb in which the bees have grown up was orientated in relation to the earth's magnetic field. Many experiments have shown that bees are influenced by the magnetic field in their orientation.

Magnetic sense will probably be revealed in far more animals than we can now imagine. In 1981, orientation experiments were made which indicated that sharks, rays, salmon fry, salamanders and field mice are affected by the magnetic field. In addition, traces of magnetic material have been found in the heads of dolphins and in mice, voles and monarch butterflies. A real revolution in our existing knowledge of the directional and magnetic sense of animals seems to lie before us. It has even been put forward that man himself might also possess such a 'sixth sense', even though we are not conscious of it. The English scientist, Robin Baker, recently did some orientation experiments with his students. They were taken off in buses blindfolded and then had to indicate the direction 'home' to the university from the 'release site' a few tens of kilometres away. Most were quite successful in their assessment, despite the fact that they had no idea where they were. Robin Baker considered that the results got worse if the students had bar magnets near their head or if they were wearing helmets with built-in circuits which produce a field of magnetic disturbance through the head. The results shown, however, are difficult to interpret, and the sources of error are so many that for the present we must I am sure take Robin Baker's assertion that there might be a human magnetic sense with a pinch of salt. Nevertheless, Robin Baker has since continued his research into man's magnetic sense with fervour and ingenuity, and supporting (as well as some conflicting) replicate experiments have been reported by independent researchers (Baker 1989). Indeed, Baker may have a case, although the issue of human magnetoreception is still highly controversial.

A research report informs us that magnetic material has been traced in human adrenal-gland tissue. This may possibly be a question of slag matter containing iron which has been stored up in the tissue. Iron forms a part of many of the chemical compounds of animals and is perhaps deposited in the form of magnetic slag products. The mere presence of magnetic material in animals is therefore far from good enough reason to conclude that they would have a magnetic sense; convincing orientation experiments are required in addition.

The 1980s have passed without any major physiological discovery being reported concerning the structure and function of a possible magnetic receptor system based on magnetite crystals (cf. Kirschvink *et al.* 1985). Hence, scepticism is spreading over whether biogenic magnetite is after all involved as a basis for the well-documented magnetic sensitivity of birds and other animals. Other alternatives of magnetoreception have attracted increased attention, especially the possibility that birds' magnetic sense is based on biochemical processes that are influenced by weak magnetic fields. One such candidate process is the light absorption by rhodopsin molecules in the photoreceptors of the eye's retina. Neurophysiological experiments have revealed that weak magnetic fields lead to different kinds of neural response in the visual system of the pigeon provided that the eyes are illuminated. This raises the possibility that the birds in some way may be able to 'see' the geomagnetic field (Semm & Demaine 1986). Hence, the birds' magnetic receptor system still remains to be elucidated – is it linked to the visual process or based on magnetite particles or what?

5.2   Different compasses

For the birds to be able to orientate successfully naturally requires of them the ability to distinguish different directions efficiently and that they stick constantly to the flight course they have chosen. In other words they must have a well-developed directional or compass sense. The compass sense of birds is not founded on one basic principle only; it is made up of several different compasses, of which the three most important ones are the sun-compass, the star-compass and the magnetic compass. I shall begin by describing each of these compasses separately, before discussing how they are co-ordinated in the birds' sense of direction.

The sun-compass

The sun-compass enables a bird to determine the compass direction using the sun's position and an inbuilt biological 24-hour clock. Since the earth rotates, the birds, like us, can see how during the course of the day the sun moves on its path from east to west. The sun's lateral speed of movement is on average about 15° per hour. The bird's inner clock compensates for this angular movement so that it can correctly determine compass direction at all times of the day.

Due south is therefore deduced by the birds roughly as follows (in the northern hemisphere; + signifies to the right, − to the left): 06.00 hours = sun's direction + 90°; 09.00 hours = sun's direction + 45°; 12.00 hours = sun's direction ± 0°; 13.00 hours − sun's direction − 15°; 15.00 hours = sun's direction − 45°; and so on. If the birds' inner clock is one hour fast or slow, the directional error is accordingly 15°. If the clock is four minutes out, the direction is affected by only 1°.

What sort of biological clock is it that forms an integral part of the sun-compass? Like other animals, including man, birds have an internal daily rhythm which persists even during isolation from the outer world. In us humans, such daily patterns in body temperature, hormone secretion, activity and sleep have been shown through experiments in isolated caves or bunkers, where the natural changes between day and night could not be perceived. These rhythms are called circadian (from the Latin *circa* = approximately and *dies* = day) since they are not exactly 24 hours when they are disengaged from the natural 24-hour rotations. Man's activity rhythm, then, is instead about 25 hours. The free-running activity rhythm of birds varies between 22 and 26 hours for different species. Under natural conditions the free-running rhythm is 'captured' and regulated in relation to the 24-hour day using the natural rotations between day and night. The transitions between dark and light and between light and dark at dawn and dusk serve as important time signals for the exact setting of the daily rhythm in both man and birds. When changes between day and night are suddenly altered, it takes four to five days before the adjustment of the internal daily rhythm to the new circumstances manages to take over. The stresses involved in adjusting to a new daily rhythm are well known to air travellers and shiftworkers.

The clock that makes up part of the birds' sun-compass thus ticks along in time with a biological daily rhythm of this kind.

The sun-compass of birds was discovered about 1950 by a research team, led by Gustav Kramer, during experiments with Starlings in orientation cages. It was found that the birds were able without too much trouble to learn to seek food in a fixed compass direction in the cage, independent of the time of day. That Starlings which learned a compass direction were using the sun was demonstrated convincingly by changing the direction of the sun with the use of mirrors; the birds then changed their direction to an equivalent degree.

The Starlings' orientation deteriorated drastically when the sun was not visible, but they could then be made to orientate in relation to a powerful lamp which was moved in a manner that imitated the sun's path. The most elegant experiment consisted in deceiving the Starlings by keeping this artificial 'sun' still. It was then revealed how the birds' choice of direction gradually shifted anti-clockwise at an angular speed equivalent to the motion of the sun, 15° per hour. A fine example of how the compass clock ticks forth directional compensation for the sun's lateral movement across the heavens!

The fact that homing pigeons use the sun-compass has been confirmed by countless release experiments. The standard procedure for demonstrating the pigeons' sun-compass consists in 're-setting' their internal clock and finding out whether they then misorientate in the expected direction. In these cases the pigeons are isolated for at least five days in artificial conditions of day/night cycles which differ from those of the natural day. If their internal clock is set six hours fast (so that the pigeons' clock is at 15.00 hours although they were in fact released at 09.00 hours), the birds ought to fly 90° wrongly to the left; if the clock is set back six hours, then they should instead fly 90° wrongly to the right. This is precisely what happens when orientation tests are made with homing pigeons in sunny weather (figure 128). Even when the pigeons are released as little as only 1 km away from the pigeon loft, in terrain over which they have flown every day throughout their life, they can be made to fly off in the wrong direction and by that means be shown to be orientating with the sun-compass. This does not, however, apply when the pigeons can see the actual pigeon loft directly from the site where they are released.

Not only Starlings and homing pigeons have been proven to have a sun-compass but so have Mallard (through release experiments) and various passerines which normally migrate at night, for example White-throated Sparrow, Red-backed Shrike and Barred Warbler (through learning experiments in cages). It is quite likely that a sun-compass exists in all birds.

Despite the fact that homing pigeons regularly use the sun-compass, this is not, it must be remembered, essential for orientation. Tests with well-trained birds show that pigeons orientate correctly even in totally overcast weather. At such times they choose the right direction independent of whether their internal clock has been re-set or not. This proves that they are not using their sun-compass. The American scientist William T. Keeton arrived at this answer. In 1969 he presented his investigation in an article that attracted much attention: 'Orientation by pigeons; is the sun necessary?'. Up to this point in time ideas had been focused largely on the role of the sun in the homing pigeons' orientation and navigation systems. William Keeton's results showed that the sun could not provide a complete explanation of the problem. This became the real starting signal for the following decade's dizzy hunt for new sensory powers in birds.

Wolfgang and Roswitha Wiltschko in Germany studied in detail the sun-compass of homing pigeons and showed, again in an elegant way, that it does not fulfil any absolutely essential function. They allowed young pigeons to grow up without ever seeing the sun. The young birds had to live in a closed room without windows and were released to fly around the pigeon loft only on totally overcast days. When they were tested, their orientation towards the pigeon loft was shown to be good, both in cloudy and in sunny weather. If their clocks were re-set, then they flew with exactly the same accuracy – even when it was sunny!

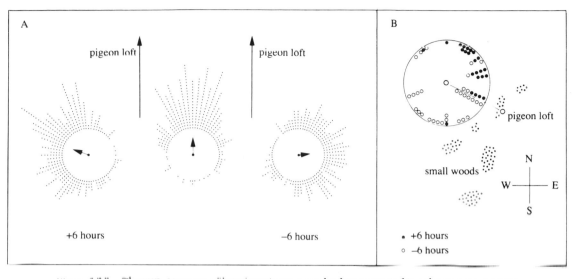

Figure 128   The sun-compass of homing pigeons can be demonstrated in release experiments where the birds' internal clock is re-set. A. Vanishing directions of homing pigeons whose internal clock has been put forward six hours, left as it is or put back six hours. The arrows show the mean direction for the different categories. Six hours' displacement corresponds to a directional error of approximately 90° to left or right when the internal clock is fast or slow. The mean directions agree well with these expectations. Note the wide scatter among individual pigeons' vanishing directions (which moreover is greater for pigeons with re-set clocks than for those with clocks left as normal). Because of this scatter, orientation experiments must always cover a large number of releases, for otherwise inadequate basic data are obtained for statistical evaluation. B. Homing pigeons take their bearings using the sun-compass even when they are released only a kilometre or so from the pigeon loft, provided that the target is not directly visible from the release site. The figure shows how the re-setting of the clock clearly influences the pigeons' choice of direction when the pigeon loft is hidden behind a clump of trees. Based on Graue (1963) and Schmidt-Koenig (1979).

Even though pigeons with incorrectly set clocks initially fly off in the wrong direction from the site where they are released, they often soon discover their mistake; many return to the pigeon loft almost as quickly as those which have a correct sun-compass. The pigeons obviously 'disconnect' their sun-compass when they notice that orientation has gone wrong and then fly home using other means. Based on Graue (1963) and Schmidt-Koenig (1979).

Pigeons which had never before seen the sun thus did not use the sun-compass when they were released in sunshine.

The sun-compass is not innate but is learned. The Wiltschko couple made a series of different tests which reveal something of how the learning process works. Homing pigeons are ready for the first tentative release tests at only eight weeks of age. By re-setting their internal clock it has been shown that they do not normally make use of the sun-compass until approximately 12 weeks of age. Their progress can be accelerated through practice releases made often in sunny weather. The young pigeons' orientation, however, is no poorer before the development of their sun-compass than it is afterwards. Thus they have the ability to orient home even before they have learned to use the sun-compass. Amazingly enough, the orientation of the youngest is, despite this, poorer in cloudy weather than in sunny weather. The explanation is probably that pigeons which have seen the sun while growing up use this right at the start as a point of reference in order to be able to maintain direction in a simple way once they have taken their bearings without using the sun-compass. In cloudy weather, therefore, the scatter of the young pigeons' flight directions increases though the mean direction still points towards the pigeon loft.

Comparative cage experiments with young Starlings suggest that these do not learn the sun-compass so quickly as the homing pigeons. Attempts to demonstrate a sun-compass in juvenile Finnish Starlings throughout their first autumn failed. Not until the following spring-migration season was clear evidence found that the sun-compass had developed.

The homing pigeons' learning of the sun-compass is also illustrated by an experiment with pigeons which were brought up without ever seeing the morning sun. During the morning the pigeons were kept in an illuminated room. Only after midday were they let out into an outdoor aviary or allowed to make exercise flights around the pigeon loft. They were therefore able to see the sun only when it was on the downward part of its track. When their internal clock was set back six hours and they were then tried out in orientation experiments in the afternoon, they still flew in exactly the right direction to the pigeon loft. Since the pigeons' internal clock showed morning time when they were released, the most likely explanation is that they did not use the sun-compass. They had of course never learnt that there was a movement of the sun in the morning with which they could determine direction.

Even though the sun's lateral movement across the sky is on average $15°$ per hour, the angular speed varies during different times of the day at different latitudes and during different seasons. Only at the poles is the angular speed exactly the same throughout the day. The nearer one gets to the latitude where the sun passes through zenith, the greater the lateral angular movement is at noon in relation to the hours at sunrise and sunset. At $50° N$ the sun moves at midsummer approximately $11°$ laterally during the hour after or before sunrise and sunset; during the noon hour, however, the lateral movement amounts to $30°$. (At the equator an extreme situation exists at the vernal and autumnal equinoxes, when the sun stands directly in the east throughout the morning, passes through zenith at noon and lies in the west throughout the afternoon.) Detailed studies in Germany and the United States of the sun-compass of homing pigeons indicate that to a certain extent pigeons compensate for these irregularities in the sun's angular speed. We may accordingly assume that birds learn the sun-compass by carefully studying the path of the sun and that pigeons from pigeon lofts at different latitudes therefore have slightly different sun-compasses. One would expect that homing pigeons and other birds at the equator learn a sun-compass of a highly divergent type.

Homing pigeons can be made to learn a false sun-compass. In one experiment, pigeons were raised under a time shift that was constant. The natural day was displaced by approximately six hours. The pigeons' night, in a dark, enclosed room, thus lasted right up to noon, when they were released into the pigeon loft. In the afternoons they were practised in orientation flights together with the other pigeons. At sunset they were returned to their room, which was kept lit until the time came for the delayed night. These pigeons coped with orientation equally as well as the other pigeons when they were released. The 'false' sun-compass that they had learned, with the morning sun in the south and the midday sun in the west, functioned admirably as a means of orientation – so long as they stuck to their daily rhythm with its time shift. When they had adapted themselves to the natural daily rhythm and began to live a totally normal life together with the other pigeons in the loft, they then flew the wrong way, about $90°$ to the left of the correct homing direction, in fresh orientation tests; they were therefore using the false sun-compass which they had learnt while growing up. The mistake in orienting gradually disappeared; this was due mainly to the fact that the pigeons stopped using the sun-compass. Their ability to 're-learn' or their confidence in their sun-compass was obviously poor.

The star-compass

Nocturnal migrants that are kept in cages are restless at night during the migration season. They hop and flutter in the natural direction of migration, to the north in spring and to the south in autumn. Through studying their night-time restlessness, Gustav Kramer discovered that small birds, Red-backed Shrikes and Blackcaps for example, make use of the starry sky for orientating: they aim fairly unambiguously in a fixed direction so long as the stars are visible. A clouded night sky, however, results in increased scatter in the choice of direction and subdued migratory restlessness. During the 1950s Franz Sauer carried on further research on the basis of these findings and was able to confirm that birds possess a star-compass. This he showed by studying caged Blackcaps, Garden Warblers and Lesser Whitethroats in a planetarium (where the pattern of the stars was projected on to the dome-shaped roof by means of an astral projector). The birds moved in their natural migration direction in relation to the stars. When the astral projector was switched over so that what the stars showed as north was moved to the opposite side of the planetarium, the birds reversed their orientation.

The most detailed planetarium studies were made at the end of the 1960s and the beginning of the 1970s by Stephen T. Emlen in the United States. He studied the Indigo Bunting, which breeds in eastern North America and winters in Central America and the West Indies. In orientation tests in the planetarium the birds were placed in small funnel-shaped cages where they had a clear view upwards only, facing the projection of the firmament. At the bottom of the cages was a wet ink pad and the walls were covered by blotting-paper. When the buntings hopped in the cages, they transmitted ink on to the paper with their feet. By studying the paper's ink pattern afterwards, it was possible to determine their choice of direction without difficulty. These simple and ingenious orientation cages have since come to be known as 'Emlen funnels'.

The Indigo Buntings' orientation, like that of the European warblers, altered in parallel with the adjusting of the night sky. The question now was how their star-compass functioned in more detail. How can the starry sky be utilised for direction-finding? Since the earth spins on its axis one revolution per 24 hours, we can see that the firmament rotates during the course of the night. The earth's axis points north in the direction of the Pole Star, which thus forms the rotational centre in the northern sky. If we look at the Pole Star, which always lies due north, we can see that the entire firmament revolves anti-clockwise around it at a rotational speed of approximately 15° per hour (one revolution per day). Stars situated near the Pole Star therefore move in a tight circle, which means that the lateral angular deviation in relation to direction north at the horizon is small. The stars at the celestial equator (among which, incidentally, the sun may be classed as a 'day star') on the other hand describe a maximum lateral angular movement, 15° an hour. The internal clock in the birds' sun-compass has the function of compensating for the sun's lateral movement. Birds which, in their star-compass, make use of those stars that move in the area around the path of the sun should consequently use a similar clock. If instead they act on stars nearer the Pole Star, they ought to have an entirely different clock, one that compensates for a smaller lateral angular movement.

How birds compensate for the shifting positions of the stars during the course of the night can be easily investigated in a planetarium, where the night sky can be moved in an instant forwards or back in relation to normal time. Such tests show that Indigo Buntings do not make use of any internal clock at all in their star-compass! They maintain their normal direction independently of whether the firmament is rotated three, six or 12 hours forwards or back (figure 129). The explanation is that they take their bearings in relation to the direction of the Pole Star, which of course lies constantly due north. By screening off various constellations or parts of the night sky, it has been shown that Indigo Buntings achieve excellent orientation even though several constellations might be missing and even if they cannot see the Pole Star itself. The most important thing for the buntings' star-compass seems to be the pattern of stars nearest the Pole Star. The birds are evidently able to use constellations such as Ursa Major (The Plough), Ursa Minor (the Little Bear), Draco, Cepheus or Cassiopeia to work out the direction north. The Indigo Buntings' striking knowledge of the star pattern helps them to utilise their star-compass even in weather that is only fair, when parts of the sky may happen to be covered by cloud.

That the star-compass is not dependent on any internal clock has also been shown to apply to the Mallard. Where the European *Sylvia* warblers are concerned, the situation is more uncertain. Franz Sauer carried out experiments in which the planetarium sky was rotated forwards or back in time. The results were scanty and difficult to interpret but suggested that the warblers' orientation was affected by these time shifts. The results if anything point to the birds using a time-compensated star-compass which is based on stars near the celestial equator. Such a star-compass may possibly be of advantage to species which migrate a long way to the south, near to the equator, where the Pole Star is no longer visible. Indigo Buntings, which spend their whole life in the northern hemisphere, can on the other hand always use the night sky's rotational pivot at the Pole Star as a directional clue. More detailed investigations are required if we are to be able to decide whether these speculations may be justified. Franz Sauer himself interpreted his results as support for the idea that birds use the stars in combination with their sense of time, not as a simple compass but in order to determine their position on the earth's surface, a possibility which I shall come back to in the section on the navigation system of birds.

Stephen Emlen was able to show how the Indigo Bunting's star-compass develops in young birds by rearing juveniles in the laboratory. One group of young birds was never allowed to see any stars while growing up. The others were kept in the planetarium on some nights. Of the 'planetarium buntings', one group was allowed to see the 'natural' night sky, which rotated around the Pole Star at a rate of one revolution per day; the other group was shown a very much displaced night sky, in which the star Betelgeuse in the Orion constellation, which is normally to be found near the celestial equator, replaced the Pole Star as the pivot around which the sky rotated. Orientation was tested under a normal night sky when the autumn-migration season had begun. The birds which had never seen any stars showed by far the poorest ability to orientate; those which had experienced a normal night sky oriented unequivocally in the normal migration direction, i.e. due south; the birds which belonged in the third group oriented in an equally clear-cut manner directly away from Betelgeuse! The buntings obviously associate the rotational pivot of the night sky with north. By learning the constellations, they acquire at the same time the ability to determine the direction north from a quick look at a part of the sky without necessarily needing to see the Pole Star itself and without henceforth needing to watch the rotation.

The birds' orientation is consequently not affected in any way by the change which the slope of the earth's axis undergoes, over a period of about 26 000 years. The change means, among other things, that the rotational centre of the northern night sky will in 30 000 years' time be located next to the star Vega, 43° from today's Pole Star. Despite this, the birds' star-compass will function just as well!

The astral day (or the astronomical day) is 23 hours and 56 minutes, on average four

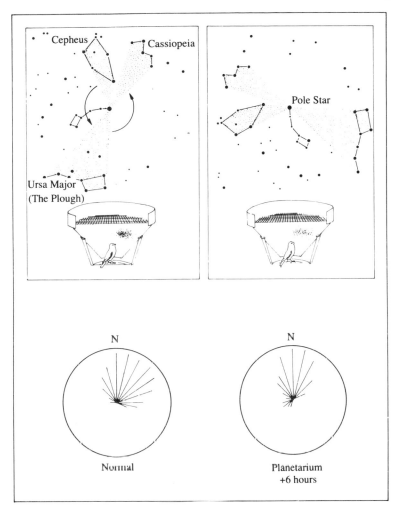

Figure 129   That Indigo Buntings do not use an internal clock in their star-compass has been demonstrated by rotating the night sky in a planetarium six hours forward, as shown in the figure. Despite this time shift the birds in the Emlen funnels continue to orient in the correct direction (northward) for spring migration, as is shown by the directional scatters in the lower part of the figure. Guided by the pattern of the stars, birds can determine the direction north towards the Pole Star and use this as a reference direction in their star-compass independently of the time of night. Based on Emlen (1975).

minutes shorter than the solar day. The reason for this difference is that the astral day is determined solely by the earth's rotation on its axis. The solar day is influenced in addition by the earth's yearly orbit around the sun. A star which stands directly below the Pole Star at a given point of time at night will therefore return the next night to the same position four minutes earlier. After a month the star returns to the position two hours earlier. In this way a successive time shift around the clock takes place until on the same night the next year the star again stands directly beneath the Pole Star at the original point of time. The appearance of the starry sky at a certain time of the night can therefore provide information on the time of the year. When birds in a planetarium are suddenly subjected to the night sky being moved forward by six hours (a quarter of a day), this is at the same time the equivalent of turning back the time of the year by three months (a quarter of a year).

Stephen Emlen investigated whether Indigo Buntings used the night sky in order to determine if it was spring or autumn, whether they should migrate north or south. He got a hint that this was not the case when the night sky was rotated 12 hours, the equivalent of a seasonal shift of half a year, without the birds altering their orientation. The decisive experiment was carried out in the following manner. A number of Indigo Buntings were caught as juveniles during the summer and had to live in the laboratory until the subsequent spring. Half of the buntings lived in a room where the day length was the whole time adjusted to the day length the buntings would have experienced under normal conditions in the wild. In September the buntings moulted into the brown winter plumage, started to put on fat and began to show migratory restlessness. The migratory restlessness decreased in November, at which time the buntings normally reach their winter quarters in Central America. The day length in the laboratory was adjusted accordingly. When the day length was increased in the spring, the birds were stimulated to moult into their blue summer plumage in March, and in April and May they put on fat reserves and came into spring-migration mood.

The other group of buntings was treated in the same manner up to December. By then they were in the brown winter plumage and had just concluded the autumn-migration period. In the middle of December the day length was suddenly increased from eleven to 15 hours. This stimulated the buntings to change into summer plumage as early as January and February and to exhibit migratory restlessness. As soon as this abnormally early spring-migration restlessness had petered out (around 1 March), the day length was gradually decreased. The buntings reacted by moulting into winter plumage, storing fat and showing autumn-migration restlessness as early as April and May. Now was the time to compare the orientation of the 'spring buntings' and the 'autumn buntings' in the planetarium under the normal spring night sky. Would the 'autumn buntings' detect, with the help of the stars, that in actual fact it was spring? No, they used their star-compass to orient south while the spring buntings at the same time and in the same planetarium oriented north!

Other investigations have shown that birds determine the season through an internal annual rhythm, known as a circannual rhythm. In step with this rhythm is a changing hormone balance which controls moult, fat deposition, spring and autumn migratory restlessness, and breeding so that they happen at the right time of year. It is mainly the decrease in day length in autumn and its increase in spring that act as external time signals for the circannual rhythm.

It is not only the stars in the sky that birds observe. Certain tests with Mallards indicate that they sometimes, when the moon is up, use a time-compensated 'moon-compass' (functioning in roughly the same manner as the sun-compass) in order to hold their course.

The star-compass is not essential for correct orientation. Despite an increased variation in direction, Indigo Buntings as well as other small birds are capable of orienting in the normal direction of migration without visible directional signs. To explain this we must take into consideration the birds' magnetic compass.

The magnetic compass
During the late 1950s a German research group studied the Robin's orientation in cages and found that the normal migration direction was maintained even in a dimly lit closed room, totally without any visible directional cues. Barely ten years later the possibility that the birds oriented from the earth's magnetic field was tested by turning the magnetic field with large

electrical coils placed around the orientation cages. When the direction of the magnetic field was altered, the Robins' orientation changed correspondingly – the magnetic compass had been discovered! This discovery was a great achievement by the researchers Friedrich W. Merkel and Wolfgang Wiltschko. Their research findings at first met with solid scepticism from other research workers; they quite simply refused to believe that birds were sensitive to the weak geomagnetic field or they failed in their attempts to corroborate the results. One reason for these failures was shown to be the different constructions of the orientation cages. Birds show a poor or non-existent reaction to the magnetic field in cages where the recording perches are positioned tangentially. In the 'Merkel cages', radial perches allow the birds to go around quickly and to pick up the magnetic field. When the sceptics borrowed the 'right' cages, the complications could be cleared up and Merkel and Wiltschko's discovery confirmed.

Through continued systematic cage tests, in which the magnetic field was adjusted in different ways, Wolfgang and Roswitha Wiltschko were able in 1972 to demonstrate how the Robin's magnetic compass, the so-called dip-compass (inclination compass), works. Before going into these results I shall just say a few words on the earth's magnetic field (figure 130). The earth's magnetic field is probably generated by electric currents in the outer, semi-molten part of the earth's core. The plane of the magnetic field lies at an angle of about $11°$ to the earth's axis. This means that the magnetic and the geographical poles do not coincide with each other. The magnetic north pole is today situated in northernmost Canada and the magnetic south pole on the coast of the Antarctic (in a strictly physical sense it is the other way around, which does not, however, have any significance for my subsequent discussion). The declination is equal to the angular difference between magnetic and geographical north. It is of course different at different points on the earth's surface (it is near zero in Scandinavia). The magnetic field extends far out into space and there it is influenced by the radiation from electrically charged particles from the sun, by the so-called 'solar wind'. At times of powerful solar eruptions the earth's magnetic field is disturbed – what are known as magnetic storms occur. At the earth's surface the magnetic field points obliquely upwards in the southern hemisphere and obliquely downwards in the northern hemisphere; at the magnetic equator the field lines run horizontally. The angle at which the magnetic field points upwards or downwards in relation to the horizontal plane is known as the inclination or dip. In the northern hemisphere the inclination increases northwards to the magnetic north pole, where it is $+90°$; the magnetic field is thus directed vertically downwards towards the earth's surface. In Scandinavia the inclination is between $+70°$ and $+80°$. The total magnetic field intensity is significantly greater at the magnetic poles, 0.6–0.7 gauss, than at the magnetic equator; it is smallest in southern Brazil, about 0.25 gauss. The earth's magnetic field exhibits considerable regional and local deviations owing to the presence of magnetic minerals in the earth's crust. I shall come back to this 'magnetic landscape' in the next section (see figure 136).

The polarity (direction) of the geomagnetic field is reversed on average a few times every million years. The last time that the magnetic north pole and the magnetic south pole changed places was probably about 700 000 years ago. The actual reversal itself, during which the magnetic field decreases in intensity, takes a relatively short time. After only a couple of thousand years a stable field with reverse polarity is again built up.

Now back to the Robin's magnetic compass (figure 130). The reason why it is called the dip-compass is that the Robin does not seem to perceive the polarity of the magnetic field but its angle of dip instead. The bird presumably determines this in relation to the vertical gravitational direction. North for the Robin is accordingly the direction in which the

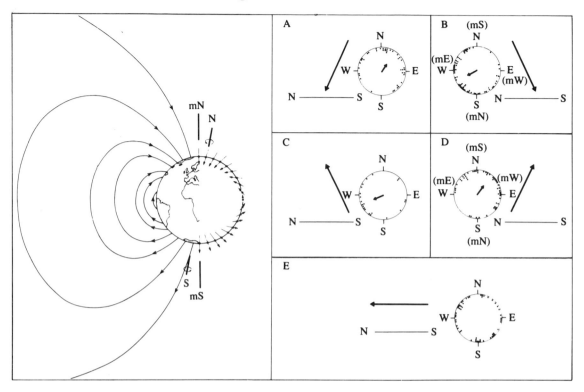

Figure 130   The earth's magnetic field extends far out into space, as shown diagrammatically by the field lines on the left-hand half of the earth. The magnetic field's direction and strength at the earth's surface are indicated by the arrows that are on the right-hand side. Note that the magnetic field is at an angle in relation to the earth's rotational axis (mN denotes magnetic north pole and mS magnetic south pole). The figures on the right show how the orientation in spring of caged Robins is affected by changes in the magnetic field. These changes have been brought about by means of large electrical coils around the cages, while at the same time the total field strength has been kept constant (0.46 gauss = the natural field strength in central Germany, where the angle of dip is 65°). The heavy arrows show the magnetic field direction (from the side with geographical north pointing to the left). The circular diagrams show the Robins' mean direction on separate nights in relation to what is north as seen geographically. The arrows inside the circles show the mean direction (omitted in fig. E, where the directional scatter is random). The orientation tests were carried out in windowless rooms, entirely without visible cues. The Robins orient in their normal migration direction, i.e. to the northeast, both when the magnetic field is normal (A) and when its polarity is totally reversed (D). When the magnetic field's horizontal or vertical components are reversed, the birds' orientation changes to the southwest, i.e. directly opposite to the normal direction (B and C). With a horizontally set magnetic field the birds lose the ability to orientate (E). Based on Wiltschko & Wiltschko (1972).

inclination of the magnetic field comes closest to the gravitational vector. When the magnetic field is horizontal, the dip-compass no longer works as an aid to orientation for the Robins, as is clear from figure 130. The dip-compass cannot therefore be used near the magnetic equator.

In the light of recent discoveries of magnetite crystals in the heads of birds, we should hardly find it odd that birds are not able to determine the polarity of the magnetic field but have to use the angle-of-dip principle. The nerves around the magnetite crystals probably pick up only forces that influence the crystals to arrange themselves parallel with the surrounding magnetic field and cannot 'sense' which part of a crystal is north pole and which is south pole.

The Robins' magnetic compass is continuously finely adjusted to the immediately surrounding magnetic field and constantly works with the best effectiveness. When caged Robins were tested in weakened (0.34 gauss) or intensified (0.60 gauss) magnetic fields in relation to the normal intensity at the actual locality (0.46 gauss in central Germany), they lost their bearings! After an acclimatisation period of a few days, however, they were able to use the modified magnetic field to orientate correctly again. After acclimatisation, Robins were able to orientate quite normally even using as weak a magnetic field as 0.16 gauss or as strong a field as 1.5 gauss.

Cage experiments have demonstrated a magnetic compass not only in the Robin but also in various *Sylvia* species, such as Blackcap, Whitethroat, Garden Warbler and Subalpine Warbler, and in Pied Flycatcher and Indigo Bunting. (Several further species have been added to this list, cf. Wiltschko & Wiltschko 1988.) In addition, the homing pigeon has a magnetic compass, this having been confirmed in many subtle release experiments. Following William Keeton's research breakthrough, when he demonstrated the ability of well-practised pigeons to orient perfectly in cloudy weather, with absolutely no help from the sun, the field was opened up for investigations of other possible means of orientation. William Keeton himself was the first to release homing pigeons with small magnets near their head. Their orientation was compared with that of control pigeons, which instead of magnets wore brass platelets wholly lacking magnetic properties. Both groups oriented correctly in sunny weather, when the sun-compass could be used. In cloudy weather, however, the pigeons which were fitted with disturbance magnets lost their bearings, while the pigeons wearing brass platelets flew in the right direction, towards the pigeon loft. Homing pigeons therefore rely on a magnetic compass in cloudy weather when the sun-compass cannot be used.

The pigeons' magnetic compass can be disturbed in an even more ingenious manner by attaching small coils around their heads and linking them to a small mercury battery which is fitted to the birds' backs. The current in the coils produces a magnetic field through the pigeons' heads. By connecting up the current in one or the other direction, the direction of the disturbing magnetic field can be fixed and in cloudy weather the pigeons can be made to orient either homeward or directly in the opposite direction. This indicates, on closer analysis, that homing pigeons, like Robins, cannot work out the polarity of the magnetic field and that they have a dip-compass.

Homing pigeons lose the ability to orientate correctly when they are released at places which have aberrant magnetic fields, known as magnetic anomalies. Amazingly enough, the ability to orientate is lost even when the birds are released in full sunshine. This was revealed by studies in the United States; at the same time it was demonstrated that the pigeons regain their orientation powers and choose the right course as soon as they have left the area of disturbance.

When magnetic minerals, mainly iron ore, are present in the earth's crust, the geomagnetic field is greatly intensified. The magnetic field strength over the Swedish iron-ore mountains of Kirunavaara and Lousavaara is amplified to twice the normal intensity, and an ordinary compass cannot be relied on there. Only a few kilometres away from these anomalies are areas where the magnetic field is perfectly normal. One of the greatest anomalies is found in Russia south of Moscow. Along a zone that is nearly 250 km in length, the field strength is everywhere intensified, to a maximum of 1.9 gauss, almost four times the normal total intensity. At similar large iron-ore deposits the magnetic landscape shows particularly strong 'undulations'. This 'landscape', however, is highly varied almost everywhere, even though the differences in field strength are more modest. Figure 136 in the following section shows how the presence of basalt and diabase gives the magnetic landscape its character over a small area in Scania, south Sweden; here the greatest anomalies involve an intensification of the normal field strength by between 4% and 8%.

When homing pigeons were released at anomalies in the United States, this was done at localities where the deviations in the magnetic field were precisely within this order of magnitude. Although an ordinary magnetic needle is hardly affected, the pigeons' magnetic sense is, to judge from the loss of orientation abilities, very sensitive to disturbance.

Is there a risk that young birds which hatch and grow up in areas of magnetic anomalies will acquire an incorrect compass? If this does actually happen, it could explain why some juvenile birds set out on autumn migration on entirely the wrong course (see section 4.9). Have the Pallas's Warblers that come to west Europe in the autumn perhaps hatched in nests which lay at the foot of Siberian mountains that are rich in iron ore?

Magnetic storms of moderate intensity produce a variation of only a few tenths of a percent in the total magnetic field strength at the earth's surface. This seems, amazingly enough, sufficient for the birds to be affected in their orientation. A study of nocturnal migrants made in the southern United States using ceilometers showed that the scatter of migration directions increased during hours of magnetic storms. At the same time it was noted that the mean direction of the migration had a tendency to shift to the left, something which has also been recorded in radar studies of the bird migration over Puerto Rico. This slight leftward shift in the migration direction would have been dismissed as an insignificant coincidence had it not been the case that released homing pigeons also bear more and more to the left as the intensity of the magnetic storm increases. The shift in the pigeons' mean direction amounts to between $10°$ and $40°$ at moderate to strong magnetic storm strength. This result was demonstrated when pigeons were released in clear weather so that they could see the sun (just as the nocturnal migrants could see the stars). In addition, statistical analyses from many years of pigeon-racing have shown that the pigeons' orientation ability is poorest at times when the number of sun-spots is exceptionally high. A large number of sun-spots generally coincides with magnetic storms on the earth. We may assume that it is the magnetism that has affected the pigeons.

The extremely fine sensitivity to magnetism that the birds exhibit at local anomalies or in magnetic storms suggests the possibility that their magnetic sense not only functions as an ordinary compass but also has some other important function. Do birds perhaps use their magnetic sense in order to fix position and determine which way to move in relation to the magnetic landscape at the earth's surface? Whether birds can read such a magnetic map or not is discussed further in the next section.

Co-ordination between the compasses

The fact that the directional sense of birds is based on at least three different compasses raises several questions. If birds have a magnetic compass the functioning of which is independent of time and weather, why then are the sun- and star-compasses needed? Which compass is most important? What happens if the different compasses simultaneously give different information on direction? There are still no comprehensive answers to these questions. Enough pieces of the jigsaw have been gathered, however, to enable us to try to put them together to form a picture of how the compasses are co-ordinated in the birds' sense of direction. I shall comment on the connections between the different compasses in turn, and begin with the relationship between the sun-compass and the magnetic compass.

William Keeton's finding that disturbance magnets cause homing pigeons to lose orientation in cloudy weather but not in sun on the face of it lends support to the conclusion that pigeons primarily use the sun-compass and make use of the magnetic compass only when this is impossible. A closer analysis shows that the situation is not so simple, for the pigeons' orientation in sunny weather is not unaffected by the magnetic field. Detailed measurements reveal that the mean direction shifts several degrees to the left when the pigeons are fitted with disturbance magnets. In similar fashion it has been shown that orientation also changes in sunny weather when the pigeons have coils around the head, even though to an extent that is slight and difficult to explain. The effects of magnetic storms and anomalies help to underline further the fact that the orientation of homing pigeons is by no means independent of the magnetic field in sunny weather. What co-operation exists between the magnetic compass and the sun-compass on these occasions?

Wolfgang and Roswitha Wiltschko have suggested that what happens is that birds use their magnetic compass in order to *determine* the flight direction and that the sun-compass (or at night the star-compass) then helps them to *maintain* this direction. The birds accordingly 'plug in' the magnetic compass from time to time to determine direction; in between they use the sun-/star-compass, which works as an 'automatic-pilot system'. The Wiltschkos further believe that the birds learn the sun-compass and calibrate it in relation to the magnetic compass.

This view is supported by the fact that homing pigeons normally have not learnt their sun-compass until they are 12 weeks old. Even before this they are able to orient homewards, but lose their bearings completely, even in sunshine, if they are released wearing disturbance magnets. Pigeons which have grown up without ever having seen the sun, and which have thus not learnt the sun-compass, in the same way lose the ability to orientate when they are wearing disturbance magnets. This shows that the magnetic compass is fully developed before the pigeons begin to learn to use the sun-compass.

One interesting result, which suggests that the magnetic compass is of more fundamental importance than the sun-compass, is that it is markedly difficult to induce homing pigeons which have had to 're-learn' their sun-compass once to misorientate in sunshine by shifting their internal clock. Their orientation is easily upset, however, by disturbance magnets. These pigeons evidently no longer rely on the sun-compass but confine themselves to using their magnetic sense.

The Wiltschko team have carried out a series of interesting cage experiments with small birds during the autumn and spring migration periods in Spain in order to ascertain the relationship between the star-compass and the magnetic compass. The tests were made on starlit nights. On certain nights the magnetic field was deflected by means of large coils around the test cages. On other occasions the magnetic horizontal component was neutralised so that the magnetic field became unusable for determining direction. How do the birds behave when the magnetic field is deflected? Do they rely mainly on their star-compass and keep orientation unchanged? Or do they adjust themselves to the magnetic compass? The results showed that the birds adjust their orientation to the magnetic compass! When the horizontal magnetic field is set to zero, they orient in the direction they have had during the last preceding test with a usable magnetic field. The birds clearly adjust their star-compass to their magnetic sense and then stick firmly to the last adjustment if the magnetic compass cannot be used. An important difference was revealed between on the one hand the reaction of Garden Warblers and Whitethroats and on the other that of Robins. The two tropical

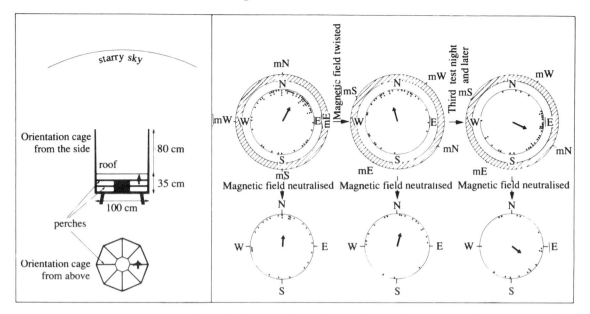

Figure 131   Robins which oriented northwards during the spring in Spain were tested in orientation cages under the natural night sky and with the natural magnetic field. When the magnetic field is deflected so that magnetic north (mN) points east-southeast, the Robins' orientation does not change during the first nights of tests. Not until the third night's testing and later do the birds seem to detect the deflection of the magnetic field and to alter their orientation accordingly. Robins evidently make use of their magnetic compass only at long time intervals and in between use the starry sky to maintain orientation. This is borne out by the fact that the birds' orientation remains unchanged when the horizontal magnetic field is set to zero. Based on Wiltschko & Wiltschko (1974, 1975).

migrants changed their orientation immediately, on the first night, when the magnetic field was twisted. By contrast, the Robins seemed not to notice that the magnetic field had been turned until after about three nights of tests (figure 131).

The star-compass seems thus to play a greater part for the Robin, a short-distance migrant, than for the Whitethroat and the Garden Warbler. The Robin does not 'plug in' its magnetic compass more than about every third night of migration – for the rest it relies on its 'automatic pilot', the star-compass. The two warblers, however, use the magnetic compass on every night of migration. The reason for this difference may possibly be that the Robin continuously has a good picture of the northern night sky, including the area around the Pole Star. For the other species the starry sky alters to a greater extent during the long-distance migration; they cannot see the Pole Star at all at more southerly latitudes, and this makes it additionally important for them to repeat the adjusting of the star-compass, using their magnetic sense, at brief intervals of time. The Indigo Bunting, like the Robin, is a short-distance migrant. According to Stephen Emlen's planetarium experiments, its star-compass also plays an independent role for several migration nights in succession.

The results described above support the view that the magnetic compass is the dominant one and that the star-compass is adjusted in relation to it. This is also confirmed by the fact that Robins can be taught to orientate correctly in relation to an artificial night sky with only 16 'stars' (points of light on the cage roof) in an arbitrary pattern. After the magnetic field has been neutralised, the birds, using the artificial night sky, maintain the orientation that they have previously selected using the magnetic compass.

A comparison between cage tests in which the birds have orientated by means of the magnetic field alone and cage tests in which they have also been able to use the night sky in addition to the magnetic field reveals interesting differences. The birds' mean direction is exactly the same in both types of test, but the orientation is appreciably more concentrated when a starry sky (real or artificial) is visible. The increase in the concentration of directions which occurs when the birds can see the starry sky is considerably greater for the Robin than for the Whitethroat and the Garden Warbler. This shows that the star-compass is of comparatively great importance to the Robin's sense of orientation. The results at the same time lend weighty support to the supposition that the magnetic compass is used for basic direction-fixing and the star-compass for helping the birds effectively to maintain this direction.

For the birds to be able to take a reading from their magnetic compass presumably requires that they move in varying directions, perhaps preferably in a circle. The magnetic compass is therefore difficult to use for keeping their course. The star-compass or the sun-compass is much better suited for this purpose.

The hours of dawn and dusk are interesting from a point of view of orientation. How do birds react during the hours when day turns into night and night turns into day, when they have the opportunity to co-ordinate sun and stars as direction signs for their migration? This question is of special interest because the majority of diurnal migrants set off precisely at dawn and the majority of nocturnal migrants precisely at dusk. I have earlier described the collective and abrupt departure of night-migrating small birds. This takes place about 30 minutes after sunset (section 4.7). Do they set off at this stroke of the clock because they can at that moment find their bearings with exceptional precision? This was asserted as early as the 1950s by the Dutch ornithologist D. A. Vleugel, who considered that the birds fixed their migration direction in relation to the direction of sunrise (diurnal migrants) or the direction of sunset (nocturnal migrants). The choice of direction is made at the start of migration; later in the day, or the night, the birds maintain the migration heading by making use of landmarks or by flying at a constant angle in relation to the wind.

The points of the compass at which the sun rises or sets are useful direction-markers; they vary to quite a small degree at different places during the autumn and spring migration seasons. At spring and autumn equinoxes, the sun rises exactly in the east and sets exactly in the west everywhere on earth.

Vleugel's suggestion that migrants might have a 'sunrise-compass' or a 'sunset-compass' has received renewed interest in connection with radar studies of nocturnal migrants in the United States. White-throated Sparrows which were released at high altitudes from a balloon oriented as quickly and accurately when they were set free so that they could see the sunset, even though it was otherwise cloudy, as when they were freed under a clear starry sky.

Cage experiments with Savannah Sparrows (*Passerculus sandwichensis*) have produced further interesting results. The birds, which are nocturnal migrants, showed the best orientation on those nights when they were made to see first the sunset and thereafter a starlit sky following the onset of darkness. The concentration of directions became only slightly poorer if they saw the sunset but as a consequence of cloudy weather were deprived of the chance to see the starry night sky. Most remarkable of all was the fact that the Savannah Sparrows seemed to lose their bearings completely if they were not placed in the cages until after the onset of dark and saw only the starry sky. The results indicate that the Savannah Sparrows fix their migration direction at the beginning of the night by means of the sunset and that they then use the pattern of the stars to maintain that direction. Unlike the Indigo Bunting or the Robin, the Savannah Sparrow does not seem to be able to choose the right direction with the aid of the stars alone. What part the magnetic compass plays in the Savannah Sparrow's orientation is unknown.

Although the pigeon's sensitivity to polarised light was demonstrated quite some time ago (see section 5.1), experiments designed to investigate whether birds may use the skylight polarisation pattern for their migratory orientation were not conducted until the 1980s. And these experiments gave a clear decision! By covering the birds' orientation cages with sheet polaroids, and shifting the axis of polarisation, it was elegantly and distinctly demonstrated how the birds, White-throated Sparrows and different species of American warblers, changed their orientation at dusk and dawn depending on the manipulated polarisation pattern (cf. Able 1989). Surely, it may well turn out that the skylight polarisation pattern has a very important role in the birds' orientation systems, in combination with other celestial and magnetic cues.

Anybody who had expected that the sense of direction of birds would prove to be uncomplicated and to be founded solely on a single reliable 'compass' must be disillusioned and confused by all the particulars that have emerged on the importance of the sun, the stars (possibly the moon), and the magnetic field. Birds seem to avail themselves of most possibilities that exist at all in order to get an impression of the points of the compass. In actual fact this redundance of 'directional feel' is perhaps not so strange. Birds transport themselves more frequently, faster and farther than all other animals. Unlike man, they cannot put their trust in technical gadgets but must take advantage of all the various orientation possibilities that the world around them offers in order to reach their goal safely.

To explain in detail how the directional sense of birds is pieced together, and how and why it differs among different species, is a great challenge for ornithological research.

5.3   How do the birds find the right migration route?

Birds can be considered to use two different principal methods for finding the right migration route: orientational migration and navigational migration. By orientational migration I mean that the birds have an innate directional instinct which leads them the right way between summer quarters and winter quarters. Navigational migration on the other hand implies that they can determine their position in relation to the migration goal towards which they navigate the whole time. Of course a combination of these two methods can be imagined, for example so that the birds cover the main part of the journey by orientational migration and swap over to navigational migration when they are approaching their final destination.

One way of investigating which method the birds use is to make displacement experiments while migration is in progress. If the birds correct for the involuntary displacement and return to their normal migration goal, then this confirms that they are using navigational migration. If, on the other hand, they continue in the normal migration direction without 'detecting' that they have been displaced, this points to orientational migration. The most comprehensive displacement experiments have been made by the Dutch researcher A. C. Perdeck (figure 132). Over a ten-year period he had more than 11 000 Starlings caught in the region of the Hague during the autumn-migration period. The birds were ringed and transported express by air approximately 600 km to the south-southeast, to Switzerland, where they were released at one of the airports in Basle, Zürich or Geneva. At the same time control birds were ringed so as to give detailed information on the normal wintering area of

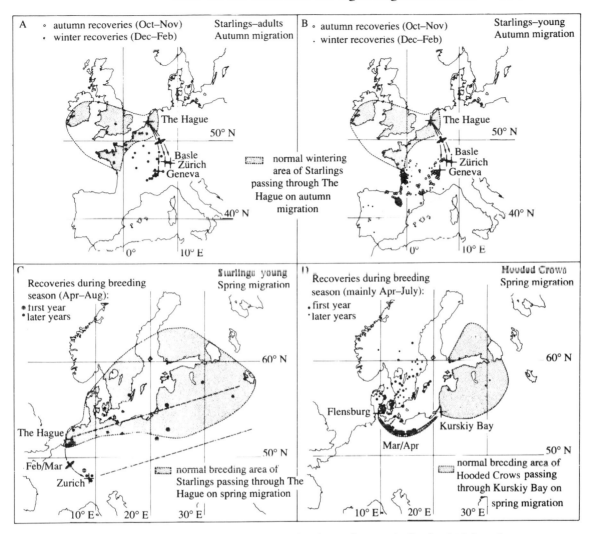

Figure 132   Ringing recoveries (more than 50 km from release site) of birds which have been subjected to displacement experiments. Map A shows recoveries of adult and map B of young Starlings during the first autumn and winter after displacement. Many adult Starlings return towards their normal winter quarters, while the young birds maintain an unchanged migration direction and reach new areas. Based on Perdeck (1958). C. Most young Starlings return to the normal breeding area when they are transported elsewhere while on spring migration. Based on Perdeck (1974, with later additions). D. Only very few Hooded Crows that were transported by train from the Kurskiy Bay region of the USSR to Flensburg returned to their normal breeding area. Most made off in the typical migration direction, northeast, to Denmark and Sweden. A total of 900 crows was freighted to Flensburg during the years 1935–1939. Based on Rüppell (1944).

the Starlings that pass through the Hague on autumn migration. The ringing recoveries of the displaced birds pointed to a clear distinction between adult and young (first-year) Starlings, independent of whether, after being air-freighted, they were released together, in mixed flocks, or not. Many adult Starlings altered their migration direction and flew northwest, i.e. towards the normal winter quarters, to which some of them even succeeded in returning for the winter. The young Starlings, however, seemed not to detect the displacement but carried on in the normal migration direction, to the west-southwest or southwest, to southwest France and Spain, some even as far as Portugal, areas which lie completely outside the population's regular winter distribution. Many of the young Starlings returned later as adults to these new wintering areas.

Perdeck drew the conclusion that Starlings over one year old use navigational migration to return to their traditional wintering area. Young Starlings, which have never before been in the winter quarters, by contrast use orientational migration. After they have, during their first winter, established themselves within a suitable area and learnt its placing on their map image of the world around them, they navigate back to the same area in subsequent years. Site fidelity (*Ortstreue*) in the wintering area is presumably explained by the advantage that it means for a bird to live in terrain that is well known to it; the fact that the bird has come through its first winter there safe and sound demonstrates at the same time that it is suitable as a place to stay. Site fidelity is not, however, absolute, for some of the displaced young Starlings were recovered in later years within the normal wintering area in north France or in southern Britain. Had they perhaps not been 'satisfied' with the first winter quarters and during their second autumn again used orientational migration, which this time took them to the normal winter quarters?

Which migration method do young Starlings use in the spring when they make their way towards the breeding area? Do they navigate back towards the place where they hatched the previous summer or do they migrate along a constant compass course? To get an answer to this question Perdeck transported about 3000 ringed young Starlings from the Hague to Zürich in early spring. The recoveries suggest more than anything that the young Starlings use navigational migration (figure 132C). The picture is not clear-cut, though, for some of the transported Starlings established themselves as breeders in Switzerland. Furthermore, certain recoveries from other displacements, from the Netherlands to Spain, suggest that some young Starlings use orientational migration during the spring as well.

In their entirety, Perdeck's various displacement experiments suggest that the Starling's method of migration works as follows. The birds as a rule use navigational migration when returning to their previous wintering and summering localities. Young birds which have never been in the winter quarters, however, use orientational migration in their first autumn. This state of affairs does not apply universally, however, to all birds. Experiments with displacement of Hooded Crows in spring from Kurskiy Bay to Flensburg suggest that adult as well as young crows primarily use orientational migration (figure 132D). The ringing recoveries show that most crows, independent of age, established themselves in Denmark or in southern Sweden instead of returning to the normal breeding area in the Baltic States and in southern Finland. Nevertheless, a small minority did return there; all of them were adults.

The obvious site fidelity which most migratory birds show to breeding and wintering localities, sometimes also to staging sites along the migration path, points to the fact that they navigate to these goals. Whether this navigational migration is resorted to only near the destination areas or whether it is going on throughout the whole migratory flight is impossible to know in the absence of further displacement experiments. Perhaps there is a distance limit from the goal beyond which navigational migration does not function effectively? The displacement experiments could be interpreted in the following way: when the crows were transported from Kurskiy Bay to Flensburg, the majority of them went beyond such a limit; by contrast, the majority of the Starlings that were taken from the Hague to Switzerland did not do so.

It may well be that the greater part of the birds' migration is in actual fact based on the orientational method. It is perfectly conceivable that a Willow Warbler for example which leaves Africa in spring migrates all the way up to north Europe guided only by its compass sense, before, in the final phase of the migration, it switches over to navigating back towards the place where it bred in the preceding summer.

How does orientational migration work? This problem has been analysed by a German research group headed by Eberhard Gwinner. It has been demonstrated that birds have hormonal annual rhythms (circannual rhythms) which govern not only processes such as coming into breeding condition, moult and fat deposition but also the migratory urge. The diminishing day length in autumn and the lengthening of the day in spring act as the timing device or *Zeitgeber* (= time-giver) which exactly synchronises these rhythms with the 12 months of the year. Nocturnal migrants in cages show intensive restlessness at night during the migration period.

So far as the total 'amount' of such migratory restlessness during a migration season is concerned, the various species differ considerably from one another in proportion to the length of the typical migration journey (figure 133). The amount of this migratory activity that is included in the birds' annual rhythm can in actual fact be considered to constitute a sort of innate 'programme' which governs the length of migration: when the migratory restlessness peters out the birds have got just the right distance, to the normal summer or winter range.

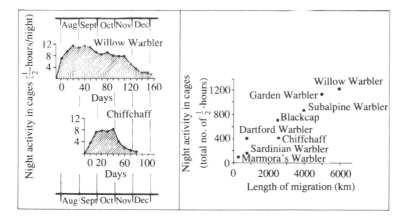

Figure 133 Migratory urge in nocturnal migrants in cages is revealed when the birds, instead of sleeping, become active at night. This nocturnal activity occurs at the same time as the species' natural migration period and is on a larger scale for long-distance migrants than for short-distance ones. The graphs show night-time activity in cages during the autumn of German Willow Warblers and Chiffchaffs. The Willow Warbler begins its autumn migration early and migrates to tropical Africa; the Chiffchaff migrates later in the autumn and moves only the short distance to the Mediterranean region and North Africa. The diagram on the right shows a directly proportional association between the total night-time activity and the length of the migration for different species. Based on Berthold (1973) and Gwinner (1969, 1977).

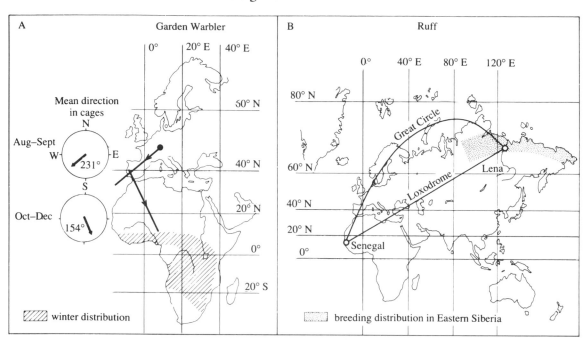

Figure 134   A. Garden Warblers from southern Germany change their mean direction in orientation cages during the course of the autumn. This suggests that the birds have an innate programme for direction which ensures that they migrate first towards Spain and Portugal and then turn southeast, towards Central Africa. The birds' mean directions during the first and the latter parts of the autumn have been drawn in on the map. Based on Gwinner & Wiltschko (1978).

B. Great Circle migration puts particularly high demands on the birds' orientation system since the direction of travel is changing the whole time. By flying in a series of different compass headings which supersede each other at frequent intervals, birds would be able to follow migration paths resembling a great circle. Ruffs from eastern Siberia migrate to West Africa in the autumn adhering closely to the Great Circle route.

Is there also an innate programme for the direction of the migration? An experiment which reinforces this possibility is illustrated in figure 134A. Garden Warblers from south Germany were kept under constant ambient conditions and were tested in orientation cages (in closed rooms, where only the magnetic compass could be used) at regular intervals during the course of the autumn. The mean direction of the birds' nocturnal migratory restlessness pointed southwest during the first part of the autumn and later shifted to south-southeast. This accords well with the ringing recoveries of German Garden Warblers, which show that the birds first migrate to the Iberian Peninsula and from there, after putting on fat, cross the Sahara in the direction of Central Africa.

The research work consequently leads us to suspect that birds are, in their internal annual rhythm, pre-programmed for the migratory journey with regard to both distance and direction(s). A migratory bird follows its inborn instinct, or can we perhaps call it an automatic-pilot system, which gives information on compass heading and degree of migratory activity as the internal seasonal clock ticks on. Can such a system really give sufficient precision and flexibility to guide the birds correctly on their travels? Some researchers, such as Jørgen Rabøl from Denmark, find this hard to believe. Rabøl argues instead that migratory birds have a very sophisticated innate 'programme', with built-in information on map co-ordinates and with instructions on navigating towards various target areas, displaced in relation to each other along the migration path, during the course of the season. So far, however, there is no evidence for Rabøl's suggestion for an innate navigation system.

If an innate orientation programme contained a sequence of different compass headings which covered the entire duration of the migration period, it could direct the birds to follow a more or less complicated and winding migration path, perhaps even a Great Circle route. The shortest route between two points on the earth's surface does not run in a straight compass heading between the points (known as the loxodrome or rhumb line) but follows the Great Circle. A Great Circle divides the globe into two equal-sized halves. The difference in distance between the Great Circle and the loxodrome is greatest in the east–west direction at latitudes near the poles. Here, then, migrants have much to save in flight energy by following the route along the Great Circle; on the other hand, they are then faced with great demands on orientation, for the compass heading is continuously changing along the Great Circle. In figure 134B examples are given of the Great Circle route and the loxodrome between the Lena River in Siberia (70° N, 125° E) and the Senegal River in West Africa (15° N, 15° W). The routes are drawn in on a map in the Mercator projection, where the loxodrome forms a straight line. This projection greatly enlarges the earth's surface towards the poles and neither in linear distance nor in surface area is it accurate.

Large numbers of Ruffs migrate from eastern Siberia to winter at the floodlands around the Senegal and Niger Rivers in Africa (see section 3.1). To follow the Great Circle route between the Lena and the Senegal they must leave the Lena on a northwesterly course (322°) and gradually change flight direction until they arrive at the Senegal on a course of between south and south-southwest (193°); the total flight distance then becomes 10 060 km. The loxodrome would allow the birds to hold a constant course towards Senegal (239°), but the distance is 11 850 km, i.e. 18% longer than the Great Circle route. During the autumn many Siberian Ruffs pass through northwest Europe, which indicates that the migration path runs close to the Great Circle. It would be interesting to learn whether the Ruffs fly out over the Arctic Ocean in order to follow the Great Circle route in detail.

In the spring the Ruffs return to Siberia by a route closer to the loxodrome. At this time the main stream passes through the Mediterranean region and central Europe. The longer itinerary in spring is probably determined by the availability of suitable staging sites; at this period the northern tundra is still frozen and has little to offer, while the winter rains provide suitable stop-over environments in the Mediterranean and the Middle East.

In the course of evolution the migratory routes of birds have been adapted to many different factors: the length of the journey, the likelihood of favourable migration weather, and the availability of suitable stop-over sites. The resulting flight path can become quite complicated and place great demands on the birds' inbuilt system for orientational migration.

Without underestimating orientational migration, the hardest mystery to explain is the navigational powers of birds. Certainly we have a fragile knowledge of just how much navigation goes into bird migration, but the striking fidelity to home area shown by migratory birds and the ability of homing pigeons to find their way home clearly demonstrate that birds really do possess a great skill in navigating. How do they go about it?

To be able to navigate it will not do just to have a reliable sense of direction; to give an idea of the position in relation to the goal a map sense is required in addition. Consider how we as human beings keep a map image in our mind, even though vague and fragmentary, when we move from place to place in the countryside or in the city. When exceptionally precise demands are placed on our navigation (for example in visual flying) or when we are moving over partially unfamiliar terrain (such as in the sport of orienteering), we use precise geographical maps in combination with a compass.

Birds must have a map sense that is much more efficient than our human one if we can judge from their ability quickly to find the right course home from distant localities. What features in the countryside do they use as a basis for their map image?

The most likely possibility is that birds, like we humans, build up their map sense by learning landmarks. Birds have an additional advantage in that while flying they can get a wide general view of the landscape. The range of vision is limited by the curvature of the earth's surface, but at only 1000 m altitude it is over 100 km. At this altitude it is therefore possible to survey over 30 000 square kilometres, an area greater than the size of Belgium. In actual fact for good landmarks the visual range is even farther – for it is not dependent only on the altitude of the observer but also on the height of the sight mark. From 1000 m up, it is thus possible in ideal visibility to see a mountain peak of similar height at a distance of more than 200 km.

Despite the fact that birds consequently have extraordinarily good chances for using landmarks to find their way home, they seem as a rule to make use of this opportunity only when they are very close to their goal. By using an aeroplane to track homeward-flying pigeons fitted with radio transmitters, it has been found that the pigeons maintain an unaltered direction of travel, even though it points a bit to the side of the goal, all the way until they have got as near as 5–10 km from the pigeon loft; only then do they make an abrupt change of course and fly straight home. Going by this reaction, we may suspect that they do not take any great notice of landmarks until they have got so close to their goal that they recognise the details in the landscape and can see exactly where the goal is. A good piece of evidence to show how important the landmarks indicating the pigeon loft's exact position are was obtained by the pigeon-rearer who erected a 30-m-high tower with a large golden sphere on the top, as he found that after this the pigeons returned home considerably more quickly than they had done before.

Some researchers into homing pigeons doubt that pigeons pay any great attention at all to landmarks when they are really close to the loft where they have made daily exercise flights throughout their life. To justify these doubts the experiments illustrated in figure 128B are used; these show that by re-setting the birds' internal clock one can trick them into flying off course when they are as close to home as 1000 m, provided that the pigeon loft itself is not directly visible from the release site. I find it almost inconceivable to imagine that at these small distances from the loft the pigeons would not immediately know their way about by means of landmarks. How would we react ourselves in a similar situation if we were equipped with a compass which, unbeknown to us, showed completely the wrong readings and we were left to make straight for home from a place a kilometre or so away? We would straightaway know our way about and be aware that the place where we found ourselves lay in a certain quarter from home. If we used our compass to go in the direction homeward, we would end up on totally the wrong course. Naturally we would soon discover that something was not right and would re-check the direction we had chosen, just as the pigeons do which fly off in the wrong direction but after only a short distance discover their mistake and correct the direction towards the pigeon loft.

The fact that pigeons can be tricked into flying off course when they are near the pigeon loft cannot therefore, in my opinion, be taken as confirmation that they totally neglect to use their knowledge of landmarks. The experiment does, though, show one thing clearly. The pigeons do *not* abandon the using of the sun-compass in well-known terrain so as to find their way home by means of various landmarks alone.

With an elegant experimental method it can be established that homing pigeons cope amazingly well with navigating without seeing any landmarks at all. The method consists in fitting frosted contact lenses or spectacles to the pigeons! The frosted glass erases all contours, and the horizon dissolves and becomes simply a diffuse transition between the light of the sky and the darker landscape below it. Since the birds can see differences in light intensity through the frosted glass, they are able to determine the sun's position. By re-setting the birds' internal daily clock, it has been shown that they use the sun-compass even when flying with the frosted lenses. In figure 135 a couple of examples are depicted of the pigeon's ability to navigate in such conditions. Judging from these and other similar results, they are able to navigate to the pigeon loft without the help of landmarks with a margin of error of only a few kilometres or less. The experiments with frosted eye lenses provide interesting evidence of how pigeons at times effectively ringed in their target and circled only a few hundred metres from the loft without being able or daring to land owing to their defective vision. Finally they tried landing by descending cautiously in what almost resembled helicopter flight and ended

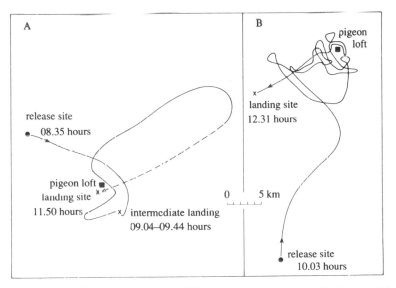

Figure 135   Flight paths of two different homing pigeons with lenses of frosted glass over their eyes. The pigeons were fitted with radio transmitters and tracked with a positional accuracy of ± 150 m from a following aeroplane. Both pigeons immediately set off in approximately the correct direction for home. The pigeon in A passed north of the goal, made a wide sweep back and made an intermediate landing for a while; it then headed off south of the pigeon loft, suddenly made a substantial course correction towards the northeast, followed by a swing to the northwest, and then missed the loft by less than 500 m. After this it flew far out of its way; the plane had to abandon the pursuit at 10.45 hours in order to refuel. After refuelling, the plane found the pigeon again one hour later perched on a log in the woods less than 1 km from the loft. On the map, the position at 10.45 hours is connected to the landing site by a broken line, as the exact flight path between these points is not known. The pigeon in B quickly made its way ahead until it arrived near the pigeon loft, where it flew around for nearly two hours. On two occasions it circled within 500 m of the loft; after having got lost away from the goal, it finally landed. Based on Schmidt-Koenig & Walcott (1978).

up on fields or in clumps of woodland, where they were located by means of the radio transmitters.

These experiments provide fairly clear information: landmarks are not a necessary condition of the homing pigeons' ability to navigate. The landscape picture seems to be of assistance mainly during the very final phase of the navigation process, when there are only a few kilometres remaining to the goal. We must look further, for other means which, judging from various homing-pigeon experiments, allow of navigation over distances of at least 1000 km with a precision, without the aid of landmarks, of plus or minus a couple of kilometres from the goal.

A classic idea is that birds use the position of the celestial bodies, i.e. the sun and the stars, to navigate. The idea was developed during the 1950s by the English ornithologist G. V. T. Matthews, who advocated the view that homing pigeons determine their position in relation to the pigeon loft by means of the sun. By extrapolating the path of the sun and comparing the sun's altitude when at its highest with the same altitude at the home base, the pigeons theoretically have a means of determining whether they are south or north of their loft; at the same time they can work out their position in an east–west direction by comparing the estimated time when the sun is at its highest with the corresponding time at the home base. This consequently requires that the pigeons have a detailed memory of the sun's path at the pigeon loft during the time of the year in question and at the same time that they have an accurate internal clock which shows the right time at the home base.

Let me give an example of how navigation by reference to the sun could be thought to work if pigeons were transported from their home in, say, Helsinki (c. 60° N, 25° E) southwest to Hamburg (54° N, 10° E). By comparing the observed (or extrapolated) altitude of the sun when at its highest point in Hamburg with the 'mental picture' of the corresponding altitude at the pigeon loft, the pigeons would find that it was approximately 6° higher than at the home base. From this they could conclude that they had been moved an equivalent number of degrees of latitude to the south. Let us further assume, for simplicity's sake, that the pigeons were released in the middle of the day at noon local Hamburg time, when the sun was at its highest due south. The pigeons' internal navigation clock would, however, follow Helsinki time and show 13.00 hours, i.e. the time one hour after the sun had reached its highest point. The pigeons would find that the sun had not moved so far on its daily course as at their home base. More exactly, the observed sun culminates one hour later, corresponding to approximately 15° of lateral movement. The pigeons would conclude from this that they had landed up an equal number of degrees of longitude to the west. (Pigeons of course 'draw their conclusions' in an instinctive way, quite differently from us with our rationalised human reasoning.) On the basis of these observations, the pigeons could then select the right course home towards the northeast.

Matthews's model for navigation using the sun places great demands on the birds' ability, after watching the sun a short while, to estimate by means of extrapolation both the sun's altitude and the time when it is at its highest point in the sky. An alternative method is that solar navigation is based on gauging of the sun's altitude and of the rate at which the sun's altitude changes; another that the gauging of the sun's altitude is combined with estimation of the angle of inclination of the sun's track. These models instead call for a great ability to remember in detail how the values in question vary during different times of the day at the home base.

Irrespective of the exact method, navigation by reference to the sun requires high precision in estimating of angle and gauging of time. An error of 1° in the sun's altitude or of five minutes in time is the equivalent in north European latitudes of a north–south and an east–west positional error of around 100 km.

In seafaring during recent centuries, right up to modern times, man has made use of solar navigation. After reliable mechanical clock workings had been devised, Greenwich Mean Time (zero meridian) and Universal Time were introduced in the 1700s in order to facilitate efficient navigation. With a sextant, an accurate chronometer and nautical almanacs, we humans can achieve a precision of around 10 km in position-fixing.

Is there anything to suggest that homing pigeons also navigate by means of the sun? In actual fact, most experiments that have been made in order to test Matthews's theory have produced negative results. Orientation researchers have, for example, failed to demonstrate any biological clock which the birds would be able to use to relate their observations at the place where they are released to the point of time at the home base. Birds certainly have an internal daily clock which is an integral part of the sun-compass, but it seems to be totally independent of the capacity for position-fixing. If this clock had been used for navigational purposes, an adjustment, for example of six hours forward, ought to be interpreted by the pigeons as a removal a long way to the west; they should therefore always head off eastwards when released. In reality, however, they fly off to the east, south, west and north, depending on whether they were released north, east, south or west of the pigeon loft. Instead of misinterpreting their position in the east–west direction, the pigeons are taken in regarding the compass direction to the sun so that their sun-compass consistently leads them 90° off course (the equivalent of six hours' lateral movement of the sun).

It is of course conceivable that, apart from the internal clock which is used for the sun-compass, pigeons have a further biological clock, a special navigating clock which shows the exact time at the home base. In an attempt to demonstrate a navigating clock of this sort, homing pigeons were subjected to completely irregular changes between day and night and at the same time were given heavy water to drink. Chemical processes take place more slowly when ordinary water is replaced by heavy water, which leaves its marks in all known types of biological clocks, from the daily movements of flower petals to the activity rhythm of animals. The free-running daily rhythm of birds is set back by heavy water by about one hour per day. The ungenial treatment which the pigeons had to endure would, it was hoped, destroy their supposed navigation clock and its synchronisation with the time at the home base with the result that after the treatment the birds would fail to navigate home by means of the sun. Instead they should, it was expected, re-synchronise their navigation clock with the daily rhythm in a new pigeon loft, more than 400 km to the west, to which they were transported after the treatment and where they were made to live for a period of a week, could see the sun and could get into the normal way of things. It was an exciting occasion when, one sunny day, the pigeons were brought out into the open countryside and released. Would they navigate back to the old pigeon loft or to the new one? They all made their way towards their old home and arrived there after normal flying time. The treatment they had received therefore resulted in no effect whatsoever on their powers of navigation; the attempt to demonstrate solar navigation had failed.

William Keeton's demonstration that well-practised pigeons cope splendidly with navigating even in cloudy weather and the Wiltschko couple's findings that young pigeons find their way home even though they have never once been allowed to see the sun while growing up are real deathblows to the notion of solar navigation. They prove that the sun is not necessary for the pigeons to navigate.

Nevertheless, we cannot exclude the fact that the sun provides a certain additional help to the birds in position-fixing, at least when it is a matter of working out their displacement in a north–south direction. A researcher with homing pigeons in England has carried out learning experiments which suggest such a possibility. Experiments were made at two places, 100 km south of and 100 km north of the pigeon loft. With reward experiments the pigeons, which had been allowed to see only sky and sun but no landmarks whatsoever, learned to determine the direction home. When the learning was completed, a pigeon accordingly chose the direction north when it found itself at the southern experimental station and south when it found itself at the northern station. Using a periscope arrangement of mirrors, the elevation of the sun's path, as the bird saw it from its cage, was altered. When the path of the sun at the southern experimental station was lowered by means of the mirrors so that it corresponded to the natural situation 200 km to the north, the pigeons chose exactly the opposite direction to before; the same thing happened when the sun's path at the northern station was raised! That the birds were helped by the sun's elevation in distinguishing between the two experiment sites can hardly be doubted; on the other hand, the original learning need not necessarily have anything to do with the direction to the pigeon loft and by that nor with the ability to navigate either. The pigeons, without having 'thought about' the home base, may have learnt to choose the direction south at the place with the lower sun height and north at the place with the higher sun height.

The stars, too, present opportunities for navigational assistance. The angle at which the Pole Star stands above the horizon immediately reveals the latitude in the northern hemisphere at which one happens to be. The longitude can be determined by means of the time displacement in the night sky's rotation at different places. Franz Sauer interpreted the results from his planetarium studies with Lesser Whitethroats and Blackcaps as indicating that the birds navigated by means of the starry sky and an internal clock. This interpretation, however, is questionable since the results are not clear-cut. Planetarium experiments in later years with Indigo Buntings are, like combined magnetic and stellar-orientation experiments with Robins, Garden Warblers and Whitethroats, contradictory to the birds using stellar navigation. The fact that bird passage is often intensive even on cloudy nights, when the starry sky is not visible, also gives us to assume that the star pattern is not of decisive importance to the navigation of nocturnal migrants. As described earlier, the northern night sky's rotational centre and the nearest surrounding constellations occupy a central role in the Indigo Bunting's star-compass. It would not, therefore, be surprising if the buntings took notice of the Pole Star's elevation in order to get an additional pointer on how far to the north or south they were.

So far as the importance of the celestial bodies is concerned, we are therefore forced to conclude that neither sun nor stars are essential for the birds to be able to find their way home. At best they provide only some of the clues to the process of navigation.

Let us instead reflect on how the magnetic field may be of use in the birds' navigation. The most surprising thing of all about the magnetic sense of birds is its extremely delicate sensitivity. The birds' orientation is affected by magnetic storms and magnetic anomalies, despite the fact that the disturbances do not amount to any more than tenths of a percent of the magnetic field's normal intensity. The alteration in the direction of the field with such small disturbances is of course absolutely negligible – less than one degree. Despite this, birds fly several tens of degrees off course or lose orientation altogether at some anomalies.

Do birds perhaps have two magnetic systems, a magnetic compass and a magnetic navigational sense? If so, it would be the latter that was confused by magnetic disturbances. To obtain a reliable direction with its magnetic compass a bird need hardly use more than a few thousand of its magnetite crystals. If, on the other hand, it 'plugged in' all the 10 million or 100 million magnetite crystals which according to theoretical calculations are to be found in its head, a sensitivity of one-hundredth of a percent (0.01%) or even less of the earth's magnetic field strength would be achievable!

We are now, as everyone realises, into highly speculative areas, but let us nevertheless continue and see where they lead. The earth's magnetic field offers interesting possibilities for position-fixing on both a global and a local scale (figure 136). By measuring the total field strength in combination with the magnetic field's angle of dip, for example, we can determine our position at all places on the earth. The magnetic total intensity increases in the northern hemisphere by almost one-hundredth of a percent per kilometre northwards. If birds really do possess the enormously fine sensitivity that the theories predict, they should consequently be able, using the magnetic field, to ring around their target to an accuracy of 1–2 km. Is this what homing pigeons do when flying with frosted-glass lenses over their eyes?

The global gradients in the magnetic field are not perfectly uniform. Irregularities are caused by local occurrences of magnetic minerals in the earth's crust. The local magnetic landscape has 'peaks' with increased magnetic strength and 'dales' with weakened magnetic field. This 'landscape' can, just as well as the ordinary topographical landscape, be of assistance to the birds when they are navigating near to the home grounds. On the other hand, it can of course also create confusion when homing pigeons are released at unfamiliar magnetic anomalies. As soon as the pigeons have left the area of the anomaly and come down to the 'magnetic flat country', where the opportunity exists to determine their position in the large-scale magnetic landscape, they correct their course and turn homewards.

How are we to interpret the experiments with disturbing magnets (of approximately the same strength as the earth's magnetic field) that cause homing pigeons to lose their bearings altogether in cloudy weather but which produce only minor deviations in orientation in sunny conditions? Do the disturbing magnets perhaps affect only the birds' magnetic compass? And does the magnetic navigational sense, calibrated for an entirely different level of minimal magnetic gradient-effects, perhaps hardly suffer at all? If this explanation is correct, it means that the pigeons managed on the whole to fix their position faultlessly despite the fact that they were wearing disturbing magnets; it was not until the direction homeward had to be gauged on the compass that mistakes arose. When the pigeons made use of the sun-compass everything went well, but not when they used the magnetic compass, which because of the disturbing magnets showed serious error.

Despite this argument being still at the speculative stage, it has to be admitted that the exceedingly sensitive magnetic sense of birds provides what is so far the most exciting explanation of their powers of navigation. If future research results were to show that the magnetic field really does have a key role in the navigation of birds and in that case presumably also of other animals – well, it will then mean a complete 'magnetic' revolution in biology!

So far in this chapter I have assumed that birds make use of what we may call site-based navigation, that they determine their position in relation to the goal by using as a guide the conditions at the place where they happen to be at the time. Another possibility is that they use as a basis information collected on the way between the home area and the place in question. Discoveries have recently been made which indicate that as an aid to their navigation pigeons make use of clues during the transportation from the pigeon loft to the place where they are released. By mere chance it was noticed that young pigeons which were transported in the back of a Volkswagen 'Variant', in other words immediately above the engine, where the magnetic field is exposed to violent disturbances, showed abnormally poor navigational ability. This has since become known in the scientific literature as 'the VW effect'.

Through further experiments under more controlled procedures, in which pigeons were shipped in solid-iron cases with greatly distorted and erratic magnetic fields or in cages that were surrounded by large electrical coils which generated 'misdirected' magnetic fields, it has

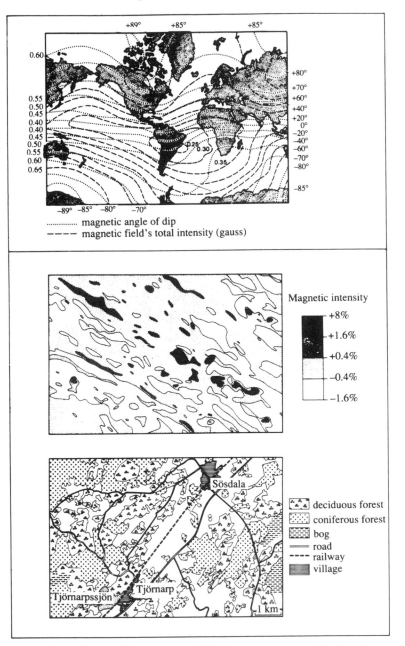

Figure 136   The earth's magnetic field shows systematic global differences, among other things in total intensity and angle of dip, as shown in the upper map. Measuring of these components can serve as a basis for a navigation system. In addition, local variations in the magnetic field occur everywhere; this results in a highly detailed magnetic landscape. This can be of use for local navigation. The lower maps show a comparison between the magnetic landscape, as mapped using field-strength measurements from aircraft, and the topographical landscape within a 54-square-kilometre area in central Scania, south Sweden. The magnetic landscape here acquires its character from scattered presence of basalt (mineral residues from minor volcanoes) combined with veins of diabase in a northwest–southeast direction. Based on Swedish geological research.

been confirmed that deviations in the magnetic field during transportation lead to various degrees of misdirection and uncertainty in navigating. Young pigeons are exceptionally susceptible in this respect. By dividing homing pigeons into groups and conveying them by different roundabout routes to one and the same release site, it has occasionally been possible to detect minor differences in the different groups' orientation.

For navigating, therefore, the pigeons seem to accept the help of some clues on the direction in which they have been moved which they manage to pick up while being transported. In this connection the magnetic sense comes into use, and perhaps also other senses too. Certain 'detour experiments' have for example been interpreted as support for the possibility that homing pigeons notice olfactory characteristics (see below regarding navigation by smell).

Nevertheless, we can probably confidently maintain that the adult pigeons' navigation is based primarily on cues perceived by the birds at the release site. By preventing pigeons from perceiving various kinds of clues while being transported, it has been established that the latter do not represent an *essential* prerequisite of perfect navigation. Homing pigeons have for example been transported in constantly rotating barrels and they have been drugged to the point of losing consciousness without this having inhibited their ability to navigate. In a series of experiments several methods of this sort were used in combination to ruin the pigeons' chances of perceiving anything useful during transportation to the release site. The pigeons were placed in compartments in rapidly rotating sealed drums and the direction of rotation was altered at irregular intervals every one or every other minute. To prevent smells from the surrounding country from reaching the birds, the drums were completely airtight; air was supplied from special air tanks. As a result of surrounding electrical coils the magnetic field could be turned at random and at frequent, irregular intervals in eight different directions. In this fashion the birds were transported between one and six hours away by road from the pigeon loft. When they had reached their destinations they had an opportunity to rest during the afternoon and night before being released the following morning. The poor pigeons were well and truly 'seasick' when the drums were open – many had vomited during the journey and were unsteady in their movements. Nevertheless, they recovered remarkably quickly. The next morning they navigated home with precisely the same accuracy and speed as the control pigeons which had been transported in open cages with an unobstructed view.

This experiment means, incidentally, that we can virtually rule out the possibility that birds use what is known as inertial navigation. The principle of this method is that the birds would record, by means of the balancing organ of the inner ear, all forces of acceleration in the lateral and longitudinal planes during their removal from the home base. Guided by these, it is theoretically possible to calculate direction and distance back to the goal. Charles Darwin was already onto these lines of thought in an article as early as 1873. Extremely sensitive accelerometers are used today in a variety of automatic inertial-navigation systems for guiding air missiles, aircraft, ships and submarines with a high degree of precision. These systems allow efficient correction for all unanticipated deviations from course caused for example by winds or ocean currents. According to experimental measurements that have been made, however, the sensitivity of the balancing organ of animals seems to be wholly inadequate for effective inertial navigation.

In Italy a research group headed by Floriano Papi has been working together in an attempt to prove that homing pigeons use navigation by smell. The idea, perhaps inspired by the ability of salmon to find the home river again by means of characteristic 'odorous substances' in the water, was presented at the beginning of the 1970s. The argument is that pigeons would, with the help of the wind, learn the 'landscape of odours' around the pigeon loft.

When the winds are in the north they bring with them particular scents; when they blow from other quarters the scent picture is a different one. When the pigeons are removed to an unfamiliar release site, they perhaps notice some unusually strong smell of the type regularly carried to the pigeon loft on northerly winds. In that case they are nearer the northerly source of the smell and with that should fly south in order to get home. The pigeons could obtain further guidance from smells which they pick up while being taken to the release site.

The Italian research team has carried out a long series of different experiments to illustrate this idea. In order to upset or destroy the pigeons' sense of smell, they have for example cut off the olfactory nerve, placed plastic tubes into the nasal passages so that the air does not reach the olfactory mucous membrane, dropped strong-smelling substances into the nostrils in order to mask other smells, or placed plugs in the nasal openings during transportation and instead supplied air 'by bottle'. In addition, a number of detour experiments have been made and the pigeons taken through widely differing 'odour landscapes'. In all cases results have been obtained which have been interpreted as supporting navigation by smell.

How these results are to be assessed, however, is uncertain, since it has been possible to substantiate them only partially, and in some cases not at all, in repeat tests made by other research groups, in the United States and in West Germany. A major complication involves the difficulty in deciding whether an inferior performance by pigeons whose sense of smell has been interfered with is due to navigational difficulties or quite simply to lack of motivation to fly far and fast when smell and breathing are out of gear. An attempt to clarify this matter was made in the United States. Homing pigeons with plastic tubes in their nasal passages were tracked from aeroplanes by means of radiotelemetry. The misgivings that the birds were, as a result of being interfered with, in poor flying spirit were overridingly confirmed. The birds landed very soon after they had been released; most flew less than 4 km, and the one that flew farthest got only 11 km before it landed. The vanishing direction from the release site, however, showed that they made off in the correct direction towards the pigeon loft. The American researchers therefore concluded that only the will to fly and not the ability to navigate had been upset.

Even if we consider only the pigeons' vanishing directions where the 'smell experiments' are concerned, there are still significant differences in results when the same experiment has been carried out in Italy and when it has been performed in the United States. 'Do American and Italian pigeons rely on different homing mechanisms?': this question constituted the title of an interesting paper in 1978 by American and Italian researchers (Papi *et al.*). An exchange of young homing pigeons between Italy and Germany showed that birds from separate regions react in a slightly dissimilar way when they are released, perhaps as a result of their placing different importance on various means of navigation. Maybe Italian pigeons, unlike American and German ones, make considerable use of olfactory information for navigating.

Confusing and conflicting results have not, however, been the outcome of all of the attempts to test the significance of smells. A particularly interesting and imaginative experiment was made in 1974 in Italy with pigeon lofts that were fitted with glass screens which acted as wind-deflectors. Pigeons were raised in aviaries which allowed free play to the wind on all sides. Large glass screens were fixed to some aviaries so that the wind, when it reached into the cage, was deflected approximately 90° clockwise or anti-clockwise (figure 137A). Pigeons in aviaries with wind screens ought, then, to learn a miscast 'olfactory map' and in release tests to fly about 90° off line, to the right or to the left. This is precisely what happens!

These findings have been fully borne out in the United States, where it has also been found

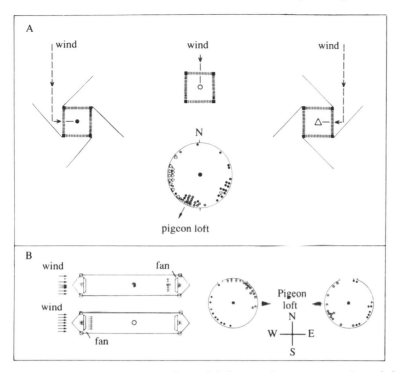

Figure 137   A. Experiments with wind-deflecting glass screens on pigeon lofts. Pigeons fly off in the wrong direction, to the right or to the left, from the release site if they have been kept in pigeon lofts where the wind is deflected clockwise or anti-clockwise by the screens. The broken lines show examples of how pigeons in aviaries with clockwise and anti-clockwise glass screens perceive a wind from the north. The aviaries without screens are control aviaries. Based on Baldaccini *et al.* (1975).

B. Experiments with fans which blow with or against the wind in corridor-style cages. Pigeons which are exposed to an easterly air stream from the fans when the wind is westerly and a westerly air stream when the wind is in the east fly off in the wrong direction at release sites east or west of the pigeon loft. Pigeons in cages where the airflow from the fans is parallel with the wind direction, however, orientate normally. Based on Ioalé *et al.* (1978).

that pigeons need to spend only one week in wind-deflected aviaries to fly the wrong way. This was demonstrated by building two such aviaries 4 km from the normal pigeon loft; in one the wind was deflected clockwise, and in the other anti-clockwise. Pigeons that were fully grown and experienced in flying were split into two groups which were simultaneously made to spend seven days each in the different deflector-aviaries. When they were then shipped off and released, they all flew home to their common loft. When they made off, however, the pigeons in group 1, which had lived in the aviary with clockwise deflector screens, flew off track to the right and the pigeons in group 2 off track to the left. A few days later the experiment was repeated, this time with the two groups each switched to the other aviary. On being released a week later, the pigeons had altered their orientation: the birds in group 1 now flew off to the left of the correct direction and the birds in group 2 to the right. To be on the safe side the experiment was taken up for a third time, with the same placing in the aviaries as the first time. On this occasion the pigeons once more showed the same directional error as in the first experiment. If the pigeons do learn a scent map, then obviously this is constantly re-examined and renewed.

A similar experiment which sheds light on the importance of winds and smells was reported from Italy (figure 137B). Homing pigeons were made to live in corridor-like glass cages with large electric fans at each end. The cages were sealed on all sides so that the pigeons were prevented from making out the wind direction; at the top, however, there were openings so that scents and smells from the surrounding air reached the birds. The cages were set up in an east–west direction. When the wind blew from either of these directions the fans were switched on, in one cage so that the air always blew in the same direction as the wind and in the other cage so that the air current consistently blew in immediately the opposite direction to the wind. In the latter cage, therefore, the pigeons were deceived into thinking that the smells were coming from the direction straight opposite that from which they actually did come. These 'duped' pigeons headed off in the wrong direction, away from the pigeon loft, when they were released west or east of the home district, just as we could expect birds that had learnt an inverted olfactory map would do. Additional support for the hypothesis of olfactory navigation in homing pigeons comes from further imaginative experiments on 'olfactory deception', e.g. with pigeons exposed to air from places other than the release site (cf. Papi 1986).

We may ask ourselves whether it really is windborne smells that provide the correct explanation for the birds' reactions in these experiments. Despite the fact that pigeons which have been exposed to deflected winds leave the place where they are released in the wrong direction, they very soon correct themselves and return home almost as quickly as normal. During some of the American tests, when the birds were kept in aviaries with deflector screens, winds never blew from the eastern sector. The pigeons' scent map in that direction should therefore have been incomplete – yet they flew away from easterly situated release sites in the typical wrong direction. In Germany a local anaesthetic was applied to the olfactory mucous membrane of pigeons which had lived in aviaries with deflector screens before the birds were released, and despite this the deflection in orientation remained.

Is there something other than smells in the wind that the pigeons could make a mental note of? Wind screens affect among other things the birds' perception of the 'sound landscape', something we should definitely take into account considering that pigeons have the capacity to hear infra-sound. In addition, the glass screens probably interfere with the pigeons' perception of the ultra-violet light and its polarisation pattern.

Another possible reason for the effect of the wind screens may be that homing pigeons use a wind-compass. Maybe pigeons make use of the wind direction prevailing on the same day as they are moved so as to be able quickly to orientate themselves at the release site in relation to the points of the compass. One may ask whether we humans would be influenced by the wind and take our bearings by it at an unfamiliar place if we had made a note beforehand of the wind direction in the home area. As soon as the 'wind-deflected' pigeons check their orientation with the sun-compass or the magnetic compass, they discover that they have been mistaken about the wind direction.

Further possible clues to the birds' powers of navigation which have been discussed in the specialist literature are differences in the gravitational force at different places on earth and differences in the Coriolis force at different latitudes. To be able to perceive these minimal forces the birds' sensitivity must be so delicately tuned that they can perceive one-hundredth of a percent of the total gravitational acceleration on the earth. These high demands on sensitivity are hardly reason enough for us to rule out the possibility that birds could perceive these forces (compare the delicate sensitivity to magnetic fields that birds seem to possess). On the other hand, it has to be admitted that we have no indications at present, with the exception of the remarkable lunar rhythm in the orientation of homing pigeons (see section 5.1), that these factors might be involved in the birds' navigation.

A bewildering multiplicity of factors has been mentioned in this chapter, all of which are conceivable foundations for the birds' map sense: landmarks, positions of the celestial bodies, magnetic fields, smells, infra-sound, and inertial, gravitational and Coriolis forces. Is bird navigation perhaps based on a combination of several different types of clues? A conclusion such as this is a very likely one when we consider all the various experiments which have been made without any one of them having provided any definitive information about any key factor in the navigation of birds. We are compelled to resign ourselves to continuing uncertainty and confusion over how the birds find the right migration route.

So, this book therefore ends with an unsolved mystery. Actually I do not think that this is any major shortcoming. It might be hoped that we humans would gain a greater degree of inspiration from unsolved mysteries than from what we believe we know – in great things as in small. May the birds continue to fly over the earth, and may mankind wonder and investigate.

Bibliography

Chapter 2. The rotating world of migratory birds

Lamb, H. H. 1972. *Climate: Present, Past and Future.* Vol. 1. Fundamentals and Climate Now. Methuen, London.

Lieth, H. & Whittaker, R. H. (eds.) 1975. *Primary Productivity of the Biosphere.* Springer-Verlag, New York.

Liljequist, G. H. 1970. *Klimatologi.* Generalstabens Litografiska Anstalt, Stockholm.

Ulfstrand, S. & Högstedt, G. 1976. Hur många fåglar häckar i Sverige? *Anser* **15**: 1–32.

Walter, H. 1973. *Vegetation of the Earth.* Springer-Verlag, New York.

Young, L. B. 1977. *Earth's Aura.* Knopf, New York.

Chapter 3. Summer and winter quarters

Section 3.1. Birds in wetlands

Atkinson, N. K., Davies, M. & Prater, A. J. 1978. The winter distribution of Purple Sandpipers in Britain. *Bird Study* **25**: 223–8.

Bagg, A. M. 1967. Factors affecting the occurrence of the Eurasian Lapwing in eastern North America. *Living Bird* **6**: 87–121.

Curry, P. J. & Sayer, J. A. 1979. The inundation zone of the Niger as an environment for Palearctic migrants. *Ibis* **121**: 20–40.

Davidson, N. C., Strann, K.-B., Crockford, N. J., Evans, P. R., Richardson, J., Standen, L. J., Townshend. D. J., Uttley, J. D., Wilson, J. R. & Wood, A. G. 1986. The origins of Knots *Calidris canutus* in arctic Norway in spring. *Ornis Scand.* **17**: 175–9.

Dean, W. R. J. 1977. Moult of Little Stints in South Africa. *Ardea* **65**: 73–9.

Dick, W. J. A., Pienkowski, M. W., Waltner, M. & Minton, C. D. T. 1976. Distribution and geographical origins of Knot *Calidris canutus* wintering in Europe and Africa. *Ardea* **64**: 22–47.

Elliott, C. C. H., Waltner, M., Underhill, L. G., Pringle, J. S. & Dick, W. J. A. 1976. The migration system of the Curlew Sandpiper *Calidris ferruginea* in Africa. *Ostrich* **47**: 191–213.

Fretwell, S. D. 1972. *Populations in a Seasonal Environment.* Princeton University Press, Princeton.

Glue, D. & Morgan, R. 1977. Recovery of bird rings in pellets and other prey traces of owls, hawks and falcons. *Bird Study* **24**: 111–13.

Glutz von Blotzheim, U. N., Bauer, K. M. & Bezzel, E. 1975. *Handbuch der Vögel Mitteleuropas.* Vol. 6. Akademische Verlagsgesellschaft, Wiesbaden.

Glutz von Blotzheim, U. N., Bauer, K. M. & Bezzel, E. 1977. *Handbuch der Vögel Mitteleuropas.* Vol. 7. Akademische Verlagsgesellschaft, Wiesbaden.

Hale, W. G. 1973. The distribution of the Redshank *Tringa totanus* in the winter range. *Zool. J. Linn. Soc.* **53**: 177–236.

Hale, W. G. 1980. *Waders.* Collins, London.

Imboden, C. 1974. Zug, Fremdansiedlung und Brutperiode des Keibitz *Vanellus vanellus* in Europa. *Orn. Beob.* **71**: 5–134.

McClure, H. E. 1974. *Migration and Survival of the Birds of Asia.* US Army Medical Component. Bangkok, Thailand.

Middlemiss, E. 1961. Biological aspects of *Calidris minuta* while wintering in south-west Cape. *Ostrich* **32**: 107–21.

Moreau, R. E. 1970. Changes in Africa as a wintering area for Palearctic birds. *Bird Study* **17**: 95–103.

Moreau, R. E. 1972. *The Palaearctic–African Bird Migration Systems.* Academic Press, London.

Morel, G. 1973. The Sahel zone as an environment for Palaearctic migrants. *Ibis* **115**: 413–17.

Morrison, R. I. G. 1975. Migration and morphometrics of European Knot and Turnstone on Ellesmere Island, Canada. *Bird-banding* **46**: 290–301.

Morrison, R. I. G. 1977. Migration of arctic waders wintering in Europe. *Polar Record* **18**: 475–86.

Pearson, D. J., Phillips, J. H. & Backhurst, G. C. 1970. Weights of some Palaearctic waders wintering in Kenya. *Ibis* **112**: 199–208.

Persson, C. 1979. Isländska rödbenor *Tringa totanus robusta* i Sydskåne. *Dansk Orn. Foren. Tidsskr.* **73**: 281–5.

Pienkowski, M. W., Knight, P. J., Stanyard, D. J. & Argyle, F. B. 1976. The primary moult of waders on the Atlantic coast of Morocco. *Ibis* **118**: 347–65.

Prater, A. J. 1976. The distribution of coastal waders in Europe and North Africa. *Proc. Int. Conf. on Conservation of Wetlands and Waterfowl* (Heiligenhafen, 1974, Int. Waterfowl Res. Bureau): 255–71.

Prater, A. J. 1981. *Estuary Birds of Britain and Ireland.* Poyser, Berkhamsted.

Roos, G. 1969. Ringmärkningsverksamheten vid Falsterbo fågelstation 1965–1967. *Vår Fågelvärld* **28**: 18–44.

Salomonsen, F. 1955. The evolutionary significance of bird migration. *Dan. Biol. Medd.* **22**, no. 6.

Salomonsen, F. 1967. *Fuglene på Grönland.* Rhodos, Copenhagen.

Salomonsen, F. 1967. *Fugletraekket og dets gåder* (2nd edition). Munksgaard, Copenhagen.

Smit, C. J. & Piersma, T. 1989. Numbers, midwinter distribution, and migration of wader populations using the East Atlantic flyway. *IWRB Special publ.* No. **9**: 24–63.

Summers, R. W., Cooper, J. & Pringle, J. S. 1977. Distribution and numbers of coastal waders (*Charadrii*) in the southwestern Cape, South Africa, summer 1975–76. *Ostrich* **48**: 85–97.

Wilson, J. R., Czajkowski, M. A. & Pienkowski, M. W. 1980. The migration through Europe and wintering in West Africa of Curlew Sandpipers. *Wildfowl* **31**: 107–22.

Section 3.2. Birds which forage on lake and sea bottom

Andersson, S. & Wester, S. 1975. Studier av strömstare i Norge 1968–1972. *Fauna och flora* **70**: 253–65.

Andersson, S. & Wester, S. 1976. Långåterfynd av nordiska strömstarar *Cinclus c. cinclus. Vår Fågelvärld* **35**: 279–86.

Atkinson-Willes, G. L. 1976. The numerical distribution of ducks, swans and coots as a guide in assessing the importance of wetlands in midwinter. *Proc. Int. Conf. on the Conservation of Wetlands and Waterfowl* (Heiligenhafen 1974, Int. Waterfowl Res. Bureau): 199–254.

Bauer, K. M. & Glutz von Blotzheim, U. N. 1968 and 1969. *Handbuch der Vögel Mitteleuropas.* Vols. 2 and 3. Akademische Verlagsgesellschaft, Frankfurt a.M.

Cramp, S. & Simmons, K. E. L. 1977. *The Birds of the Western Palearctic.* Vol. 1. Oxford University Press, Oxford.

Curry, P. J. & Sayer, J. A. 1979. The inundation zone of the Niger as an environment for Palearctic migrants. *Ibis* **121**: 20–40.

Dau, C. P. & Kistchinski, S. A. 1977. Seasonal movements and distribution of the Spectacled Eider. *Wildfowl* **28**: 65–75.

Impekoven, M. 1964. Zugwege und Verbreitung der Knäkente, *Anas querquedula;* eine Analyse der europäischen Beringungsresultate. *Orn. Beob.* **61**: 1–34.

Joensen, A. H. 1973. Moult migration and wing-feather moult of seaducks in Denmark. *Danish Review of Game Biology* **8**, no. 4.

Joensen, A. H. 1974. Waterfowl populations in Denmark 1965–1973. *Danish Review of Game Biology* 9, no. 1.

Johnson, A. & Hafner, H. 1970. Winter wildfowl counts in south-east Europe and western Turkey. *Wildfowl* 21: 22–36.

King, J. G. 1973. A cosmopolitan duck moulting resort; Takslesluk Lake, Alaska. *Wildfowl* 24: 103–9.

Kistchinski, S. A. 1973. Waterfowl in north-east Asia. *Wildfowl* 24: 88–102.

Mathiasson, S. 1970. Numbers and distribution of Long-tailed Ducks wintering in northern Europe. *Brit. Birds* 63: 414–24.

Monval, J.-Y. & Pirot, J.-Y. 1989. Results of the IWRB International Waterfowl Census 1967–1986. *IWRB Special Publ.* No. 8, 145 pp.

Nilsson, L. 1975. Midwinter distribution and numbers of Swedish Anatidae. *Ornis Scand.* 6: 83–107.

Nilsson, L. 1980. Wintering diving duck population and available food resources in the Baltic. *Wildfowl* 31: 131–43.

Persson, L. E. 1977. Pegelundersökningar i Östersjön: Hanöbukten. Utbredning, abundans och biomassa 1975 och 1976. Sydlänens Kustundersökningar. Report no. 36, Lund.

Roux, F. 1973. Censuses of Anatidae in the Central Delta of the Niger and the Senegal Delta – January 1972. *Wildfowl* 24: 63–80.

Roux, F. 1976. The status of wetlands in the West African Sahel: their value for waterfowl and their future. *Proc. Int. Conf. on the Conservation of Wetlands and Waterfowl* (Heiligenhafen 1974, Int. Waterfowl Res. Bureau): 272–87.

Roux, F., Jarry, G., Mahéo, R. & Tamisier, A. 1976 and 1977. Importance, structure et origine des populations d'Anatidés hivernant dans le delta du Sénégal. *L'Oiseau et R.F.O.* 46: 299–336 and 47: 1–24.

Salomonsen, F. 1967. *Fugletraekket og dets gåder* (2nd edition). Munksgaard, Copenhagen.

Salomonsen, F. 1967. *Fuglene på Grönland*. Rhodos, Copenhagen.

Salomonsen, F. 1968. The moult migration. *Wildfowl* 19: 5–24.

Wolff, W. J. 1966. Migration of Teal ringed in the Netherlands. *Ardea* 54: 230–70.

Section 3.3. Birds which feed on terrestrial plants

Ahlén, I. 1977. *Faunavård*. Skogshögskolan Naturvårdsverket, Stockholm.

Bauer, K. M. & Glutz von Blotzheim, U. N. 1968. *Handbuch der Vögel Mitteleuropas*. Vol. 2. Akademische Verlagsgesellschaft, Frankfurt a.M.

Cramp, S. & Simmons, K. E. L. 1977. *The Birds of the Western Palearctic*. Vol. 1. Oxford University Press, Oxford.

Dementev, G. P. & Gladkov, N. A. 1967. *Birds of the Soviet Union*. Vol. IV. Israel Program for Scientific Translations, Jerusalem.

Doude van Troostwijk, W. J. 1974. Ringing data on White-fronted Geese *Anser a. albifrons* in the Netherlands, 1953–1968. *Ardea* 62: 98–110.

Folkestad, A. O. 1975. Wetland bird migration in central Norway. *Ornis Fenn.* 52: 49–56.

Haftorn, S. 1971. *Norges fugler*. Universitetsforlaget, Oslo.

IWRB (International Waterfowl and Wetlands Research Bureau). 1989. Flyways and reserve networks for waterbirds (eds. Boyd, H. & Pirot, J.-Y.). *IWRB Special Publ.* No. 9, 111 pp.

Larsson, K., Forslund, P., Gustafsson, L. & Ebbinge, B. S. 1988. From the high Arctic to the Baltic: the successful establishment of a Barnacle Goose *Branta leucopsis* population on Gotland, Sweden. *Ornis Scand.* 19: 182–9.

Mathiasson, S. 1963. The Bean Goose, *Anser fabalis*, in Skåne, Sweden, with remarks on occurrence and migration through northern Europe. *Acta Vertebratica* 2: 417–533.

Nilsson, L. 1976. Kanadagåsens *Branta canadensis* utbredning i södra Sverige under vinterhalvåret. *Anser* 15: 241–6.

Nilsson, L. 1979. Observationer under vårflyttningen 1978 av sädgäss märkta i Skåne. *Anser* 18: 1–4.

Nilsson, L. 1984. Migrations of Fennoscandian Bean Geese *Anser fabalis*. *Swedish Wildlife Res.* 13: 83–106.

Nilsson, L. & Persson, H. 1978. Antal och rörelser hos övervintrande gäss i Skåne 1977–1978. *Anser* **17**: 139–45.

Ogilvie, M. A. 1978. *Wild Geese.* T. & A. D. Poyser, Berkhamsted.

Owen, M. 1980. *Wild Geese of the World.* Batsford, London.

Salomonsen, F. 1958. The present status of the Brent Goose (*Branta bernicla*) in western Europe. *Vidensk. Medd. dansk naturh. Foren.* **120**: 43–80.

Salomonsen, F. 1967. *Fuglene på Grönland.* Rhodos, Copenhagen.

Summers, R. W. & Underhill, L. G. 1987. Factors related to breeding production of Brent Geese *Branta b. bernical* and waders (Charadrii) on the Taimyr Peninsula. *Bird Study* **34**: 161–71.

Section 3.4. Birds which feed on fish

Ashmole, N. P. 1971. Sea Bird Ecology and the Marine Environment. *Avian Biology* (Academic Press, New York) Vol. 1: 223–86.

Blomqvist, S. & Peterz, M. 1984. Cyclones and pelagic seabird movements. *Mar. Ecol. Prog. Ser.* **20**: 85–92.

Cooke, F. & Mills, E. L. 1972. Summer distribution of pelagic birds off the coast of Argentine. *Ibis* **114**:245–51.

Cramp, S. & Simmons, K. E. L. 1977. *The Birds of the Western Palearctic.* Vol. 1. Oxford University Press, Oxford.

Grimes, L. G. 1977. A radar study of tern movements along the coast of Ghana. *Ibis* **119**: 28–36.

Harris, M. P. & Hansen, L. 1974. Sea-bird transects between Europe and Rio Plate, South America, in autumn 1973. *Dansk Orn. Foren. Tidsskr.* **68**: 117–37.

Jönsson, P. E. & Peterz, M. 1976. Havsfåglar vid Kullen 1970–1974. *Anser* **15**: 51–64.

Langham, N. P. E. 1971. Seasonal movements of British terns in the Atlantic Ocean. *Bird Study* **18**: 155–75.

Meltofte, H. & Overlund, E. 1974. Forekomsten av Suler *Sula bassana* ved Blåvandshuk 1963–1971. *Dansk Orn. Foren. Tidsskr.* **68**: 43–8.

Muus, B. J. & Dahlström, P. 1969. *Havsfisk och fiske.* Norstedts, Stockholm.

Nelson, B. 1978. *The Gannet.* T. & A. D. Poyser, Berkhamsted.

Nelson, B. 1980. *Seabirds – their Biology and Ecology.* Hamlyn, London.

Österlöf, S. 1977. Migration, wintering areas, and site tenacity of the European Osprey *Pandion h. haliaetus. Ornis Scand.* **8**: 61–78.

Perrins, C. M. & Brooke, M. de L. 1976. Manx Shearwaters in the Bay of Biscay. *Bird Study* **23**: 295–9.

Perrins, C. M., Harris, M. P. & Britton, C. K. 1973. Survival of Manx Shearwaters *Puffinus puffinus. Ibis* **115**: 535–48.

Peterz, M. 1978. Havsfåglar vid Kullen och i Kattegatt hösten 1977. *Anser* **17**: 154–60.

Pettersson, G. & Unger, U. 1972. Havsfågelstudier på Västkusten under tioårsperioden 1960–1969. *Vår Fågelvärld* **31**: 229–36.

Phillips, J. H. 1963. The pelagic distribution of the Sooty Shearwater *Procellaria grisea. Ibis* **105**: 340–53.

Stresemann, E. & Stresemann, V. 1966. Die Mauser der Vögel. *J. Orn.* **107**: Sonderheft.

Thomson, A. L. 1965. The transequatorial migration of the Manx Shearwater. *L'Oiseau et R.F.O.* **35**: 130–40.

Voous, K. H. & Wattel, J. 1963. Distribution and migration of the Greater Shearwater. *Ardea* **51**: 143–57.

Section 3.5. Birds which obtain food at the water's surface

Andersson, M. 1976. Predation and kleptoparasitism by skuas in a Shetland seabird colony. *Ibis* **118**: 208–17.

Ashmole, N. P. 1971. Sea Bird Ecology and the Marine Environment. *Avian Biology* (Academic Press, New York) Vol. 1: 223–86.

Coulson, J. C. 1966. The movements of the Kittiwake. *Bird Study* **13**: 107–15.

Devillers, P. 1977. The skuas of the North American Pacific coast. *Auk* **94**: 417–29.

Furness, R. 1978. Movements and mortality rates of Great Skuas ringed in Scotland. *Bird Study* **25**: 229–38.

Furness, R. W. 1987. *The Skuas*. T. & A. D. Poyser, Calton.

Glutz von Blotzheim, U. N., Bauer, K. & Bezzel, E. 1977. *Handbuch der Vögel Mitteleuropas*. Vol. 7. Akademische Verlagsgesellschaft, Wiesbaden.

Rudebeck, G. 1957. Studies on Some Palaearctic and Arctic Birds in their Winter Quarters in South Africa. *South African Animal Life* (Almquist & Wiksell, Stockholm) Vol. 4: 459–507.

Salomonsen, F. 1967. Migratory movements of the Arctic Tern (*Sterna paradisaea*) in the Southern Ocean. *Biol. Medd. Dan. Vid. Selsk.* **24**: 1.

Salomonsen, F. 1967. *Fuglene på Grönland*. Rhodos, Copenhagen.

Salomonsen, F. 1976. The South Polar Skua *Stercorarius maccormicki* in Greenland. *Dansk Orn. Foren. Tidsskr.* **70**: 81–9.

Schiemann, H. 1972. Über Winterquartiere nordeuropäischer Odinshünchen. *Vogelwarte* **26**: 329–36.

Thomson, A. L. 1966. An analysis of recoveries of Great Skuas ringed in Shetland. *Brit. Birds* **59**: 1–15.

Tuck, L. M. 1971. The occurrence of Greenland and European birds in Newfoundland. *Bird-banding* **42**: 184–209.

Young, E. C. 1963. Feeding habits of the South Polar Skua *Catharacta maccormicki*. *Ibis* **105**: 301–18.

Section 3.6. Birds of prey

Bernis, F. 1975. Migracion de Falconiformes y Ciconia spp. por Gibraltar. *Ardeola* **21**.

Broekhuysen, G. J. & Siegfried, W. R. 1970. Age and moult in the Steppe Buzzard in southern Africa. *Ostrich* suppl. **8**: 223–37.

Broekhuysen, G. J. & Siegfried, W. R. 1971. Dimensions and weight of the Steppe Buzzard in southern Africa. *Ostrich* suppl. **9**: 31–9.

Brown, L. 1970. *African Birds of Prey*. Collins, London.

Brown, L. & Amadon, D. 1968. *Eagles, Hawks and Falcons of the World*. Country Life Books, Middlesex.

Dementev, G. P. & Gladkov, N. A. 1966. *Birds of the Soviet Union*. Israel Program for Scientific Translations, Jerusalem.

Glutz von Blotzheim, U. N., Bauer, K. M. & Bezzel, E. 1971. *Handbuch der Vögel Mitteleuropas*. Vol. 4. Akademische Verlagsgesellschaft, Frankfurt a.M.

Hörnfeldt, B. 1978. Synchronous population fluctuations in voles, small game, owls and tularemia in northern Sweden. *Oecologia* **32**: 141–52.

Lack, D. 1971. *Ecological Isolation in Birds*. Blackwell, Oxford.

Löfgren, O., Hörnfeldt, B. & Carlsson, B.-G. 1986. Site tenacity and nomadism in Tengmalm's Owl (*Aegolius funereus*) in relation to cyclic food production. *Oecologia* **69**: 321–6.

Lundberg, A. 1979. Ecology of owls (Strigidae), especially the Ural Owl *Strix uralensis*, in central Sweden. PhD thesis, Uppsala University.

Lundgren, U. 1979. Fjällvråkens *Buteo lagopus* uppträdande i Sverige hösten 1978. *Vår Fågelvärld* **38**: 95–100.

Moreau, R. E. 1972. *The Palaearctic–African Bird Migration Systems*. Academic Press, London.

Nielsen, B. P. 1977. Danske musvågers *Buteo buteo* traekforhold og spredning. *Dansk Orn. Foren. Tidsskr.* **71**: 1–3.

Nordström, G. 1963. Einige Ergebnisse der Vogelberingung in Finland in den Jahren 1913–1962. *Ornis Fenn.* **40**: 81–124.

Olsson, V. 1958. Dispersal, migration, longevity and death causes of *Strix aluco*, *Buteo buteo*, *Ardea cinerea* and *Larus argentatus*. *Acta Vertebratica* **1**, no. 2.

Roos, G. 1978. Sträckräkningar och miljöövervaktning: långsiktiga förändringar i höststräckets numerär vid Falsterbo 1942–1977. *Anser* **17**: 133–8.

Roos, G. 1979. Sträcksummor och årliga populationsfluktuationer hos ormvråk *Buteo buteo*. *Anser* **18**: 48–50.

Rudebeck, G. 1963. Studies on Some Palaearctic and Arctic Birds in their Winter Quarters in South

Africa. 4. Birds of Prey (Falconiformes). *South African Animal Life.* (Almqvist & Wiksell, Stockholm) Vol. 9: 418–53.

Saurola, P. 1977. Suomalaisten hiirihaukkojen muuttoreitit. *Lintumies* 12: 45–53.

Sylvén, M. 1978. Interspecific relations between sympatrically wintering Common Buzzards *Buteo buteo* and Rough-legged Buzzards *Buteo lagopus. Ornis Scand.* 9: 197–206.

Sylvén, M. 1982. Reproduction and survival in Common Buzzards *Buteo buteo*, illustrated by the seasonal allocation of energy expenses. PhD thesis, Lund University.

Ulfstrand, S. & Högstedt, G. 1976. Hur många fåglar häckar i Sverige. *Anser* 15: 1–32.

Vande Weghe, J.-P. 1978. Les rapaces paléarctiques au Rwanda. *Gerfaut* 68: 493–517.

Voous, K. H. 1960. *Atlas of European Birds.* Nelson, London.

Section 3.7. Insect-eaters

Alerstam, T. 1975. Redwing (*Turdus iliacus*) migration towards southeast over southern Sweden. *Vogelwarte* 28: 2–17.

Alerstam, T. 1976. Nocturnal migration of thrushes (*Turdus* spp.) in southern Sweden. *Oikos* 27: 457–75.

Ashmole, M. J. 1962. The migration of European thrushes: a comparative study based on ringing records. *Ibis* 104: 314–46, 522–59.

Askenmo, C., von Brömssen, A., Ekman, J. & Jansson, C. 1977. Impact of some wintering birds on spider abundance in spruce. *Oikos* 28: 90–4.

Backhurst, G. C. & Pearson, D. J. 1979. Southward migration at Ngulia, Tsavo, Kenya 1978/79. *Scopus* 3: 19–25.

Backhurst, G. C. & Pearson, D. J. 1981. Ringing and migration at Ngulia, Tsavo, November 1980–January 1981. *Scopus* 5: 28–30.

Berndt, R. & Henss, M. 1967. Die Kohlmeise, *Parus major*, als Invasionsvogel. *Vogelwarte* 24: 17–37.

Berthold, P. 1988. Evolutionary aspects of migratory behavior in European warblers. *J. Evol. Biol.,* 1: 195–209.

Deshler, W. 1974. An examination of the extent of fire in the grassland and savanna of Africa along the southern side of the Sahara. *Int. Symp. Remote Sensing Environment*, Ann Arbor: 23–30.

Elgood, J. H., Sharland, R. E. & Ward, P. 1966. Palaearctic migrants in Nigeria. *Ibis* 108: 84–116.

Gibb, J. 1958. Predation by tits and squirrels on the eucosmid *Ernarmonia conicolana. J. Anim. Ecol.* 27: 375–96.

Jansson, C., Ekman, J. & von Brömssen, A. 1981. Winter mortality and food supply in tits *Parus* spp. *Oikos* 37: 313–22.

Källander, H. 1981. The effects of provision of food in winter on a population of the Great Tit *Parus major* and the Blue Tit *P. caeruleus. Ornis Scand.* 12: 244–8.

Källander, H. & Karlsson, J. 1981. Population fluctuations of some north European bird species in relation to winter temperatures. *Proc. Second Nordic Congr. Ornithol.*: 111–17.

Lack, D. 1966. *Population Studies of Birds.* Clarendon Press, Oxford.

Lindholm, C.-G. 1978. Talgoxens sträck över Östersjön höstarna 1975 och 1976. *Anser*, suppl. 3: 145–53.

Lindskog, H. & Roos, G. 1979. Höststräckets förlopp hos blåmes *Parus caeruleus* och talgoxe *Parus major* vid Falsterbo 1973–1978. *Anser* 18: 171–88.

Mathiasson, S. 1971. Untersuchungen an Klappergrasmücken (*Sylvia curruca*) in Niltal in Sudan. *Vogelwarte* 26: 212–21.

Moreau, R. E. 1972. *The Palaearctic–African Bird Migration Systems.* Academic Press, London.

Morel, G. 1973. The Sahel zone as an environment for Palaearctic migrants. *Ibis* 115: 413–17.

Mork, K. 1974. Ringmerkningsresultat for raudvengtrast, *Turdus iliacus*, i Norge. *Sterna* 13: 77–107.

Otvos, I. S. 1979. The effects of insectivorous bird activities in forest ecosystems: an evaluation. In: *The Role of Insectivorous Birds in Forest Ecosystems* (ed. J. G. Dickson *et al.* Academic Press, New York).

Paevskii, V. A. 1973. Atlas of bird migration according to banding data at the Courland Spit. In: *Bird Migrations, Ecological and Physiological Factors* (ed. B. E. Bykhovskii, John Wiley, Jerusalem).

Pearson, D. J. 1973. Moult of some Palaearctic warblers wintering in Uganda. *Bird Study* 20: 24–36.

Pearson, D. J. & Backhurst, G. C. 1976. The southward migration of Palaearctic birds over Ngulia, Kenya. *Ibis* **118**: 78–105.

Perrins, C. M. 1966. The effect of beech crops on Great Tit populations and movements. *Brit. Birds* **59**: 419 32.

Perrins, C. M. 1979. *British Tits*. Collins, London.

Price, T. 1981. The ecology of the Greenish Warbler *Phylloscopus trochiloides* in its winter quarters. *Ibis* **123**: 131–44.

Roos, G. 1984. Migration, wintering and longevity of birds ringed at Falsterbo (1947–1980). *Anser*, supplement 13.

Rudebeck, G. 1957. Studies on Some Palaearctic and Arctic Birds in their Winter Quarters in South Africa. *South African Animal Life* (Almqvist & Wiksell, Stockholm), Vol. 4: 459–507.

Simms, E. 1978. *British Thrushes*. Collins, London.

Sinclair, A. R. E. 1978. Factors affecting the food supply and breeding season of resident birds and movements of Palaearctic migrants in a tropical African savannah. *Ibis* **120**: 480–97.

Southwood, T. R. E. 1978. The components of diversity *Royal Entomological Soc. Symp.* **9**: 19–40.

Svensson, S. 1981. Populationsfluktuationer hos mesar *Parus*, nötväcka *Sitta europaea* och trädkrypare *Certhia familiaris* i södra Sverige. *Proc. Second Nordic Congr. Ornithol.*: 9–18.

Ulfstrand, S. 1962. On the nonbreeding ecology and migratory movements of the Great Tit (*Parus major*) and the Blue Tit (*Parus caeruleus*) in Southern Sweden. *Vår Fågelvärld*, suppl. **3**: 145 pp.

Winkler, R. 1974. Der Herbstdurchzug von Tannenmeise, Blaumeise und Kohlmeise (*Parus ater, caeruleus* und *major*) auf dem Col de Bretolet (Wallis). *Orn. Beob.* **71**: 135–52.

Zink, G. 1973. *Der Zug Europäischer Singvögel*. 1. Lieferung. (Vogelwarte Radolfzell).

Section 3.8. Seed-eaters

Andersson, M. 1981. Födohamstring hos fåglar. *Vår Fågelvärld* **40**: 177–84.

Bock, C. E. & Lepthien, L. W. 1976. Synchronous eruptions of boreal seed-eating birds. *Am. Nat.* **110**: 559–71.

Bossema, I. 1979. Jays and oaks: an eco-ethological study of a symbiosis. *Behaviour* **70**: 1–117.

Dementev, G. P. *et al.* 1954. *Birds of the Soviet Union*. Vol. V. Israel Program for Scientific Translations, Jerusalem 1970.

Enemar, A. 1969. Gråsiskan *Carduelis flammea* i Ammarnäs-området, Lycksele lappmark, år 1968. *Vår Fågelvärld* **28**: 230–5.

Enemar, A. & Nyström, B. 1981. Om gråsiskans *Carduelis flammea* beståndsväxlingar, föda och häckning i fjällbjörkskog, södra Lappland. *Vår Fågelvärld* **40**: 409–26.

Grodzinski, W. & Sawicka-Kapusta, K. 1970. Energy values of tree seeds eaten by small mammals. *Oikos* **21**: 52–8.

Harper, J. L. 1977. *Population Biology of Plants*. Academic Press, London.

Hora, B. (ed.) 1981. *The Oxford Encyclopedia of Trees of the World*. Oxford University Press, Oxford.

Janzen, D. H. 1976. Why bamboos wait so long to flower. *Ann. Rev. Ecol. Syst.* **7**: 347–91.

John, A. W. G. & Roskell, J. 1985. Jay movements in autumn 1983. *Brit. Birds* **78**: 611–37.

Källander, H. 1978. Hoarding in the Rook *Corvus frugilegus*. *Anser*, suppl. **3**: 124–8.

Meltofte, H. 1983. Arrival and pre-nesting period of the Snow Bunting *Plectrophenax nivalis* in East Greenland. *Polar Research* **1**: 185–98.

Mirov, N. T. 1967. *The Genus Pinus*. The Ronald Press Co., New York.

Nethersole-Thompson, D. 1966. *The Snow Bunting*. Oliver & Boyd, Edinburgh.

Newton, I. 1972. *Finches*. Collins, London.

Nilsson, S. 1858. *Skandinavisk Fauna. Foglarna*. Vol. 1. Gleerups förlag, Lund.

Salomonsen, F. 1967. *Fuglene på Grönland*. Rhodos, Copenhagen.

Salomonsen, F. 1971, 1979. Fra Zoologisk Museum XXIV, XXV. Tolvte, Trettende forelöbige lista over genfundne grönlandske ringfugle. *Dansk Orn. Foren. Tidsskr.* **65**: 11–19, **73**: 191–206.

Sjörs, H. 1971. *Ekologisk Botanik*. Almqvist & Wiksell, Stockholm.

Snow, D. W. 1971. Evolutionary aspects of fruit-eating by birds. *Ibis* **113**: 194–202.

Sorensen, A. E. 1981. Interactions between birds and fruit in a temperate woodland. *Oecologia* **50**: 242–9.

Stiles, E. W. 1980. Patterns of fruit presentation and seed dispersal in bird-disseminated woody plants in the eastern deciduous forest. *Am. Nat.* **116**: 670–88.

Svärdson, G. 1957. The 'invasion' type of bird migration. *Brit. Birds* **50**: 314–43.

Swanberg, P. O. 1975. Nötkråkans *Nucifraga c. caryocatactes* vana att upplägga vinterförråd. *Grus* **2**: 10–18.

Temple, S. A. 1977. Plant–animal mutualism: coevolution with Dodo leads to near extinction of plant. *Science* **197**: 885–6.

Ulfstrand, S. 1963. Ecological aspects of irruptive bird migration in northwestern Europe. *Proc. 13 Int. Orn. Congr.* 780–94.

Vander Wall, S. B. & Balda, R. P. 1977. Coadaptations of the Clark's Nutcracker and the Piñon Pine for efficient seed harvest and dispersal. *Ecological Monographs* **47**: 89–111.

Voous, K. H. 1960. *Atlas of European Birds*. Nelson, London.

Section 3.9. Omnivorous birds

Andersson, Å. 1970. Gråtrutar och soptippar. *Hygienisk Revy* **9**: 410–15.

Baker, R. R. 1980. The significance of the Lesser Black-Backed Gull to models of bird migration. *Bird Study* **27**: 41–50.

Busse, P. 1969. Results of ringing of European *Corvidae*. *Acta Ornith.* **11**: 263–328.

Haftorn, S. 1971. *Norges fugler*. Universitetsforlaget, Oslo.

Johansson, L. & Lundberg, A. 1977. Kråkans *Corvus corone cornix* höstflyttning över Ålands hav. *Vår Fågelvärld* **36**: 229–37.

Rendahl, H. 1961. Die Zugverhältnisse der schwedischen Rabenvögel. *Arkiv f. Zoologi*, **12**: 421–510.

Salomonsen, F. 1967. *Fugletraekket og dets gåder* (2nd edition). Munksgaard, Copenhagen.

Section 3.10. The evolution of bird migration

Alerstam, T. & Enckell, P. H. 1979. Unpredictable habitats and evolution of bird migration. *Oikos* **33**: 228–32.

Alerstam, T. & Högstedt, G. 1982. Bird migration and reproduction in relation to habitats for survival and breeding. *Ornis Scand.* **13**: 25–37.

Britton, P. L. & Rathbun, G. B. 1978. Two migratory thrushes and the African Pitta in coastal Kenya. *Scopus* **2**: 11–17.

Elgood, J. H., Fry, C. H. & Dowsett, R. J. 1973. African migrants in Nigeria. *Ibis* **115**: 1–45, 375–411.

Fretwell, S. 1980. Evolution of migration in relation to factors regulating bird numbers. In: *Migrant Birds in the Neotropics* (ed. A. Keast & E. S. Morton, Smithsonian Institution Press, Washington): 517–27.

Haffer, J. 1969. Speciation in Amazonian forest birds. *Science* **165**: 131–7.

Haffer, J. 1974. Pleistozäne Differenzierung der amazonischen Vogelfauna. *Bonner zool. Beitr.* **25**: 87–117.

Karr, J. R. 1976. On the relative abundance of migrants from the North Temperate Zone in tropical habitats. *Wilson Bull.* **88**: 433–58.

Karr, J. R. 1980. Patterns in the migration system between the North Temperate Zone and the Tropics. In: *Migrant Birds in the Neotropics* (ed. A. Keast & E. S. Morton, Smithsonian Institution Press, Washington): 529–43.

Königsson, L.-K. & Frängsmyr, T. 1977. *Istid – nutid – istid*. Natur och Kultur, Stockholm.

Lamb, H. H. 1977. *Climate – Present, Past and Future*. Vol. 2, Climatic History and the Future. Methuen & Co., London.

Moreau, R. E. 1966. *The Bird Faunas of Africa and Its Islands*. Academic Press, London.

Moreau, R. E. 1972. *The Palaearctic–African Bird Migration Systems*. Academic Press, London.

Williams, M. A. J. & Faure, H. 1980. *The Sahara and the Nile. Quaternary Environments and Prehistoric Occupation in Northern Africa*. A. A. Balkema, Rotterdam.

Chapter 4. The migratory journey

Section 4.1. Methods of studying bird migration

Alerstam, T. 1975. Crane *Grus grus* migration over sea and land. *Ibis* **117**: 489–95.

Alerstam, T., Bauer, C. A. & Roos, G. 1974. Spring migration of Eiders *Somateria mollissima* in southern Scandinavia. *Ibis* **116**: 194–210.

Alerstam, T. & Ulfstrand, S. 1974. A radar study of the autumn migration of Wood Pigeons *Columba palumbus* in southern Scandinavia. *Ibis* **116**: 522–42.

Bergman, G. 1974. The spring migration of the Long-tailed Duck and the Common Scoter in western Finland. *Ornis Fenn.* **51**: 129–45.

Bergman, G. & Donner, K. O. 1964. An analysis of the spring migration of the Common Scoter and the Long-tailed Duck in southern Finland. *Acta Zool. Fenn.* **105**: 1–59.

Cochran, W. W. 1972. *Long-distance Tracking of Birds*. I: Animal Orientation and Navigation (NASA SP-262, Washington): 39–59.

Dorka, V. 1966. Das jahres- und tageszeitliche Zugmuster von Kurz- und Langstreckenziehern nach Beobachtungen auf den Alpenpassen Cou-Bretolet. *Orn. Beob.* **63**: 165–223.

Eastwood, E. 1967. *Radar Ornithology*. Methuen, London.

Edelstam, C. 1972. The visible migration of birds at Ottenby, Sweden. *Vår Fågelvärld*, suppl. 7.

Gauthreaux, S. A. 1969. A portable ceilometer technique for studying low level nocturnal migration. *Bird-banding* **40**: 309–20.

Lind, G. 1980. Änglar – finns dom? Om radar och fågelekon. *Lundaforskare föreläser* **12**: 38–45.

Lowery, G. H. & Newman, R. J. 1966. A continentwide view of bird migration on four nights in October. *Auk* **83**: 547–86.

Nisbet, I. C. T. 1959. Calculation of flight directions of birds observed crossing the face of the moon. *Wilson Bull.* **71**: 237–43.

Nisbet, I. C. T. 1963. Quantitative study of migration with 23-centimetre radar. *Ibis* **105**: 435–60.

Richardson, W. J. 1979. Radar techniques for wildlife studies. *Nat. Wildl. Fed. Sci. Tech.* Ser. 3: 171–9.

Rudebeck, G. 1950. Studies on bird migration. *Vår Fågelvärld*, suppl. **1**.

Ulfstrand, S., Roos, G., Alerstam, T. & Österdahl, L. 1974. Visible bird migration at Falsterbo, Sweden. *Vår Fågelvärld*, suppl. **8**.

Section 4.2. Flight speed

Bilo, D. & Nachtigall, W. 1977. Biophysics of bird flight: questions and results. *Fortschritte der Zoologie* **24**: 217–34.

Brown, R. M. J. 1963. The flight of birds. *Biol. Rev.* **38**: 460–89.

Bruderer, B. 1971. Radarbeobachtungen über den Frühlingszug im Schweizerischen Mittelland. *Orn. Beob.* **68**: 89–158.

Bruderer, B. & Weitnauer, E. 1972. Radarbeobachtungen über Zug und Nachtflüge des Mauerseglers (*Apus apus*). *Rev. Suisse de Zoologie* **79**: 1190–1200.

Gatter, W. 1979. Unterschiedliche Zuggeschwindigkeit nahe verwandter Vogelarten. *J. Orn.* **120**: 221–5.

Greenewalt, C. H. 1975. The flight of birds. *Trans. Am. Phil. Soc.*, New Series 65, part 4.

Kipp, F. A. 1959. Der Handflügel-Index als flugbiologisches Mass. *Vogelwarte* **20**: 77–86.

Kokshaysky, N. V. 1979. Tracing the wake of the flying bird. *Nature* **279**: 146–8.

Nachtigall, W. 1975. Vogelflügel und Gleitflug. *J. Orn.* **116**: 1–38.

Nachtigall, W. 1979. Der Taubenflügel in Gleitflugstellung: Geometrische Kenngrössen der Flügelprofile und Luftkrafterzeugung. *J. Orn.* **120**: 30–40.

Noer, H. 1979. Speeds of migrating waders *Charadriidae*. *Dansk Orn. Foren. Tidsskr.* **73**: 215–24.

Pennycuick, C. J. 1969. The mechanics of bird migration. *Ibis* **111**: 525–56.

Pennycuick, C. J. 1972. *Animal Flight*. Edward Arnold, London.

Pennycuick, C. J. 1975. Mechanics of Flight. *Avian Biology* (Academic Press, New York) Vol. 5: 1–75.

Pennycuick, C J. 1989. *Bird Flight Performance*. Oxford University Press, Oxford.

Rayner, J. M. V. 1979. A new approach to animal flight mechanics. *J. Exp. Biol.* **80**: 17–54.

Rüppell, G. 1977. *Bird Flight*. Van Nostrand Reinhold Co., New York.

Schnell, G. D. & Hellack, J. J. 1979. Bird flight speeds in nature: optimized or a compromise? *Am. Nat.* **113**: 53–66.

Torre-Bueno, J. R. & Larochelle, J. 1978. The metabolic cost of flight in unrestrained birds. *J. Exp. Biol.* **75**: 223–9.

Tucker, V. A. 1968. Respiratory exchange and evaporative water loss in the flying Budgerigar. *J. Exp. Biol.* **48**: 67–87.

Tucker, V. A. 1969. The energetics of bird flight. *Scientific American* **220**, no. 5: 70–8.

Section 4.3. Migration in flocks

Badgerow, J. P. & Hainsworth, F. R. 1981. Energy savings through formation flight? A re-examination of the Vee formation. *J. Theor. Biol.* **93**: 41–52.

Balcomb, R. 1977. The grouping of nocturnal passerine migrants. *Auk* **94**: 479–88.

Bruderer, B. & Steidinger, P. 1972. *Methods of Quantitative and Qualitative Analysis of Bird Migration With a Tracking Radar*. I: Animal Orientation and Navigation (NASA SP-262, Washington): 151–67.

D'Arms, E. & Griffin, D. R. 1972. Balloonists' reports of sounds audible to migrating birds. *Auk* **89**: 269–79.

Gauthreaux, S. A. 1972. Behavioural responses of migrating birds to daylight and darkness: a radar and direct visual study. *Wilson Bull.* **84**: 136–48.

Gould, L. L. & Heppner, F. 1974. The Vee formation of Canada Geese. *Auk* **91**: 494–506.

Higdon, J. J. L. & Corrsin, S. 1978. Induced drag of a bird flock. *Am. Nat.* **112**: 727–44.

Hummel, D. 1973. Die Leistungsersparnis beim Verbandsflug. *J. Orn.* **114**: 259–82.

Hummel, D. 1978. Die Leistungsersparnis in Flugformationen von Vögeln mit Unterschieden in Grösse, Form und Gewicht. *J. Orn.* **119**: 52–73.

Keeton, W. T. 1970. Comparative orientational and homing performances of single pigeons and small flocks. *Auk* **87**: 797–9.

Kjellén, N. & Sylvén, M. 1978. Artsammansättningen i flockar av flyttande änder – några spekulationer. *Anser* **17**: 35–40.

Lissaman, P. B. S. & Shollenberger, C. A. 1970. Formation flight of birds. *Science* **168**: 1003–5.

May, R. M. 1979. Flight formations in geese and other birds. *Nature* **282**: 778–80.

Nisbet, I. C. T. 1963. Quantitative study of migration with 23-centimetre radar. *Ibis* **105**: 435–60.

Tamm, S. 1980. Bird orientation: single homing pigeons compared to small flocks. *Behav. Ecol. Sociobiol.* **7**: 319–22.

Wagner, G. 1975. Zur Frage des Flugführens im heimkehrenden Brieftaubengruppen. *Z. Tierpsychol.* **39**: 61–74.

Williams, T. C., Williams, J. M. & Klonowski, T. J. 1974. Flight patterns of birds studied with an 'Ornithar'. In: *A Conference on the Biological Aspects of the Bird/Aircraft Collision Problem* (ed. S. A. Gauthreaux, Clemson University): 477–90.

Section 4.4. Soaring flight

Bernis, F. 1975. Migracion de Falconiformes y *Ciconia* spp. por Gibraltar. parte II, IV. *Ardeola* **21** Espécial.

Hürzeler, E. 1950. Kranichzug 4300 m hoch über den Kanal. *Orn. Beob.* **47**: 172.

Kerlinger, P. 1989. *Flight Strategies of Migrating Hawks*. University of Chicago Press, Chicago.

Newton, I. 1979. *Population Ecology of Raptors*. T. & A. D. Poyser, Berkhamsted.

Pennycuick, C. J. 1972. Soaring behaviour and performance of some East African birds, observed from a motor-glider. *Ibis* **114**: 178–218.

Pennycuick, C. J. 1975. Mechanics of Flight. *Avian Biology* (Academic Press, New York) Vol 5: 1–75.

Pennycuick, C. J., Alerstam, T. & Larsson, B. 1979. Soaring migration of the Common Crane *Grus grus* observed by radar and from an aircraft. *Ornis Scand.* **10**: 241–51.

Peterz, M. & Rönnertz, T. 1980. Iakttagelser rörande flyktsättet hos några havsfåglar. *Anser* **19**: 167–73.

Porter, R. & Willis, I. 1968. The autumn migration of soaring birds at the Bosphorus. *Ibis* **110**: 520–36.

Richardson, W. J. 1975. Autumn hawk migration in Ontario studied with radar. *Proc. of North American Hawk Migration Conf.* (ed. M. Harwood): 47–58.

Roos, G. 1978. Sträckräkningar och miljöövervakning: långsiktiga förändringar i höststräckets numerär vid Falsterbo 1942–1977. *Anser* **17**: 133–8.

Rudebeck, G. 1950. Studies on bird migration. *Vår Fågelvärld*, suppl. **1**.

Rudebeck, G. 1972. Falsterbo – god fågellokal året runt. Falsterboguide. *Svenska Naturskyddsföreningen*: 17–53.

Rüppell, G. 1977. *Bird Flight.* Van Nostrand Reinhold Co., New York.

Smith, N. G. 1980. Hawk and vulture migrations in the Neotropics. *Migrant Birds in the Neotropics* (ed. A. Keast & E. S. Morton, Smithsonian Inst., Washington): 51–65.

Wallington, C. E. 1977. *Meteorology for Glider Pilots* (3rd edition). John Murray, London.

Welch, A. & L. & Irving, F. 1977. *New Soaring Pilot* (3rd edition). John Murray, London.

Wilson, I. A. 1975. Sweeping flight and soaring by albatrosses. *Nature* **257**: 307–8.

Wood, C. J. 1973. The flight of albatrosses (a computer simulation). *Ibis* **115**: 244–56.

Section 4.5. Flight altitude

Alerstam, T. 1977. Fågelsträckets höjd. *Anser* **16**: 189–202.

Bellrose, F. C. 1971. The distribution of nocturnal migrants in the air space. *Auk* **88**: 397–424.

Bruderer, B. 1971. Radarbeobachtungen über den Frühlingszug im Schweizerischen Mittelland. *Orn. Beob.* **68**: 89–158.

Elkins, N. 1979. High altitude flight by swans. *Brit. Birds* **72**: 238–9.

Gauthreaux, S. A. 1972. Behavioural responses of migrating birds to daylight and darkness: a radar and direct visual study. *Wilson Bull.* **84**: 136–48.

Gustafson, T., Lindkvist, B. & Kristiansson, K. 1973. New method for measuring the flight altitude of birds. *Nature* **244**: 112–13.

Gustafson, T., Lindkvist, B., Gotborn, L. & Gyllin, R. 1977. Altitudes and flight times for Swifts *Apus apus*. *Ornis Scand.* **8**: 87–95.

Laybourne, R. C. 1974. Collision between a vulture and an aircraft at an altitude of 37 000 feet. *Wilson Bull.* **86**: 461–2.

Manville, R. M. 1963. Altitude record for Mallard. *Wilson Bull.* **75**: 92.

Martens, J. 1971. Zur Kenntnis des Vogelzuges in nepalischen Himalaya. *Vogelwarte* **26**: 113–28.

Myres, M. T. 1964. Dawn ascent and re-orientation of Scandinavian thrushes (*Turdus* spp) migrating at night over the northeastern Atlantic Ocean in autumn. *Ibis* **106**: 7–51.

Pennycuick, C. J. 1978. Fifteen testable predictions about bird flight. *Oikos* **30**: 165–76.

Richardson, W. J. 1976. Autumn migration over Puerto Rico and the western Atlantic: a radar study. *Ibis* **118**: 309–32.

Richardson, W. J. 1979. Southeastward shorebird migration over Nova Scotia and New Brunswick in autumn: a radar study. *Can. J. Zool.* **57**: 107–24.

Roos, G. 1975. Falsterbonytt: juli–oktober 1975. *Anser* **14**: 237–46.

Schmidt-Nielsen, K. 1975. *Animal Physiology.* Cambridge University Press, Cambridge.

Stewart, A. G. 1978. Swans flying at 8000 metres. *Brit. Birds* **71**: 459–60.

Swan, L. W. 1970. Goose of the Himalayas. *Natural History* **79** (Dec.): 68–75.

Thiollay, J.-M. 1979. La migration des Grues à travers l'Himalaya et la prédation par les Aigles Royaux. *Alauda* **47**: 83–92.

Torre-Bueno, J. R. 1978. Evaporative cooling and water balance during flight in birds. *J. Exp. Biol.* **75**: 231–6.

Tucker, V. A. 1968. Respiratory physiology of House Sparrows in relation to high-altitude flight. *J. Exp. Biol.* **48**: 55–66.

Section 4.6. Fat as flight fuel

Ash, J. S. 1969. Spring weights of trans-Saharan migrants in Morocco. *Ibis* **111**: 1–10.

Bibby, C. J. & Green, R. E. 1981. Autumn migration strategies of Reed and Sedge Warblers. *Ornis Scand.* **12**: 1–12.

Biebach, H., Friedrich, W. & Heine, G. 1986. Interaction of bodymass, fat, foraging and stopover period in trans-Sahara migrating passerine birds. *Oecologia* **69**: 370–9.

Blyumental, T. I. 1973. Development of fall migratory state in some wild passerine birds (bioenergetic aspect). In: *Bird Migrations. Ecological and Physiological Factors* (ed. B. E. Bykhovski, John Wiley, Chichester): 125–218.

Dolnik, V. R. & Gavrilov, V. M. 1973. Energy metabolism during flight of some passerines. In: *Bird Migrations* (as above): 288–96.

Dowsett, R. J. & Fry, C. H. 1971. Weight losses of trans-Saharan migrants. *Ibis* **113**: 531–3.

Evans, P. R. 1966. Migration and orientation of passerine night migrants in northeast England. *J. Zool.* **150**: 319–69.

Fogden, M. P. L. 1972. Premigratory dehydration in the Reed Warbler *Acrocephalus scirpaceus* and water as a factor limiting migratory range. *Ibis* **114**: 548–52.

Fry, C. H., Ash, J. S. & Ferguson-Lees, I. J. 1970. Spring weights of some Palaearctic migrants at Lake Chad. *Ibis* **112**: 58–82.

Fry, C. H., Ferguson-Lees, I. J. & Dowsett, R. J. 1972. Flight muscle hypertrophy and ecophysiological variation of Yellow Wagtail *Motacilla flava* races at Lake Chad. *J. Zool.* **167**: 293–306.

Gladwin, T. W. 1963. Increases in the weights of Acrocephali. *Bird Migration* **2**: 319–24.

Hussel, D. J. 1969. Weight loss of birds during nocturnal migration. *Auk* **86**: 75–83.

Hussel, D. J. & Lambert, A. B. 1980. New estimates of weight loss in birds during nocturnal migration. *Auk* **97**: 547–58.

Jones, P. J. & Ward, P. 1977. Evidence of pre-migratory fattening in three tropical granivorous birds. *Ibis* **119**: 200–3.

Larkin, R. P., Griffin, D. R., Torre-Bueno, J. R. & Teal, J. 1979. Radar observations of bird migration over the western North Atlantic Ocean. *Behav. Ecol. Sociobiol.* **4**: 225–64.

McClintock, C. P., Williams, T. C. & Teal, J. M. 1978. Autumnal bird migration observed from ships in the western North Atlantic Ocean. *Bird-banding* **49**: 262–77.

McNeil, R. & Burton, J. 1977. Southbound migration of shorebirds from the Gulf of St Lawrence. *Wilson Bull.* **89**: 167–71.

McNeil, R. & Cadieux, F. 1972. Fat content and flight range capabilities of some adult spring and fall migrant North American shorebirds in relation to migration routes on the Atlantic coast. *Naturaliste Can.* **99**: 589–605.

Moreau, R. E. 1961. Problems of Mediterranean–Saharan migration. *Ibis* **103**: 373–427, 580–623.

Moreau, R. E. 1969. Comparative weights of some trans-Saharan migrants at intermediate points. *Ibis* **111**: 621–4.

Moreau, R. E. 1972. *The Palaearctic–African Bird Migration Systems.* Academic Press, London.

Nisbet, I. C. T. 1970. Autumn migration of the Blackpoll Warbler: evidence for long flight provided by regional survey. *Bird-banding* **41**: 207–40.

Nisbet, I. C. T., Drury, W. H. & Baird, J. 1963. Weight loss during migration. *Bird-banding* **34**: 107–59.

Odum, E. P., Connell, C. E. & Stoddard, H. L. 1961. Flight energy and estimated flight ranges of some migratory birds. *Auk* **78**: 515–27.

Pearson, D. J. 1971. Weights of some Palaearctic migrants in southern Uganda. *Ibis* **113**: 173–84.

Pearson, D. J., Backhurst, G. C. & Backhurst, D. E. G. 1979. Spring weights and passage of Sedge Warblers *Acrocephalus schoenobaenus* in central Kenya. *Ibis* **121**: 8–19.

Pennycuick, C. J. 1975. Mechanics of Flight. *Avian Biology* (Academic Press, London) Vol. 5: 1–75.

Pienkowski, M. W., Lloyd, C. S. & Minton, C. D. T. 1979. Seasonal and migrational weight changes in Dunlins. *Bird Study* **26**: 134–48.

Rappole, J. H. & Warner, D. W. 1976. Relationships between behaviour, physiology and weather in avian transients at a migration stopover site. *Oecologia* **26**: 193–212.

Raveling, D. G. & Lefebre, E. A. 1967. Energy metabolism and theoretical flight range of birds. *Bird-banding* **38**: 97–113.

Richardson, W. J. 1979. Southeastward shorebird migration over Nova Scotia and New Brunswick in autumn: a radar study. *Can. J. Zool.* **57**: 107–24.

Smith, G. A. 1979. Spring weights of selected trans-Saharan migrants in north west Morocco. *Ringing and Migration* **2**: 151–5.

Smith, V. W. 1966. Autumn and spring weights of some Palaearctic migrants in central Nigeria. *Ibis* **108**: 492–512.

Summers, R. W. & Waltner, M. 1979. Seasonal variation in the mass of waders in southern Africa, with special reference to migration. *Ostrich* **50**: 21–37.

Thompson, M. C. 1974. Migratory patterns of Ruddy Turnstones in the central Pacific region. *Living Bird* **12**: 5–23.

Ward, P. 1971. The migration patterns of *Quelea quelea* in Africa. *Ibis* **113**: 275–97.

Ward, P. & Jones, P. J. 1977. Pre-migratory fattening in three races of the Red-billed Quelea *Quelea quelea*, an intra-tropical migrant. *J. Zool.* **181**: 43–56.

Williams, T. C. & Williams, J. M. 1978. An oceanic mass migration of land birds. *Scientific American* **239** (Oct.): 138–45.

Zink, G. 1975. *Der Zug Europäischer Singvögel*. 2. Lieferung. (Vogelzug-Verlag, Möggingen).

Section 4.7. Diurnal and nocturnal migration

Alerstam, T. 1978. Reoriented bird migration in coastal areas: dispersal to suitable resting grounds? *Oikos* **30**: 405–8.

Bramley, A. 1979. White Storks migrating at night by utilising oil flares. *Brit. Birds* **72**: 229.

Evans, P. R. 1966. An approach to the analysis of visible migration and a comparison with radar observations. *Ardea* **54**: 14–44.

Hebrard, J. J. 1971. The nightly initiation of passerine migration in spring: a direct visual study. *Ibis* **113**: 8–18.

Section 4.8. Weather and wind

Alerstam, T. 1976. Bird migration in relation to wind and topography. PhD thesis, Lund University.

Alerstam, T. 1978. Analysis and theory of visible bird migration. *Oikos* **30**: 273–349.

Alerstam, T. 1979. Optimal use of wind by migrating birds: combined drift and overcompensation. *J. Theor. Biol.* **79**: 341–53.

Alerstam, T. & Pettersson, S.-G. 1976. Do birds use waves for orientation when migrating across the sea? *Nature* **259**: 205–7.

Koskimies, J. 1950. The life of the Swift, *Micropus apus*, in relation to the weather. *Annales Acad. Scient. Fennicae*. series A, IV, no. 15, 152 pp.

Lack, D. 1956. *Swifts in a Tower*. Methuen, London.

Lack, D. 1960. The influence of weather on passerine migration. A review. *Auk* **77**: 171–209.

Liljequist, G. H. 1970. *Klimatologi*. Generalstabens Litografiska Anstalt, Stockholm.

Lowery, G. H. 1945. Trans-Gulf spring migration of birds and the coastal hiatus. *Wilson Bull.* **57**: 92–121.

Nisbet, I. C. T. & Drury, W. H. 1968. Short-term effects of weather on bird migration: a field study using multivariate statistics. *Anim. Behav.* **16**: 496–530.

Richardson, W. J. 1978. Timing and amount of bird migration in relation to weather: a review. *Oikos* **30**: 224–72.

Richardson, W. J. 1979. Southeastward shorebird migration over Nova Scotia and New Brunswick in autumn: a radar study. *Can. J. Zool.* **37**: 107–24.

Roos, G. 1962. Vinterfåglar på Falsterbonäset. *Fauna och Flora* **57**: 249–73.

Roos, G. 1975. De arktiska vadarnas flyttning över Falsterbo sommaren 1974 enligt tre olika registreringsmetoder. *Anser* **14**: 79–92.

Roos, G. 1977. Sträckräkningar vid Falsterbo hösten 1975. *Anser* **16**: 169–88.

Roos, G. 1978. Sträckräkningar vid Falsterbo hösten 1976. *Anser* **17**: 1–22.

Roos, Gunnar, publishes consecutive reports and analyses of the annual migration observations at Falsterbo in the journal *Anser*. These papers give a unique insight, unparalleled anywhere in the world, into the migratory course of events and its dependence on weather in various species.

Rudebeck, G. 1950. Studies on bird migration. *Vår Fågelvärld.* Suppl. 1.

Swanberg, P. O. 1948. Ett tillfälligt återsträck av bofinkar den 3 april 1947. *Vår Fågelvärld* 7: 125–6.

Svärdson, G. 1951. Swift (*Apus apus*) movements in summer. *Proc. 10 Int. Orn. Congress* (Uppsala 1950): 335–8.

Williams, T. C. & Williams, J. M. 1978. An oceanic mass migration of land birds. *Scientific American* 239 no. 4: 138–45.

Section 4.9. Dangers during the migration

Baker, K. 1977. Westward vagrancy of Siberian passerines in autumn 1975. *Bird Study* 24: 233–42.

Bourne, W. R. P. 1980. The midnight descent, dawn ascent and reorientation of land birds migrating across the North Sea in autumn. *Ibis* 122: 536–40.

Clapham, C. S. 1964. The birds of the Dahlac Archipelago. *Ibis* 106: 376–88.

Diamond, J. M. 1982. Mirror-image navigational errors in migrating birds. *Nature* 295: 277–8.

Durand, A. L. 1963. A remarkable fall of American land birds on the 'Mauretania', New York to Southampton, October 1962. *Brit. Birds* 56: 157–64.

Durand, A. L. 1972. Landbirds over the North Atlantic: unpublished records 1961–65 and thoughts a decade later. *Brit. Birds* 65: 428–42.

Elkins, N. 1979. Nearctic landbirds in Britain and Ireland: a meteorological analysis. *Brit. Birds* 72: 417–33.

Flach, B. 1959. Höstobservationer i Libanon. *Fauna och Flora* 54: 161–80.

Gibson, D. D. 1981. Migrant birds at Shemya Island, Aleutian Islands, Alaska. *Condor* 83: 65–77.

Haag, W. & Beck, P. 1979. Zum Frühjahrszug paläarktischer Vögel über die westliche Sahara. *J. Orn.* 120: 237–46.

Hope Jones, P. 1980. The effect on birds of a North Sea gas flare. *Brit. Birds* 73: 547–55.

Hørring, R. 1933. *Aethia cristatella* skudt ved Island. *Dansk Orn. Foren. Tidsskr.* 27: 103–5.

Iso-Iivari, L. 1976. Check-list of the birds of Northern Europe. *Ornis Fenn.* 53: 57–63.

Jany, E. 1960. An Brutplätzen des Lannerfalken (*Falco biarmicus erlangeri*) in einer Kieswüste der inneren Sahara (Nordrand des Serir Tibesti) zur Zeit des Frühjahrzuges. *Proc. 12 Int. Orn. Congr.* (Helsinki 1958): 343–52.

Karlsson, J. 1977. Fågelkollisioner med master och andra byggnadsverk. *Anser* 16: 203–16.

MacDonald, S. M. & Mason, C. F. 1973. Predation of migrant birds by gulls. *Brit. Birds* 66: 361–3.

Moreau, R. E. 1972. *The Palaearctic–African Bird Migration Systems.* Academic Press, London.

Rabøl, J. 1976. The orientation of Pallas's Leaf Warbler *Phylloscopus proregulus* in Europe. *Dansk Orn. Foren. Tidsskr.* 70: 5–16.

Rudebeck, G. 1950 and 1951. The choice of prey and modes of hunting of predatory birds with special reference to their selective effect. *Oikos* 2: 67–88, 3: 200–31.

Sharrock, J. T. R. 1974. *Scarce Migrant Birds in Britain and Ireland.* T. & A. D. Poyser, Berkhamsted.

Sharrock, J. T. R. & Sharrock, E. M. 1976. *Rare Birds in Britain and Ireland.* T. & A. D. Poyser, Berkhamsted.

Ullman, M. 1989. Why are northern Yellow-browed Warblers, *Phylloscopus inornatus*, and Pallas's Warblers, *Ph. proregulus*, earlier than southern? (in Swedish with English summary). *Vår Fågelvärld* 48: 467–75.

Wahlgren, F. 1867. *Phaleris psittacula* funnen i Sverige. *Sv. jägarförb. nya tidsskr.* 5: 108–11.

Walter, H. 1968. Zur Abhängigkeit des Eleonorafalken (*Falco eleonorae*) vom mediterranen Vogelzug. *J. Orn.* 109: 323–65.

Walter, H. 1979. *Eleonora's Falcon.* The University of Chicago Press, Chicago.

Chapter 5. Orientation and navigation

The following gives only general reviews which have served as a basis for all the three ensuing sections. Other reference literature is mentioned separately for each section.

Emlen, S. T. 1975. Migration: Orientation and Navigation. *Avian Biology* (Academic Press, New York) Vol. 5: 129–219.

Keeton, W. T. 1979. Avian orientation and navigation: a brief overview. *Brit. Birds* **72**: 451–70.

Matthews, G. V. T. 1968. *Bird Navigation* (2nd edition). Cambridge University Press.

Schmidt-Koenig, K. 1979. *Avian Orientation and Navigation.* Academic Press, London.

Schmidt-Koenig, K. & Keeton, W. T. (eds.) 1978. *Animal Migration, Navigation and Homing.* Springer-Verlag, Berlin.

Section 5.1. The sensory world of birds

Århem, P. 1979. Några problem kring fågelögats fysiologi. *Vår Fågelvärld* **38**: 73–82.

Baker, R. R. 1981. *Human Navigation and the Sixth Sense.* Hodder and Stoughton, London.

Baker, R. R. 1989. *Human Navigation and Magnetoreception.* Manchester University Press, Manchester.

Bookman, M. A. 1977. Sensitivity of the homing pigeon to an earth-strength magnetic field. *Nature* **267**: 340–2.

Ehinger, B. 1977. Vikingarnas navigationsinstrument 1000 år före sin tid. *Forskning och Framsteg* no. 8/77: 39–43.

Keeton, W. T. 1974. The mystery of pigeon homing. *Scientific American* **231**, no. 6: 96–107.

Kirschvink, J. L. 1981. Ferromagnetic crystals (magnetite?) in human tissue. *J. Exp. Biol.* **92**: 333–5.

Kirschvink, J. L. & Gould, J. L. 1981. Biogenic magnetite as a basis for field detection in animals. *BioSystems* **13**: 181–201.

Kirschvink, J. L., Jones, D. S. & MacFadden, B. J. (eds.) 1985. *Magnetite Biomineralization and Magnetoreception in Organisms.* Plenum Press, New York.

Kreithen, M. L. 1978. Sensory mechanisms for animal orientation – can any new ones be discovered? In: *Animal Migration, Navigation and Homing* (eds. K. Schmidt-Koenig & W. T. Keeton, Springer-Verlag, Berlin): 25–34.

Kreithen, M. L. & Keeton, W. T. 1974. Detection of polarized light by the Homing Pigeon *Columba livia. J. Comp. Physiol.* **89**: 83–92.

Larkin, T. & Keeton, W. T. 1978. An apparent lunar rhythm in the day-to-day variations in initial bearings of Homing Pigeons. In: *Animal Migration, Navigation and Homing* (as above): 92–106.

Pettigrew, J. D. 1978. A role for the avian pecten oculi in orientation to the sun? In: *Animal Migration, Navigation and Homing* (as above): 42–54.

Presti, D. & Pettigrew, J. D. 1980. Ferromagnetic coupling to muscle receptors as a basis for geomagnetic field sensibility in animals. *Nature* **285**: 99–101.

Quine, D. B. & Kreithen, M. L. 1981. Frequency shift discrimination: can Homing Pigeons locate infrasounds by Doppler shifts? *J. Comp. Physiol.* **141**: 153–5.

Semm, P. & Demaine, C. 1986. Neurophysiological properties of magnetic cells in the pigeon's visual system. *J. Comp. Physiol* **A159**: 619–25.

Walcott, C., Gould, J. L. & Kirschvink, J. L. 1979. Pigeons have magnets. *Science* **205**: 1027–9.

Section 5.2. Different compasses

Able, K. P. 1989. Skylight polarization patterns and the orientation of migratory birds. *J. Exp. Biol.* **141**: 241–56.

Emlen, S. T. 1975. The stellar-orientation system of a migratory bird. *Scientific American* **233**. no. 2: 102–11.

Graue, L. C. 1963. The effect of phase shifts in the day–night cycle on pigeon homing at distances of less than one mile. *Ohio Journ. of Science* **63**: 214–17.

Moore, F. R. 1977. Geomagnetic disturbance and the orientation of nocturnally migrating birds. *Science* **196**: 682–4.

Moore, F. R. 1978. Sunset and the orientation of a nocturnal migrant bird. *Nature* **274**: 154–6.

Walcott, C. 1978. Anomalies in the earth's magnetic field increase the scatter of pigeons' vanishing bearings. In: *Animal Migration, Navigation and Homing* (eds. K. Schmidt-Koenig & W. T. Keeton, Springer-Verlag, Berlin): 143–51.

Wiltschko, R. 1980 and 1981. Die Sonnenorientierung der Vögel. *J. Orn.* **121**: 121–43, **122**: 1–22.

Wiltschko, R. & Wiltschko, W. 1978. Relative importance of stars and the magnetic field for the accuracy of orientation in night migrating birds. *Oikos* **30**: 195–206.

Wiltschko, R., Nohr, D. & Wiltschko, W. 1981. Pigeons with a deficient sun compass use the magnetic compass. *Science* **214**: 343–5.

Wiltschko, W. & Wiltschko, R. 1972. Magnetic compass of European Robins. *Science* **176**: 62–4.

Wiltschko, W. & Wiltschko, R. 1974. Bird orientation under different sky sectors. *Z. Tierpsychol.* **35**: 536–42.

Wiltschko, W. & Wiltschko, R. 1975. The interaction of stars and magnetic field in the orientation system of night migrating birds. *Z. Tierpsychol.* **37**: 337–55, **39**: 265–82.

Wiltschko, W. & Wiltschko, R. 1976. Die Bedeutung des Magnetkompasses für die Orientierung der Vögel. *J. Orn.* **117**: 362–87.

Wiltschko, W. & Wiltschko, R. 1988. Magnetic orientation in birds. *Current Ornithology*, Vol. **5**: 67–121.

Section 5.3. How do birds find the right migration route?

Baldaccini, N. E., Benvenuti, S., Fiaschi, V. & Papi, F. 1975. Pigeon navigation: effects of wind deflection at home cage on homing behaviour. *J. Comp. Physiol.* **99**: 177–86.

Berthold, P. 1973. Relationships between migratory restlessness and migration distance in six *Sylvia* species. *Ibis* **115**: 594–9.

Gould, J. L. 1980. The case for magnetic sensitivity in birds and bees (such as it is). *Amer. Scientist* **68**: 256–67.

Gwinner, E. 1969. Untersuchungen zur Jahresperiodik von Laubsängern. *J. Orn.* **110**: 1–21.

Gwinner, E. 1977. Circannual rhythms in bird migration. *Ann. Rev. Ecol. Syst.* **8**: 381–405.

Gwinner, E. & Wiltschko, W. 1978. Endogenously controlled changes in migratory direction of the Garden Warbler *Sylvia borin*. *J. Comp. Physiol.* **125**: 267–73.

Ioalé, P., Papi, F., Fiaschi, V. & Baldaccini, N. E. 1978. Pigeon navigation: effects upon homing behaviour by reversing wind direction at the loft. *J. Comp. Physiol.* **128**: 285–95.

Moore, B. R. 1980. Is the homing pigeon's map geomagnetic? *Nature* **285**: 69–70.

Papi, F. 1986. Pigeon navigation: solved problems and open questions. *Monitore Zool. Ital.* **20**: 471–517.

Papi, F., Keeton, W. T., Brown, A. I. & Benvenuti, S. 1978. Do American and Italian pigeons rely on different homing mechanisms? *J. Comp. Physiol.* **128**: 303–17.

Perdeck, A. C. 1958. Two types of orientation in migrating Starlings *Sturnus vulgaris* and Chaffinches *Fringilla coelebs*, as revealed by displacement experiments. *Ardea* **46**: 1–37.

Perdeck, A. C. 1974. An experiment on the orientation of juvenile Starlings during spring migration. *Ardea* **62**: 190–5.

Rüppell, W. 1944. Versuche über Heimfinden ziehender Nebelkrähen nach Verfrachtung. *J. Orn.* **92**: 106–32.

Schmidt-Koenig, K. & Walcott, C. 1978. Tracks of pigeons homing with frosted lenses. *Anim. Behav.* **26**: 480–6.

Waldvogel, J. A., Phillips, J. B., McCorkle, D. F. & Keeton, W. T. 1980. Short-term residence in deflector lofts alters initial orientation of homing pigeons. *Behav. Ecol. Sociobiol.* **7**: 207–11.

Wallraff, H. G. 1980. Does pigeon homing depend on stimuli perceived during displacement? *J. Comp. Physiol.* **139**: 193–208.

Whiten, A. 1972. Operant study of sun altitude and pigeon navigation. *Nature* **237**: 405–6.

Index